LINEAR AND
NONLINEAR WAVES

LINEAR AND NONLINEAR WAVES

G. B. WHITHAM F. R. S.

Professor of Applied Mathematics
California Institute of Technology

A WILEY-INTERSCIENCE PUBLICATION

JOHN WILEY & SONS
New York • Chichester • Brisbane • Toronto • Singapore

Library of Congress Cataloging in Publication Data:

Whitham, Gerald Beresford, 1927–
   Linear and nonlinear waves.

   (Pure and applied mathematics)
   "A Wiley-Interscience publication."
   Bibliography: p.
   1. Wave-motion, Theory of. 2. Waves. I. Title.

QA927.W48   531'.1133   74-2070
ISBN 0-471-94090-9

Printed in the United States of America

10

PREFACE

This is an expanded version of a course given for a number of years at the California Institute of Technology. It was designed for applied mathematics students in the first and second years of graduate study; it appears to have been equally useful for students in engineering and physics.

The presentation is intended to be self-contained but both the order chosen for the topics and the level adopted suppose previous experience with the elementary aspects of linear wave propagation. The aim is to cover all the major well-established ideas but, at the same time, to emphasize nonlinear theory from the outset and to introduce the very active research areas in this field. The material covered is outlined in detail in Chapter 1. The mathematical development of the subject is combined with considerable discussion of applications. For the most part previous detailed knowledge of a field of application is not assumed; the relevant physical ideas and derivation of basic equations are given in depth. The specific mathematical background required is familiarity with transform techniques, methods for the asymptotic expansion of integrals, solutions of standard boundary value problems and the related topics that are usually referred to collectively as "mathematical methods."

Parts of the account are drawn from research supported over the last several years by the Office of Naval Research. It is a pleasure to express my gratitude to the people there, particularly to Leila Bram and Stuart Brodsky.

My special thanks to Vivian Davies and Deborah Massey who typed the manuscript and cheerfully put up with my constant rewrites and changes.

G. B. WHITHAM

Pasadena, California
December 1973

v

CONTENTS

6 Gas Dynamics   143

LINEAR AND
NONLINEAR WAVES

CHAPTER 1

Introduction and General Outline

Wave motion is one of the broadest scientific subjects and unusual in that it can be studied at any technical level. The behavior of water waves and the propagation characteristics of light and sound are familiar from everyday experience. Modern problems such as sonic booms or moving bottlenecks in traffic are necessarily of general interest. All these can be appreciated in a descriptive way without any technical knowledge. On the other hand they are also intensively studied by specialists, and almost any field of science or engineering involves some questions of wave motion.

There has been a correspondingly rich development of mathematical concepts and techniques to understand the phenomena from the theoretical standpoint and to solve the problems that arise. The details in any particular application may be different and some topics will have their own unique twists, but a fairly general overall view has been developed. This book is an account of the underlying mathematical theory with emphasis on the unifying ideas and the main points that illuminate the behavior of waves. Most of the typical techniques for solving problems are presented, but these are not pursued beyond the point where they cease to give information about the nature of waves and become exercises in "mathematical methods," difficult and intriguing as these may be. This applies particularly to linear wave problems. Important and fundamental properties of linear theory which are basic to the understanding of waves must be covered. But one could then fill volumes with solutions and techniques for specific problems. This is not the purpose of the book. Although the basic material on linear waves is included, some previous experience with linear theory is assumed and the emphasis is on the conceptually more difficult nonlinear theory. The study of nonlinear waves started over a hundred years ago with the pioneering work of Stokes (1847) and Riemann (1858), and it has proceeded at an accelerating pace, with considerable development in recent years. The purpose here is to give a unified treatment of this body of material.

The mathematical ideas are liberally interspersed with discussion of

1

specific cases and specific physical fields. Particularly in nonlinear problems this is essential for stimulation and illumination of the correct mathematical arguments, and, in any case, it makes the subject more interesting. Many of these topics are related to some branch of fluid mechanics, or to examples such as traffic flow which are treated in analogous fashion. This is unavoidable, since the main ideas of nonlinear waves were developed in these subjects, although it doubtless also reflects personal interest and experience. But the account is not written specifically for fluid dynamicists. The ideas are presented in general, and topics for application or motivation are chosen with a general reader in mind. It is assumed that flood waves in rivers, waves in glaciers, traffic flow, sonic booms, blast waves, ocean waves from storms, and so on, are of universal interest. Other fields are not excluded, and detailed discussion is given, for example, of nonlinear optics and waves in various mechanical systems. On the whole, though, it seemed better in applications to concentrate in a nontrivial way on representative areas, rather than to present superficial applications to sets of equations merely quoted from every conceivable field.

The book is divided into two parts, the first on hyperbolic waves and the second on dispersive waves. The distinction will be explained in the next section. In Part I the basic ideas are presented in Chapters 2, 5, 7, while in Part II they appear in Chapters 11, 14, 15, 17. The intervening chapters amplify the general ideas in specific contexts and may be read in full or sampled according to the reader's interests. It should also be possible to proceed directly to Part II from Chapter 2.

1.1 The Two Main Classes of Wave Motion

There appears to be no single precise definition of what exactly constitutes a wave. Various restrictive definitions can be given, but to cover the whole range of wave phenomena it seems preferable to be guided by the intuitive view that a wave is any recognizable signal that is transferred from one part of the medium to another with a recognizable velocity of propagation. The signal may be any feature of the disturbance, such as a maximum or an abrupt change in some quantity, provided that it can be clearly recognized and its location at any time can be determined. The signal may distort, change its magnitude, and change its velocity provided it is still recognizable. This may seem a little vague, but it turns out to be perfectly adequate and any attempt to be more precise appears to be too restrictive; different features are important in different types of wave.

Nevertheless, one can distinguish two main classes. The first is formulated mathematically in terms of hyperbolic partial differential equations, and such waves will be referred to as *hyperbolic*. The second class cannot be characterized as easily, but since it starts from the simplest cases of dispersive waves in linear problems, we shall refer to the whole class as *dispersive* and slowly build up a more complete picture. The classes are not exclusive. There is some overlap in that certain wave motions exhibit both types of behavior, and there are certain exceptions that fit neither.

The prototype for hyperbolic waves is often taken to be the wave equation

$$\varphi_{tt} = c_0^2 \nabla^2 \varphi, \tag{1.1}$$

although the equation

$$\varphi_t + c_0 \varphi_x = 0 \tag{1.2}$$

is, in fact, the simplest of all. As will be seen, there is a precise definition for hyperbolic equations which depends only on the form of the equations and is independent of whether explicit solutions can be obtained or not. On the other hand, the prototype for dispersive waves is based on a type of solution rather than a type of equation. A linear dispersive system is any system which admits solutions of the form

$$\varphi = a \cos (\kappa x - \omega t), \tag{1.3}$$

where the frequency ω is a definite real function of the wave number κ and the function $\omega(\kappa)$ is determined by the particular system. The phase speed is then $\omega(\kappa)/\kappa$ and the waves are usually said to be "dispersive" if this phase speed is not a constant but depends on κ. The term refers to the fact that a more general solution will consist of the superposition of several modes like (1.3) with different κ. [In the most general case a Fourier integral is developed from (1.3).] If the phase speed ω/κ is not the same for all κ, that is, $\omega \neq c_0 \kappa$ where c_0 is some constant, the modes with different κ will propagate at different speeds; they will disperse. It is convenient to modify the definition slightly and say that (1.3) is dispersive if $\omega'(\kappa)$ is not constant, that is, $\omega''(\kappa) \neq 0$.

It should be noted that (1.3) is also a solution of the hyperbolic equation (1.2) with $\omega = c_0 \kappa$, or of (1.1) with $\omega = \pm c_0 \kappa$. But these cases are excluded from the dispersive classification by the condition $\omega'' \neq 0$. However, it is not hard to find cases of genuine overlap in which the equations are hyperbolic and yet have solutions (1.3) with nontrivial dispersion relations $\omega = \omega(\kappa)$. One such example is the Klein-Gordon equation

$$\varphi_{tt} - \varphi_{xx} + \varphi = 0. \tag{1.4}$$

It is hyperbolic and yet (1.3) is a solution with $\omega^2 = \kappa^2 + 1$. This dual behavior is limited to relatively few instances and should not be allowed to obscure the overall differences between the two main classes. It does perhaps contribute to a fairly common misunderstanding, particularly encouraged in mathematical books, that wave motion is synonymous with hyperbolic equations and (1.3) is a less sophisticated approach to the same thing. The true emphasis should probably be the other way round. Rich and various as the class of hyperbolic waves may be, it is probably fair to say that the majority of wave motions fall into the dispersive class. The most familiar of all, ocean waves, is a dispersive case governed by Laplace's equation with strange boundary conditions at the free surface!

The first part of this book is devoted to hyperbolic waves and the second to dispersive waves. The theory of hyperbolic waves enters again into the study of dispersive waves in various curious ways, so the second part is not entirely independent of the first. The remainder of this chapter is an outline of the various themes, most of which are taken up in detail in the remainder of the book. The purpose is to introduce the material, but at the same time to give an overall view which is extracted from the detailed account.

1.2 Hyperbolic Waves

The wave equation (1.1) arises in acoustics, elasticity, and electromagnetism, and its basic properties and solutions were first developed in these areas of classical physics. In all cases, however, this is not the whole story.

In acoustics, one starts with the equations for a compressible fluid. Even if viscosity and heat conduction are neglected, this is a set of nonlinear equations in the velocity vector $\mathbf{u}$, the density ρ, and pressure p. Acoustics refers to the approximate linear theory in which all the disturbances are assumed to be small perturbations to an ambient constant state in which $\mathbf{u} = 0$, $\rho = \rho_0$, $p = p_0$. The equations are linearized by retaining only first order terms in the small quantities $\mathbf{u}$, $\rho - \rho_0$, $p - p_0$, that is, all powers higher than the first and all products of small quantities are omitted. It can then be shown that each component of $\mathbf{u}$ and the perturbations $\rho - \rho_0$, $p - p_0$ satisfy the wave equation (1.1). Once this has been solved for the appropriate boundary conditions or initial conditions that provide the source of the sound, it is natural to ask various questions about how this solution relates to the original nonlinear equations. Even for such weak perturbations, are the linear results accurate and are any important qualitative features lost in the approximation? If the disturbances are not weak, as in explosions or in the disturbances caused by high speed supersonic aircraft and missiles, what progress can be made directly on the original nonlinear equations? What are the modifying effects of viscosity and heat conduction? The answers to these questions in gas dynamics led to most of the fundamental ideas in nonlinear hyperbolic waves. The most outstanding new phenomenon of the nonlinear theory is the appearance of shock waves, which are abrupt jumps in pressure, density, and velocity: the blast waves of explosions and the sonic booms of high speed aircraft. But the whole intricate machinery of nonlinear hyperbolic equations had to be developed for their prediction, and a full understanding required analysis of the viscous effects and some aspects of kinetic theory.

In this way a set of basic ideas became clear within the context of gas dynamics, although one should add that the investigation of more complicated cases and the search for deeper understanding of the kinetic theory aspects, for example, are still active fields. The basic mathematical theory, developed in gas dynamics, is appropriate for any system governed by nonlinear hyperbolic equations, and it has been used and refined in many other fields.

In elasticity, the classical wave theory is also obtained after linearization. Even with the linear theory, the situation is more complicated because the system of equations leads to essentially two wave equations of the form (1.1) with two functions φ_1, φ_2 and two wave speeds, c_1, c_2, which are associated with the different modes of propagation for compression waves and shear waves. The two functions φ_1 and φ_2 are coupled through the appropriate boundary conditions, and generally the problem is much more complicated than merely solving the wave equation (1.1). At a free surface of an elastic body, there is further complication in that surface waves, so-called Rayleigh waves, are possible; these are perhaps more akin to dispersive waves and travel at a speed smaller than both c_1 and c_2. Because of these extra complications, the nonlinear theory has not been developed as fully as in gas dynamics.

In electromagnetism there is also the complication that while different components of the electric and magnetic fields satisfy (1.1), they are coupled by additional equations and by the boundary conditions. Although the classical Maxwell equations are posed in linear form from the outset, there is much present interest in "nonlinear optics," since devices such as lasers produce intense waves and various media react nonlinearly.

The corresponding mathematical theme started from the study of solutions of (1.1). The one dimensional equation for plane waves,

$$\varphi_{tt} - c_0^2 \varphi_{xx} = 0, \qquad (1.5)$$

is particularly simple. It can be rewritten in terms of new variables

$$\alpha = x - c_0 t, \qquad \beta = x + c_0 t, \tag{1.6}$$

as

$$\varphi_{\alpha\beta} = 0. \tag{1.7}$$

This is immediately integrated to show that the general solution is

$$\varphi = f(\alpha) + g(\beta)$$

$$= f(x - c_0 t) + g(x + c_0 t), \tag{1.8}$$

where f and g are arbitrary functions.

The solution is a combination of two waves, one with shape described by the function f moving to the right with speed c_0, and the other with shape g moving to the left with speed c_0. It would be even simpler if there were only one wave. The required equation corresponds to factoring (1.5) as

$$\left(\frac{\partial}{\partial t} - c_0 \frac{\partial}{\partial x} \right) \left(\frac{\partial}{\partial t} + c_0 \frac{\partial}{\partial x} \right) \varphi = 0 \tag{1.9}$$

and retaining only one of the factors. If we retain only

$$\varphi_t + c_0 \varphi_x = 0, \tag{1.10}$$

the general solution is

$$\varphi = f(x - c_0 t). \tag{1.11}$$

This is the simplest hyperbolic wave problem. Although the classical problems led to (1.5), many wave motions have now been studied which do in fact lead to (1.10). Examples are flood waves, waves in glaciers, waves in traffic flow, and certain wave phenomena in chemical reactions. We shall start with these in Chapters 2 and 3. Just as in the classical problems, the original formulations lead to nonlinear equations and the simplest is

$$\varphi_t + c(\varphi) \varphi_x = 0, \tag{1.12}$$

where the propagation speed $c(\varphi)$ is a function of the local disturbance φ. The study of this deceptively simple-looking equation will provide all the main concepts for nonlinear hyperbolic waves. We follow the ideas which were developed first in gas dynamics, but now we develop them in the simpler mathematical context. The main nonlinear feature is the breaking of waves into shock waves, and the corresponding mathematical theory is the theory of characteristics and the special treatment of shock waves. This is all presented in detail in Chapter 2. The theory is then applied and supplemented in Chapter 3 in a full discussion of the topics of flood waves and similar waves noted earlier.

The first order equation (1.12) is called *quasi-linear* in that it is nonlinear in φ but is linear in the derivatives φ_t, φ_x. The general nonlinear first order equation for $\varphi(x,t)$ is any functional relation between φ, φ_t, φ_x. This more general case as well as the extension to first order equations in n independent variables is included in Chapter 2.

In the framework of (1.12), shock waves appear as discontinuities in φ. However, the derivation of (1.12) usually involves approximations which are not strictly valid when shock waves arise. In gas dynamics the corresponding approximation is the omission of viscous and heat conduction effects. Again, the same mathematical effects can be seen in examples simpler than gas dynamics, even though the appropriate ideas were first explored there. These effects are included in Chapters 2 and 3. The simplest case is the equation

$$\varphi_t + \varphi \varphi_x = \nu \varphi_{xx}. \qquad (1.13)$$

It was particularly stressed by Burgers (1948) as being the simplest one to combine typical nonlinearity with typical heat diffusion, and it is usually referred to as Burgers' equation. It was probably introduced first by Bateman (1915). It acquired even more interest when it was shown by Hopf (1950) and Cole (1951) that the general solution could be obtained explicitly. Various questions can be investigated in great detail on this typical example, and then used with confidence in other cases where the full solution is not available and one must resort to special or approximate methods. Chapter 4 is devoted to Burgers' equation and its solution.

For two independent variables, usually the time and one space dimension, the general system corresponding to (1.12) is

$$A_{ij}\frac{\partial u_j}{\partial t} + a_{ij}\frac{\partial u_j}{\partial x} + b_i = 0, \qquad i = 1,\ldots,n, \qquad (1.14)$$

for n unknowns $u_i(x,t)$. (The usual convention is used that summation $j = 1,\ldots,n$ is to be understood for the repeated subscript j.) For linear systems, the matrices A_{ij}, a_{ij} are independent of $\mathbf{u}$, and the vector b_i is a linear expression

$$b_i = b_{ij}u_j \qquad (1.15)$$

in $\mathbf{u}$; (1.5) can be written in this form. When A_{ij}, a_{ij}, b_i are functions of $\mathbf{u}$ but not of its derivatives, the system is quasi-linear. Chapter 5 starts with a discussion of the conditions necessary for (1.14) to be hyperbolic (and hence to correspond to hyperbolic waves), and it then turns to the general theory of characteristics and shocks for such hyperbolic systems.

Gas dynamics is the subject that provided the basis for this material and is its most fruitful physical context. Chapter 6 is a fairly detailed account of gas dynamics for both unsteady problems and supersonic flow. Problems of cylindrical and spherical explosions are included, since they also reduce to two independent variables.

For genuine two or three dimensional problems, we turn in Chapter 7 to a more comprehensive discussion of solutions of the wave equation (1.1). It is perhaps a novelty in a book on wave propagation to delay this so long, and to give such an extensive discussion of nonlinear effects first. This is due to an ordering based on the number of dimensions rather than the difficulty of the concepts or the availability of mathematical techniques. Chapter 7 includes the aspects of solutions to (1.1) which reveal information about the nature of the wave motion involved and which offer the possibility of generalization to other wave systems. The prime example of this is the theory of geometrical optics, which extends to linear waves in nonhomogeneous media and is the basis for similar developments related to shock propagation in nonlinear problems. No attempt is made to give even a relatively brief account of the huge areas of diffraction and scattering theory, nor of the special features of elastic or electromagnetic waves. These are all too extensive to be adequately treated in a book that has such a broad range of topics already.

Chapters 8 and 9 devoted to shock dynamics and propagation problems related to sonic booms build on all this material and show how it can be brought to bear on difficult nonlinear problems. In these two chapters, intuitive ideas and approximations based on physical arguments are used to surmount the mathematical difficulties. Although these problems are drawn from fluid mechanics, it is hoped that the results and the style of thinking will be useful in other fields.

The final chapter on hyperbolic waves concerns those situations where waves of different orders are present simultaneously. A typical example is the equation

$$\eta\left(\varphi_{tt} - c_0^2\varphi_{xx}\right) + \varphi_t + a_0\varphi_x = 0. \qquad (1.16)$$

This is hyperbolic with characteristic velocities $\pm c_0$ determined from the second order wave operator. Yet if η is small, the lower order wave operator $\varphi_t + a_0\varphi_x = 0$ should be a good approximation in some sense, and this predicts waves with speed a_0. It turns out that both kinds of wave play important roles, and there are important interaction effects between the two. The higher order waves carry the "first signal" with speed c_0, but the "main disturbance" travels with the lower order waves at speed a_0. In the nonlinear counterparts to (1.16) this has important bearing on properties of shocks and their structure. This general topic is taken up in Chapter 10.

1.3 Dispersive Waves

Dispersive waves are not classified as easily as hyperbolic waves. As explained in connection with (1.3), the discussion stems from certain types of oscillatory solution representing a train of waves. Such solutions are obtained from a variety of partial differential equations and even certain integral equations. One rapidly realizes that it is the dispersion relation, written

$$\omega = W(\kappa), \tag{1.17}$$

connecting the frequency ω and the wave number κ, which characterizes the problem. The source of this relation in the particular system of equations governing the problem is of subsidiary importance. Some of the typical examples are the beam equation

$$\varphi_{tt} + \gamma^2 \varphi_{xxxx} = 0, \qquad \omega = \pm \gamma \kappa^2, \tag{1.18}$$

the linear Korteweg-deVries equation

$$\varphi_t + c_0 \varphi_x + \nu \varphi_{xxx} = 0, \qquad \omega = c_0 \kappa - \nu \kappa^3, \tag{1.19}$$

and the linear Boussinesq equation

$$\varphi_{tt} - \alpha^2 \varphi_{xx} = \beta^2 \varphi_{xxtt}, \qquad \omega = \pm \alpha \kappa (1 + \beta^2 \kappa^2)^{-1/2}. \tag{1.20}$$

Equations 1.19 and 1.20 appear in the approximate theories of long water waves. The general equations for linear water waves require more detail to explain, but the upshot is a solution (1.3) for the displacement of the surface with

$$\omega = \pm (g\kappa \tanh \kappa h)^{1/2}, \tag{1.21}$$

where h is the undisturbed depth and g is the gravitational acceleration. Another example is the classical theory for the dispersive effects of electromagnetic waves in dielectrics; this leads to

$$(\omega^2 - \nu_0^2)(\omega^2 - c_0^2 \kappa^2) = \omega^2 \nu_p^2, \tag{1.22}$$

where c_0 is the speed of light, ν_0 is the natural frequency of the oscillator, and ν_p is the plasma frequency.

For linear problems, solutions more general than (1.3) are obtained by superposition to form Fourier integrals, such as

$$\varphi = \int_0^\infty F(\kappa) \cos(\kappa x - Wt) \, d\kappa, \qquad (1.23)$$

where $W(\kappa)$ is the dispersion function (1.17) appropriate to the system. Formally, at least, this is a solution for arbitrary $F(\kappa)$, which is then chosen to fit the boundary or initial conditions, with use of the Fourier inversion theorem.

The solution in (1.23) is a superposition of wavetrains of different wave numbers, each traveling with its own phase speed

$$c(\kappa) = \frac{W(\kappa)}{\kappa}. \qquad (1.24)$$

As time evolves, these different component modes "disperse," with the result that a single concentrated hump, for example, disperses into a whole oscillatory train. This process is studied by various asymptotic expansions of (1.23). The key concept that comes out of the analysis is that of the *group velocity* defined as

$$C(\kappa) = \frac{dW}{d\kappa}. \qquad (1.25)$$

The oscillatory train arising from (1.23) does not have constant wavelength; the whole range of wave numbers κ is still present. In a sense to be explained, the different values of wave number propagate through this oscillatory train and the speed of propagation is the group velocity (1.25). In a similar sense it is found that energy also propagates with the group velocity. For genuinely dispersive waves, the case $W \propto \kappa$ is excluded so that the phase velocity (1.24) and the group velocity (1.25) are not the same. And it is the group velocity which plays the dominant role in the propagation.

In view of its great importance, and with an eye to nonuniform media and nonlinear waves, it is desirable to find direct ways of deriving the group velocity and its properties without the intermediary of the Fourier analysis. This can be done very simply on an intuitive basis, which can be justified later. Assume that the nonuniform oscillatory wave is described approximately in the form

$$\varphi = a \cos\theta, \qquad (1.26)$$

where a and θ are functions of x and t. The function $\theta(x,t)$ is the "phase" which measures the point in the cycle of $\cos\theta$ between its extreme values of ± 1, and $a(x,t)$ is the amplitude. The special uniform wavetrain has

$$a = \text{constant}, \quad \theta = \kappa x - \omega t, \quad \omega = W(\kappa). \qquad (1.27)$$

In the more general case, we define a *local* wave number $k(x,t)$ and a *local* frequency $\omega(x,t)$ by

$$k(x,t) = \frac{\partial \theta}{\partial x}, \qquad \omega(x,t) = -\frac{\partial \theta}{\partial t}. \qquad (1.28)$$

Assume now that these are still related by the dispersion relation

$$\omega = W(k). \qquad (1.29)$$

This is then an equation for θ:

$$\frac{\partial \theta}{\partial t} + W\left(\frac{\partial \theta}{\partial x}\right) = 0, \qquad (1.30)$$

and its solution determines the kinematic properties of the wavetrain. It is more convenient to eliminate θ from (1.28) to obtain

$$\frac{\partial k}{\partial t} + \frac{\partial \omega}{\partial x} = 0, \qquad (1.31)$$

and to work with the pair of relations (1.29) and (1.31). Replacing ω by $W(k)$ in (1.31), we have

$$\frac{\partial k}{\partial t} + C(k)\frac{\partial k}{\partial x} = 0, \qquad (1.32)$$

where $C(k)$ is the group velocity defined in (1.25). This equation for k is just the simplest nonlinear hyperbolic equation given in (1.12)! It may be interpreted as a wave equation for the propagation of k with speed $C(k)$. In this rather subtle way, hyperbolic phenomena are hidden in dispersive waves. This may be exploited to bring the methods of Part I to bear on dispersive wave problems.

The more intuitive analysis of group velocity indicated here is readily extended to more dimensions and to nonuniform media where the exact solutions are either inconvenient or unobtainable. The results then usually may be justified directly as the first term in an asymptotic solution. These basic questions with emphasis on the understanding of group velocity arguments are studied in Chapter 11.

Once the group velocity arguments are established, they provide a surprisingly simple yet powerful method for deducing the main features of any linear dispersive system. A wide variety of such cases is given in Chapter 12.

It is easy to show asymptotically from the Fourier integral (1.23) that energy ultimately propagates with the group velocity. For purposes of generalization, it is again important to have direct approaches to this basic result. Some of these are explained in Chapter 11, but until recently there was no wholly satisfactory approach. In the last few years, the problem has been resolved as an offshoot of the investigation of the corresponding questions for nonlinear waves. The nonlinear problems required a more powerful approach altogether, and eventually the possibility of using variational principles was realized. These appear to provide the correct tools for all these questions in both linear and nonlinear dispersive waves. Judging from its recent success, this variational approach has led to a completely fresh view of the subject. It is taken up for linear waves in preliminary fashion in Chapter 11 and the full nonlinear version is described in Chapter 14.

The intermediate Chapter 13 is on the subject of water waves. This is perhaps the most varied and fascinating of all the subjects in wave motion. It includes a wide range of natural phenomena in the oceans and rivers, and suitably interpreted it applies to gravity waves in the atmosphere and other fluids. It has provided the impetus and background for the development of dispersive wave theory, with much the same role that gas dynamics has played for hyperbolic waves. In particular, the fundamental ideas for nonlinear dispersive waves originated in the study of water waves.

1.4 Nonlinear Dispersion

In 1847 Stokes showed that the surface elevation η in a plane wavetrain on deep water could be expanded in powers of the amplitude a as

$$\eta = a\cos\left(\kappa x - \omega t\right) + \tfrac{1}{2}\kappa a^2 \cos 2\left(\kappa x - \omega t\right)$$

$$+ \tfrac{3}{8}\kappa^2 a^3 \cos 3\left(\kappa x - \omega t\right) + \cdots, \tag{1.33}$$

where

$$\omega^2 = g\kappa\left(1 + \kappa^2 a^2 + \cdots\right). \tag{1.34}$$

The linear result would be the first term in (1.33) in agreement with (1.3) and the dispersion relation would be

$$\omega^2 = g\kappa, \tag{1.35}$$

in agreement with (1.21) since one takes the limiting form $\kappa h \rightarrow \infty$ for deep water. There are two key ideas here. First, there exist periodic wavetrain solutions in which the dependent variables are functions of a phase $\theta = \kappa x - \omega t$, but the functions are no longer sinusoidal; (1.33) is the Fourier series expansion of the appropriate function $\eta(\theta)$. The second crucial idea is that the dispersion relation (1.34) also involves the amplitude. This introduces a qualitatively new feature and the nonlinear effects are not merely slight corrections.

In 1895 Korteweg and deVries showed that long waves, in water of relatively shallow depth, could be described approximately by a nonlinear equation of the form

$$\eta_t + (c_0 + c_1 \eta)\eta_x + \nu \eta_{xxx} = 0, \tag{1.36}$$

where c_0, c_1, and ν are constants. A linearization of this for very small amplitudes would drop the term $c_1 \eta \eta_x$; the resulting linear equation has solutions

$$\eta = a \cos(\kappa x - \omega t),$$
$$\omega = c_0 \kappa - \nu \kappa^3. \tag{1.37}$$

One could improve on this by Stokes-type expansions in the amplitude. But one can do better: Korteweg and deVries showed that periodic solutions

$$\eta = f(\theta), \qquad \theta = \kappa x - \omega t$$

of (1.36) could be found in closed form, and without further approximation, in terms of Jacobian elliptic functions. Since $f(\theta)$ was found in terms of the elliptic function $cn\theta$, they named the solutions *cnoidal* waves. This work endorses the general conclusions of Stokes' work. First, the existence of periodic wavetrains is demonstrated explicitly. Second, $f(\theta)$ contains an arbitrary amplitude a, and the solution includes a specified dispersion relation between ω, κ, and a, the most important nonlinear effect being again the inclusion of the amplitude in this relation.

But even more was found. One limit of $cn\theta$ (as the modulus tends to 1) is the sech function. Either by taking this limit or directly from (1.36), the special solution

$$\eta = a \operatorname{sech}^2 \left\{ \left(\frac{c_1 a}{12\nu} \right)^{1/2} (x - Ut) \right\}, \tag{1.38}$$

$$U = c_0 + \frac{1}{3} c_1 a \tag{1.39}$$

may be established. In this limit the period has become infinite and (1.38) represents a single hump of positive elevation. It is the "solitary wave," discovered experimentally by Scott Russell (1844), and previously analyzed on an approximate basis by Boussinesq (1871) and Rayleigh (1876). The inclusion of the solitary wave with the periodic wavetrains in the same analysis was an important step. Equation 1.39 for the velocity of propagation U in terms of the amplitude is the remnant of the dispersion relation in this nonperiodic case.

Although the equation originated in water waves, it was subsequently realized that the Korteweg-deVries equation is one of the simplest prototypes that combines nonlinearity and dispersion. In this respect it is analogous to Burgers' equation, which combines nonlinearity with diffusion. It has now been derived as a useful equation in other fields.

In recent years other simple equations have been derived in various fields and also used as prototypes to develop and test ideas. Notable among these are the equation

$$\varphi_{tt} - \varphi_{xx} + V'(\varphi) = 0, \tag{1.40}$$

a natural generalization of the linear Klein-Gordon equation, and

$$i\psi_t + \psi_{xx} + |\psi|^2 \psi = 0, \tag{1.41}$$

a generalization of Schrödinger's equation. We return to comment on these later.

First we must consider the question of how to build further on Stokes' general result, confirmed by many other examples, that the existence of periodic wavetrains is a typical feature of nonlinear dispersive systems. These solutions are the counterparts of (1.3) but one cannot proceed by simple Fourier superposition. However, the eventual description of many important results in linear theory is in terms of the group velocity for modulated wavetrains as described following (1.26). These ideas are not crucially dependent on the Fourier synthesis and a theory of *nonlinear* group velocities can be developed. The appropriate analysis can be put in a general, concise form using the variational techniques already referred to. The theory is given in Chapter 14. The dependence of the dispersion relation on the amplitude introduces a number of new phenomena (for example, there are two group velocities) and these are discussed in general terms in Chapter 15. In addition to the original problems of water waves, one of the main fields of application is the new, rapidly expanding field of nonlinear optics. A selection of applications to both fields is given in Chapter 16.

One of the most interesting topics in nonlinear optics is the self-focusing of beams, and (1.41) arises in this context. Equation 1.40, particularly in the so-called Sine-Gordon case of

$$\varphi_{tt} - \varphi_{xx} + \sin \varphi = 0, \tag{1.42}$$

arises in a number of areas. Both of these equations share with the Korteweg-deVries equation in having solitary wave solutions as limiting cases. Solitary waves were always of obvious interest, since they are strictly nonlinear phenomena with no counterparts in linear dispersive theory. But until recently little further was known. Now, stemming from the remarkable work of Gardner, Greene, Kruskal, and Miura (1967) on the Korteweg-deVries equation and Perring and Skyrme (1962) and Lamb (1967, 1971) on the Sine-Gordon equation, families of exact solutions representing interacting solitary waves have been found. The surprising result is that solitary waves retain their individuality under interaction and eventually emerge with their original shapes and speeds. These solutions are only one class obtained in a more general attack on the equations, with further results on the solutions for arbitrary initial conditions being fairly complete. Zakharov and Shabat (1972) extended the methods of Gardner et al. to the cubic Schrödinger equation (1.41) and found similar results. An account of these important and ingenious investigations is given in Chapter 17.

Part I

HYPERBOLIC WAVES

CHAPTER 2

Waves and First Order Equations

We start the detailed discussion of hyperbolic waves with a study of first order equations. As noted in Chapter 1, the simplest wave equation is

$$\rho_t + c_0 \rho_x = 0, \qquad c_0 = \text{constant.} \qquad (2.1)$$

When this equation arises, the dependent variable is usually the density of something so we now use the symbol ρ rather than the all-purpose symbol φ of the introduction. The general solution of (2.1) is $\rho = f(x - c_0 t)$, where $f(x)$ is an arbitrary function, and the solution of any particular problem consists merely of matching the function f to initial or boundary values. It clearly describes a wave motion since an initial profile $f(x)$ would be translated unchanged in shape a distance $c_0 t$ to the right at time t. At two observation points a distance s apart, exactly the same disturbance would be recorded with a time delay of s/c_0.

Although this linear case is almost trivial, the nonlinear counterpart

$$\rho_t + c(\rho)\rho_x = 0, \qquad (2.2)$$

where $c(\rho)$ is a given function of ρ, is certainly not and a study of it leads to most of the essential ideas for nonlinear hyperbolic waves. As remarked earlier, many of the classical examples of wave propagation are described by second or higher order equations such as the wave equation $c_0^2 \nabla^2 \varphi = \varphi_{tt}$, but a surprising number of physical problems do lead directly to (2.2) or extensions of it. Examples will be given after a preliminary discussion of the solution. Even in higher order problems, one often searches for special solutions or approximations that involve (2.2).

2.1 Continuous Solutions

One approach to the solution of (2.2) is to consider the function $\rho(x, t)$ at each point of the (x, t) plane and to note that $\rho_t + c(\rho)\rho_x$ is the total

19

derivative of ρ along a curve which has slope

$$\frac{dx}{dt} = c(\rho) \tag{2.3}$$

at every point of it. For along any curve in the (x,t) plane, we may consider x and ρ to be functions of t, and the total derivative of ρ is

$$\frac{d\rho}{dt} = \frac{\partial\rho}{\partial t} + \frac{dx}{dt}\frac{\partial\rho}{\partial x}.$$

The total derivative notation should be sufficient to indicate when x and ρ are being treated as functions of t on a certain curve; the introduction of new symbols each time this is done eventually becomes confusing. We now consider a curve $\mathcal{C}$ in the (x,t) plane which satisfies (2.3). Of course such a curve cannot be determined explicitly in advance since the defining equation (2.3) involves the unknown values of ρ on the curve. However, its consideration will lead us to a simultaneous determination of a possible curve $\mathcal{C}$ and the solution ρ on it. On $\mathcal{C}$ we deduce from the total derivative relation and from (2.2) that

$$\frac{d\rho}{dt} = 0, \qquad \frac{dx}{dt} = c(\rho). \tag{2.4}$$

We first observe that ρ remains constant on $\mathcal{C}$. It then follows that $c(\rho)$ remains constant on $\mathcal{C}$, and therefore that the curve $\mathcal{C}$ must be a straight line in the (x,t) plane with slope $c(\rho)$. Thus the general solution of (2.2) depends on the construction of a family of straight lines in the (x,t) plane, each line with slope $c(\rho)$ corresponding to the value of ρ on it. This is easily done in any specific problem.

Let us take for example the initial value problem

$$\rho = f(x), \quad t = 0, \quad -\infty < x < \infty,$$

and refer to the (x,t) diagram in Fig. 2.1. If one of the curves $\mathcal{C}$ intersects $t = 0$ at $x = \xi$ then $\rho = f(\xi)$ on the whole of that curve. The corresponding slope of the curve is $c(f(\xi))$, which we will denote by $F(\xi)$; it is a known function of ξ calculated from the function $c(\rho)$ in the equation and the given initial function $f(\xi)$. The equation of the curve then is

$$x = \xi + F(\xi)t.$$

This determines one typical curve and the value of ρ on it is $f(\xi)$. Allowing ξ to vary, we obtain the whole family:

$$\rho = f(\xi), \qquad c = F(\xi) = c(f(\xi)) \tag{2.5}$$

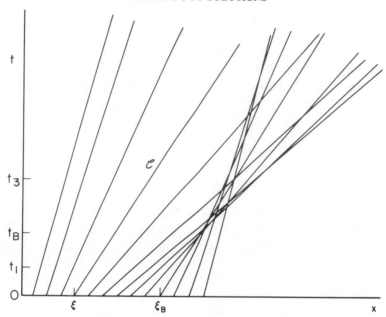

Fig. 2.1. Characteristic diagram for nonlinear waves.

on

$$x = \xi + F(\xi)t. \qquad (2.6)$$

We may now change the emphasis and use (2.5) and (2.6) as an analytic expression for the solution, free of the particular construction. That is, ρ is given by (2.5) where $\xi(x, t)$ is defined implicitly by (2.6). Let us check that this gives the solution. From (2.5),

$$\rho_t = f'(\xi)\xi_t, \qquad \rho_x = f'(\xi)\xi_x,$$

and from the t and x derivatives of (2.6),

$$0 = F(\xi) + \{1 + F'(\xi)t\}\xi_t,$$

$$1 = \{1 + F'(\xi)t\}\xi_x.$$

Therefore

$$\rho_t = -\frac{F(\xi)f'(\xi)}{1 + F'(\xi)t}, \qquad \rho_x = \frac{f'(\xi)}{1 + F'(\xi)t}, \qquad (2.7)$$

and we see that

$$\rho_t + c(\rho)\rho_x = 0$$

since $c(\rho) = F(\xi)$. The initial condition $\rho = f(x)$ is satisfied because $\xi = x$ when $t = 0$.

The curves used in the construction of the solution are the *characteristic curves* for this special problem. Similar characteristics play an important role in all problems involving hyperbolic differential equations. In general, characteristic curves do not have the property that the solution remains constant along them. This happens to be true in the special case of (2.2); it is not the defining property of characteristics. The general definitions will be considered later, but it will be convenient now to refer to the curves defined by (2.3) as characteristics.

The basic idea of wave propagation is that some recognizable feature of the disturbance moves with a finite velocity. For hyperbolic equations, the characteristics correspond to this idea. Each characteristic curve in (x, t) space represents a moving wavelet in x space, and the behavior of the solution on a characteristic curve corresponds to the idea that information is carried by that wavelet. The mathematical statement in (2.4) may be given this type of emphasis by saying that different values of ρ "propagate" with velocity $c(\rho)$. Indeed, the solution at time t can be constructed by moving each point on the initial curve $\rho = f(x)$ a distance $c(\rho)t$ to the right; the distance moved is different for the different values of ρ. This is shown in Fig. 2.2 for the case $c'(\rho) > 0$; the corresponding time levels are indicated in Fig. 2.1. The dependence of c on ρ produces the typical nonlinear distortion of the wave as it propagates. When $c'(\rho) > 0$, higher values of ρ propagate faster than lower ones. When $c'(\rho) < 0$, higher values of ρ propagate slower and the distortion has the opposite tendency to that shown in Fig. 2.2. For the linear case, c is constant and the profile is translated through a distance ct without any change of shape.

It is immediately apparent from Fig. 2.2 that the discussion is far from complete. Any compressive part of the wave, where the propagation velocity is a decreasing function of x, ultimately "breaks" to give a

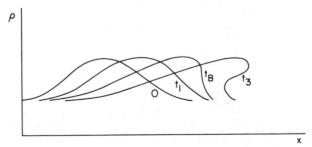

Fig. 2.2. Breaking wave: successive profiles corresponding to the times 0, t_1, t_B, t_3 in Fig. 2.1.

triple-valued solution for $\rho(x,t)$. The breaking starts at the time indicated by $t=t_B$ in Fig. 2.2, when the profile of ρ first develops an infinite slope. The analytic solution (2.7) confirms this and allows us to determine the breaking time t_B. On any characteristic for which $F'(\xi)<0$, ρ_x and ρ_t become infinite when

$$t=-\frac{1}{F'(\xi)}.$$

Therefore breaking first occurs on the characteristic $\xi=\xi_B$ for which $F'(\xi)<0$ and $|F'(\xi)|$ is a maximum; the time of first breaking is

$$t_B=-\frac{1}{F'(\xi_B)}. \tag{2.8}$$

This development can also be followed in the (x,t) plane. A compressive part of the wave with $F'(\xi)<0$ has converging characteristics; since the characteristics are straight lines, they must eventually overlap to give a region where the solution is multivalued, as in Fig. 2.1. This region may be considered as a fold in the (x,t) plane made up of three sheets, with different values of ρ on each sheet. The boundary of the region is an envelope of characteristics. The family of characteristics is given by (2.6) with ξ as parameter. The condition that two neighboring characteristics ξ, $\xi+\delta\xi$ intersect at a point (x,t) is that

$$x=\xi+F(\xi)t$$

and

$$x=\xi+\delta\xi+F(\xi+\delta\xi)t$$

hold simultaneously. In the limit $\delta\xi\to0$, these give

$$x=\xi+F(\xi)t \quad\text{and}\quad 0=1+F'(\xi)t$$

for the implicit equations of an envelope. The second of these relations shows that an envelope is formed in $t>0$ by those characteristics for which $F'(\xi)<0$. The minimum value of t on the envelope occurs for the value of ξ for which $-F'(\xi)$ is maximum. This is the first time of breaking in agreement with (2.8). If $F''(\xi)$ is continuous, the envelope has a cusp at $t=t_B$, $\xi=\xi_B$, as shown in Fig. 2.1.

An extreme case of breaking arises when the initial distribution has a discontinuous step with the value of $c(\rho)$ behind the discontinuity greater than that ahead. If we have the initial functions

$$f(x)=\begin{cases} \rho_1, & x>0 \\ \rho_2, & x<0 \end{cases}$$

and

$$F(x) = \begin{cases} c_1 = c(\rho_1), & x > 0 \\ c_2 = c(\rho_2), & x < 0 \end{cases}$$

with $c_2 > c_1$, then breaking occurs immediately. This is shown in Fig. 2.3 for the case $c'(\rho) > 0$, $\rho_2 > \rho_1$. The multivalued region starts right at the origin and is bounded by the characteristics $x = c_1 t$ and $x = c_2 t$; the boundary is no longer a cusped envelope since F and its derivatives are not continuous. Nevertheless, the result may be considered as the limit of a series of smoothed-out steps, and the breaking point moves closer to the origin as the initial profile approaches the discontinuous step.

On the other hand, if the initial step function is expansive with $c_2 < c_1$, there is a perfectly good continuous solution. It may be obtained as the limit of (2.5) and (2.6) in which all the values of F between c_2 and c_1 are taken on characteristics through the origin $\xi = 0$. This corresponds to a fan of characteristics in the (x,t) plane as in Fig. 2.4. Each member of the fan has a different slope F but the same ξ. The function F is a step function but we use all the values of F between c_2 and c_1 on the face of the step and take them all to correspond to $\xi = 0$. In the fan, the solution (2.5), (2.6) then reads

$$c = F, \qquad x = Ft, \qquad \text{for } c_2 < F < c_1,$$

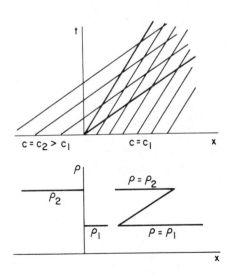

Fig. 2.3. Centered compression wave with overlap.

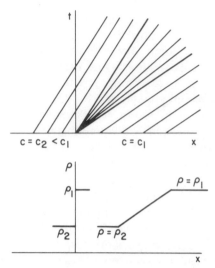

Fig. 2.4. Centered expansion wave.

and by elimination of F we have the simple explicit solution for c:

$$c = \frac{x}{t}, \qquad c_2 < \frac{x}{t} < c_1.$$

The complete solution for c is

$$c = \begin{cases} c_1, & c_1 < \frac{x}{t}, \\ \dfrac{x}{t}, & c_2 < \dfrac{x}{t} < c_1, \\ c_2, & \dfrac{x}{t} < c_2. \end{cases} \qquad (2.9)$$

The relation $c = c(\rho)$ can be solved to determine ρ. For the compressive step, $c_2 > c_1$, the fan in the (x,t) plane is reversed to produce the overlap shown in Fig. 2.3.

In most physical problems where this theory arises, $\rho(x,t)$ is just the density of some medium and is inherently single-valued. Therefore when breaking occurs (2.2) must cease to be valid as a description of the physical problem. Even in cases such as water waves where a multivalued solution for the height of the surface could at least be interpreted, it is still found that (2.2) is inadequate to describe the process. Thus the situation is that some assumption or approximate relation in the formulation leading to (2.2) is no longer valid. In principle one must return to the physics of the problem, see what went wrong, and formulate an improved theory. However, it turns out, as we shall see, that the foregoing solution can be saved by allowing discontinuities into the solution; there is then a single-valued solution with a simple jump discontinuity to replace the multivalued continuous solution. This requires some mathematical extension of what we mean by a "solution" to (2.2), since strictly speaking the derivatives of ρ will not exist at a discontinuity. It can be done through the concept of a "weak solution." But it is important to appreciate that the real issue is not just a mathematical question of extending the solution of (2.2). The breakdown of the continuous solution is associated with the breakdown of some approximate relation in the physics, and the two aspects must be considered together. It is found, for example, that there are several possible families of discontinuous solutions, all satisfactory mathematically; the nonuniqueness can be resolved only by appeal to the physics.

Clearly then, we cannot proceed further without discussion of some physical problems. The prototype is the nonlinear theory of waves in a gas and the formation of shock waves. When viscosity and heat conduction are ignored, the equations of gas dynamics have breaking solutions similar to the preceding ones. As the gradients become steep, just before breaking, the effects of viscosity and heat conduction are no longer negligible. These effects can be included to give an improved theory and waves no longer break in that theory. There is a thin region, a shock wave, in which viscosity and heat conduction are crucially important; outside the shock wave, viscosity and heat conduction may still be neglected. The flow variables change rapidly in the shock. This shock region is idealized into a discontinuity in the "extended" inviscid theory, and only shock conditions relating the jumps of the flow variables across the discontinuity need to be added to the inviscid theory.

We will study all these various aspects in detail. However, gas dynamics is not the simplest example, since it involves higher order equations, and we shall discuss the essential ideas first in the context of the simpler first order problems. It should be remembered, though, that these ideas were developed for gas dynamics, and we are reversing the chronological order. The basic ideas were elucidated by Poisson (1807), Stokes (1848), Riemann (1858), Earnshaw (1858), Rankine (1870), Hugoniot (1889), Rayleigh (1910), Taylor (1910)—a most impressive list. The time required indicates that putting the different aspects together was quite a complicated affair.

2.2 Kinematic Waves

In many problems of wave propagation there is a continuous distribution of either material or some state of the medium, and (for a one dimensional problem) we can define a density $\rho(x,t)$ per unit length and a flux $q(x,t)$ per unit time. We can then define a flow velocity $v(x,t)$ by

$$v = \frac{q}{\rho}.$$

Assuming that the material (or state) is conserved, we can stipulate that the rate of change of the total amount of it in any section $x_1 > x > x_2$ must be balanced by the net inflow across x_1 and x_2. That is,

$$\frac{d}{dt} \int_{x_2}^{x_1} \rho(x,t)\, dx + q(x_1,t) - q(x_2,t) = 0. \qquad (2.10)$$

If $\rho(x,t)$ has continuous derivatives, we may take the limit as $x_1 \to x_2$ and obtain the conservation equation

$$\frac{\partial \rho}{\partial t} + \frac{\partial q}{\partial x} = 0. \qquad (2.11)$$

The simplest wave problems arise when it is reasonable, on either theoretical or empirical grounds, to postulate (in a first approximation!) a functional relation between q and ρ. If this is written as

$$q = Q(\rho), \qquad (2.12)$$

(2.11) and (2.12) form a complete system. On substitution we have

$$\rho_t + c(\rho)\rho_x = 0 \qquad (2.13)$$

where

$$c(\rho) = Q'(\rho). \qquad (2.14)$$

This leads to our (2.2) and a typical solution is given by (2.5) to (2.6). The breaking requires us to reconsider both the mathematical assumption that ρ and q have derivatives and the physical assumption that $q = Q(\rho)$ is a good approximation. To fix ideas for the further development of the theory some specific examples are noted briefly here. We shall return to them in Chapter 3 for a more detailed discussion after the theoretical ideas are complete.

An amusing case (which is also important) concerns traffic flow. It is reasonable to suppose that some essential features of fairly heavy traffic flow may be obtained by treating a stream of traffic as a continuum with an observable density $\rho(x,t)$, equal to the number of cars per unit length, and a flow $q(x,t)$, equal to the number of cars crossing the position x per unit time. For a stretch of highway with no entries or exits, cars are conserved! So we stipulate (2.10). For traffic it also seems reasonable to argue that the traffic flow q is determined primarily by the local density ρ and to propose (2.12) as a first approximation. Such functional relations have been studied and documented to some extent by traffic engineers. We can then apply the theory. But it is clear in this case that when breaking occurs there is no lack of possible explanations for some breakdown in the formulation. Certainly the assumption $q = Q(\rho)$ is a very simplified view of a very complicated phenomenon. For example, if the density is changing rapidly (as it is near breaking), one expects the drivers to react to more than the local density and one also expects that there will be a time lag before they respond adequately to the changing conditions. One might also question the continuum assumption itself.

Another example is flood waves in long rivers. Here ρ is replaced by the cross-sectional area of the channel, A, and this varies with x and t as the level of the river rises. If q is the volume flux across the section, then (2.10) between A and q expresses the conservation of water. Although the fluid flow is extremely complicated, it seems reasonable to start with a functional relation $q = Q(A)$ as a first approximation to express the increase in flow as the level rises. Such relations have been plotted from empirical observations on various rivers. But it is again clear that this assumption is an oversimplification which may well have to be corrected if troubles arise in the theory.

A similar example, proposed and studied extensively by Nye (1960), is the example of glacier flow. The flow velocity is expected to increase with the thickness of the ice, and it seems reasonable to assume a functional dependence between the two.

In chromatography and in similar exchange processes studied in problems of chemical engineering, the same theory arises. The formulation is a little more complicated. The situation is that a fluid carrying dissolved substances or particles or ions flows through a fixed bed and the material being carried is partially adsorbed on the fixed solid material in the bed. The fluid flow is idealized to have a constant velocity V. Then if ρ_f is the density of the material carried in the fluid, and ρ_s is the density deposited on the solid,

$$\rho = \rho_f + \rho_s, \qquad q = V\rho_f.$$

Hence the conservation equation (2.11) reads

$$\frac{\partial}{\partial t}(\rho_f + \rho_s) + \frac{\partial}{\partial x}(V\rho_f) = 0.$$

A second relation concerns the rate of deposition on the solid bed. The exchange equation

$$\frac{\partial \rho_s}{\partial t} = k_1 (A - \rho_s)\rho_f - k_2 \rho_s (B - \rho_f)$$

is apparently the simplest equation with the required properties. The first term represents deposition from the fluid to the solid at a rate proportional to the amount in the fluid, but limited by the amount already on the solid up to a capacity A. The second term is the reverse transfer from the solid to the fluid. (In some processes, the second term is just proportional to ρ_s; this is the limit $B \to \infty$, $k_2 B$ finite.) In equilibrium, the right hand side of the equation vanishes and ρ_s is a definite function of ρ_f. In slowly varying conditions, with relatively large reaction rates k_1 and k_2, we may take a first approximation in which the right hand side still vanishes ("quasi-equilibrium") and we have

$$\rho_s = A \frac{k_1 \rho_f}{k_2 B + (k_1 - k_2)\rho_f}.$$

Thus ρ_s is a function of ρ_f; hence q is a function of ρ. When changes become rapid, just before breaking, the term $\partial \rho_s / \partial t$ in the rate equation can no longer be neglected.

As a different type of example, the concept of group velocity can be fitted into this general scheme. In linear dispersive waves, as already noted following (1.26), there are oscillatory solutions with a local wave number $k(x,t)$ and a local frequency $\omega(x,t)$. Thus k is the density of the waves—the number of wave crests per unit length—and ω is the flux—number of wave crests crossing the position x per unit time. If we expect that wave crests will be conserved in the propagation, we have, in differential form, the conservation equation

$$\frac{\partial k}{\partial t} + \frac{\partial \omega}{\partial x} = 0.$$

In addition, k and ω are related by the dispersion relation

$$\omega = \omega(k).$$

Hence

$$\frac{\partial k}{\partial t} + \omega'(k) \frac{\partial k}{\partial x} = 0.$$

We have a wave propagation for the variations of the local wave number of the "carrier" wavetrain, and the propagation velocity is $d\omega/dk$. This is the group velocity. These ideas will be considered in full detail in the later discussion of dispersive waves.

The wave problems listed here depend primarily on the conservation equation (2.11), and for this reason they were given the name *kinematic waves* (Lighthill and Whitham, 1955) in contrast to the usual acoustic or elastic waves which depend strongly on how the acceleration is determined through the laws of dynamics.

After this review of some of the physical problems, we return to the study of breaking and shock waves in order to complete the theory. Further details of the physical problems are pursued in Chapter 3.

2.3 Shock Waves

When breaking occurs we question the assumption $q = Q(\rho)$ in (2.12) and also the differentiability of ρ and q in (2.11). But, provided the continuum assumption is adequate, we still insist on the conservation equation (2.10).

Consider first the mathematical question of whether discontinuities are possible. Certainly a simple jump discontinuity in ρ and in q is feasible as far as (2.10) is concerned; all the expressions in (2.10) have a meaning. Does (2.10) provide any restriction? To answer this, suppose there is a discontinuity at $x = s(t)$ and that x_1 and x_2 are chosen so that $x_1 > s(t) > x_2$. Suppose ρ and q and their first derivatives are continuous in $x_1 \geqslant x > s(t)$ and in $s(t) > x \geqslant x_2$, and have finite limits as $x \rightarrow s(t)$ from above and below. Then (2.10) may be written

$$q(x_2,t) - q(x_1,t) = \frac{d}{dt}\int_{x_2}^{s(t)}\rho(x,t)\,dx + \frac{d}{dt}\int_{s(t)}^{x_1}\rho(x,t)\,dx$$

$$= \rho(s^-,t)\dot{s} - \rho(s^+,t)\dot{s} + \int_{x_2}^{s(t)}\rho_t(x,t)\,dx + \int_{s(t)}^{x_1}\rho_t(x,t)\,dx,$$

where $\rho(s^-,t)$, $\rho(s^+,t)$ are the value of $\rho(x,t)$ as $x \rightarrow s(t)$ from below and above, respectively, and $\dot{s} = ds/dt$. Since ρ_t is bounded in each of the intervals separately, the integrals tend to zero in the limit as $x_1 \rightarrow s^+$, $x_2 \rightarrow s^-$. Therefore

$$q(s^-,t) - q(s^+,t) = \{\rho(s^-,t) - \rho(s^+,t)\}\dot{s}.$$

A conventional notation is to use a subscript 1 for the values ahead of the shock and a subscript 2 for values behind. Then if U is the shock velocity, $\dot{s}$,

$$q_2 - q_1 = U(\rho_2 - \rho_1). \tag{2.15}$$

The condition may also be written in the form

$$-U[\rho] + [q] = 0, \tag{2.16}$$

where the brackets indicate the jump in the quantity. This form gives a nice correspondence between the shock condition and the differential equation (2.11), the correspondence being

$$\frac{\partial}{\partial t} \leftrightarrow -U[\quad], \qquad \frac{\partial}{\partial x} \leftrightarrow [\quad]. \tag{2.17}$$

We can now extend our solutions of (2.10) to allow such discontinuities. In any continuous part of the solution, (2.11) will still be satisfied and the assumption (2.12) may be retained. Since $q = Q(\rho)$ in the continuous parts, we have $q_2 = Q(\rho_2)$ and $q_1 = Q(\rho_1)$ on the two sides of any shock, and the shock condition (2.15) may be written

$$U = \frac{Q(\rho_2) - Q(\rho_1)}{\rho_2 - \rho_1}. \tag{2.18}$$

The problem then reduces to fitting shock discontinuities into the solution (2.5), (2.6) in such a way that (2.18) is satisfied and multivalued solutions are avoided.

The simplest case is the problem

$$\left. \begin{array}{ll} \rho = \rho_1, & c = c(\rho_1) = c_1, \quad x > 0, \\ \rho = \rho_2, & c = c(\rho_2) = c_2, \quad x < 0, \end{array} \right\} \quad t = 0,$$

with $c_2 > c_1$. The breaking solution was indicated in Fig. 2.3. Now a single-valued solution is possible which is just a shock moving with velocity (2.18):

$$\rho = \rho_1, \quad x > Ut,$$
$$\rho = \rho_2, \quad x < Ut.$$

This is represented schematically in Fig. 2.5.

A popular way to derive the shock condition is to view this particular solution from a frame of reference in which the shock is at rest, as shown

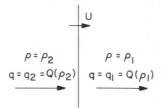

Fig. 2.5. Flow quantities for moving shock.

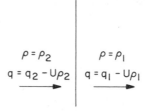

Fig. 2.6. Flow quantities relative to stationary shock.

in Fig. 2.6. The relative flows become $q_1 - U\rho_1$ and $q_2 - U\rho_2$. The conservation law may be stated immediately in the form

$$q_1 - U\rho_1 = q_2 - U\rho_2,$$

and (2.15) follows.

Before proceeding with the general problem of shock fitting, we consider the alternative view that the differential equation (2.11) is adequate but that the assumed relation (2.12) is insufficient.

2.4 Shock Structure

As a particular case, we need to find and examine a more accurate description of the simple discontinuous solution represented in Fig. 2.5. This is the problem of finding the "shock structure."

In many problems of kinematic waves, it would be a better approximation to suppose that q is a function of the density gradient ρ_x as well as ρ. A simple assumption is to take

$$q = Q(\rho) - \nu\rho_x, \tag{2.19}$$

where ν is a constant. In traffic flow, for example, we may argue that drivers will reduce their speed to account for an increasing density ahead, and conversely. This argument would propose a positive value for ν, and we see below that the sign is important. If ν is small, in some suitable dimensionless measure, (2.12) is a good approximation provided ρ_x is not relatively large. At breaking, ρ_x becomes large and the correction term becomes crucial, however small ν may be. Now in *this* formulation, consider continuous solutions. From (2.11) and (2.19), they satisfy

$$\rho_t + c(\rho)\rho_x = \nu\rho_{xx}, \qquad c(\rho) = Q'(\rho). \tag{2.20}$$

The term $c(\rho)\rho_x$ in (2.20) leads to steepening and breaking. On the other hand, the term $\nu\rho_{xx}$ introduces diffusion typical of the heat equation

$$\rho_t = \nu\rho_{xx}.$$

For the heat equation, the solution of the initial step function problem

$$\left.\begin{array}{ll} \rho = \rho_1, & x > 0, \\ \rho = \rho_2, & x < 0, \end{array}\right\} \quad t = 0$$

is

$$\rho = \rho_2 + \frac{\rho_1 - \rho_2}{\sqrt{\pi}} \int_{-\infty}^{x/\sqrt{4\nu t}} e^{-\zeta^2} d\zeta.$$

This represents a smoothed-out step approaching values ρ_1, ρ_2 as $x \to \pm\infty$, and with slope decreasing like $(\nu t)^{-1/2}$. The two opposite tendencies of nonlinear steepening and diffusion are combined in (2.20). The significance of $\nu > 0$ can be seen from the heat equation; solutions are unstable if $\nu < 0$.

We now look within the framework of this more accurate theory for the solution to replace the one shown in Fig. 2.5. One obvious idea is to look for a steady profile solution in which

$$\rho = \rho(X), \qquad X = x - Ut,$$

where U is a constant still to be determined. Then from (2.20),

$$\{c(\rho) - U\}\rho_X = \nu\rho_{XX}.$$

Integrating once, we have

$$Q(\rho) - U\rho + A = \nu\rho_X, \tag{2.21}$$

where A is a constant of integration. An implicit relation for $\rho(X)$ is obtained in the form

$$\frac{X}{\nu} = \int \frac{d\rho}{Q(\rho) - U\rho + A}, \tag{2.22}$$

but the qualitative behavior is more readily seen directly from (2.21). We are interested in the possibility of a solution which tends to constant states $\rho \to \rho_1$ as $X \to +\infty$, $\rho \to \rho_2$ as $X \to -\infty$. If such a solution exists with $\rho_X \to 0$ as $X \to \pm\infty$, the arbitrary parameters U, A must satisfy

$$Q(\rho_1) - U\rho_1 + A = Q(\rho_2) - U\rho_2 + A = 0.$$

In particular,

$$U = \frac{Q(\rho_2) - Q(\rho_1)}{\rho_2 - \rho_1}. \tag{2.23}$$

In such a solution, the relation between the velocity U and the two states at $\pm\infty$ is exactly the same as in the shock condition!

The values ρ_1, ρ_2 are zeros of $Q(\rho) - U\rho + A$, and in general they are simple zeros. As $\rho \to \rho_1$ or ρ_2 in (2.22), the integral diverges and $X \to \pm\infty$ as required. If $Q(\rho) - U\rho + A < 0$ between the two zeros, and if ν is positive, we have $\rho_X < 0$ and the solution is as shown in Fig. 2.7 with ρ increasing monotonically from ρ_1 at $+\infty$ to ρ_2 at $-\infty$. If $Q(\rho) - U\rho + A > 0$ and $\nu > 0$, the solution increases from ρ_2 at $-\infty$ to ρ_1 at $+\infty$. It is clear from (2.21) that if ρ_1, ρ_2 are kept fixed (so that U, A are fixed), a change in ν can be absorbed by a change in the X scale. As $\nu \to 0$, the profile in Fig. 2.7 is compressed in the X direction and tends in the limit to a step function increasing ρ from ρ_1 to ρ_2 and traveling with the velocity given by (2.23). This is exactly the discontinuous shock solution seen in Fig. 2.5. For small nonzero ν the shock is a rapid but continuous increase taking place over a narrow region. The breaking due to the nonlinearity is balanced by the diffusion in this narrow region to give a steady profile.

One very important point is the sign of the change in ρ. A continuous wave carrying an increase of ρ will break forward and require a shock with $\rho_2 > \rho_1$ if $c'(\rho) > 0$; it will break backward and require a shock with $\rho_2 < \rho_1$ if $c'(\rho) < 0$. The shock structure given by (2.21) must agree. As remarked above, ν is always positive for stability, so the direction of increase of ρ depends on the sign of $Q(\rho) - U\rho + A$ between the two zeros ρ_1 and ρ_2. But

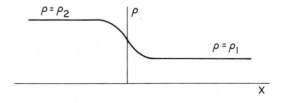

Fig. 2.7. Shock structure.

$c'(\rho) = Q''(\rho)$. Hence when $c'(\rho) > 0$, $Q(\rho) - U\rho + A < 0$ between zeros and the solution is as seen in Fig. 2.7 with $\rho_2 > \rho_1$ as required. If $c'(\rho) < 0$, the step is reversed and $\rho_2 < \rho_1$. The breaking argument and the shock structure agree.

In the special case of a quadratic expression for $Q(\rho)$, taken as

$$Q(\rho) = \alpha\rho^2 + \beta\rho + \gamma, \tag{2.24}$$

the integral in (2.22) is easily evaluated. The sign of α determines the sign of $c'(\rho) = Q''(\rho)$ and we consider $\alpha > 0$, for definiteness. We may write

$$Q - U\rho + A = -\alpha(\rho - \rho_1)(\rho_2 - \rho),$$

where

$$U = \beta + \alpha(\rho_1 + \rho_2), \qquad A = \alpha\rho_1\rho_2 - \gamma.$$

Then (2.22) becomes

$$\frac{X}{\nu} = -\int \frac{d\rho}{\alpha(\rho - \rho_1)(\rho_2 - \rho)} = \frac{1}{\alpha(\rho_2 - \rho_1)} \log \frac{\rho_2 - \rho}{\rho - \rho_1}. \tag{2.25}$$

As $X \to \infty$, $\rho \to \rho_1$ exponentially, and as $X \to -\infty$, $\rho \to \rho_2$ exponentially. There is no precise thickness to the transition region, but we can introduce various measures of the scale, such as the length over which 90% of the change occurs or $(\rho_2 - \rho_1)$ divided by the maximum slope $|\rho_X|$. Clearly all such measures of thickness are proportional to

$$\frac{\nu}{\alpha(\rho_2 - \rho_1)}. \tag{2.26}$$

If this is small compared with other typical lengths in the problem, the rapid shock transition is satisfactorily approximated by a discontinuity. We confirm that the thickness tends to zero as $\nu \to 0$ for fixed ρ_1, ρ_2, but it also should be noted that sufficiently weak shocks with $(\rho_2 - \rho_1)/\rho_1 \to 0$ ultimately become thick for fixed ν, however small. For weak shocks $Q(\rho)$ can always be approximated by a suitable quadratic over the range ρ_1 to ρ_2, so that (2.25) applies. Even for moderately strong shocks it is a good overall approximation to the shape.

The shock structure is only one special solution of (2.20), but from it we might expect in general that when $\nu \to 0$ in some suitable nondimensional form, solutions of (2.20) tend to solutions of

$$\rho_t + c(\rho)\rho_x = 0$$

together with discontinuous shocks satisfying

$$U = \frac{Q(\rho_2) - Q(\rho_1)}{\rho_2 - \rho_1}.$$

This is true when the solutions are compared at fixed (x, t) with $\nu \to 0$. However, the fact that the shock transition becomes very wide as $(\rho_2 - \rho_1)/\rho_1 \to 0$, for fixed ν, means that in any problem where the shocks ultimately tend to zero strength as $t \to \infty$, there may be some final stage with extremely weak shocks when the discontinuous theory will be invalid. This is often a very uninteresting stage, since the shocks must be very weak.

Otherwise, we can say that the two alternative ways of improving on the unacceptable multivalued solutions agree. The use of discontinuous shocks is the easier analytically and can be carried further in more complicated problems.

Confirmation in more detail would require some explicit solutions of (2.20) which involve shocks of varying strength. Although solutions are not known for a general $Q(\rho)$, it turns out that (2.20) can be solved explicitly when $Q(\rho)$ is once again a quadratic in ρ. If (2.20) is multiplied by $c'(\rho)$, it may be written

$$c_t + cc_x = \nu c'(\rho)\rho_{xx}$$

$$= \nu c_{xx} - \nu c''(\rho)\rho_x^2. \tag{2.27}$$

If $Q(\rho)$ is quadratic, $c(\rho)$ is linear in ρ, then $c''(\rho) = 0$ and we have

$$c_t + cc_x = \nu c_{xx}. \tag{2.28}$$

This is Burgers' equation and it can be solved explicitly. The main results are given in Chapter 4. For the present, we accept the evidence for pursuing discontinuous solutions of (2.2) bearing in mind that for extremely weak shocks it will not be appropriate. For the extremely weak shocks, $Q(\rho)$ can be approximated by a quadratic and Burgers' equation can be used.

The arguments in this section depend strongly on $\nu > 0$. As noted previously, this is required for stability of the problem. Interesting cases of instability do occur, however, in traffic flow and flood waves. They are discussed in Chapter 3.

2.5 Weak Shock Waves

In a number of situations the shocks are weak in that $(\rho_2 - \rho_1)/\rho_1$ is small, but they are not so extremely weak that they may no longer be treated as discontinuities. It is useful to note some approximations for such cases.

The shock velocity

$$U = \frac{Q(\rho_2) - Q(\rho_1)}{\rho_2 - \rho_1}$$

tends to the characteristic velocity

$$c(\rho) = \frac{dQ}{d\rho}$$

in the limit as the shock strength $(\rho_2 - \rho_1)/\rho_1 \to 0$. For weak shocks the expression for the shock velocity U may be expanded in a Taylor series in $(\rho_2 - \rho_1)/\rho_1$ as

$$U = Q'(\rho_1) + \frac{1}{2}(\rho_2 - \rho_1)Q''(\rho_1) + O(\rho_2 - \rho_1)^2.$$

The propagation velocity $c(\rho_2) = Q'(\rho_2)$ may also be expanded as

$$c(\rho_2) = c(\rho_1) + (\rho_2 - \rho_1)Q''(\rho_1) + O(\rho_2 - \rho_1)^2.$$

Therefore

$$U = \frac{1}{2}(c_1 + c_2) + O(\rho_2 - \rho_1)^2, \tag{2.29}$$

where $c_1 = c(\rho_1)$ and $c_2 = c(\rho_2)$. To this approximation, the shock velocity is the mean of the characteristic velocities on the two sides of it. In the (x, t) plane the shock curve bisects the angle between the characteristics which meet on the shock. This property is useful for sketching in the shocks, but it also simplifies the analytic determination of shock positions. Clearly the relation is exact when $Q(\rho)$ is a quadratic.

2.6 Breaking Condition

A continuous wave breaks and requires a shock if and only if the propagation velocity c decreases as x increases. Therefore when the shock is included we have

$$c_2 > U > c_1, \tag{2.30}$$

where all velocities are measured positive in the direction of x increasing and the subscript 1 refers to the value of c just ahead of the shock (i.e., greater value of x) and the subscript 2 refers to the value of c just behind the shock. A shock produces an increase in c, and it is supersonic viewed from ahead and subsonic viewed from behind. As regards the jump condition alone, it would be feasible to fit in discontinuities with $c_2 < c_1$. However, shocks with $c_2 < c_1$ could never be formed from a continuous wave and they are never required; they are excluded from consideration for this reason.

One question about this argument is the point that the solution represented in Fig. 2.5 might be set up with $c_2 < c_1$ (by some complicated but probably highly unrealistic device). Of course we have already noted in (2.9) and in Fig. 2.6 a satisfactory continuous solution for such initial conditions. Still, to be particularly awkward, one might insist that Fig. 2.5 gives an alternative solution. The answer is that this proposed solution is unstable. That is, small perturbations would change the flow into some- thing quite different—the expansion fan solution of (2.9). This is a "disintegration argument" which is complementary to the "formation argument." The instability will not be considered in detail in this chapter since the formation argument is convincing and unambiguous. For higher order equations the shock formation becomes harder to study and the instability arguments sometimes give an easier method to decide whether a particular shock satisfying the shock conditions is really possible.

For gas dynamic shocks, the inequality corresponding to (2.30) is equivalent to the condition that the entropy of the gas *increases* as the gas passes through the shock. The entropy condition was the first argument for the irreversibility of shock waves, that is, that the shock transition goes only one way. However, conditions like (2.30) are more general. In some problems there is no obvious counterpart to entropy; in others, such as magnetogasdynamics, the entropy condition does not rule out some inad- missible shocks.

An alternative view of these criteria is that any acceptable discon- tinuous shock must have a satisfactory shock structure when described by more accurate equations. This is a more satisfactory point of view, since it appeals to a more realistic description of the phenomenon. However, the analysis may become prohibitive and one often resorts to the indirect arguments in the framework of the simpler theory.

This alternative approach was checked in the discussion of shock structure in Section 2.4. When $c'(\rho) > 0$ we found only a shock structure for $\rho_2 > \rho_1$; since $c'(\rho) > 0$, this is equivalent to $c_2 > c_1$. When $c'(\rho) < 0$, we found $\rho_2 < \rho_1$, but the change in sign of $c'(\rho)$ means that $c_2 > c_1$. Since $c(\rho) = Q'(\rho)$, the shock velocity lies between the values c_1 and c_2 by Rolle's theorem.

2.7 Note on Conservation Laws and Weak Solutions

Mathematically, the composite solution composed of continuously differentiable parts satisfying

$$\frac{\partial \rho}{\partial t} + \frac{\partial Q(\rho)}{\partial x} = 0 \tag{2.31}$$

together with jump discontinuities satisfying

$$-U[\rho] + [Q(\rho)] = 0 \tag{2.32}$$

can be considered a weak solution of (2.31). Briefly, the idea is as follows. *Associate* with (2.31) the equation

$$-\iint_R \{\rho \phi_t + Q(\rho)\phi_x\} \, dx \, dt = 0, \tag{2.33}$$

where R is an arbitrary rectangle in the (x, t) plane, and ϕ is an arbitrary "test" function with continuous first derivatives in R and $\phi = 0$ on the boundary of R. If ρ and $Q(\rho)$ are continuously differentiable, (2.31) and (2.33) are equivalent. On the one hand, if (2.31) is multiplied by ϕ and integrated over R, we may deduce (2.33) after integration by parts. On the other hand, integration by parts on (2.33) leads to

$$\iint_R \left\{ \frac{\partial \rho}{\partial t} + \frac{\partial Q(\rho)}{\partial x} \right\} \phi \, dx \, dt = 0,$$

and, since this must hold for all arbitrary continuous ϕ, (2.31) follows. However, (2.33) allows more general possibilities, since the admissible functions $\rho(x, t)$ need not have derivatives. Functions $\rho(x, t)$ which satisfy (2.33) for all test functions ϕ are called "weak solutions" of (2.31).

We now investigate what this extended meaning of solution has achieved. Consider the possibility of a weak solution $\rho(x, t)$, that is, one satisfying (2.33), which is continuously differentiable in two parts R_1 and R_2 of R, but with a simple jump discontinuity across the dividing boundary, S, between R_1 and R_2. We may integrate by parts in each of the separate regions R_1, R_2, and deduce from (2.33) that

$$\iint_{R_1} \left\{ \frac{\partial \rho}{\partial t} + \frac{\partial Q(\rho)}{\partial x} \right\} \phi \, dx \, dt + \iint_{R_2} \left\{ \frac{\partial \rho}{\partial t} + \frac{\partial Q(\rho)}{\partial x} \right\} \phi \, dx \, dt$$

$$+ \int_S \{ [\rho] l + [Q(\rho)] m \} \phi \, ds = 0,$$

where (l, m) is the normal to S and $[\rho]$, $[Q(\rho)]$ denote the jumps across S. The line integral on S consists of the two contributions from the boundary terms of R_1 and R_2 obtained in the integration by parts. Since this equation must hold for all test functions ϕ, we deduce that (2.31) holds inside each of the regions R_1 and R_2, but in addition we deduce

$$[\rho]l + [Q(\rho)]m = 0 \quad \text{on} \quad S.$$

This is the shock condition (2.32), since $U = -l/m$. Thus weak solutions of this type would satisfy (2.31) at points of continuity and allow jump discontinuities satisfying the shock condition. Just what we want!

At first sight, the weak solution concept appears to bypass the more involved and less precise discussion of the real physical processes. But it is not really so. Corresponding to the differential equation

$$\frac{\partial \rho}{\partial t} + c(\rho) \frac{\partial \rho}{\partial x} = 0$$

there are an infinite number of conservation equations

$$\frac{\partial f(\rho)}{\partial t} + \frac{\partial g(\rho)}{\partial x} = 0. \tag{2.34}$$

Any choice which satisfies

$$g'(\rho) = f'(\rho)c(\rho) \tag{2.35}$$

will do. For differentiable functions $\rho(x, t)$, these are all equivalent. However, their integrated forms are *not* equivalent and lead to different jump conditions. The weak solution of (2.34) will require the shock condition

$$-U[f(\rho)] + [g(\rho)] = 0; \tag{2.36}$$

different choices of f and g lead to different relations between ρ_1, ρ_2, and U. Therefore a discussion of the physical processes is still necessary in order to pick out which weak solution is relevant to the particular physical problem at hand.

From the differential equation 2.34 we can propose a candidate for a conservation equation in integrated form:

$$\frac{d}{dt} \int_{x_2}^{x_1} f(\rho) \, dx + [g(\rho)]_{x_2}^{x_1} = 0. \tag{2.37}$$

But whether this holds for nondifferentiable ρ can be decided only by returning to the original formulation of the problem. In Section 2.2 we argued in the correct order: first (2.10) then (2.11). The reverse order, going from an equivalent partial differential equation to an integrated form, introduces the lack of uniqueness.

If (2.37) is the true conservation equation, then (2.36) may be deduced as the shock condition by the same argument that was used in Section 2.3. Thus the correct choice of weak solution is made on the basis of which quantities are really conserved across the shock. In view of the lack of uniqueness and possible confusion, it is felt that the weak solution concept is not particularly valuable in this context and it is better to stress that physical problems are first formulated in the basic integrated forms from which both the partial differential equations and the appropriate jump conditions follow.

A looser form of the weak solution idea is sometimes useful in a preliminary look at a problem. If, for example, we ask whether (2.34) might admit moving discontinuities as part of the solution, we might try

$$f(\rho) = f_0(x) H(x - Ut) + f_1,$$

$$g(\rho) = g_0(x) H(x - Ut) + g_1,$$

where $H(x)$ is the Heaviside step function and f_1, g_1 are continuous functions. On substitution in the equation we obtain δ function terms

$$(-Uf_0 + g_0)\delta(x - Ut)$$

plus less singular terms. We deduce that

$$-Uf_0 + g_0 = 0,$$

and this is the shock condition (2.36), since $f_0 = [f]$, $g_0 = [g]$. This does not avoid the lack of uniqueness, of course, and it also uses δ functions in a slightly dubious way. The use of δ functions in nonlinear problems usually is excluded because there is no satisfactory meaning to powers and products of such generalized functions; we have retained an artificial linearity by expressing $f(\rho)$ and $g(\rho)$ separately, rather than using a single expression for ρ. Of course, the justification of the δ function argument is via the weak solution.

In contrast, consider the same question of the possibility of moving discontinuities for (2.20) written in the form

$$\frac{\partial \rho}{\partial t} + \frac{\partial Q(\rho)}{\partial x} = \nu \frac{\partial^2 \rho}{\partial x^2}.$$

If ρ and $Q(\rho)$ are expressed in terms of $H(x - Ut)$, the term $\partial^2\rho/\partial x^2$ will have a term $[\rho]\delta'(x - Ut)$ and there is no other term as singular as $\delta'(x - Ut)$ to balance it. We conclude that $[\rho] = 0$ and that discontinuities are not possible. This is clearly a useful tool for a preliminary assessment.

2.8 Shock Fitting; Quadratic $Q(\rho)$

After discussion of these various points of view we now turn to the analytic problem of fitting discontinuous shocks satisfying

$$U = \frac{Q(\rho_2) - Q(\rho_1)}{\rho_2 - \rho_1} \tag{2.38}$$

into the continuous solution

$$\rho = f(\xi),$$
$$x = \xi + F(\xi)t. \tag{2.39}$$

Any multivalued part of the wave profile must be replaced by an appropriate discontinuity, as shown in Fig. 2.8.* The correct position for the discontinuity may be determined by the followng ingenious argument. Both the multivalued curve and the discontinuous curve satisfy conservation. Therefore $\int \rho \, dx$ under each curve must be the same; hence the discontinuity must cut off lobes of equal area, shown shaded in Fig. 2.8.

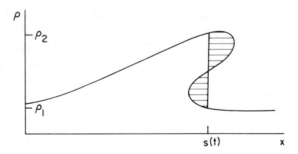

Fig. 2.8. Equal area construction for the position of the shock in a breaking wave.

*The figure is drawn for the case $c'(\rho) > 0$ but all the formulas in this section are correct for either case.

This determination, although quite general, is not in a convenient form for analytic work. The general case gets complicated and it is worthwhile to do a special case first. The special case is again a quadratic expression for $Q(\rho)$. This includes the case of weak disturbances about a value $\rho = \rho_0$, since $Q(\rho)$ can then be approximated by

$$Q = Q(\rho_0) + Q'(\rho_0)(\rho - \rho_0) + \frac{1}{2} Q''(\rho_0)(\rho - \rho_0)^2,$$

and for this reason it has considerable generality.

We consider

$$Q(\rho) = \alpha \rho^2 + \beta \rho + \gamma.$$

Then

$$c(\rho) = Q'(\rho) = 2\alpha \rho + \beta$$

and the shock velocity (2.38) becomes

$$U = \frac{1}{2}(c_1 + c_2),$$

where $c_1 = c(\rho_1)$, $c_2 = c(\rho_2)$.

The simplicity of this case is that the whole problem can be written in terms of c. The continuous solution is

$$c = F(\xi),$$
$$x = \xi + F(\xi)t, \tag{2.40}$$

and shocks must be fitted in such that

$$U = \frac{1}{2}(c_1 + c_2) = \frac{1}{2}\{F(\xi_1) + F(\xi_2)\}, \tag{2.41}$$

where ξ_1 and ξ_2 are the values of ξ on the two sides of the shock. Since ρ and c are linearly related, the conservation of ρ implies conservation of c; that is, $\int c\,dx$ is conserved in the solution. Therefore for this special case the shock construction for the (ρ, x) curve in Fig. 2.8 applies equally well to the (c, x) curve.

It is convenient to note for future reference that this solution in terms of c solves the equation

$$c_t + cc_x = 0, \tag{2.42}$$

with weak solutions chosen to satisfy the conservation law

$$c_t + \left(\tfrac{1}{2}c^2\right)_x = 0 \tag{2.43}$$

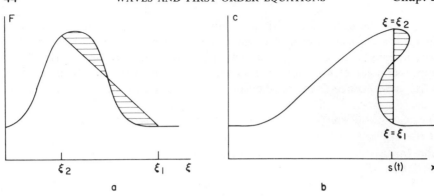

Fig. 2.9. Equal area construction: (*a*) on the initial profile; (*b*) on the transformed breaking profile.

so that the shock condition is

$$U = \frac{\tfrac{1}{2}c_2^2 - \tfrac{1}{2}c_1^2}{c_2 - c_1} = \frac{1}{2}(c_1 + c_2). \tag{2.44}$$

Equation 2.42 is true for general $Q(\rho)$, since it is $c'(\rho)$ times $\rho_t + c(\rho)\rho_x = 0$; (2.44) is always a possible weak solution, but it is the *correct* choice only when $Q(\rho)$ is quadratic or approximated by a quadratic since it is only in that case that the integrated form of (2.43) holds across discontinuities.

    The shock construction can now be combined with the continuous solution (2.40). Since we now work with c the awkward distinction between the two cases $c'(\rho) \gtrless 0$ does not arise. According to (2.40) the solution at time t is obtained from the initial profile $c = F(\xi)$ by translating each point a distance $F(\xi)t$ to the right, as shown in Fig. 2.9. The shock cuts out the

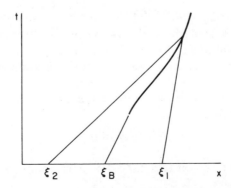

Fig. 2.10. The (x,t) diagram associated with the shock construction in Fig. 2.9.

part corresponding to $\xi_2 \geqslant \xi \geqslant \xi_1$. If the discontinuity line in Fig. 2.9b is also mapped back as in Fig. 2.9a, it is a straight line chord between the points $\xi = \xi_1$ and $\xi = \xi_2$ on the curve $F(\xi)$. Moreover, since areas are preserved under the mapping, the equal area property still holds in Fig. 2.9a; the chord on the F curve cuts off lobes of equal area. The shock determination can then be described entirely on the fixed $F(\xi)$ curve by drawing all the chords with the equal area property. The pairs $\xi = \xi_1$, $\xi = \xi_2$ at the ends of each chord relate characteristics which meet on the shock. The (x, t) plane is shown in Fig. 2.10. The equal area property can be written analytically as

$$\frac{1}{2} \left\{ F(\xi_1) + F(\xi_2) \right\} (\xi_1 - \xi_2) = \int_{\xi_2}^{\xi_1} F(\xi) \, d\xi, \qquad (2.45)$$

since the left hand side is the area under the chord and the right hand side is the area under the F curve. If the shock is at $x = s(t)$ at time t, we also have

$$s(t) = \xi_1 + F(\xi_1) t \qquad (2.46)$$

$$s(t) = \xi_2 + F(\xi_2) t \qquad (2.47)$$

from the second of (2.40). The three equations (2.45)–(2.47) determine the three functions $s(t)$, $\xi_1(t)$, and $\xi_2(t)$. The determination of $s(t)$ is implicit involving the two additional functions $\xi_1(t)$ and $\xi_2(t)$ which determine the characteristics meeting the shock at time t. The values of c on the two sides of the shock are $c_1 = F(\xi_1)$ and $c_2 = F(\xi_2)$; the values of ρ are obtained from c.

Since the shock determination (2.45)–(2.47) was obtained geometrically, it is interesting to check directly that it does indeed satisfy the shock condition (2.41). We may write this verification as an independent derivation of the result in (2.45). We have to find three functions $s(t)$, $\xi_1(t)$, $\xi_2(t)$ which satisfy (2.46), (2.47), and

$$\dot{s}(t) = \frac{1}{2} \left\{ F(\xi_1) + F(\xi_2) \right\}. \qquad (2.48)$$

(Dots denote t derivatives.) From (2.46) and (2.47), we have

$$t = - \frac{\xi_1 - \xi_2}{F(\xi_1) - F(\xi_2)}, \qquad (2.49)$$

and

$$\dot{s}(t) = \{1 + tF'(\xi_1)\}\dot{\xi}_1 + F(\xi_1),$$

$$\dot{s}(t) = \{1 + tF'(\xi_2)\}\dot{\xi}_2 + F(\xi_2).$$

If we take the mean of these last two expressions for $\dot{s}$ in order to preserve symmetry, substitute for t from (2.49), and then substitute in (2.48), we obtain

$$\frac{1}{2}\left\{F'(\xi_1)\dot{\xi}_1 + F'(\xi_2)\dot{\xi}_2\right\}(\xi_1 - \xi_2) + \frac{1}{2}\left\{F(\xi_1) + F(\xi_2)\right\}\left(\dot{\xi}_1 - \dot{\xi}_2\right)$$

$$= F(\xi_1)\dot{\xi}_1 - F(\xi_2)\dot{\xi}_2.$$

This may be integrated to (2.45); the constant of integration may be dropped since the starting point of the shock, $\xi_1 = \xi_2$, must be a solution.

The expression (2.49) for the time can be used to follow the development of the shock. Since $t > 0$, all the relevant chords in Fig. 2.9a must have negative slope. Since $\xi_1 > \xi_2$ by the choice of notation, $F(\xi_2) > F(\xi_1)$, that is, $c_2 > c_1$ as we decided from the breaking condition. The earliest time for the shock corresponds to the steepest chord. This is the limit when the chord is tangent at the point of inflexion $\xi = \xi_B$, say. Then $F(\xi_1) = F(\xi_2)$ so the shock starts with zero strength and the time is

$$t_B = -\frac{1}{F'(\xi_B)}.$$

This all fits with the conditions for the first point of breaking discussed in (2.8). For an F curve like Fig. 2.9a, the chords tend to the horizontal as $t \to \infty$, with $F(\xi_2) - F(\xi_1) \to 0$; hence $c_2 - c_1 \to 0$ and the shock strength tends to zero as $t \to \infty$.

Single Hump.

To study the shock in detail, we suppose first that $F(\xi)$ is equal to a constant c_0 outside the range $0 < \xi < L$, and $F(\xi) > c_0$ in the range. Equation 2.45 may be written as

$$\frac{1}{2}\left\{F(\xi_1) + F(\xi_2) - 2c_0\right\}(\xi_1 - \xi_2) = \int_{\xi_2}^{\xi_1}\left\{F(\xi) - c_0\right\}d\xi.$$

As time goes on, ξ_1 increases and eventually exceeds L. At this stage

$F(\xi_1)=c_0$ and the shock is moving into the constant region $c=c_0$. The function $\xi_1(t)$ can then be eliminated, for we have

$$\frac{1}{2}\{F(\xi_2)-c_0\}(\xi_1-\xi_2)=\int_{\xi_2}^{L}\{F(\xi)-c_0\}\,d\xi, \qquad t=\frac{\xi_1-\xi_2}{F(\xi_2)-c_0}.$$

Therefore

$$\frac{1}{2}\{F(\xi_2)-c_0\}^2 t=\int_{\xi_2}^{L}\{F(\xi)-c_0\}\,d\xi.$$

At this stage, the shock position and the value of c just behind the shock are given by

$$s(t)=\xi_2+F(\xi_2)t,$$
$$c=F(\xi_2),$$

(2.50)

where $\xi_2(t)$ satisfies

$$\frac{1}{2}\{F(\xi_2)-c_0\}^2 t=\int_{\xi_2}^{L}\{F(\xi)-c_0\}\,d\xi.$$

As $t\to\infty$, we have $\xi_2\to0$ and $F(\xi_2)\to c_0$; hence the equation for $\xi_2(t)$ takes the limiting form

$$\frac{1}{2}\{F(\xi_2)-c_0\}^2 t\sim A,$$

where

$$A=\int_{0}^{L}\{F(\xi)-c_0\}\,d\xi$$

is the area of the hump above the undisturbed value c_0. We have $\xi_2\to0$, $F(\xi_2)\sim c_0+\sqrt{2A/t}$. Therefore the asymptotic formulas for $s(t)$ and c in (2.50) are

$$s\sim c_0 t+\sqrt{2At},$$
$$c-c_0\sim\sqrt{\frac{2A}{t}},$$

(2.51)

at the shock. The shock curve is asymptotically parabolic and the shock strength $(c-c_0)/c_0$ tends to zero like $t^{-1/2}$.

The solution behind the shock is given by (2.40) with $0<\xi<\xi_2$. Since

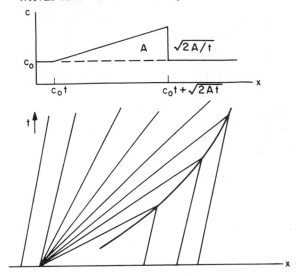

Fig. 2.11. The asymptotic triangular wave.

$\xi_2 \to 0$ as $t \to 0$, all the relevant values of ξ also tend to zero and the asymptotic form is

$$c \sim \frac{x}{t}, \qquad c_0 t < x < c_0 t + \sqrt{2At} . \tag{2.52}$$

The asymptotic solution and the corresponding (x, t) diagram are shown in Fig. 2.11. Notice that the details of the initial distribution are lost; only $A = \int_0^L \{F(\xi) - c_0\} d\xi$ appears in the ultimate asymptotic behavior.

N Wave.

Other problems can be worked out in a similar way. One important case is when $F(\xi)$ has a positive and a negative phase about an undisturbed value c_0, as in Fig. 2.12. There are now two shocks, corresponding to the two compression phases at the front and at the back where $F'(\xi) < 0$. The families of chords for each are shown in the figure. As $t \to \infty$, the pair (ξ_2, ξ_1) for the front shock approach $(0, \infty)$, whereas for the rear shock, (ξ_2, ξ_1) approach $(-\infty, 0)$. Asymptotically the front shock is

$$s \sim c_0 t + \sqrt{2At}$$

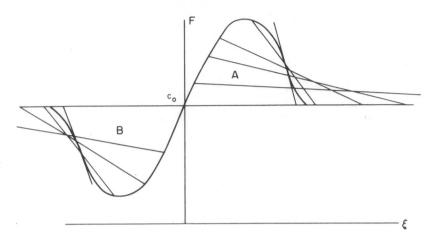

Fig. 2.12. Shock construction for an N wave.

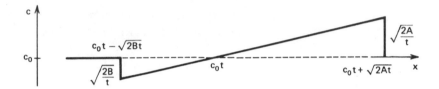

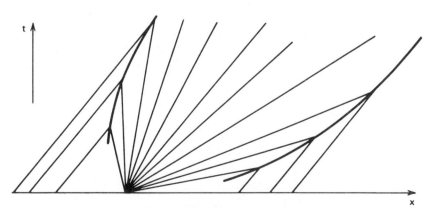

Fig. 2.13. The asymptotic N wave.

and the jump of c is

$$c - c_0 \sim \sqrt{\frac{2A}{t}} \, ,$$

where A is the area of the F curve above $c = c_0$. The rear shock has

$$x \sim c_0 t - \sqrt{2Bt} \, ,$$

$$c - c_0 \sim -\sqrt{\frac{2B}{t}} \, ,$$

where B is the area below $c = c_0$. The solution between the shocks is again asymptotically

$$c \sim \frac{x}{t}, \qquad c_0 t - \sqrt{2Bt} < x < c_0 t + \sqrt{2At} \, . \tag{2.53}$$

The asymptotic form and the (x, t) diagram are shown in Fig. 2.13. Because of the shape of the wave profile, it is known as the N wave.

Periodic Wave.

Another interesting problem is that of an initial distribution

$$c = F(\xi) = c_0 + a \sin \frac{2\pi\xi}{\lambda}. \tag{2.54}$$

In this case, the shock equations (2.45)–(2.47) simplify considerably for all times t. Consider one period $0 < \xi < \lambda$ as in Fig. 2.14. Relation 2.45 becomes

$$(\xi_1 - \xi_2) \sin \frac{\pi}{\lambda} (\xi_1 + \xi_2) \cos \frac{\pi}{\lambda} (\xi_1 - \xi_2) = \frac{\lambda}{\pi} \sin \frac{\pi}{\lambda} (\xi_1 - \xi_2) \sin \frac{\pi}{\lambda} (\xi_1 + \xi_2),$$

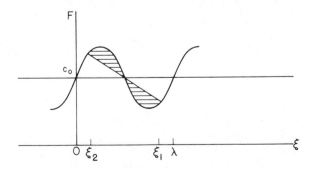

Fig. 2.14. Shock construction for a periodic wave.

and the relevant choice is the trivial one

$$\sin \frac{\pi(\xi_1 + \xi_2)}{\lambda} = 0, \quad \text{that is,} \quad \xi_1 + \xi_2 = \lambda.$$

From the difference and sum of (2.46) and (2.47), we have

$$t = \frac{\xi_1 - \xi_2}{2a \sin \frac{\pi}{\lambda}(\xi_1 - \xi_2)},$$

$$s = c_0 t + \frac{\lambda}{2},$$

respectively. The discontinuity in c at the shock is

$$c_2 - c_1 = a \sin \frac{2\pi\xi_1}{\lambda} - a \sin \frac{2\pi\xi_2}{\lambda}$$

$$= 2a \sin \frac{\pi}{\lambda}(\xi_1 - \xi_2).$$

If we introduce

$$\xi_1 - \xi_2 = \frac{\lambda\theta}{\pi}, \qquad \xi_1 + \xi_2 = \lambda,$$

we have

$$t = \frac{\lambda}{2\pi a} \frac{\theta}{\sin\theta},$$

$$s = c_0 t + \frac{\lambda}{2}, \tag{2.55}$$

$$\frac{c_2 - c_1}{c_0} = \frac{2a}{c_0} \sin \theta.$$

The shock has constant velocity c_0 and this result could have been deduced in advance from the symmetry of the problem. The shock starts with zero strength corresponding to $\theta = 0$ at time $t = \lambda/2\pi a$. It reaches a maximum strength of $2a/c_0$ for $\theta = \pi/2$, $t = \lambda/4a$, and decays ultimately with $\theta \to \pi$, $t \to \infty$,

$$\frac{c_2 - c_1}{c_0} \sim \frac{\lambda}{c_0 t}. \tag{2.56}$$

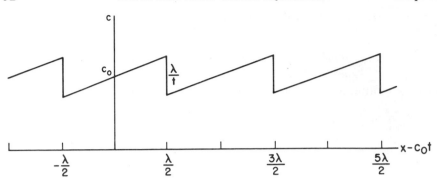

Fig. 2.15. Asymptotic form of a periodic wave.

It is interesting that the final decay formula does not even depend explicitly on the amplitude a. However, the condition for its applicability is $t \gg \lambda/a$. For any periodic $F(\xi)$, sinusoidal or not, $\xi_1 - \xi_2 \to \lambda$ as $t \to \infty$; hence from (2.49)

$$\frac{c_2 - c_1}{c_0} = \frac{F(\xi_2) - F(\xi_1)}{c_0} \sim \frac{\lambda}{c_0 t}.$$

Between successive shocks, the solution for c is linear in x with slope $1/t$ as before, and the asymptotic form of the entire profile is the sawtooth shown in Fig. 2.15.

Confluence of Shocks.

When a number of shocks are produced it is possible in general for one of them to overtake the shock ahead; they then combine and continue as a single shock. This is also described by our shock solution. Consider the F curve in Fig. 2.16. Two shocks are formed corresponding to the points of inflexion P and Q with families of equal area chords typified by $P_1 P_2$ and $Q_1 Q_2$. As time goes on the points Q_1 and P_2 approach each other until the stage in Fig. 2.16b is reached where a common chord cuts off lobes of equal area for both humps. At this stage the characteristics corresponding to P_2' and Q_1' are the same, and therefore the shocks have just combined into one as shown in the (x, t) diagram Fig. 2.17. All the characteristics between Q_2' and P_1' have now been absorbed by one or other of the shocks; a single shock proceeds using chords $P_1'' Q_2''$ as in Fig. 2.16c, counting only *total* areas above and below the chord in the equal area construction.

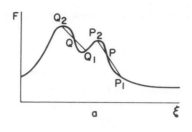

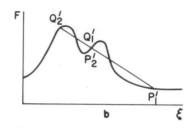

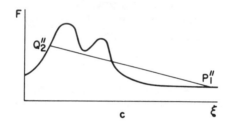

Fig. 2.16. Construction for merging shocks.

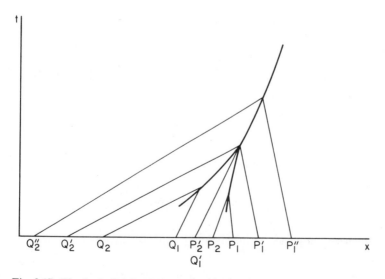

Fig. 2.17. The (x, t) diagram for merging shocks corresponding to Figs. 2.16.

2.9 Shock Fitting; General $Q(\rho)$

For the general dependence of q on ρ, the shock determination can be put into an analytic form similar to (2.45)–(2.47). The complication is the nonlinear relation between c and ρ, so that the construction in Fig. 2.8 is correct for ρ but not for c. Accordingly, we must work with ρ. But if we then plot (ρ, x) curves similar to Fig. 2.9, the discontinuity line does not map into a straight chord because the translation is proportional to c and not to ρ. Thus the mapping back onto the initial curve is no particular advantage.

However, we can proceed as follows. Introduce the function $\xi(\rho)$, which is the inverse of

$$\rho = f(\xi),$$

and introduce also the function $X(\rho, t)$, which is the inverse of the function $\rho = \rho(x, t)$ in the multivalued solution. That is, we fix attention on a particular value of ρ and note where it is now, $X(\rho, t)$, and where it was initially, $\xi(\rho)$. From the equation for the characteristics we have

$$X(\rho, t) = c(\rho)t + \xi(\rho). \tag{2.57}$$

Consider the shock at $s(t)$ and let ρ_1 and ρ_2 be the values ahead and behind the shock, respectively. The equal area construction in Fig. 2.8 may be written

$$\int_{\rho_2}^{\rho_1} X(\rho, t)\, d\rho = (\rho_1 - \rho_2)s(t).$$

[This is true for either case $c'(\rho) \gtrless 0$. We always take ρ_1 to be the value ahead of the shock and ρ_2 to be the value behind. If $c'(\rho) > 0$, then $\rho_2 > \rho_1$; if $c'(\rho) < 0$, then $\rho_2 < \rho_1$.] Hence from (2.57),

$$\int_{\rho_2}^{\rho_1} \{ c(\rho)t + \xi(\rho) \}\, d\rho = (\rho_1 - \rho_2)s(t).$$

Since $c(\rho) = Q'(\rho)$, this may be written

$$(q_1 - q_2)t - (\rho_1 - \rho_2)s(t) = -\int_{\rho_2}^{\rho_1} \xi(\rho)\, d\rho. \tag{2.58}$$

The right hand side can be integrated by parts and rewritten as

$$-\rho_1\xi_1 + \rho_2\xi_2 + \int_{\xi_2}^{\xi_1} \rho(\xi)\, d\xi.$$

The shock position $s(t)$ is given by

$$s(t) = \xi_1 + c_1 t,$$
$$s(t) = \xi_2 + c_2 t; \tag{2.59}$$

these may be solved for $s(t)$ and t and substituted in (2.58). Finally, (2.58) becomes

$$\{(q_2 - q_1) - (\rho_2 c_2 - \rho_1 c_1)\} \frac{\xi_1 - \xi_2}{c_1 - c_2} = \int_{\xi_2}^{\xi_1} \rho \, d\xi, \tag{2.60}$$

where ρ, q, and c are all to be evaluated as functions of ξ through the relations

$$\rho = f(\xi), \qquad q = Q(f(\xi)), \qquad c = Q'(f(\xi)) \tag{2.61}$$

and subscripts indicate values for $\xi = \xi_1$ and $\xi = \xi_2$. [This makes it a little clearer than using $f(\xi)$ for ρ, $F(\xi)$ for c and introducing a new symbol for q as a function of ξ.] Equations 2.59 and 2.60 give three relations for $s(t)$, $\xi_1(t)$, $\xi_2(t)$. Again it may be verified directly by differentiation that the shock condition

$$\dot{s} = \frac{q_2 - q_1}{\rho_2 - \rho_1}$$

is satisfied. When q is quadratic in ρ, it is easily verified that (2.60) reduces to (2.45). The problems like a single hump or an N wave can be analyzed as before and are qualitatively similar. The asymptotic formulas (2.51), (2.52), and (2.53) still apply with the modification that

$$A = c'(\rho_0) \int_0^L (\rho - \rho_0) \, d\xi,$$

and B is changed similarly. The expressions for ρ may be deduced from those for c, since the disturbance is weak in the asymptotic limit and $\rho - \rho_0 = (c - c_0)/c'(\rho_0)$ to first order.

2.10   Note on Linearized Theory

When disturbances are weak, nonlinear equations are often "linearized" by neglecting all but the first order powers of the perturbations. For weak disturbances with $(c - c_0)/c_0 \ll 1$, the equation

$$c_t + cc_x = 0$$

would be linearized to

$$c_t + c_0 c_x = 0.$$

As noted earlier the solution of this equation is $c - c_0 = f(x - c_0 t)$. The breaking effect and the formation of shocks are completely absent, yet we see from Figs. 2.11, 2.13, and 2.15 that these become crucial after a sufficient time, *however weak the initial disturbance may be*. Thus it is clear from comparison of the answers that the linearized approximation cannot be uniformly valid as $t \to \infty$.

This may also be seen directly, by looking at the linear theory as the first term in a naive expansion in powers of a small parameter. Suppose ϵ measures the maximum initial value of $(c - c_0)/c_0$, and a solution is sought in the form

$$c = c_0 + \epsilon c_1(x, t) + \epsilon^2 c_2(x, t) + \cdots.$$

When this expansion is substituted in $c_t + c c_x = 0$ and coefficients of ϵ^n are equated to zero, we have a hierarchy of equations starting with

$$c_{1t} + c_0 c_{1x} = 0,$$

$$c_{2t} + c_0 c_{2x} = -c_1 c_{1x},$$

$$c_{3t} + c_0 c_{3x} = -c_2 c_{1x} - c_1 c_{2x}.$$

These are easily solved successively since at each step we have

$$\phi_t + c_0 \phi_x = \Phi(x, t),$$

where Φ is known from the previous step. If we introduce the characteristic coordinate $y = x - c_0 t$, this may be written

$$\left(\frac{\partial \phi}{\partial t} \right)_{y = \text{const.}} = \Phi(y + c_0 t, t).$$

Therefore

$$\phi = \int_0^t \Phi(y + c_0 \tau, \tau) \, d\tau + \Psi(y).$$

The initial condition on c may be written

$$c = c_0 + \epsilon P(x) \qquad \text{at } t = 0,$$

and it is satisfied by

$$c_1 = P(x), \quad c_n = 0 \quad (n > 1) \qquad \text{at } t = 0.$$

Hence the complementary functions $\Psi(y)$ are zero in the solutions for the c_n, $n > 1$. The first three c_n are found to be

$$c_1 = P(y),$$

$$c_2 = -tP(y)P'(y),$$

$$c_3 = \frac{t^2}{2}(P^2P')'.$$

It is clear that in general c_n will contain a term of the form $t^{n-1}R_n(y)$. Therefore the successive terms in the assumed series for c are of order $\epsilon^n t^{n-1}$, and the series is not uniformly valid as $t \to \infty$.

The failure of the linearized theory, brought out strongly in the solution for the higher order terms, is that it approximates the characteristics as lines $x - c_0 t = \text{constant}$. The slight inclination of the true characteristic lines, relative to each other, accumulates to a large displacement as $t \to \infty$. The correct solution may be written as a telescoping function:

$$c = c_0 + \epsilon P(x - ct),$$

$$= c_0 + \epsilon P(x - [c_0 + \epsilon P]t),$$

and so on. The naive perturbation expansion can then be obtained by an inadvisable use of Taylor series!

2.11 Other Boundary Conditions; The Signaling Problem

The solution for the initial value problem has been given in great detail. Other boundary value problems can be solved in similar fashion. It is clear from the characteristic form (2.4) that the solution is determined once the value of ρ is given on any curve that intersects each characteristic once. Such a boundary value provides the initial conditions for integrating the two ordinary equations in (2.4) along the characteristic through that point. In principle, this is repeated at each point of the boundary curve to build up the solution in the whole region covered by the characteristics through the boundary curve. If the curve intersects characteristics twice, as does curve ABC in Fig. 2.18, the data can only be posed on AB or BC, otherwise the integration starting from AB, say, will conflict with the data on arrival at BC. Since the characteristics may depend on the solution, the region covered and the admissibility of the boundary curve cannot always be decided in advance.

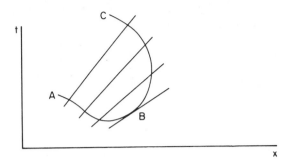

Fig. 2.18. Characteristics and initial data.

A standard boundary value problem is the so-called signaling problem for which

$$\rho = \rho_0 \quad \text{for } x > 0,\ t = 0,$$

$$\rho = g(t) \quad \text{for } t > 0,\ x = 0,$$

and the solution is required in $x > 0$, $t > 0$. Of course, this problem only arises in the case $c = Q'(\rho) > 0$. The (x, t) diagram is shown in Fig. 2.19. Characteristics start from the positive x axis and the positive t axis. Those from the t axis have $\rho = \rho_0$, $c = c(\rho_0) = c_0$ and are straight lines $x - c_0 t$ = constant. Thus they predict

$$\rho = \rho_0, \quad c = c_0 \quad \text{in } x > c_0 t. \tag{2.62}$$

For the characteristics starting from the t axis, let a typical one start at $t = \tau$. Then

$$\rho = g(\tau),$$
$$x = G(\tau)(t - \tau), \tag{2.63}$$

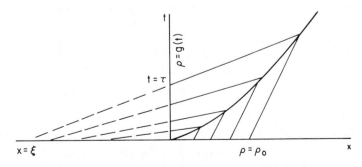

Fig. 2.19. The (x, t) diagram for the signaling problem.

where $G(\tau) = c\{g(\tau)\}$. This gives the solution implicitly in terms of $\tau(x,t)$.

The solution can be related to the solution of the initial value problem in two ways. The first method is to note that the two solutions agree if

$$\xi = -\tau G(\tau), \qquad f(\xi) = g(\tau), \qquad F(\xi) = G(\tau). \tag{2.64}$$

This corresponds to continuing the characteristic through $t = \tau$, $x = 0$, back to the x axis and denoting the point of intersection as $x = \xi$; in this way the signaling problem is formulated as an initial value problem. The alternative method is to interchange the roles of x and t, and of q and ρ; in the formulas, $dp/dq = 1/c$ will appear in place of $dq/d\rho = c$.

Any multivalued overlap in the solution (2.63) has to be resolved by shocks. If $G(0+) > c_0$, there will be an overlap immediately since the first characteristic $x = tG(0+)$ of the disturbed region is ahead of the last characteristic $x = c_0 t$ of the undisturbed region. In that case a shock of finite strength will start from the origin. The shock determination can be taken from the results for the initial value problem using either of the above methods, or it can be developed independently. If characteristics $\tau_1(t)$ and $\tau_2(t)$ meet the shock at the time t, then from (2.63),

$$s(t) = (t - \tau_1)c_1, \qquad c_1 = G(\tau_1),$$
$$\tag{2.65}$$
$$s(t) = (t - \tau_2)c_2, \qquad c_2 = G(\tau_2),$$

and the formula corresponding to (2.60) may be written

$$\{(q_2 - \rho_2 c_2)c_1 - (q_1 - \rho_1 c_1)c_2\}\frac{\tau_2 - \tau_1}{c_2 - c_1} = -\int_{\tau_1}^{\tau_2} q(\tau)\,d\tau. \tag{2.66}$$

Equations 2.65 and 2.66 provide the three implicit equations for the functions $\tau_1(t)$, $\tau_2(t)$, $s(t)$. The most important case is that of a front shock formed at the origin [i.e., $G(0+) > c_0$] and propagating into the undisturbed region. Then we have $\rho_1 = \rho_0$, $c_1 = c_0$, $q_1 = q_0$, and τ_1 can be eliminated from (2.65) and (2.66). At the same time we drop the subscript 2 and reduce the shock relation (2.66) to

$$\{(q - q_0) - (\rho - \rho_0)c\}(t - \tau) = -\int_0^{\tau}\{q(\tau') - q_0\}\,d\tau'. \tag{2.67}$$

Here ρ, q, and c are functions of τ determined from

$$\rho = g(\tau), \qquad q = Q(g(\tau)), \qquad c = Q'(g(\tau));$$

they are all known functions of each other and when one is prescribed as a function of τ the others follow. Equations 2.63 determine the solution in the disturbed region behind the shock; (2.67) determines the appropriate value $\tau(t)$ at the shock and on substitution in (2.63) we have both the position of the shock and the value of ρ just behind it.

In the initial motion of the shock, the value of $\tau(t)$ in (2.67) is small and we have

$$\{(q_i - q_0) - (\rho_i - \rho_0)c_i\}(t - \tau) = -(q_i - q_0)\tau + O(\tau^2),$$

where ρ_i, q_i, and c_i are the initial values on $x = 0$; that is, $\rho_i = g(0+)$, and so on. Therefore

$$\tau = \left\{1 - \frac{q_i - q_0}{(\rho_i - \rho_0)c_i}\right\}t + O(t^2).$$

From (2.63), the shock position is

$$x = (t - \tau)c_i + O(t^2)$$

$$= \frac{q_i - q_0}{\rho_i - \rho_0}t + O(t^2).$$

The shock starts with velocity $(q_i - q_0)/(\rho_i - \rho_0)$, and this result can be seen directly from the shock condition. If $g(\tau)$ remains constant and equal to ρ_i, this is exact for all t and the solution is a shock of constant velocity separating the two uniform regions $\rho = \rho_0$ and $\rho = \rho_i$.

If $g(\tau)$ returns to ρ_0, the shock ultimately decays. For a single positive phase with $g(\tau)$ returning to ρ_0 at $\tau = T$, the asymptotic behavior corresponds to $\tau \to T$, $t \to \infty$, $\rho \to \rho_0$ in (2.67). In this limit, (2.67) becomes

$$\frac{1}{2}c'(\rho_0)(\rho - \rho_0)^2 t \sim \int_0^T \{q(\tau') - q_0\}\,d\tau',$$

and the expression for the shock position in (2.63) becomes

$$x \sim c_0 t + c'(\rho_0)(\rho - \rho_0)t.$$

Therefore at the shock we have

$$\rho - \rho_0 \sim \frac{1}{c'(\rho_0)}\sqrt{\frac{2A}{t}}, \qquad c - c_0 \sim \sqrt{\frac{2A}{t}},$$

$$x \sim c_0 t + \sqrt{2At},$$

(2.68)

where

$$A = c'(\rho_0) \int_0^T (q - q_0) \, d\tau.$$

In the region behind the shock,

$$c \sim \frac{x}{t}, \qquad c_0 t < x < c_0 t + \sqrt{2At} \,,$$

$$\rho - \rho_0 \sim \frac{(c - c_0)}{c'(\rho_0)} \sim \frac{1}{c'(\rho_0)} \frac{x - c_0 t}{t}.$$

(2.69)

These results are very similar to those for the initial value problem. Other cases may be studied in the same way. If the positive phase is followed by a negative phase, there is a second shock whose asymptotic behavior is given by (2.68) with the modifications that A is replaced by the corresponding integral over the negative phase and the signs are changed appropriately. The ultimate form is an N wave with the formulas (2.69) extended back to the rear shock.

2.12 More General Quasi-Linear Equations

The general quasi-linear equation of first order is linear in ρ_t and ρ_x but may also have an undifferentiated term. The coefficients of ρ_t, ρ_x and the undifferentiated term may be any functions of ρ, x, t. If the coefficient of ρ_t is nonzero, the equation may be divided by this quantity and written in the form

$$\rho_t + c\rho_x = b, \tag{2.70}$$

where b and c are functions of ρ, x, and t. Such equations can again be reduced to the integration of ordinary differential equations along characteristic curves by writing (2.70) in the characteristic form

$$\frac{d\rho}{dt} = b(\rho, x, t), \qquad \frac{dx}{dt} = c(\rho, x, t). \tag{2.71}$$

In particular the initial value problem with initial data

$$\rho = f(x), \qquad t = 0$$

is solved by integrating the coupled ordinary differential equations in

(2.71) subject to the initial conditions

$$\rho = f(\xi), \quad x = \xi, \quad \text{at } t = 0.$$

Each choice of ξ leads to the determination of the characteristic through $x = \xi$ and the value of ρ along it. The solution in a whole region is obtained by varying the parameter ξ.

When $b \neq 0$, ρ is not constant along the characteristics and generally the characteristics are not straight lines. But the method of determination is qualitatively the same. Again waves may break with the characteristics overlapping in the (x, t) plane. Again the multivalued solutions may be avoided by including suitable discontinuities.

Some interesting cases concerning breaking arise and we will consider two examples here.

Damped Waves.

Consider as a first example the case

$$c_t + cc_x + ac = 0, \tag{2.72}$$

where a is a positive constant. In characteristic form it is written

$$\frac{dc}{dt} = -ac, \qquad \frac{dx}{dt} = c. \tag{2.73}$$

If we take the initial value problem, the first equation may be integrated to

$$c = e^{-at}f(\xi). \tag{2.74}$$

Then the second equation is

$$\frac{dx}{dt} = e^{-at}f(\xi),$$

and we require $x = \xi$ at $t = 0$. The solution is

$$x = \xi + \frac{1 - e^{-at}}{a}f(\xi). \tag{2.75}$$

The nonlinearity gives the typical distortion of the wave profile, but simultaneously the wave is damped due to the presence of the un-differentiated term in the equation.

Consider now the question of breaking. This is most easily investigated by seeing whether the characteristic curves (2.75) have an envelope. An envelope of these curves satisfies the derivative of (2.75) with respect to the parameter ξ:

$$0 = 1 + \frac{1 - e^{-at}}{a} f'(\xi). \tag{2.76}$$

Since $a > 0$, $t > 0$, this is possible if and only if

$$f'(\xi) < -a. \tag{2.77}$$

Thus breaking occurs if and only if the initial curve has a large enough negative slope; the damping may prevent breaking if the compressive phase is not steep enough.

Although the appropriate equations are more complicated than those just considered (see Chapter 3), this type of inequality determines whether the tidal variation propagating up a river will be strong enough to produce breaking into a bore, or whether the friction will dominate. For most rivers the frictional effects dominate. However, those famous rivers that have a bore have high enough tides at the mouth and additional reinforcement from rapid narrowing of the river to overcome the various frictional effects. This theory has been discussed and applied by Abbott (1956). It will be referred to again in Section 5.7.

Waves Produced by a Moving Source.

If b is independent of ρ in (2.70), it may be interpreted as an external source of the fluid. A particularly interesting case is when the source distribution moves with constant velocity V. There is a recent example, in the more complicated context of magnetogasdynamics, where a wave motion is produced by applying a moving force to the fluid (Hoffman, 1967). We can examine some of the qualitative effects in our simple model. We take

$$b = B(x - Vt),$$

where V is constant and $B(x)$ is a positive function tending rapidly to zero as $|x| \to \infty$. We assume that ρ has a constant value $\rho = \rho_0$ at $t = 0$. If $c_0 = c(\rho_0)$, there are important differences depending on whether the source moves supersonically with $V > c_0$ or subsonically with $V < c_0$.

The surprising result is that a supersonic source need not produce a shock, whereas a subsonic source always does. This can be seen quite simply by looking for a steady profile solution with

$$\rho = \rho(X), \qquad X = x - Vt. \tag{2.78}$$

Since we are only looking at models anyway, with the aim of showing qualitative effects, let us take the special case

$$c_t + cc_x = B(x - Vt). \tag{2.79}$$

Then in the steady profile solution

$$(c - V)c_X = B(X),$$

$$\tfrac{1}{2}(V - c)^2 - \tfrac{1}{2}(V - c_0)^2 = - \int_X^\infty B(y)\, dy.$$

In the supersonic case, $V > c_0$, the solution for c is

$$c = V - \left\{ (V - c_0)^2 - 2\int_X^\infty B(y)\, dy \right\}^{1/2}. \tag{2.80}$$

If

$$V - c_0 > \left\{ 2\int_{-\infty}^\infty B(y)\, dy \right\}^{1/2}, \tag{2.81}$$

(2.80) is a satisfactory single-valued solution for all X and no shock is required. The criterion (2.81) is an inequality between the speed $V - c_0$ and the total source strength

$$\int_{-\infty}^\infty B(y)\, dy.$$

We can get a feeling for the result by the following argument. If the source moves with a large supersonic speed, the only shock that could keep up with it would be strong. But if the source is relatively small, a strong shock cannot be produced and is not required.

When the inequality (2.81) does not hold, (2.80) breaks down for $X \leqslant X_0$, where

$$V - c_0 = \left\{ 2\int_{X_0}^\infty B(y)\, dy \right\}^{1/2}.$$

At $X = X_0$, $c = V$ and transients from the starting conditions can and do overtake the wave. The solution cannot be completed without a detailed discussion of the transients. Similarly, in the subsonic case the solution cannot be established without full discussion of the transients. In both these cases shocks are found to occur. A detailed discussion is given by Hoffman (1967).

2.13 Nonlinear First Order Equations

The discussion of quasi-linear equations has raised many questions that require further consideration. Before proceeding, however, we note briefly that similar constructions using characteristics go through in the general case of fully nonlinear first order equations. These results will also be needed later.

It will be useful to have the characteristic form for an equation in n independent variables $(x_1,\ldots,x_n)$. We consider, then, a function $\phi(x_1,\ldots,x_n)$ which satisfies a differential equation

$$H(\mathbf{p},\phi,\mathbf{x})=0, \tag{2.82}$$

where $\mathbf{p}$ and $\mathbf{x}$ denote the vectors with components p_i and x_i, $i=1,\ldots,n$, and

$$p_i = \frac{\partial \phi}{\partial x_i}. \tag{2.83}$$

We may motivate the characteristic form by asking whether there are curves in x space with special properties akin to the characteristics of the quasi-linear equations. Any curve $\mathcal{C}$ in x space may be written in parametric form

$$\mathbf{x}=\mathbf{x}(\lambda).$$

The total derivative of ϕ along the curve $\mathcal{C}$ is*

$$\frac{d\phi}{d\lambda} = \frac{\partial \phi}{\partial x_j}\frac{dx_j}{d\lambda} = p_j \frac{dx_j}{d\lambda}.$$

Is there a choice of the direction vector $dx_j/d\lambda$ which has special significance for the solution of (2.82)? In the quasi-linear case where we have

---

*We use the summation convention that a repeated subscript is automatically summed over $1,\ldots,n$.

$H \equiv c_j(\phi, \mathbf{x})p_j - b(\phi, \mathbf{x})$, we choose

$$\frac{dx_j}{d\lambda} = c_j(\phi, \mathbf{x})$$

so that $d\phi/d\lambda = c_j p_j$; we then use the equation to obtain

$$\frac{d\phi}{d\lambda} = b(\phi, \mathbf{x}).$$

But generally the p_i cannot be eliminated in the expression for $d\phi/d\lambda$ whatever the choice of the $dx_j/d\lambda$. We do not have an ordinary differential equation for ϕ alone; the p_i are involved. However, consider in addition the total derivatives of the p_i on $\mathcal{C}$. We have

$$\frac{dp_i}{d\lambda} = \frac{d}{d\lambda}\left(\frac{\partial\phi}{\partial x_i}\right) = \frac{\partial^2\phi}{\partial x_i \partial x_j}\frac{dx_j}{d\lambda}, \tag{2.84}$$

and the x_i derivative of (2.82) yields

$$\frac{\partial^2\phi}{\partial x_i \partial x_j}\frac{\partial H}{\partial p_j} + \frac{\partial H}{\partial\phi}\frac{\partial\phi}{\partial x_i} + \frac{\partial H}{\partial x_i} = 0. \tag{2.85}$$

Comparing the two, we see the special advantage in choosing curves $\mathcal{C}$ defined by

$$\frac{dx_i}{d\lambda} = \frac{\partial H}{\partial p_i}. \tag{2.86}$$

For in that case (2.84) may be calculated from (2.85) as

$$\frac{dp_i}{d\lambda} = -p_i\frac{\partial H}{\partial\phi} - \frac{\partial H}{\partial x_i}. \tag{2.87}$$

Then if we add

$$\frac{d\phi}{d\lambda} = p_j\frac{\partial H}{\partial p_j}, \tag{2.88}$$

(2.86)–(2.88) are a complete set of $(2n+1)$ ordinary differential equations for determining a "characteristic curve" $x_i(\lambda)$ and the values of ϕ and p_i along it. In principle, the solution in a whole region can be obtained by integrating these characteristic equations along the characteristics covering the region.

In the special case of the quasi-linear equation, $H \equiv c_i(\phi, \mathbf{x})p_i - b(\phi, \mathbf{x})$, (2.86) and (2.88) reduce to

$$\frac{dx_i}{d\lambda} = c_i(\phi, \mathbf{x}),$$

$$\frac{d\phi}{d\lambda} = p_j c_j = b(\phi, \mathbf{x}),$$

which may be solved independently of (2.87). In the earlier discussion one of the x_i was the time t, the corresponding c_i was unity and the parameter λ was t itself.

CHAPTER 3

Specific Problems

In this chapter the basic ideas developed so far are applied in more detail to the particular cases raised in Section 2.2. At the same time, the general ideas can be taken further on the basis of specific sets of equations.

3.1 Traffic Flow

The application of these ideas to traffic flow was formulated and discussed independently by Lighthill and Whitham (1955) and Richards (1956). It is clear in this case that the flow velocity

$$V(\rho) = \frac{Q(\rho)}{\rho}$$

must be a decreasing function of ρ which starts from a finite maximum value at $\rho = 0$ and decreases to zero as $\rho \to \rho_j$, the value for which the cars are bumper to bumper. Thus $Q(\rho)$ is zero at both $\rho = 0$ and $\rho = \rho_j$, and has a maximum value q_m at some intermediate density ρ_m. It has the general convex form shown in Fig. 3.1. Actual observations of traffic flow indicate that typical values for a single lane are $\rho_j \sim 225$ vehicles per mile, $\rho_m \sim 80$ vehicles per mile, $q_m \sim 1500$ vehicles per hour. It appears to be roughly correct to multiply these values by the number of lanes for multilane highways. It is interesting that, according to these figures, the maximum flow rate q_m is attained at a low velocity in the neighborhood of 20 miles per hour.

The propagation velocity for the waves is

$$c(\rho) = Q'(\rho) = V(\rho) + \rho V'(\rho).$$

Since $V'(\rho) < 0$, the propagation velocity is less than the car velocity; waves propagate backward through the stream of traffic and drivers are warned of disturbances ahead. The velocity c is the slope of the (q, ρ) curve so the

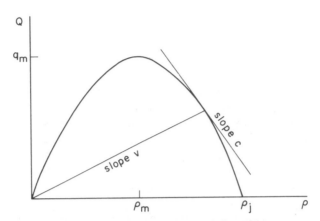

Fig. 3.1. Flow-density curve in traffic flow.

waves move forward or backward relative to the road depending on whether $\rho < \rho_m$ or $\rho > \rho_m$. At the maximum flow rate, $\rho = \rho_m$, the waves are stationary relative to the road, so the propagation velocity relative to the cars is then the same as $q_m/\rho_m \sim 20$ mph.

Near $\rho = \rho_j$, we can make a rough estimate on the basis of a simple reaction time argument. If we assume that a driver and his car take a time δ to react to any change ahead, then the gap between cars should be kept at $V\delta$ for safety. If h is the headway, defined as the distance between the front ends of successive cars, and L is the typical car length, this leads to

$$V = \frac{h - L}{\delta}.$$

Since $h = 1/\rho$, $L = 1/\rho_j$, we have

$$V(\rho) = \frac{L}{\delta}\left(\frac{\rho_j}{\rho} - 1\right), \qquad Q(\rho) = \frac{L}{\delta}(\rho_j - \rho).$$

One should probably interpret this as an estimate of the slope of the $Q(\rho)$ curve at ρ_j, rather than as a realistic prediction of a linear dependence on ρ. In any event, it gives $c_j = -L/\delta$ for the propagation velocity there. In the traffic flow context δ is usually estimated in the range 0.5–1.5 sec, although in other circumstances the human reaction time can be much faster. With $L = 20$ ft, $\delta = 1$ sec, we have $c_j \sim -14$ mph.

Greenberg (1959) found a good fit with data for the Lincoln Tunnel in

New York by taking

$$Q(\rho) = a\rho \log \frac{\rho_j}{\rho},$$

with $a = 17.2$ mph, $\rho_j = 228$ vpm (vehicles per mile). For this formula, the relative propagation velocity $V - c$ is equal to the constant value a at all densities. The values of ρ_m and q_m are $\rho_m = 83$ vpm, $q_m = 1430$ vph (vehicles per hour). The logarithmic formula does not give a finite value for V as $\rho \to 0$, but the theory would be on dubious ground for very light traffic so this point alone is not important. With a finite maximum V and a finite $V'(\rho)$, we have $c \to V$ as $\rho \to 0$, so one should expect $V - c$ to decrease at the lighter densities.

Since $Q(\rho)$ is convex with $Q''(\rho) < 0$, c itself is always a decreasing function of ρ. This means that a local increase of density propagates as shown in Fig. 3.2 with a shock forming at the back. Individual cars move faster than the waves, so that a driver enters such a local density increase from behind; he must decelerate rapidly through the shock but speeds up only slowly as he leaves the congestion. This seems to accord with experience. The details can be analyzed by the theory of Chapter 2. In particular the final asymptotic behavior is the triangular wave which is the last profile in Fig. 3.2. The length of the wave increases like $t^{1/2}$ and the shock decays like $t^{-1/2}$. The actual analytic expressions are

$$c \sim \frac{x}{t}, \qquad \rho - \rho_0 \sim \frac{x - c_0 t}{c'(\rho_0) t} \qquad \text{for} \quad c_0 t - \sqrt{2Bt} < x < c_0 t,$$

where

$$B = |c'(\rho_0)| \int_{-\infty}^{\infty} (\rho - \rho_0) \, dx.$$

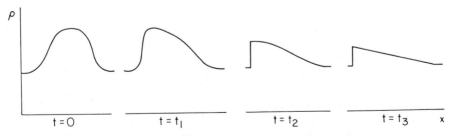

Fig. 3.2. Breaking wave in traffic flow.

The shock is at

$$x = c_0 t - \sqrt{2Bt} \ ,$$

and the jumps of c and ρ at the shock are

$$c - c_0 \sim -\sqrt{\frac{2B}{t}} \ , \qquad \rho - \rho_0 \sim \frac{1}{|c'(\rho_0)|} \sqrt{\frac{2B}{t}} \ .$$

Traffic Light Problem.

A more complicated problem is the analysis of the flow at a traffic light. We construct the characteristics in the (x,t) diagram. These are lines of constant density and their slopes $c(\rho)$ determine the corresponding values of ρ on them. So the problem is solved once the (x,t) diagram has been obtained.

Suppose first that the red period of the light is long enough to allow the incoming traffic to flow freely at some value $\rho_i < \rho_m$. Then we may start with characteristics of slope $c(\rho_i)$ intersecting the t axis in the interval AB in Fig. 3.3; AB is part of a green period. [The (x,t) diagram is plotted with x vertical and t horizontal since this is the usual practice in the references on traffic flow.] Just below the red period BC, the cars are stationary with $\rho = \rho_j$; hence the characteristics have the negative slope $c(\rho_j)$. The line of separation between the stopped queue at the traffic light and the free flow must be a shock BP, and from the shock condition its velocity is

$$-\frac{q(\rho_i)}{\rho_j - \rho_i} \ .$$

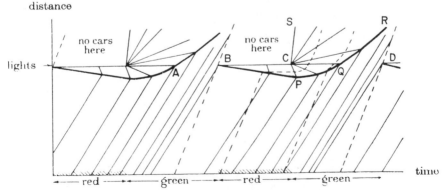

Fig. 3.3. Wave diagram for an efficient traffic light.

When the light turns green at C, the leading cars can go at the maximum speed since $\rho = 0$ ahead of them. (The finite acceleration could be allowed for roughly by extending the effective red period.) This is represented by the characteristic CS with maximum slope $c(0)$. Between CS and CP we have an expansion fan with all values of c being taken. Exactly at the intersection CQ, the slope c must be zero. But this corresponds to the maximum $q = q_m$. Therefore we have the interesting result that q attains its maximum value right at the traffic light. The shock $BPQR$ is weakened by the expansion fan and ultimately accelerates through the intersection, provided the green period is long enough. The criterion for whether the shock gets through is easily established. The total incoming flow for the time BQ is $(t_r + t_s) q_i$ where t_r is the red period BC and t_s is the part of the green period before the shock gets through. The flow across the intersection in this time is $t_s q_m$. These two must be equal; therefore

$$ t_s = \frac{t_r q_i}{q_m - q_i} . $$

For the shock to get through and the light to operate freely, the green period must exceed this critical value.

If the shock does not get through, the flow never becomes free and the notorious traffic crawl develops. It is perhaps sufficient to show the corresponding (x, t) diagram Fig. 3.4 without comment!

Higher Order Effects; Diffusion and Response Time.

There are two obvious additional effects one may wish to include in the theory. One was mentioned in Section 2.4: the dependence of q on ρ_x as well as ρ. This introduces in a rough way the drivers' awareness of conditions ahead, and it produces a *diffusion* of the waves. The simplest assumption with the correct qualitative behavior is

$$ q = Q(\rho) - \nu \rho_x, \qquad v = V(\rho) - \frac{\nu}{\rho} \rho_x, \tag{3.1} $$

and one does not have much basis for any more complicated choice.

The second effect is the time lag in the response of the driver and of his car to any changes in the flow conditions. One way to introduce this effect is to consider the expression for v in (3.1) as a desired velocity which the driver accelerates toward; therefore the equation

$$ v_t + v v_x = -\frac{1}{\tau} \left\{ v - V(\rho) + \frac{\nu}{\rho} \rho_x \right\} \tag{3.2} $$

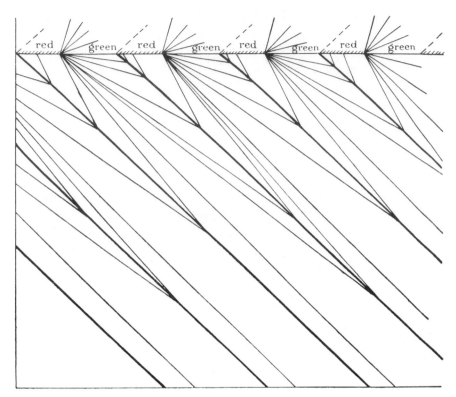

Fig. 3.4. Wave diagram for the slow crawl at an overcrowded traffic light.

may be introduced for the acceleration. The coefficient τ is a measure of the response time and is akin to the quantity δ mentioned earlier. Equation 3.2 is to be solved together with the conservation equation

$$\rho_t + (\rho v)_x = 0. \tag{3.3}$$

When v, τ are both small in a suitable nondimensional measure, (3.2) is approximated by $v = V(\rho)$, and we have the simpler theory. With the higher order terms included in (3.2), we expect shocks to appear as smooth steps and so on. This is true on the whole, but the situation turns out to be more complicated.

It is always helpful to get a first feel for a nonlinear equation by looking at the linearized theory, even though the linearization may have its own shortcomings, as we discussed in Section 2.10. If (3.2) and (3.3) are linearized for small perturbations about $\rho = \rho_0$, $v = v_0 = V(\rho_0)$, by substituting

$$\rho = \rho_0 + r, \qquad v = v_0 + w,$$

and retaining only first powers of r and w, we have

$$\tau(w_t + v_0 w_x) = -\left\{ w - V'(\rho_0) r + \frac{\nu}{\rho_0} r_x \right\},$$

$$r_t + v_0 r_x + \rho_0 w_x = 0.$$

The kinematic wave speed is

$$c_0 = \rho_0 V'(\rho_0) + V(\rho_0);$$

hence $V'(\rho_0) = -(v_0 - c_0)/\rho_0$. Introducing this expression and then eliminating w, we have

$$\frac{\partial r}{\partial t} + c_0 \frac{\partial r}{\partial x} = \nu \frac{\partial^2 r}{\partial x^2} - \tau\left(\frac{\partial}{\partial t} + v_0 \frac{\partial}{\partial x} \right)^2 r. \tag{3.4}$$

When $\nu = \tau = 0$, we have the linearized approximation to the kinematic waves: $r = f(x - c_0 t)$. The term proportional to ν introduces typical diffusion of the heat equation type. The effect of the finite response time τ is more complicated, but a quick insight can be gained as follows. In the basic wave motion governed by the left hand side, $r = f(x - c_0 t)$, so that t derivatives are approximately equal to $-c_0$ multiplied by x derivatives:

$$\frac{\partial}{\partial t} \simeq -c_0 \frac{\partial}{\partial x}. \tag{3.5}$$

If this approximation is used in the right hand side of (3.4), the equation reduces to

$$\frac{\partial r}{\partial t} + c_0 \frac{\partial r}{\partial x} = \left\{ \nu - (v_0 - c_0)^2 \tau \right\} \frac{\partial^2 r}{\partial x^2}. \tag{3.6}$$

There is a combined diffusion when

$$\nu > (v_0 - c_0)^2 \tau \tag{3.7}$$

but instability if

$$\nu < (v_0 - c_0)^2 \tau. \tag{3.8}$$

This is reasonable; for stability a driver should look far enough ahead to make up for his response time.

The stability criterion can be verified directly from the complete equation (3.4) in the traditional way. There are exponential solutions of (3.4) with

$$r \propto e^{ikx - i\omega t}$$

provided that

$$\tau(\omega - v_0 k)^2 + i(\omega - c_0 k) - \nu k^2 = 0.$$

The exponential solutions will be stable provided $\mathcal{I}\omega < 0$ for both of the roots ω. It is easily verified that the requirement for this is (3.7), so the result of the approximate procedure is confirmed and extended to all wavelengths.

Higher Order Waves.

It is important to note that the right hand side of (3.4) is itself a wave operator and we may write the equation as

$$\frac{\partial r}{\partial t} + c_0 \frac{\partial r}{\partial x} = -\tau \left(\frac{\partial}{\partial t} + c_+ \frac{\partial}{\partial x} \right) \left(\frac{\partial}{\partial t} + c_- \frac{\partial}{\partial x} \right) r, \qquad (3.9)$$

where

$$c_+ = v_0 + \sqrt{\nu/\tau}, \qquad c_- = v_0 - \sqrt{\nu/\tau}.$$

It would be expected therefore that waves traveling with speeds c_+ and c_- also play some role. It would be premature to go deeply into this question at this stage, but one remark has great significance in interpreting the stability condition. We shall see later in our discussion of higher order equations that the propagation speeds in the highest order derivatives always determine the fastest and slowest signals. Thus in the present case however small τ may be provided it is nonzero, the fastest signal travels with speed c_+ and the slowest with speed c_-. It is clear therefore that the approximation

$$\frac{\partial r}{\partial t} + c_0 \frac{\partial r}{\partial x} = 0 \qquad (3.11)$$

could only make sense if

$$c_- < c_0 < c_+. \qquad (3.12)$$

But this is exactly the stability criterion (3.7). So the flow is stable only if (3.12) holds, and then it is appropriate to approximate (3.9) by (3.11) for small τ. There is a nice correspondence between stability and wave interaction.

Equation 3.9 arises in several applications and a full discussion is given in Chapter 10.

Shock Structure.

The more complicated form of the higher order corrections introduces a new possibility in the shock structure. For the simple diffusion term used in Section 2.4 with $\nu > 0$, a continuous shock structure was obtained. We shall see now that this is not always the case when there are additional higher terms. We look for a steady profile solution of (3.2)–(3.3) with

$$\rho = \rho(X), \qquad v = v(X), \qquad X = x - Ut,$$

where U is the constant translational velocity. Equation 3.3 becomes

$$- U\rho_X + (v\rho)_X = 0 \tag{3.13}$$

and may be integrated to

$$\rho(U - v) = A, \tag{3.14}$$

where A is a constant. Equation 3.2 becomes

$$\tau\rho(v - U)v_X + \nu\rho_X + \rho v - Q(\rho) = 0. \tag{3.15}$$

Since $v = U - A/\rho$, this may be reduced to

$$\left(\nu - \frac{A^2}{\rho^2}\tau \right)\rho_X = Q(\rho) - \rho U + A. \tag{3.16}$$

For $\tau = 0$, it is the same as (2.21), as it should be. For $\tau \neq 0$, the possibility that $\nu - A^2\tau/\rho^2$ may vanish introduces the new effects.

As before we are interested in solution curves between ρ_1 at $X = +\infty$ and ρ_2 at $X = -\infty$. These values will be zeros of the right hand side of (3.16). For traffic flow $c'(\rho) = Q''(\rho) < 0$, so $\rho_2 < \rho_1$ and the right hand side of (3.16) is positive for $\rho_2 < \rho < \rho_1$. If $\nu - A^2\tau/\rho^2$ remains positive in this range, then $\rho_X > 0$ and we have a smooth profile as in Fig. 3.5. In view of (3.14), the condition for $\nu - A^2\tau/\rho^2$ to remain positive may be written

$$\nu > (v - U)^2\tau, \qquad \text{that is,} \quad v - \sqrt{\nu/\tau} < U < v + \sqrt{\nu/\tau}. \tag{3.17}$$

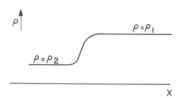

Fig. 3.5. Continuous shock structure.

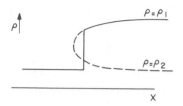

Fig. 3.6. Shock structure with an inner discontinuity.

This is similar in form to the linearized stability criterion (3.7) with v_0 replaced by the local velocity v and c_0 replaced by the shock velocity U. We might also interpret it in a way similar to (3.12) as a warning of possible complications if a shock tries to violate the higher order signal speeds. However, it is not necessarily an unstable situation. The conditions for the uniform states at $\pm \infty$ to be stable are

$$v_1 - \sqrt{\nu/\tau} < c_1 < v_1 + \sqrt{\nu/\tau} , \qquad v_2 - \sqrt{\nu/\tau} < c_2 < v_2 + \sqrt{\nu/\tau} . \quad (3.18)$$

It is possible, in general, for these to be satisfied and yet (3.17) to be violated. When this is the case, $\nu - A^2\tau/\rho^2$ changes sign in the profile, as in Fig. 3.6, and a single-valued continuous profile is no longer possible.

In most problems of shock structure, when the profile turns back on itself in this way, it is rectified by fitting in an appropriate discontinuity. The situation again corresponds, strictly speaking, to a breakdown of the assumptions for the particular level of description, but the introduction of a discontinuity, *provided it corresponds to a valid integrated form of the basic equations*, avoids an explicit discussion of yet higher order effects. In the case of (3.2) and (3.3), it is not clear which conservation forms are appropriate for the discontinuity conditions nor what additional effects should be introduced. One expects a discontinuous profile shown by the full curve in Fig. 3.6, but the precise determination of the discontinuity is not clear for this case. In other cases discussed later the details can be completed. The point to stress here is that the discontinuities in the simple theory using

$$\rho_t + c(\rho)\rho_x = 0$$

may be only partially resolved into continuous transitions in a more accurate formulation.

A Note on Car-following Theories.

Considerable work has been done on discrete models where the motion of the nth car in a line of cars is prescribed in terms of the motion of the other cars. [See, for example, Newell (1961) and the earlier references given there.] If the position of the nth car is $s_n(t)$ at time t, the assumed laws of motion usually take the form

$$\dot{s}_n(t+\Delta) = G\left\{ s_{n-1}(t) - s_n(t) \right\}, \tag{3.19}$$

between velocity $\dot{s}_n$ and headway $h_n = s_{n-1} - s_n$ with a time lag Δ to account for the driver response time. If $G(h_n)$ is chosen to be linear in h_n, or if the equation is linearized to study fluctuations about a uniform state, solutions can be obtained by Laplace transforms. In general, however, one must appeal to computer studies.

This type of model takes a more rigid view of how each individual car moves, so it is narrower in scope than the continuum theory, where the whole complicated behavior of the individuals is lumped together in the function $Q(\rho)$ and the parameters ν and τ. But each model leads to a particular form for these quantities, which may be helpful in interpreting observational data. Moreover, such models may lead to additional effects that cannot be seen in the continuum theory.

To see the correspondence of the particular car-following model in (3.19) with the continuum theory, we first note the relation of $G(h)$ to $Q(\rho)$. In a uniform stream with equal spacing h, the velocities in (3.19) are all equal and are given by the relation $v = G(h)$. Since $h = 1/\rho$, $v = q/\rho$, the function $Q(\rho)$ in the corresponding continuum equations is

$$Q(\rho) = \rho G\left(\frac{1}{\rho}\right).$$

If empirical or other information is known about $G(h)$, it may be transferred to information about $Q(\rho)$ near $\rho = \rho_j$. Of course at lower densities $Q(\rho)$ will be affected more by cars overtaking and changing lanes.

The wave propagation described by (3.19), in which the motion of a lead car is transmitted successively back through the stream, should be a typical finite difference version of the earlier continuum results with this choice of $Q(\rho)$. The finite difference form of (3.19) also introduces higher order effects equivalent to those in (3.2) and we can make a detailed comparison. If we let

$$v_n(t) = \dot{s}_n(t), \qquad s_{n-1}(t) - s_n(t) = h_n(t), \tag{3.20}$$

(3.19) is equivalent to the pair of equations

$$v_n(t+\Delta) = G(h_n), \tag{3.21}$$

$$\frac{dh_n}{dt} = v_{n-1}(t) - v_n(t). \tag{3.22}$$

In this form we introduce continuous functions $v(x,t)$ and $h(x,t)$ such that

$$v(s_n, t) = v_n(t), \tag{3.23}$$

$$h\left(\frac{s_{n-1}+s_n}{2}, t\right) = h_n(t), \tag{3.24}$$

and obtain corresponding partial differential equations in the approximations of small Δ and small h_n. Equation 3.21 may be written

$$v\{s_n(t+\Delta), t+\Delta\} = G\left\{h\left(s_n + \frac{1}{2}h_n, t\right)\right\},$$

and it may be approximated by

$$v + (v_t + vv_x)\Delta = G(h) + \frac{1}{2}hG'(h)h_x, \tag{3.25}$$

where the functions are evaluated at $x = s_n(t)$ and the errors are of order Δ^2, h^2. Equation 3.22 may be written

$$\frac{d}{dt}h\left(\frac{s_{n-1}+s_n}{2}, t\right) = v(s_{n-1}, t) - v(s_n, t)$$

and approximated by

$$h_t + vh_x = hv_x \qquad \text{at } x = \frac{s_{n-1}+s_n}{2}. \tag{3.26}$$

The error in (3.26) is *third* order in h [due to the centering of h at the midpoint $(s_{n-1}+s_n)/2$], so the equation is correct to both first and second orders. In terms of $\rho = 1/h$, $V(\rho) = G(h)$, (3.25)–(3.26) become

$$v + (v_t + vv_x)\Delta = V(\rho) + \frac{1}{2}\frac{V'(\rho)}{\rho}\rho_x, \tag{3.27}$$

$$\rho_t + (\rho v)_x = 0. \tag{3.28}$$

To lowest order in Δ and h, we would have

$$v = V(\rho), \qquad \rho_t + (\rho v)_x = 0,$$

which is just the kinematic theory. The differencing has been arranged so that the next order corrections leave the conservation equation (3.28) unchanged.

Equations 3.27 and 3.28 are identical with (3.2) and (3.3) if we take

$$\tau = \Delta, \qquad \nu = -\frac{1}{2} V'(\rho).$$

Since $V - c = -\rho V'(\rho)$, the stability criterion (3.7) may be written

$$2\rho^2 |V'(\rho)| \Delta < 1,$$

or, equivalently,

$$2G'(h)\Delta < 1.$$

This is exactly the condition found in the car-following theories (Chandler, Herman, and Montroll, 1958; Kometani and Sasaki, 1958). Similarly the shock structures discussed earlier on the basis of (3.2) should be close to those discussed by Newell (1961) on the basis of (3.19).

An effect that cannot be covered by the continuum theory is the actual collision of cars. In a queue described by (3.19), this occurs if $s_{n-1} - s_n$ ever drops to the car length L. In the special case

$$\dot{s}_n(t+\Delta) = \alpha \{ s_{n-1}(t) - s_n(t) - L \},$$

which can be solved by Laplace transforms, it may be shown that the criterion for avoiding collision is

$$\alpha\Delta < \frac{1}{e};$$

this is slightly more stringent than the stability criterion $2\alpha\Delta < 1$ found above. The analysis would take us too far afield and the reader is referred to the discussion of local stability in the paper by Herman, Montroll, Potts, and Rothery (1959).

3.2 Flood Waves

For flood waves, the "density" in the sense of the general theory presented in Chapter 2 is the cross-sectional area of the riverbed, $A(x,t)$, at position x along the river at time t. If the volume flow across the section is $q(x,t)$ per unit time, the conservation equation is

$$\frac{d}{dt} \int_{x_2}^{x_1} A(x,t)\,dx + q(x_1,t) - q(x_2,t) = 0,$$

or, in differentiated form,

$$\frac{\partial A}{\partial t} + \frac{\partial q}{\partial x} = 0. \tag{3.29}$$

Flow in a river is obviously so complicated that any flow model for the second relation between q and A must be extremely approximate, giving only qualitative effects and general order of magnitude results for propagation speeds, wave profiles, and so on. However, observations during slow changes in the river level may be used also to establish the dependence of depth and the area A on the flow q. These provide empirical curves for the function

$$q = Q(A,x) \tag{3.30}$$

in steady flows. This relation can be combined with (3.29) to give a first approximation for unsteady flows which vary slowly. Then $A(x,t)$ satisfies

$$\frac{\partial A}{\partial t} + \frac{\partial Q}{\partial A} \frac{\partial A}{\partial x} = -\frac{\partial Q}{\partial x}. \tag{3.31}$$

We have again the theory discussed in Chapter 2 with the propagation velocity

$$c = \frac{\partial Q}{\partial A} = \frac{1}{b} \frac{\partial Q}{\partial h}. \tag{3.32}$$

[The second form introduces the breadth b and depth h, and $dA = bdh$.] This is the Kleitz-Seddon formula for flood waves, apparently established first by Kleitz (1858, unpublished) and thoroughly discussed and used effectively by Seddon (1900).

Empirical relations for (3.30) can be viewed against simple theoretical models. The relation is an expression of the balance between the frictional force of the river bed and the gravitational force. In theoretical models, the frictional force is usually assumed to be proportional to v^2, where v is the average velocity

$$v = \frac{q}{A},$$

and also proportional to the wetted perimeter P of the cross-section at position x. This force is then expressed as $\rho_0 C_f P v^2$ per unit length of river, where ρ_0 is the density of water and C_f is a friction coefficient. The gravitational force is $\rho_0 g A \sin\alpha$ per unit length, where α is the angle of inclination of the surface of the river. Hence

$$v = \sqrt{\frac{A}{P} \frac{g \sin\alpha}{C_f}} \; ,$$

(3.33)

$$Q = vA = \sqrt{\frac{A^3}{P} \frac{g \sin\alpha}{C_f}} \; .$$

The wetted perimeter P is a function of A, and C_f may also be allowed to depend on A. For broad rivers P varies little with the depth of the river and may be taken to be constant. If C_f and α are also taken to be constants, (3.33) gives the Chezy law

$$v \propto A^{1/2}, \qquad Q \propto A^{3/2}.$$

Then the propagation velocity

$$c = \frac{d}{dA}(vA) = v + A\frac{dv}{dA} = \frac{3}{2}v.$$

More generally, P and C_f are functions of A, and power law dependences for these give $v \propto A^n$, $Q \propto A^{1+n}$ with other values for n. For example, a triangular cross-section gives $P \propto A^{1/2}$ and leads to $n = \frac{1}{4}$; Manning's law $C_f \propto A^{-1/3}$ leads to $n = \frac{2}{3}$. For all these power laws the propagation velocity is

$$c = (1+n)v.$$

As expected, flood waves move faster than the fluid but the propagation velocity may not be very much greater than the fluid velocity.

Seddon turns the calculation around and uses his observations of the propagation velocity to deduce the effective shape of the bed, that is, the dependence of P on A. This is a valuable idea in all kinematic wave problems: use observations of the propagation velocity c to infer the q-ρ relation.

If the dependence of Q on x is omitted, (3.31) reduces to

$$A_t + c(A)A_x = 0,$$

and the general solution may be taken from Chapter 2 with shocks fitted in as discontinuities satisfying

$$U = \frac{q_2 - q_1}{A_2 - A_1}.$$

For the power laws suggested (and this is also borne out by observations), $c(A)$ is an increasing function of A; hence waves due to an increase in height break forward, and shocks carry an increase in height, $A_2 > A_1$.

Higher Order Effects.

As in the other examples discussed, a more accurate treatment of the relation between q and A than that expressed in (3.30) involves higher derivatives. In unsteady flow the frictional and gravitational forces do not balance exactly and their difference is proportional to the acceleration of the fluid; the difference between the slope of the water surface and the slope of the bottom also makes a contribution.

It will be valuable to express the equations in conservation form so that, when necessary, appropriate discontinuity conditions can also be deduced. For simplicity, we consider the case of a broad rectangular channel of constant inclination α and work with the depth h and mean velocity v as basic variables in place of A and q. The conservation of fluid for unit breadth can then be written

$$\frac{d}{dt}\int_{x_2}^{x_1} h\,dx + [hv]_{x_2}^{x_1} = 0, \tag{3.34}$$

and we need to add a more detailed formulation of the conservation of momentum. The appropriate equation in hydraulic theory is

$$\frac{d}{dt}\int_{x_2}^{x_1} hv\,dx + [hv^2]_{x_2}^{x_1} + \left[\frac{1}{2}gh^2\cos\alpha\right]_{x_2}^{x_1}$$

$$= \int_{x_2}^{x_1} gh\sin\alpha\,dx - \int_{x_2}^{x_1} C_f v^2\,dx. \tag{3.35}$$

Apart from the common factors ρ_0 (the constant density of water) and the breadth b which have been cancelled through, the five terms in this equation are, respectively, (1) the rate of increase of momentum in the section $x_2 < x < x_1$, (2) the net transport of momentum across x_1 and x_2, (3) the net total pressure force acting across x_1 and x_2, (4) the component of the gravitational force down the incline, and (5) the frictional effects of the bottom. The pressure term requires some comment perhaps. In hydraulic theory the dependence of the velocity on the coordinate y normal to the bed is averaged out to $v(x,t)$ and the fluid acceleration in the y direction is neglected. The latter assumption means that the pressure satisfies a hydrostatic law

$$\frac{\partial p}{\partial y} = -\rho_0 g \cos \alpha.$$

Hence

$$p - p_0 = (h - y)\rho_0 g \cos \alpha$$

and the total contribution of the perturbed pressure integrated over a cross section of the river is

$$b\int_0^h (p - p_0)\, dy = \frac{1}{2}h^2 \rho_0 gb \cos \alpha;$$

this is the origin of the third term in (3.35).

Equations 3.34 and 3.35 are the two conservation equations for h and v. If h and v are assumed to be continuously differentiable, we may take the limit $x_1 - x_2 \to 0$ to obtain partial differential equations for h and v. It will be a minor saving in writing to introduce $g' = g\cos\alpha$ and the slope $S = \tan\alpha$. The equations for h and v are then

$$h_t + (hv)_x = 0,$$

$$(hv)_t + \left(hv^2 + \frac{1}{2}g'h^2\right)_x = g'hS - C_f v^2. \tag{3.36}$$

We may also use the first equation to simplify the second and take the equivalent pair

$$h_t + vh_x + hv_x = 0,$$

$$v_t + vv_x + g'h_x = g'S - C_f \frac{v^2}{h}. \tag{3.37}$$

The kinematic wave approximation to (3.37) neglects the left hand side of the second equation and takes

$$h_t + (hv)_x = 0, \qquad v = \left(\frac{g'S}{C_f}\right)^{1/2} h^{1/2}. \tag{3.38}$$

In this kinematic theory, discontinuous shocks must satisfy the shock condition

$$U = \frac{v_2 h_2 - v_1 h_1}{h_2 - h_1}.$$
                                                                        (3.39)

Stability; Roll Waves.

We now consider the consequences of the additional terms in (3.37). For simplicity, S and C_f are assumed to be constant. As in the traffic flow problem, we look first at the linearized form of the equations for small perturbations about a constant state $v = v_0$, $h = h_0$, where

$$C_f \frac{v_0^2}{h_0} = g' S.$$
                                                                        (3.40)

If we substitute

$$v = v_0 + w, \qquad h = h_0 + \eta$$

and neglect all but the first powers of w and η, we have

$$\eta_t + v_0 \eta_x + h_0 w_x = 0,$$

$$w_t + v_0 w_x + g' \eta_x + g' S \left(\frac{2w}{v_0} - \frac{\eta}{h_0} \right) = 0.$$

We may then eliminate w and write the single equation for η in the form

$$\left(\frac{\partial}{\partial t} + c_+ \frac{\partial}{\partial x} \right)\left(\frac{\partial}{\partial t} + c_- \frac{\partial}{\partial x} \right)\eta + \frac{2g' S}{v_0} \left(\frac{\partial}{\partial t} + c_0 \frac{\partial}{\partial x} \right)\eta = 0, \quad (3.41)$$

where

$$c_+ = v_0 + \sqrt{g' h_0}, \qquad c_- = v_0 - \sqrt{g' h_0}, \qquad c_0 = \frac{3 v_0}{2}. \quad (3.42)$$

The equation is now the same as in the earlier discussion of (3.4) and (3.9), with appropriate changes in the expressions for c_+, c_-, and c_0. Accordingly, the stability condition is

$$c_- < c_0 < c_+,$$
                                                                        (3.43)

and this also ensures that the lower order approximation

$$\frac{\partial \eta}{\partial t} + c_0 \frac{\partial \eta}{\partial x} = 0$$
                                                                        (3.44)

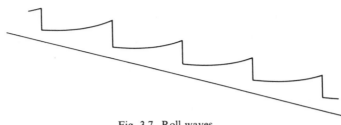

Fig. 3.7. Roll waves.

does not violate the characteristic condition. Equation 3.44 is the linearized version of (3.38), of course.

The stability conditions may also be written, using (3.42), as

$$v_0 < 2\sqrt{g'h_0} \, ,$$

or, again, from (3.40) as

$$S < 4C_f.$$

For rivers, v_0 is usually much less than $\sqrt{g'h_0}$, but spillways from dams and other man-made conduits easily exceed the critical values. The resulting flow is not necessarily completely chaotic and without structure. In favorable circumstances, it takes the form of "roll waves," as shown in Fig. 3.7, with a periodic structure of discontinuous bores separated by smooth profiles. Early observational data and photographs of the phenomenon were obtained by Cornish in 1905 and are beautifully described in his classic book (Cornish, 1934), which summarizes his observations of waves in sand and water. The most specific data refer to a stone conduit in the Alps (the Grünnbach, Merligen) with a slope of 1 in 14. On an occasion when the mean depth was approximately 3 in., the mean flow velocity was estimated as 10 ft/sec and the whole roll wave pattern moved downstream at an average speed of 13.5 ft/sec. For these figures the Froude number $v_0/\sqrt{g'h_0}$ is 3.5, considerably in excess of the critical value of 2. These values would give $S/C_f = 12.5$ and lead to $C_f \approx 0.006$.

Jeffreys (1925) proposed the instability argument and noted that for smooth cement channels (for which he performed experiments) the friction coefficient is $C_f \approx 0.0025$; this value of C_f agrees with current values. For the latter value, uniform flows should become unstable when the slope S exceeds 1 in 100. Jeffreys found his own experiments on the production of roll waves inconclusive, but he felt that long channels with slopes considerably in excess of 1 in 100 were needed. Much later Dressler (1949) took up the subject and showed how to construct nonlinear solutions of (3.36), with appropriate jump conditions, to describe the roll wave pattern. The details will be indicated after the question of steady profile waves for the stable case has been considered.

Monoclinal Flood Wave.

The structure of the shocks arising in the kinematic theory (3.38 and 3.39) is particularly important in the flood wave problem, since in reality the shock thickness is of the order of 50 miles! It is obtained as usual by searching for steady profile solutions in a more detailed description which, in this case, is provided by (3.37). We look for solutions with

$$h = h(X), \qquad v = v(X), \qquad X = x - Ut.$$

The equations may be written

$$(v - U)\frac{dv}{dX} + g'\frac{dh}{dX} = g'S - C_f\frac{v^2}{h}, \tag{3.45}$$

$$h(U - v) = B, \tag{3.46}$$

where the continuity equation has been integrated to (3.46) and B is the constant of integration. The uniform states (h_1, v_1) at $X = \infty$ satisfy

$$g'S - C_f\frac{v_1^2}{h_1} = g'S - C_f\frac{v_2^2}{h_2} = 0,$$

$$h_1(U - v_1) = h_2(U - v_2) = B.$$

If we express all flow quantities in terms of h_1 and h_2, we have

$$v_1^2 = \frac{S}{C_f}g'h_1, \qquad v_2^2 = \frac{S}{C_f}g'h_2, \tag{3.47}$$

$$B = \left(\frac{v_2 - v_1}{h_2 - h_1}\right)h_1h_2 = \left(\frac{g'S}{C_f}\right)^{1/2}\frac{h_1h_2}{h_1^{1/2} + h_2^{1/2}}, \tag{3.48}$$

$$U = \frac{v_2h_2 - v_1h_1}{h_2 - h_1} = \left(\frac{g'S}{C_f}\right)^{1/2}\frac{h_2^{3/2} - h_1^{3/2}}{h_2 - h_1}. \tag{3.49}$$

The last of these is exactly the shock condition governing discontinuities in the kinematic theory (3.39). This is the usual pattern and we expect the solutions of (3.45) and (3.46) to provide the structure of these kinematic shocks.

When v is eliminated from (3.45) and (3.46), the equation for $h(X)$ takes the form

$$\frac{dh}{dX} = -\frac{(B-Uh)^2 C_f - g'h^3 S}{g'h^3 - B^2}. \tag{3.50}$$

Since the numerator must vanish for $h = h_1$ and $h = h_2$, these two values must be roots of the cubic. Then the third root is

$$H = \frac{C_f}{S} \frac{B^2}{g'h_1 h_2}$$

$$= \frac{h_1 h_2}{\left(h_1^{1/2} + h_2^{1/2}\right)^2}.$$

Since $H < h_1, h_2$, and the solution has h between h_1 and h_2, this third root $h = H$ is never a value taken in the solution considered.

Equation 3.50 may now be written

$$\frac{dh}{dX} = -S \frac{(h_2 - h)(h - h_1)(h - H)}{h^3 - B^2/g'}, \tag{3.51}$$

and the behavior of the solution depends critically on the sign of the denominator $h^3 - B^2/g'$ and its possible change of sign in the profile. From (3.46),

$$g'h^3 - B^2 = g'h^3 - (U - v)^2 h^2$$

$$= h^2 \left\{ g'h - (U - v)^2 \right\};$$

hence the sign is positive or negative depending on whether

$$U \lessgtr v + \sqrt{g'h}.$$

[From (3.48), $B > 0$; therefore from (3.46), $U > v$ and U is always greater than $v - \sqrt{g'h}$.]

When $h_2 \to h_1$, we see from (3.49) that

$$U \to \frac{3}{2} \left(\frac{g'S}{C_f} \right)^{1/2} h_1^{1/2} = \frac{3}{2} v_1.$$

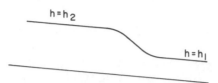

Fig. 3.8.   Structure of the monoclinal flood wave.

In the stable case, $\frac{3}{2}v_1 < v_1 + \sqrt{g'h_1}$; thus for weak waves we start the integral curve of (3.51) from $h = h_1$, $X = \infty$, with the denominator in (3.51) negative. Accordingly, $h_X > 0$, h increases, $g'h^3 - B^2$ remains positive and we have a smooth profile as shown in Fig. 3.8. This is the so-called monoclinal flood wave. The fact that $h_2 > h_1$ is required for this profile agrees with the tendency of breaking of the kinematic waves, since this is a problem with $c'(h) > 0$. A smooth profile of this type will continue to hold for the range of shock velocities

$$\frac{3v_1}{2} < U < v_1 + \sqrt{g'h_1} \ . \tag{3.52}$$

From (3.47) and (3.49) it is easily shown that this is the range

$$1 < \left(\frac{h_2}{h_1}\right)^{1/2} < \frac{1 + \left\{1 + 4(S/C_f)^{1/2}\right\}^{1/2}}{2(S/C_f)^{1/2}} \ .$$

But (3.52) is the more significant form, in view of the physical interpretation of the velocities. The shock moves faster than the lower order waves but slower than the higher order waves in the flow ahead.
    When

$$v_1 + \sqrt{g'h_1} \ < U < v_2 + \sqrt{g'h_2} \ ,$$

the denominator in (3.51) changes sign in the profile and the integral curve

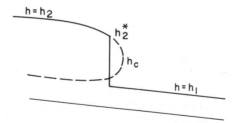

Fig. 3.9. Structure of the monoclinal flood wave with an inner discontinuity.

turns back on itself as in Fig. 3.9. A single-valued profile is recovered by fitting in a discontinuity as indicated. In contrast with the case of traffic flow, basic conservation equations in integrated form, (3.34) and (3.35), are known, and these apply whether the solution has discontinuities or not. If the same procedure developed in Section 2.3 is used on these,* the appropriate jump conditions at a discontinuity located at $x = s(t)$ are

$$-\dot{s}[h] + [hv] = 0, \tag{3.53}$$

$$-\dot{s}[hv] + \left[hv^2 + \frac{1}{2}g'h^2\right] = 0. \tag{3.54}$$

It should be especially noted that the right hand side of (3.35) makes no contribution in the final limit $x_1 \rightarrow x_2$. These discontinuity conditions go along with the equations (3.36), just as (3.39) goes along with (3.38). One must be careful to pair correctly the equations and shock conditions at each level of description. In a change of the level of description, both the equations and shock conditions change in number. The discontinuities described by (3.53) and (3.54) are in reality the turbulent bores familiar in water wave theory as "hydraulic jumps" or breakers on a beach.

In the present context, the proposal is to fit a discontinuity satisfying (3.53) and (3.54) into the steady profile solution of (3.36); hence it will also have the velocity U. In view of (3.46) any discontinuity between branches of the profile [including the lines $h = h_1$ and $h = h_2$ as possible solutions of (3.51)] will automatically have $h(U - v)$ continuous to satisfy (3.53). The second requirement (3.54) determines where it should be placed. The condition (3.54) requires

$$hv(v - U) + \frac{1}{2}g'h^2$$

to be continuous. From (3.46), this can be modified into

$$\frac{B^2}{h} + \frac{1}{2}g'h^2$$

should be continuous. If the discontinuity is chosen to take the profile from $h = h_1$ to a point $h = h^*$ on the upper branch as shown in Fig. 3.9, the

---

*Here to complete the physical problem we are requiring results from the later mathematical development in Chapters 5 and 10, but it seems better to include them here, with a minimum of explanation, rather than delaying completion of the solution.

requirement is

$$\frac{B^2}{h^*} + \frac{1}{2}g'h^{*2} = \frac{B^2}{h_1} + \frac{1}{2}g'h_1^2,$$

that is,

$$\frac{h^*}{h_1} = \frac{\{1 + 8B^2/g'h_1^3\}^{1/2} - 1}{2}.$$

This can be expressed in terms of h_1 and h_2 using (3.48). It can be verified that it meets the requirements (1) that $g'h^{*3} - B^2 > 0$ so that $h = h^*$ is on the upper branch, and (2) that $h^* < h_2$ provided $S < 4C_f$.

The overall conclusion, then, is that the original discontinuity of the kinematic theory (3.38 and 3.39) is resolved into a smooth profile when viewed from the more detailed description (3.36), *provided (3.52) is satisfied*. For stronger shocks that violate (3.52), some discontinuity remains and corresponds to a shock discontinuity in the solution of (3.36). Further interpretations of the significance of (3.52) will be given (see Chapter 10) after the theory of characteristics and shocks for higher order systems has been developed in detail.

The roll wave patterns referred to earlier are obtained by piecing together smooth sections satisfying (3.50) with discontinuous bores satisfying (3.53) and (3.54). It may be shown that $g'h^3 - B^2$ must change sign in the profile but the smooth parts are kept monotonic by demanding that the numerator of (3.50) also vanish at the critical depth. This requirement relates the two parameters B and U; one or the other may be kept as a basic parameter in the family of solutions and is determined by the total volume flow. For further details, reference should be made to Dressler's (1949) paper.

3.3 Glaciers

Nye (1960, 1963) has pointed out that these ideas on flood waves apply equally to the study of waves on glaciers and has developed the particular aspects that are most important there. He refers to Finsterwalder (1907) for the first studies of wave motion on glaciers and to independent formulations by Weertman (1958).

In view of the difficulties of collecting data on the flow curves for glaciers, due to both the inaccessibility and the slowness of the flow, more reliance is placed on semitheoretical derivations. These consider in more detail the shearing motion in two dimensional steady flow down a constant slope. Let $u(y)$ be the velocity of the layer at a distance y from the ground and let $\tau(y)$ be the shearing stress. For ice it seems to be appropriate to take the stress-strain relation as

$$\mu\frac{du}{dy} = \tau^n, \tag{3.55}$$

where $n \approx 3$ or 4. (Newtonian viscosity would be the case $n = 1$.) In addition, ice slips in its bed according to the approximate law

$$\nu u(0) = \tau^m(0), \tag{3.56}$$

where $m \approx \frac{1}{2}(n+1) \approx 2$. On the layer between y and $y + \delta y$, the difference in shearing stress must balance the gravitational force. If α is the angle of the slope and ρ is the density of ice, we have

$$\delta\tau = -\rho\,\delta y\,g\sin\alpha.$$

That is,

$$\frac{d\tau}{dy} = -\rho g\sin\alpha. \tag{3.57}$$

Since τ vanishes at the surface $y = h$, the solution for τ is

$$\tau = (h-y)\rho g\sin\alpha.$$

Then, integrating (3.55) with boundary condition (3.56), we have

$$u(y) = \frac{(\rho g\sin\alpha)^m h^m}{\nu} + \frac{1}{\mu}\frac{(\rho g\sin\alpha)^n}{n+1}\left\{h^{n+1} - (h-y)^{n+1}\right\}. \tag{3.58}$$

The flow per unit breadth is

$$Q^*(h) = \int_0^h u\,dy$$

$$= \frac{(\rho g\sin\alpha)^m h^{m+1}}{\nu} + \frac{(\rho g\sin\alpha)^n h^{n+2}}{n+2}. \tag{3.59}$$

For order of magnitude purposes, one may take

$$Q^*(h) \propto h^N,$$

with N roughly in the range 3 to 5. The propagation speed is

$$c = \frac{dQ^*}{dh} = N\upsilon$$

where υ is the average velocity Q^*/h. Thus the waves move about three to five times faster than the average flow velocity. Typical velocities are of the order of 10 to 100 meters per year.

Various problems can be solved using the results and ideas of Chapter 2. An interesting question considered by Nye is the effect of periodic accumulation and evaporation of the ice; depending on the period, this may refer either to seasonal or climatic changes. To do this a prescribed source term $f(x,t)$ is added to the continuity equation; that is, one takes

$$h_t + q_x^* = f(x,t), \qquad q^* = Q^*(h,x). \tag{3.60}$$

The consequences are determined from integration of the characteristic equations

$$\frac{dh}{dt} = f(x,t) - Q_x^*(h,x),$$

$$\frac{dx}{dt} = Q_h^*(h,x).$$

The main result is that parts of the glacier may be very sensitive, and relatively rapid local changes can be triggered by the source term.

3.4 Chemical Exchange Processes; Chromatography; Sedimentation in Rivers

The formulation of equations for exchange processes between a solid bed and a fluid flowing through it was given in Section 2.2. The exchange may involve particles or ions of some substance, or it may be heat exchange between the solid bed and the fluid. Another instance is sediment transport in rivers (see Kynch, 1952).

The equations coupling the density ρ_f in the fluid and the density ρ_s on the solid are

$$\frac{\partial}{\partial t}(\rho_f + \rho_s) + \frac{\partial}{\partial x}(V\rho_f) = 0, \tag{3.61}$$

$$\frac{\partial \rho_s}{\partial t} = k_1(A - \rho_s)\rho_f - k_2\rho_s(B - \rho_f). \tag{3.62}$$

For relatively slow changes in the densities and relatively high reaction rates k_1, k_2, the second equation is taken in the approximate quasi-equilibrium form in which the $\partial \rho_s / \partial t$ is neglected and

$$\rho_s = R(\rho_f) = \frac{k_1 A \rho_f}{k_2 B + (k_1 - k_2)\rho_f}. \tag{3.63}$$

When this relation is substituted into (3.61), we have

$$\frac{\partial \rho_f}{\partial t} + \frac{V}{1 + R'(\rho_f)} \frac{\partial \rho_f}{\partial x} = 0. \tag{3.64}$$

Thus the density changes propagate downstream at the rate

$$c = \frac{V}{1 + R'(\rho_f)}, \tag{3.65}$$

and

$$R'(\rho_f) = \frac{k_1 k_2 A B}{\left\{ k_2 B + (k_1 - k_2)\rho_f \right\}^2}. \tag{3.66}$$

If the densities concerned are small this is approximately

$$c = \frac{k_2 B}{k_1 A + k_2 B} V. \tag{3.67}$$

The propagation speed depends on the reaction rates involved, being slower for substances with larger attraction toward the solid. If a mixture of substances is present in the fluid at the entrance of the column and the components have different reactions rates, they will travel down the column at different speeds. In this way the column can be used to separate the mixture into bands of the individual components. If they are also colored, a spectrum is formed. This is the basic process of chromatography. The nonlinear effects produce heavier concentrations at the beginning or end of a band depending on the sign of $c'(\rho_f)$. Of course, the nonlinear equations for a single component apply only after the separation has taken place.

The shock structure and other aspects can be studied from the full equations 3.61 and 3.62. It is remarkable in this case that the full equations can be transformed (exactly) into a linear equation. This was shown by Thomas (1944). First, a moving coordinate system

$$\tau = t - \frac{x}{V}, \qquad \sigma = \frac{x}{V}$$

is introduced; the equations then take the form

$$\frac{\partial \rho_f}{\partial \sigma} + \frac{\partial \rho_s}{\partial \tau} = 0,$$

$$\frac{\partial \rho_s}{\partial \tau} = \alpha \rho_f - \beta \rho_s - \gamma \rho_s \rho_f. \tag{3.68}$$

The first equation is solved identically by

$$\rho_f = \psi_\tau, \qquad \rho_s = -\psi_\sigma, \tag{3.69}$$

and the second equation provides an equation for ψ:

$$\psi_{\sigma\tau} + \alpha\psi_\tau + \beta\psi_\sigma + \gamma\psi_\sigma\psi_\tau = 0. \tag{3.70}$$

If we now make the nonlinear transformation

$$\gamma\psi = \log\chi, \tag{3.71}$$

we deduce

$$\chi_{\sigma\tau} + \alpha\chi_\tau + \beta\chi_\sigma = 0; \tag{3.72}$$

the nonlinear transformation eliminates the nonlinear term. In terms of the original variables the transformation is

$$\rho_f = \frac{1}{\gamma}\frac{\chi_t}{\chi}, \qquad \rho_s = -\frac{\chi_t + V\chi_x}{\gamma\chi}, \tag{3.73}$$

$$\chi_{tt} + V\chi_{xt} + (\alpha + \beta)\chi_t + \beta V\chi_x = 0. \tag{3.74}$$

The linear equation can be solved in general by transform methods so that in this case the solutions of the approximate equation (3.64), including shocks when necessary, may be compared in detail with the exact solution. This has been investigated extensively by Goldstein (1953) and Goldstein and Murray (1959). The exact solution endorses the views and methods for including discontinuous shocks in solutions of (3.64). The details are not given here since Burgers' equation is simpler to deal with and provides the same endorsement. Some of the relevant analysis will appear in Chapter 10 in a different connection.

For exchange processes, $\alpha = k_1 A$ and $\beta = k_2 B$ are both positive. For these signs, the uniform state is always stable. This may be checked by considering perturbations in (3.74). We note that the lower order waves travel at a speed

$$c_0 = \frac{\beta V}{\alpha + \beta},$$

while the wave speeds given by the higher order terms are $c_- = 0$ and $c_+ = V$. Thus for $\alpha > 0$, $\beta > 0$, the stability criterion $c_- < c_0 < c_+$ is satisfied.

CHAPTER 4

Burgers' Equation

The simplest equation combining both nonlinear propagation effects and diffusive effects is Burgers' equation

$$c_t + cc_x = \nu c_{xx}. \tag{4.1}$$

In (2.28) we saw that this is an exact equation for waves described by

$$\rho_t + q_x = 0, \qquad q = Q(\rho) - \nu \rho_x, \tag{4.2}$$

in the case that $Q(\rho)$ is a quadratic function of ρ. In general, if the two effects are important in a problem, there is usually some way of extracting (4.1) either as a precise approximation or as a useful basis for rough estimates.

For a general $Q(\rho)$ in (4.2), for example, the equation may be written

$$c_t + cc_x = \nu c_{xx} - \nu c''(\rho)\rho_x^2, \tag{4.3}$$

where $c(\rho) = Q'(\rho)$ as usual. The ratio of $\nu c''(\rho)\rho_x^2$ to νc_{xx} is of the order of the amplitude of the disturbance, and we therefore expect that (4.1) is a good approximation for small amplitude. We are then assuming that omission of this particular small amplitude term does not produce accumulating errors (as $t \to \infty$, say) which eventually lead to nonuniform validity. We know, in contrast, that to linearize the left hand side by $c_t + c_0 c_x$, where c_0 is some constant unperturbed value, would be disastrous in this respect. But as a check, we may verify that in the shock structure solution (see Section 2.4), where the diffusion terms are greatest, the term $\nu c''(\rho)\rho_x^2$ remains of smaller order than νc_{xx} in the strength of the shock. This kind of argument can be made the basis of formal perturbation expansions in terms of appropriate precisely defined small parameters. On the other hand, the fact that the terms retained in (4.1) represent identifiable and important phenomena, whereas the term $\nu c''(\rho)\rho_x^2$ appears more as a mathematical nuisance, leads one to suggest (4.1) as a useful overall description even beyond the range of strict validity.

In a similar fashion, Burgers' equation is relevant in higher order systems such as (3.2)–(3.3), when nonlinear propagation is combined with diffusion. Of course it is limited to the stable range and to parts of the solution where the lower order waves are dominant. The appropriate form is easily recognized and again can usually be substantiated by more formal procedures. In the case of (3.2)–(3.3), we know from (3.6) that the effective diffusivity is $\nu^* = \nu - (v_0 - c_0)^2\tau$ and we would use (4.1) with this value. Indeed, (3.6) is the fully linearized Burgers' equation for this system.

Our general purpose now is to show that the exact solution of Burgers' equation endorses the ideas regarding shocks that were developed in Chapter 2. That is, we want to confirm that as $\nu \to 0$ (in appropriate dimensionless form) the solutions of (4.1) reduce to solutions of

$$c_t + cc_x = 0, \tag{4.4}$$

with discontinuous shocks which satisfy

$$U = \frac{1}{2}(c_1 + c_2), \qquad c_2 > U > c_1, \tag{4.5}$$

and the shocks are located at the positions determined in Section 2.8.

4.1 The Cole-Hopf Transformation

Independently, Cole (1951) and Hopf (1950) noted the remarkable result that (4.1) may be reduced to the linear heat equation by the nonlinear transformation

$$c = -2\nu\frac{\varphi_x}{\varphi}. \tag{4.6}$$

This is similar to Thomas' earlier transformation of the exchange equations described in Section 3.4. It is again convenient to do the transformation in two steps. First introduce

$$c = \psi_x,$$

so that (4.1) may be integrated to

$$\psi_t + \frac{1}{2}\psi_x^2 = \nu\psi_{xx}.$$

Then introduce

$$\psi = -2\nu\log\varphi$$

to obtain

$$\varphi_t = \nu \, \varphi_{xx}. \tag{4.7}$$

The nonlinear transformation just eliminates the nonlinear term. The general solution of the heat equation (4.7) is well known and can be handled by a variety of methods.

The basic problem considered in Chapter 2 is the initial value problem:

$$c = F(x) \qquad \text{at } t = 0.$$

This transforms through (4.6) into the initial value problem

$$\varphi = \Phi(x) = \exp\left\{ -\frac{1}{2\nu} \int_0^x F(\eta)\, d\eta \right\}, \qquad t = 0, \tag{4.8}$$

for the heat equation. The solution for φ is

$$\varphi = \frac{1}{\sqrt{4\pi\nu t}} \int_{-\infty}^{\infty} \Phi(\eta) \exp\left\{ -\frac{(x-\eta)^2}{4\nu t} \right\} d\eta. \tag{4.9}$$

Therefore, from (4.6), the solution for c is

$$c(x,t) = \frac{\displaystyle\int_{-\infty}^{\infty} \frac{x-\eta}{t} e^{-G/2\nu} d\eta}{\displaystyle\int_{-\infty}^{\infty} e^{-G/2\nu} d\eta}, \tag{4.10}$$

where

$$G(\eta; x, t) = \int_0^{\eta} F(\eta')\, d\eta' + \frac{(x-\eta)^2}{2t}. \tag{4.11}$$

4.2 Behavior as $\nu \to 0$

The behavior of the exact solution (4.10) is now considered as $\nu \to 0$ while x, t and $F(x)$ are held fixed. [Strictly speaking this means we consider a family of solutions with $\nu = \epsilon \nu_0$ and take the limit as $\epsilon \to 0$, holding $\nu_0, x, t, F(x)$ fixed.] As $\nu \to 0$, the dominant contributions to the integrals in (4.10) come from the neighborhood of the stationary points of G. A stationary point is where

$$\frac{\partial G}{\partial \eta} = F(\eta) - \frac{x-\eta}{t} = 0. \tag{4.12}$$

Let $\eta = \xi(x, t)$ be such a point; that is, $\xi(x, t)$ is defined as a solution of

$$F(\xi) - \frac{(x - \xi)}{t} = 0. \tag{4.13}$$

The contribution from the neighborhood of a stationary point, $\eta = \xi$, in an integral

$$\int_{-\infty}^{\infty} g(\eta) e^{-G(\eta)/2\nu} d\eta,$$

is

$$g(\xi) \sqrt{\frac{4\pi\nu}{G''(\xi)}} \; e^{-G(\xi)/2\nu};$$

this is the standard formula of the method of steepest descents.

Suppose first that there is only one stationary point $\xi(x, t)$ which satisfies (4.13). Then

$$\int_{-\infty}^{\infty} \frac{x - \eta}{t} e^{-G/2\nu} d\eta \sim \frac{x - \xi}{t} \sqrt{\frac{4\pi\nu}{G''(\xi)}} \; e^{-G(\xi)/2\nu}, \tag{4.14}$$

$$\int_{-\infty}^{\infty} e^{-G/2\nu} d\eta \sim \sqrt{\frac{4\pi\nu}{G''(\xi)}} \; e^{-G(\xi)/2\nu}, \tag{4.15}$$

and in (4.10) we have

$$c \sim \frac{x - \xi}{t}, \tag{4.16}$$

where $\xi(x, t)$ is defined by (4.13). This asymptotic solution may be rewritten

$$\begin{aligned} c &= F(\xi) \\ x &= \xi + F(\xi)t. \end{aligned} \tag{4.17}$$

It is exactly the solution of (4.4) which was discussed in (2.5) and (2.6); the stationary point $\xi(x, t)$ becomes the characteristic variable.

However, we saw that in some cases (4.17) gives a multivalued solution after a sufficient time, and discontinuities must be introduced. Yet the solution (4.10) for Burgers' equation is clearly single-valued and continuous for all t. The explanation is that when this stage is reached there are two stationary points that satisfy (4.13), and the foregoing analysis of the asymptotic behavior requires modification. If the two stationary points are denoted by ξ_1 and ξ_2 with $\xi_1 > \xi_2$, there will be contributions as shown in (4.14) and (4.15) from both ξ_1 and ξ_2. Therefore the dominant behavior will be included if we take

$$c \sim \frac{\dfrac{x - \xi_1}{t} \{G''(\xi_1)\}^{-1/2} e^{-G(\xi_1)/2\nu} + \dfrac{x - \xi_2}{t} \{G''(\xi_2)\}^{-1/2} e^{-G(\xi_2)/2\nu}}{\{G''(\xi_1)\}^{-1/2} e^{-G(\xi_1)/2\nu} + \{G''(\xi_2)\}^{-1/2} e^{-G(\xi_2)/2\nu}}. \quad (4.18)$$

When $G(\xi_1) \neq G(\xi_2)$, the accentuation by the small denominator ν in the exponents makes one or the other of the terms overwhelmingly large as $\nu \to 0$. If $G(\xi_1) < G(\xi_2)$, we have

$$c \sim \frac{x - \xi_1}{t};$$

if $G(\xi_1) > G(\xi_2)$,

$$c \sim \frac{x - \xi_2}{t}.$$

In each case (4.17) applies with either ξ_1 or ξ_2 for ξ. But the choice is now unambiguous. Both ξ_1 and ξ_2 are functions of (x, t); the criterion $G(\xi_1) \gtrless G(\xi_2)$ will determine the appropriate choice of ξ_1 or ξ_2 for given (x, t). The changeover from ξ_1 to ξ_2 will occur at those (x, t) for which

$$G(\xi_1) = G(\xi_2).$$

From (4.11), this is when

$$\int_0^{\xi_2} F(\eta') \, d\eta' + \frac{(x - \xi_2)^2}{2t} = \int_0^{\xi_1} F(\eta') \, d\eta' + \frac{(x - \xi_1)^2}{2t}. \quad (4.19)$$

Since ξ_1 and ξ_2 both satisfy (4.13), the condition may be written

$$\frac{1}{2} \{ F(\xi_1) + F(\xi_2) \} (\xi_1 - \xi_2) = \int_{\xi_2}^{\xi_1} F(\eta') \, d\eta'. \quad (4.20)$$

This is exactly the shock determination obtained in (2.45). The changeover in the choice of terms in (4.18) leads to the discontinuity in $c(x, t)$ in the limit $\nu \to 0$. All the details of Section 2.8 can be confirmed similarly. We conclude that solutions of Burgers' equation approach those described by (4.4) and (4.5) as $\nu \to 0$.

In reality ν is fixed, but it is relatively small and we expect that the limit solution for $\nu \to 0$ will often be a good approximation. For this argument, since ν is a dimensional quantity, we have to introduce a nondimensional measure of ν by comparing it with some other quantity of the same dimension. This is not hard to do. In the single hump problem, for example, where $F(x)$ is as shown in Fig. 2.9, we may introduce the parameter

$$A = \int_{-\infty}^{\infty} \{ F(x) - c_0 \} \, dx. \tag{4.21}$$

The dimensions of A and ν are both L^2/T, so that

$$R = \frac{A}{2\nu} \tag{4.22}$$

is a dimensionless number, and by "ν small" we mean $R \gg 1$. If the length of the hump is L, the number R measures the ratio of the nonlinear term $(c - c_0)c_x$ to the diffusion term νc_{xx}, *in those regions where the x scale for the derivatives is L.* (Inside shocks, for example, the x scale is of smaller order.) It will be convenient to refer to R as the Reynolds number, following the practice in viscous flow.

Even with the meaning of "small ν" decided, there are distinctions between the limit solution $\nu \to 0$ and the solution for fixed small ν. As we saw in (2.26), the shock thickness tends to infinity if the strength tends to zero. Therefore for fixed R, even if it is large, any solution that includes shock formation or a shock decaying as $t \to \infty$ will not always be well approximated by the discontinuity theory in these regions. As regards a shock formation region, the precise details are not usually important; one just wants a good estimate of *where* it forms, without details of the profile, and this is provided by the discontinuity theory. The effects of diffusion on decaying shocks as $t \to \infty$ is of more interest. We will explore these questions through typical examples in the following sections.

4.3 Shock Structure

The shock structure for (4.1) satisfies

$$- Uc_X + cc_X = \nu c_{XX}, \qquad X = x - Ut.$$

Hence

$$\frac{1}{2} c^2 - Uc + C = \nu c_X.$$

If $c \to c_1, c_2$ as $X \to \pm \infty$,

$$U = \frac{1}{2} (c_1 + c_2), \qquad C = \frac{1}{2} c_1 c_2,$$

and the equation may be written

$$(c - c_1)(c_2 - c) = -2\nu c_x.$$

The solution is

$$\frac{X}{\nu} = \frac{2}{c_2 - c_1} \log \frac{c_2 - c}{c - c_1};$$

this agrees with (2.25), since $c = 2\alpha\rho + \beta$ for quadratic $Q(\rho)$. Solving for c, we have

$$c = c_1 + \frac{c_2 - c_1}{1 + \exp \dfrac{c_2 - c_1}{2\nu}(x - Ut)}, \qquad U = \frac{c_1 + c_2}{2}. \qquad (4.23)$$

One can study how an initial step diffuses into this steady profile by taking

$$F(x) = \begin{cases} c_1, & x > 0, \\ c_2 > c_1, & x < 0, \end{cases}$$

in (4.10)–(4.11). The solution may be put in the form

$$c = c_1 + \frac{c_2 - c_1}{1 + h \exp \dfrac{c_2 - c_1}{2\nu}(x - Ut)}, \qquad U = \frac{c_1 + c_2}{2}. \qquad (4.24)$$

where

$$h = \frac{\displaystyle\int_{-(x - c_1 t)/\sqrt{4\nu t}}^{\infty} e^{-\zeta^2} d\zeta}{\displaystyle\int_{(x - c_2 t)/\sqrt{4\nu t}}^{\infty} e^{-\zeta^2} d\zeta}. \qquad (4.25)$$

For fixed x/t in the range $c_1 < x/t < c_2$, $h \to 1$ as $t \to \infty$, and the solution approaches (4.23).

4.4 Single Hump

A special solution with a single hump may be obtained by taking

$$F(x) = c_0 + A\delta(x) \qquad (4.26)$$

as the initial condition. The parameter A agrees with (4.21) and the

Reynolds number is $R = A/2\nu$. The constant c_0 may be omitted without loss of generality, since the substitution

$$c = c_0 + \tilde{c}, \qquad x = c_0 t + \tilde{x} \qquad (4.27)$$

in Burgers' equation reduces it to

$$\tilde{c}_t + \tilde{c}\tilde{c}_{\tilde{x}} = \nu \tilde{c}_{\tilde{x}\tilde{x}} \qquad (4.28)$$

Thus omission of c_0 is equivalent to viewing the solution from a frame of reference moving with velocity c_0. Accordingly we consider only

$$F(x) = A\delta(x). \qquad (4.29)$$

The lower limit in the integral in (4.11) is arbitrary since it cancels out in (4.10). Therefore we may choose it to be $0+$ and include the δ function for $\eta < 0$ but not for $\eta > 0$. Then

$$G = \begin{cases} \dfrac{(x-\eta)^2}{2t}, & \eta > 0, \\[2ex] \dfrac{(x-\eta)^2}{2t} - A, & \eta < 0. \end{cases}$$

The integrals in the numerator of (4.10) may be evaluated and those in the denominator written in terms of the complementary error function. The result is

$$c(x,t) = \sqrt{\frac{\nu}{t}} \; \frac{(e^R - 1)e^{-x^2/4\nu t}}{\sqrt{\pi} + (e^R - 1)\displaystyle\int_{x/\sqrt{4\nu t}}^{\infty} e^{-\zeta^2} d\zeta}, \qquad R = \frac{A}{2\nu}. \quad (4.30)$$

The similarity form of the solution, that is,

$$c = \sqrt{\frac{\nu}{t}} \; f\!\left(\frac{x}{\sqrt{\nu t}}; \frac{A}{\nu}\right),$$

could have been predicted by dimensional arguments. The only dimensional parameters in the problem, A and ν, both have dimensions L^2/T; there is no separate length and time with which to scale x and t separately.

As $R \to 0$ we would expect the diffusion to dominate over the nonlinearity. For $R \ll 1$ the denominator in (4.30) is $\sqrt{\pi} + O(R)$, uniformly in

x, t, ν; hence c may be approximated by

$$c(x,t) = \sqrt{\frac{\nu}{\pi t}} \; R e^{-x^2/4\nu t}$$

$$= \frac{A}{\sqrt{4\pi\nu t}} e^{-x^2/4\nu t}. \tag{4.31}$$

This is the source solution of the heat equation $c_t = \nu c_{xx}$, so the expectation is verified.

To discuss the behavior for large R it is convenient to introduce the similarity variable $z = x/\sqrt{2At}$ and to write (4.30) as

$$c = \sqrt{\frac{2A}{t}} \; g(z, R),$$

$$g(z, R) = \frac{(e^R - 1)}{2\sqrt{R}} \; \frac{e^{-z^2 R}}{\sqrt{\pi} + (e^R - 1) \int_{z\sqrt{R}}^{\infty} e^{-\zeta^2} d\zeta}, \tag{4.32}$$

$$z = \frac{x}{\sqrt{2At}}.$$

We now discuss the behavior of g as $R \to \infty$ for different ranges of z. In all cases, $e^R - 1$ may be approximated by e^R and we may use

$$g \sim \frac{1}{2\sqrt{R}} \; \frac{e^{R(1-z^2)}}{\sqrt{\pi} + e^R \int_{z\sqrt{R}}^{\infty} e^{-\zeta^2} d\zeta}. \tag{4.33}$$

If $z < 0$, the integral tends to

$$\int_{-\infty}^{\infty} e^{-\zeta^2} d\zeta = \sqrt{\pi} \; ;$$

therefore $g \to 0$ at least like $1/\sqrt{R}$. If $z > 0$, the integral becomes small and we use the asymptotic expansion

$$\int_{\eta}^{\infty} e^{-\zeta^2} d\zeta \sim \frac{e^{-\eta^2}}{2\eta} \qquad \text{as } \eta \to \infty.$$

Therefore

$$g \sim \frac{z}{1 + 2z\sqrt{\pi R} \; e^{R(z^2-1)}}, \qquad z > 0, \quad R \to \infty. \tag{4.34}$$

If $0 < z < 1$, we have

$$g \sim z, \qquad 0 < z < 1, \quad R \to \infty, \tag{4.35}$$

whereas if $z > 1$, $g \to 0$ as $R \to \infty$. Thus $g \to 0$ except in $0 < z < 1$, and in that range $g \sim z$. In the original variables, the result reads

$$c \sim \begin{cases} \dfrac{x}{t} & \text{in } 0 < x < \sqrt{2At}, \\ 0 & \text{outside.} \end{cases}$$

This is the appropriate solution of (4.4) with a shock at $x = \sqrt{2At}$. The shock velocity is $U = \sqrt{A/2t}$, and c jumps from zero to $\sqrt{2A/t}$, so the shock condition (4.5) is satisfied.

The same expression (with $c_0 = 0$ to fit our assumption here) appeared in (2.52) for the ultimate behavior of the solution of (4.4) for a general single hump. That was asymptotic in a different sense; it was the behavior as $t \to \infty$ within the description provided by (4.4). For a δ function initial condition, it is valid immediately.

The shock is located at $z = 1$, and for large but finite R (4.34) shows a rapid transition from exponentially small values in $z > 1$ to $g \sim z$ in $z < 1$. In the transition layer $z \approx 1$, (4.34) may be approximated by

$$g \approx \frac{1}{1 + 2\sqrt{\pi R}\, e^{2R(z-1)}}. \tag{4.36}$$

In the original variables this would give

$$c \approx \sqrt{\frac{2A}{t}}\; \frac{1}{1 + \exp\left\{ \dfrac{1}{2\nu}\sqrt{\dfrac{2A}{t}}\,(x - \sqrt{2At}) + \tfrac{1}{2}\log\dfrac{2\pi A}{\nu} \right\}}. \tag{4.37}$$

It agrees with the shock profile (4.23), with $c_2 - c_1 = \sqrt{2A/t}$ and the shock located at $x = \sqrt{2At}$ to first order. From (4.36), the transition layer is of thickness $O(R^{-1})$ around $z = 1$.

There is another (weaker) transition layer at $z = 0$ to smooth out the discontinuity in derivative between $g \sim 0$ in $z < 0$ to $g \sim z$ in $0 < z < 1$. It is clear from (4.33) that this transition layer occurs for

$$z = O(R^{-1/2}),$$

and for these values (4.33) may be approximated by

$$g \approx \frac{e^{-Rz^2}}{2\sqrt{R}\int_{z\sqrt{R}}^{\infty} e^{-\zeta^2}d\zeta}. \tag{4.38}$$

In the original variables we have

$$c \approx \sqrt{\frac{\nu}{t}} \; \frac{e^{-x^2/4\nu t}}{\int_{x/\sqrt{4\nu t}}^{\infty} e^{-\zeta^2}d\zeta}. \tag{4.39}$$

The form of the solution for large R is shown in Fig. 4.1, where $g(z) = c\sqrt{t/2A}$ is plotted against z. As $R \to \infty$ the shock layer becomes a discontinuity in c and the transition layer at $x = 0$ becomes a discontinuity in c_x. In the scaled variables g and z, the profile is independent of t. Therefore if the value of R provided by the initial condition is large, the shock remains relatively thin and the discontinuity theory of (4.4) is a good approximation *for all* t. This is true even though the shock strength is proportional to $\sqrt{2A/t}$ and tends to zero as $t \to \infty$.

A significant point in this connection is that the area under the profile remains constant even with diffusion included, since

$$\frac{d}{dt}\int_{-\infty}^{\infty} c \, dx = \left[\nu c_x - \frac{1}{2}c^2\right]_{-\infty}^{\infty} = 0.$$

Hence the "effective" Reynolds number defined as

$$\frac{1}{2\nu}\int_{-\infty}^{\infty} c \, dx$$

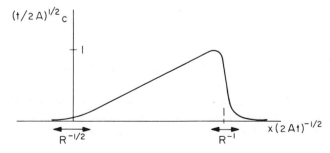

Fig. 4.1. Triangular wave solution of Burgers' equation.

remains constant for all t. The next example will show that the more usual situation is for diffusion to take over ultimately in the final decay, and that the single hump is exceptional in this respect.

4.5 *N* Wave

The final examples we consider are more easily derived by choosing appropriate solutions for φ to satisfy the heat equation (4.7) and then substituting in (4.6) to obtain c. As a rough qualitative guide to the appropriate choice, the profile for c will be something like φ_x. Thus for the single hump we could have taken the solution of φ corresponding to an initial step function. To obtain an N wave for c, we choose the source solution of the heat equation for φ:

$$\varphi = 1 + \sqrt{\frac{a}{t}}\; e^{-x^2/4\nu t}. \tag{4.40}$$

From (4.6), the corresponding solution for c is

$$c = -\frac{2\nu\,\varphi_x}{\varphi} = \frac{x}{t}\,\frac{\sqrt{a/t}\; e^{-x^2/4\nu t}}{1 + \sqrt{a/t}\; e^{-x^2/4\nu t}}. \tag{4.41}$$

Since φ has a δ function behavior as $t \to 0$, this is a little hard to interpret as an initial value problem on c. However, for any $t > 0$ it has the form shown in Fig. 4.2, with a positive and negative phase, and we may take the profile at any $t = t_0 > 0$ to be the initial profile. It should be typical of all N wave solutions.

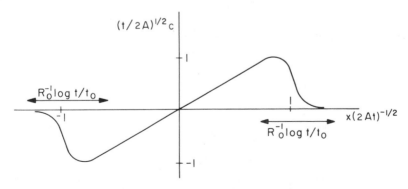

Fig. 4.2. *N* wave solution of Burgers' equation.

The area under the positive phase of the profile is

$$\int_0^\infty c\,dx = -2\nu[\log\varphi]_0^\infty$$

$$= 2\nu\log\left(1+\sqrt{\frac{a}{t}}\,\right). \qquad (4.42)$$

The magnitude of the area of the negative phase is the same. Thus in marked contrast with the previous case, the area of the positive phase tends to zero as $t\to\infty$. If the value of (4.42) at the initial time t_0 is denoted by A, we may introduce a Reynolds number

$$R_0 = \frac{A}{2\nu} = \log\left(1+\sqrt{\frac{a}{t_0}}\,\right). \qquad (4.43)$$

But as time goes on the *effective* Reynolds number will be

$$R(t) = \frac{1}{2\nu}\int_0^\infty c\,dx = \log\left(1+\sqrt{\frac{a}{t}}\,\right), \qquad (4.44)$$

and this tends to zero as $t\to\infty$. If $R_0\gg1$, we may expect the "inviscid theory" of (4.4)–(4.5) to be a good approximation for some time, but as $t\to\infty$, $R(t)\to0$ and the diffusion term will eventually become dominant. This is different from the previous example in which the effective Reynolds number defined in the same way remains equal to the initial Reynolds number. We now verify the details.

In terms of R_0 and t_0, $a = t_0(e^{R_0}-1)^2$; hence (4.41) may be written

$$c = \frac{x}{t}\left\{1+\sqrt{\frac{t}{t_0}}\,\frac{e^{x^2/4\nu t}}{e^{R_0}-1}\right\}^{-1}. \qquad (4.45)$$

For $R_0\gg1$ (corresponding to $t_0\ll a$), it may be approximated by

$$c\sim\frac{x}{t}\left\{1+\sqrt{\frac{t}{t_0}}\,e^{(x^2/2At-1)R_0}\right\}^{-1} \qquad (4.46)$$

for all x and t. Now for fixed t and $R_0\to\infty$,

$$c\sim\begin{cases}\dfrac{x}{t}, & -\sqrt{2At}<x<\sqrt{2At}\,,\\[2mm] 0, & |x|>\sqrt{2At}\,.\end{cases}$$

This is exactly the inviscid solution. However, for any fixed a and ν we see directly from (4.41) [and it may be verified also from (4.46)] that

$$c \sim \frac{x}{t}\sqrt{\frac{a}{t}}\; e^{-x^2/4\nu t} \qquad \text{as } t \to \infty. \qquad (4.47)$$

This is the dipole solution of the heat equation. The diffusion dominates the nonlinear term in the final decay. It should be remembered, though, that this final period of decay is for extremely large times; the inviscid theory is adequate for most of the interesting range.

4.6 Periodic Wave

A periodic solution may be obtained by taking for φ a distribution of heat sources spaced a distance λ apart. Then

$$\varphi = (4\pi\nu t)^{-1/2} \sum_{n=-\infty}^{\infty} \exp\left\{ -\frac{(x-n\lambda)^2}{4\nu t} \right\}, \qquad (4.48)$$

$$c = -2\nu \frac{\varphi_x}{\varphi} = \frac{\displaystyle\sum_{-\infty}^{\infty} \{(x-n\lambda)/t\} \exp\{-(x-n\lambda)^2/4\nu t\}}{\displaystyle\sum_{-\infty}^{\infty} \exp\{-(x-n\lambda)^2/4\nu t\}}. \qquad (4.49)$$

When $\lambda^2/4\nu t \gg 1$, the exponential with the minimum value of $(x-n\lambda)^2/4\nu t$ will dominate over all the others. Therefore the term with $n=m$ will dominate for $(m-\tfrac{1}{2})\lambda < x < (m+\tfrac{1}{2})\lambda$, and (4.49) is approximately

$$c \sim \frac{x-m\lambda}{t}, \qquad \left(m-\frac{1}{2}\right)\lambda < x < \left(m+\frac{1}{2}\right)\lambda.$$

This is a sawtooth wave with a periodic set of shocks a distance λ apart, and c jumps from $-\lambda/2t$ to $\lambda/2t$ at each shock. The result agrees with (2.56).

To study the final decay, $\lambda^2/4\nu t \ll 1$, we may use an alternative form of the solution. The expression (4.48) is periodic in x, and in the interval $-\lambda/2 < x < \lambda/2$,

$$\varphi \to \delta(x) \qquad \text{as } t \to 0.$$

The initial condition can be expanded in a Fourier series as

$$\Phi(x) = \frac{1}{\lambda}\left\{1 + 2\sum_{1}^{\infty}\cos\frac{2\pi n x}{\lambda}\right\},$$

and the corresponding solution of the heat equation for φ is

$$\varphi = \frac{1}{\lambda}\left\{1 + 2\sum_{1}^{\infty}\exp\left(-\frac{4\pi^2 n^2}{\lambda^2}\nu t\right)\cos\frac{2\pi n x}{\lambda}\right\}. \tag{4.50}$$

It may be verified directly that this is the Fourier series of (4.48). In this form

$$c = -\frac{2\nu\varphi_x}{\varphi} = \frac{\dfrac{8\pi\nu}{\lambda}\sum\limits_{1}^{\infty}n\exp\left(-\dfrac{4\pi^2 n^2}{\lambda^2}\nu t\right)\sin\dfrac{2\pi n x}{\lambda}}{1 + 2\sum\limits_{1}^{\infty}\exp\left(-\dfrac{4\pi^2 n^2}{\lambda^2}\nu t\right)\cos\dfrac{2\pi n x}{\lambda}}. \tag{4.51}$$

When $\nu t/\lambda^2 \gg 1$, the term with $n = 1$ dominates the series and we have

$$c \sim \frac{8\pi\nu}{\lambda}\exp\left(-\frac{4\pi^2\nu t}{\lambda^2}\right)\sin\frac{2\pi x}{\lambda}. \tag{4.52}$$

This is a solution of $c_t = \nu c_{xx}$, and again the diffusion dominates in the ultimate decay.

4.7 Confluence of Shocks

When a shock overtakes another shock, they merge into a single shock of increased strength as described for the inviscid solution ($\nu \to 0$) on the F curve in Fig 2.16. It is possible to give a simple solution of Burgers' equation that describes this process for arbitrary ν.

The solution for a single shock is given in (4.23) and the corresponding expression for φ may be written in the form

$$\varphi = f_1 + f_2, \qquad f_j = \exp\left(-\frac{c_j x}{2\nu} + \frac{c_j^2 t}{4\nu} - b_j\right). \tag{4.53}$$

In (4.23), the parameters b_1, b_2 which locate the initial position of the shock are taken to be zero. The expressions f_1, f_2 are clearly solutions of the heat equation (4.7). The expression for c is

$$c = -\frac{2\nu\varphi_x}{\varphi} = \frac{c_1 f_1 + c_2 f_2}{f_1 + f_2}. \tag{4.54}$$

For $c_2 > c_1$, f_1 dominates as $x \to +\infty$ and we have $c \to c_1$; f_2 dominates as $x \to -\infty$ to give $c \to c_2$. The center of the shock is where $f_1 = f_2$, that is, $x = \frac{1}{2}(c_1 + c_2)t$.

Now since any f_j is a solution of the heat equation, we may clearly add further terms in (4.53) and generate more general solutions of Burgers' equation. Such solutions represent interacting shocks. We consider the case

$$\varphi = f_1 + f_2 + f_3, \qquad b_1 = b_2 = 0, \qquad b_3 = \frac{c_3 - c_2}{2\nu}$$

$$c = \frac{c_1 f_1 + c_2 f_2 + c_3 f_3}{f_1 + f_2 + f_3}, \qquad c_3 > c_2 > c_1 > 0. \tag{4.55}$$

If ν is reasonably small, we can recognize shock transitions between the states c_1, c_2, c_3 by noting in which regions the corresponding f dominates. At $t = 0$, f_1 dominates in $0 < x$, f_2 in $-1 < x < 0$, f_3 in $x < -1$. Thus we have a shock transition from c_1 to c_2 centered at $x = 0$, and one from c_2 to c_3 centered at $x = -1$. For $t > 0$, the regions in which $c \simeq c_1$, $c \simeq c_2$, $c \simeq c_3$ can be found in the same way and the result is shown in Fig 4.3. For early times the transition from c_1 to c_2 occurs where $f_1 = f_2$ on

$$x = \frac{c_1 + c_2}{2} t; \tag{4.56}$$

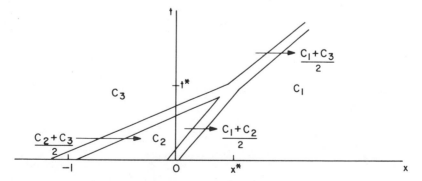

Fig. 4.3. Merging shocks.

the transition from c_2 to c_3 occurs for $f_2 = f_3$ on

$$x = \frac{c_2 + c_3}{2} t - 1. \tag{4.57}$$

Since $\frac{1}{2}(c_2 + c_3) > \frac{1}{2}(c_1 + c_2)$, the second shock overtakes the first at the point (x^*, t^*) determined by (4.56) and (4.57). At this point

$$f_1 = f_2 = f_3.$$

For $t > t^*$, there is no longer any region where f_2 dominates and the continuing solution describes a single shock transition between c_1 and c_3, moving with velocity $\frac{1}{2}(c_1 + c_3)$ on the path

$$x - x^* = \frac{c_1 + c_3}{2} (t - t^*) \tag{4.58}$$

determined by $f_1 = f_3$.

CHAPTER 5

Hyperbolic Systems

The next step in the development of the theory of hyperbolic waves is to see how the ideas and methods established so far can be extended and amplified for the study of higher order systems. Some preliminary remarks have been made in the discussion of the various modifying effects on a single basic wave motion, but the questions relating directly to the possibilities of a number of different wave modes in a system have been touched upon only in passing. We now enter into the general discussion of these questions.

Many physical problems lead to the formulation of a quasi-linear system of first order equations; such equations are linear in the first derivatives of the dependent variables, but the coefficients may be functions of the dependent variables. When these equations describe wave motion, a good understanding of many of the issues can be developed from the study of plane waves. Accordingly, we start with the case of two independent variables. The two variables are often the time and one space variable so we denote them by t and x and use corresponding terminology, but the discussion applies to any two-variable system. If the dependent variables are $u_i(x,t)$, $i = 1, \ldots, n$, the general quasi-linear first order system is

$$A_{ij} \frac{\partial u_j}{\partial t} + a_{ij} \frac{\partial u_j}{\partial x} + b_i = 0, \qquad i = 1, \cdots, n, \tag{5.1}$$

where the matrices A, a, and the vector $\mathbf{b}$ may be functions of $u_1, \ldots, u_n$, as well as x and t. (Here and throughout summation over a repeated subscript is automatic unless otherwise indicated.)

In this chapter we establish the conditions for (5.1) to be hyperbolic and discuss some of the general consequences. Some brief comments on the situation for more space dimensions are made, but the case of more dimensions is developed primarily in the context of specific problems and considerable use is made of the fact that any small region of a two or three dimensional wave behaves *locally* as if it were plane.

113

5.1 Characteristics and Classification

The key to the solution of a single first order equation as described in Chapter 2 was the use of the family of characteristic curves in the (x, t) plane; along each characteristic curve the partial differential equation could be reduced to an ordinary differential equation. In some cases the solution could then be found analytically. But at worst the partial differential equation could be reduced to a set of ordinary differential equations for stepwise numerical integration. In either event, one could proceed to build up the solution by successive "local" considerations of small regions; the whole solution did not need to be calculated at once. This, of course, corresponds to the simple ideas of wave phenomena; in any small time increment the behavior at a point can be influenced only by points near enough for their waves to arrive in time. For the system (5.1), we ask whether such local calculations are possible. If they are, the system is hyperbolic and a suitable precise definition will be framed.

In general, any one of the equations in (5.1) has different combinations of $\partial u_j / \partial t$ and $\partial u_j / \partial x$ for each u_j. That is, it couples information about the rates of change of the different u_j in different directions; one cannot deduce information about the increments of all the u_j for a step in any *single* direction. But we are at liberty to manipulate the n equations in (5.1) to see whether this information can be obtained from some combination of them. We therefore consider the linear combination

$$l_i \left(A_{ij} \frac{\partial u_j}{\partial t} + a_{ij} \frac{\partial u_j}{\partial x} \right) + l_i b_i = 0, \tag{5.2}$$

where the vector $\mathbf{l}$ is a function of $x, t, \mathbf{u}$, and investigate whether $\mathbf{l}$ can be chosen so that (5.2) takes the form

$$m_j \left(\beta \frac{\partial u_j}{\partial t} + \alpha \frac{\partial u_j}{\partial x} \right) + l_j b_j = 0. \tag{5.3}$$

If this is possible, (5.3) provides a relation between the directional derivatives of *all* the u_j in the single direction (α, β). When this is the case, it will be valuable to introduce curves in the (x, t) plane defined by the vector field (α, β). If $x = X(\eta)$, $t = T(\eta)$ is the parametric representation of a typical member of this family, the total derivative of u_j on the curve is

$$\frac{du_j}{d\eta} = T' \frac{\partial u_j}{\partial t} + X' \frac{\partial u_j}{\partial x}.$$

Without loss of generality, we may take

$$\alpha = X'(\eta), \qquad \beta = T'(\eta),$$

and write (5.3) as

$$m_j \frac{du_j}{d\eta} + l_j b_j = 0. \tag{5.4}$$

The conditions for (5.2) to be in the form (5.4) are

$$l_i A_{ij} = m_j T', \qquad l_i a_{ij} = m_j X',$$

and we may eliminate the m_j to give

$$l_i(A_{ij} X' - a_{ij} T') = 0. \tag{5.5}$$

These are n equations for the multipliers l_i and the direction (X', T'). Since they are homogeneous in the l_i, a necessary and sufficient condition for a nontrivial solution is that the determinant

$$|A_{ij} X' - a_{ij} T'| = 0. \tag{5.6}$$

This is a condition on the direction of the curve. Such a curve is said to be a *characteristic* and the corresponding equation (5.4) is said to be in *characteristic form*.

Each equation in characteristic form provides only one relation between the n derivatives of u_j along the corresponding characteristic curve. To proceed with a local construction of the solution in some small region, we shall see below that n independent equations in characteristic form are needed. This is the basis for the definition of a hyperbolic system.

First, however, a relatively mild but important restriction on the systems to be included in the definition must be noted. The restriction concerns the coefficient matrices A and a. It is not hard to see that either one, or even both, may be singular in simple situations. If the determinant $|A_{ij}| = 0$, then $T' = 0$ is a solution of (5.6) and the x direction is characteristic; if $|a_{ij}| = 0$, then $X' = 0$ is a solution and the t direction is characteristic. Clearly, it may be acceptable for the axes to be characteristics, and these possibilities should be included in the discussion. Yet in some cases where both matrices are singular, the systems are so degenerate that they must be excluded. The two situations can be distinguished by checking whether a rotation of the axes cures the difficulty or not. If the trouble is merely that the original axes coincide with characteristics, a rotation of the axes will lead to a new system with nonsingular matrices. A rotation of the axes in (5.1) replaces the original matrices by a linear combination of them. Therefore the appropriate condition is that

$$|\lambda A_{ij} + \mu a_{ij}| \neq 0, \qquad (5.7)$$

for some λ, μ, not both zero, and we do not need to carry out the transformation explicitly. If (5.7) can never be satisfied, we have the degenerate case that must be excluded. In that case, all directions are formally characteristic and the discussion is spurious. On the basis of examples given in the next section, it appears that systems which are so degenerate are unnecessarily large and can be reduced to smaller systems with coefficients that satisfy (5.7).

With this restriction the following definition is introduced.

Definition. A system (5.1), satisfying (5.7), is *hyperbolic* if n linearly independent real vectors $\mathbf{l}^{(k)}$, $k = 1, \dots, n$, can be found such that

$$l_i^{(k)} \left\{ A_{ij} \alpha^{(k)} - a_{ij} \beta^{(k)} \right\} = 0 \qquad (5.8)$$

for each k, and the corresponding directions $\{\alpha^{(k)}, \beta^{(k)}\}$ are real with $\alpha^{(k)^2} + \beta^{(k)^2} \neq 0$.

It should be noted that the emphasis is on there being n independent vectors $\mathbf{l}^{(k)}$, and it is not important that the corresponding directions $(\alpha^{(k)}, \beta^{(k)})$ be distinct. If the directions are distinct so that there are n different families of characteristics, the system is said to be totally hyperbolic; but we shall have little use for this term. As we shall see below it is possible for (5.6) to have less than n different solutions and yet n independent vectors $\mathbf{l}$ can be found.

Special Case $A_{ij} = \delta_{ij}$.

In many problems, the system (5.1) appears in the special form

$$\frac{\partial u_i}{\partial t} + a_{ij} \frac{\partial u_j}{\partial x} + b_i = 0, \qquad (5.9)$$

where the A matrix is the unit matrix. In other cases one could transform the system to this form by multiplying by A^{-1}, after a change of coordinates if the original matrix A is singular. It is seldom worthwhile to do this reduction in detail, but we may, when convenient, refer to this form without loss of generality. It is clear from (5.6) that in this form $T' \neq 0$, so characteristic curves will never have the direction of the x axis. We can therefore parametrize a characteristic curve by t itself and describe the curve by $x = X(t)$. The linear combination

$$l_i \frac{\partial u_i}{\partial t} + l_i a_{ij} \frac{\partial u_i}{\partial x} + l_i b_i = 0$$

takes the characteristic form

$$l_i \frac{du_i}{dt} + l_i b_i = 0 \quad \text{on} \quad \frac{dX}{dt} = c, \tag{5.10}$$

provided that

$$l_i a_{ij} = l_j c. \tag{5.11}$$

In particular, the *characteristic velocity* c must satisfy

$$|a_{ij} - c\delta_{ij}| = 0. \tag{5.12}$$

The possible roots c are the eigenvalues of the matrix and the vectors l are the left eigenvectors.

Two results follow from standard theorems in linear algebra:

The eigenvectors l corresponding to different eigenvalues c are linearly independent. Hence *the system is hyperbolic if (5.12) has n different real roots c*.

If a_{ij} is a real symmetric matrix, then all the roots of (5.12) are real and n independent real eigenvectors can be found. *Hence the system is hyperbolic if a is real and symmetric*.

5.2 Examples of Classification

Before proceeding with the use of the equations in characteristic form and with the further properties of characteristics, a few examples will illustrate the ideas and show some of the peculiarities that may arise in the classification of systems.

Example 1. First consider the wave equation

$$u_{tt} - \gamma u_{xx} = 0.$$

This can be written as a system by introducing $u_x = v$, $u_t = w$ and writing

$$v_t - w_x = 0,$$

$$w_t - \gamma v_x = 0.$$

The linear combination

$$l_1(v_t - w_x) + l_2(w_t - \gamma v_x) = 0$$

takes the characteristic form

$$l_1(v_t + cv_x) + l_2(w_t + cw_x) = 0,$$

provided

$$-\gamma l_2 = c l_1,$$

$$-l_1 = c l_2.$$

There are nontrivial solutions when $c^2 = \gamma$. If $\gamma > 0$, we may take

$$c = +\sqrt{\gamma}, \qquad l_1 = -\sqrt{\gamma}, \qquad l_2 = 1;$$

$$c = -\sqrt{\gamma}, \qquad l_1 = +\sqrt{\gamma}, \qquad l_2 = 1.$$

The two vectors **l** are linearly independent, hence the system is hyperbolic. If $\gamma < 0$ there are no real characteristic forms; in fact, the equation is the prototype of *elliptic* equations.

Example 2. The heat equation

$$u_t - u_{xx} = 0$$

is equivalent to the system

$$u_t - v_x = 0,$$

$$u_x - v = 0.$$

It is clear that the combination

$$l_1(u_t - v_x) + l_2(u_x - v) = 0$$

can be in characteristic form only if $l_1 = 0$. Thus the only solution is $\mathbf{l} = (0, 1)$ or a scalar multiple of this. Since there is only one vector **l** for the second order system, it is not hyperbolic. If we check the general formalism, (5.6) reduces this case to

$$\begin{vmatrix} X' & T' \\ -T' & 0 \end{vmatrix} = 0, \qquad \text{that is,} \quad T'^2 = 0.$$

Thus the x axis is a double characteristic, but there is only one characteristic form for it:

$$u_x - v = 0.$$

Example 3. The simplest second order hyperbolic equation is

$$u_{xt} = 0;$$

an equivalent system is

$$u_t - v = 0,$$

$$v_x = 0.$$

In this case both matrices A and a are singular but (5.7) is satisfied and there is no trouble. Equation 5.6 is

$$\begin{vmatrix} X' & 0 \\ 0 & -T' \end{vmatrix} = 0, \qquad \text{that is,} \quad X'T' = 0.$$

Both the t axis and x axis are characteristics, and the original equations are already in characteristic form.

Example 4. Consider now

$$u_{tt} - \gamma u_{xx} + u = 0.$$

If we introduce $u_x = v$, $u_t = w$, as in Example 1, the extra undifferentiated term in u prevents the completely obvious elimination of u and might suggest keeping three equations. If we choose

$$u_x - v = 0,$$

$$u_t - w = 0,$$

$$w_t - \gamma v_x + u = 0,$$

as an equivalent system, we have the trouble that both matrices A and a are singular, and so are all linear combinations of them. Equation 5.6 is

$$\begin{vmatrix} -T' & 0 & 0 \\ X' & 0 & 0 \\ 0 & \gamma T' & X' \end{vmatrix} = 0$$

and is clearly satisfied for all values of (X', T'). However, this system is excluded by (5.7).

At least in the case $\gamma > 0$, we can implement the suggestion noted earlier that the system is probably too big and can be reduced. We can spot the reduction by writing the equation as

$$\left(\frac{\partial}{\partial t} - \sqrt{\gamma} \, \frac{\partial}{\partial x} \right) \left(\frac{\partial}{\partial t} + \sqrt{\gamma} \, \frac{\partial}{\partial x} \right) u + u = 0.$$

Then the introduction of

$$\varphi = u_t + \sqrt{\gamma} \, u_x$$

leads to the *second order* system

$$\varphi_t - \sqrt{\gamma} \, \varphi_x + u = 0,$$

$$u_t + \sqrt{\gamma} \, u_x - \varphi = 0.$$

This has nonsingular coefficients. In fact it is already in characteristic form, and there are just two characteristics.

Example 5. An alternative system that might be proposed for the equation

$$u_{tt} - \gamma u_{xx} + u = 0$$

is

$$u_t - w = 0,$$

$$v_t - w_x = 0,$$

$$w_t - \gamma v_x + u = 0.$$

This differs from Example 4 in that the equation $v_t - w_x = 0$, obtained by eliminating u, has been substituted for $u_x - v = 0$. Now A is the unit matrix and we should have no trouble. The condition (5.6) is found to be

$$\begin{vmatrix} X' & 0 & 0 \\ 0 & X' & T' \\ 0 & \gamma T' & X' \end{vmatrix} = 0, \qquad \text{that is,} \quad X'(X'^2 - \gamma T'^2) = 0.$$

Two of the roots $X' = \pm \sqrt{\gamma} \, T'$ are clearly the characteristics of the original equation, but why has an extra characteristic $X' = 0$ arisen? The system is not too large as a system, but it is no longer equivalent to the original equation. It is in fact equivalent to

$$\frac{\partial}{\partial t}(u_{tt} - \gamma u_{xx} + u) = 0.$$

The extra characteristic corresponds to the extra t derivative.

Example 6.   The system

$$u_t + C(u,v)u_x = 0,$$

$$v_t + C(u,v)v_x = u,$$

is clearly an example with one characteristic on which $dx/dt = C$, but with two independent characteristic forms. Hence it is hyperbolic.

Example 7.   The system

$$u_t + C(u)u_x = 0,$$

$$v_t + C(u)v_x + C'(u)vu_x = 0,$$

occurs in dispersive waves. The only possible characteristic form is the first equation as it stands. Hence the system is not hyperbolic. Yet because of the exceptional case that the first equation can be solved independently of the second, we can integrate the first equation along the characteristics $dx/dt = C$. Then once u is known in a *whole region* , u_x can be calculated and the second equation can be integrated along the same characteristics to find v. For these purposes it is like a hyperbolic system with a double characteristic, yet formally it would be classified as parabolic.

In Examples 2 to 7 the classification is not completely straightforward. We now add a few nonlinear examples where there is no problem in the classification but they are typical and rather well known. We list pertinent information with a minimum of explanation.

Example 8: Gas Dynamics.   In the compressible inviscid flow of gas with velocity u, pressure p, density ρ, and entropy S, the equations (see Chapter 6) are

$$\rho_t + u\rho_x + \rho u_x = 0,$$

$$u_t + uu_x + \frac{1}{\rho}p_x = 0,$$

$$S_t + uS_x = 0,$$

where $p = p(\rho, S)$. The characteristic equations are

$$\frac{dp}{dt} \pm \rho a \frac{du}{dt} = 0 \qquad \text{on} \quad \frac{dx}{dt} = u \pm a,$$

$$\frac{dS}{dt} = 0 \qquad \text{on} \quad \frac{dx}{dt} = u,$$

where $a^2 = (\partial p / \partial \rho)_{S = \text{constant}}$. For a gas with constant specific heats $p = \kappa \rho^\gamma e^{S/c_v}$ and $a^2 = \gamma p / \rho$.

Example 9: River Waves and Shallow Water Theory. The equations were given in (3.37); the characteristic forms are

$$\frac{d}{dt}\left(v \pm 2\sqrt{g'h}\right) = g'S - C_f \frac{v^2}{h} \qquad \text{on} \quad \frac{dx}{dt} = v \pm \sqrt{g'h} .$$

Example 9′. The reduced kinematic approximation (3.38) has a single characteristic form

$$\frac{dh}{dt} = 0 \qquad \text{on} \quad \frac{dx}{dt} = \frac{3}{2}\left(\frac{g'S}{C_f}\right)^{1/2} h^{1/2}.$$

Example 10: Magnetogasdynamics. For a conducting gas in a magnetic field the equations (using standard notation) are sometimes taken to be

$$\rho_t + u\rho_x + \rho u_x = 0,$$

$$\rho(u_t + uu_x) + p_x = jB,$$

$$\frac{1}{\gamma - 1}(p_t + up_x) - \frac{\gamma}{\gamma - 1}\frac{p}{\rho}(\rho_t + u\rho_x) = \frac{j^2}{\sigma},$$

$$B_t + E_x = 0,$$

$$\epsilon_0 E_t + \frac{1}{\mu}B_x + j = 0,$$

where $j = \sigma(E - uB)$. The characteristic velocities are $\pm(\epsilon_0 \mu)^{-1/2}$, $u \pm a, u$.

Example 10′. When the conductivity σ is very high the following reduced set may be derived and is often used as an adequate approximation:

$$\rho_t + u\rho_x + \rho u_x = 0,$$

$$B_t + uB_x + Bu_x = 0,$$

$$\rho(u_t + uu_x) + p_x + \frac{1}{\mu} BB_x = 0,$$

$$\frac{1}{\gamma - 1}(p_t + up_x) - \frac{\gamma}{\gamma - 1}\frac{p}{\rho}(\rho_t + u\rho_x) = 0.$$

The characteristic velocities are now $u \pm (a^2 + B^2/\mu\rho)^{1/2}$, u, u.

Example 11: Nonlinear Effects for Electromagnetic Waves. In a simple but probably unrealistic formulation of the effects in nonlinear optics, one may take

$$\frac{\partial B}{\partial t} + \frac{\partial E}{\partial x} = 0,$$

$$\frac{\partial D}{\partial t} + \frac{1}{\mu}\frac{\partial B}{\partial x} = 0,$$

with $D = D(E)$. The characteristic equations are

$$\frac{dB}{dt} \pm \frac{1}{c(E)}\frac{dE}{dt} = 0 \qquad \text{on} \quad \frac{dx}{dt} = \pm c(E)$$

where $c(E) = \{\mu D'(E)\}^{-1/2}$. Dispersive effects usually make the relation $D = D(E)$ inadequate.

Example 12: Nonlinear Elastic Waves in a Bar. The one dimensional equation for waves in a bar may be formulated in terms of the displacement $\xi(x,t)$ of a section initially at position x and the stress $\sigma(x,t)$, as

$$\rho_0 \xi_{tt} = \sigma_x,$$

where ρ_0 is the initial density in the unstrained state. If the strain $\epsilon = \xi_x$ and velocity $u = \xi_t$ are introduced, the equivalent pair

$$\rho_0 u_t - \sigma_x = 0,$$

$$\epsilon_t - u_x = 0,$$

may be used. The linear theory takes $\sigma \propto \epsilon$, but nonlinear effects may be included by taking σ as a more general function $\sigma = \sigma(\epsilon)$. The characteristic velocities are $\pm \{\sigma'(\epsilon)/\rho_0\}^{1/2}$. For the appropriate choices of $\sigma(\epsilon)$, there are interesting effects in the wave propagation; in particular, it is a little surprising perhaps to find that shocks are produced in the *unloading* phase of a disturbance. An account is included in Courant and Friedrichs (1948, p. 235).

5.3 Riemann Invariants

Each equation in characteristic form introduces a particular linear combination of the derivatives. For simplicity we consider the reduced form (5.10), where the linear combination concerned is $l_i du_i / dt$. In a linear problem, the vector $\mathbf{l}$ is independent of $\mathbf{u}$ so that a new variable $r = l_i u_i$ simplifies the form of the equation to

$$\frac{dr}{dt} + f(x, t, \mathbf{u}) = 0.$$

In nonlinear problems, however, $\mathbf{l}$ may depend on $\mathbf{u}$ and it is not always possible to achieve this form. It would be necessary to find λ and r such that

$$l_i \, du_i = \lambda \, dr,$$

or, equivalently,

$$l_i = \lambda \frac{\partial r}{\partial u_i}. \tag{5.13}$$

(Here x and t are held fixed; the differential dr refers only to changes in $\mathbf{u}$.) This is a special case of Pfaff's problem for the integrability of differential forms. For $n = 2$ we may eliminate r and find an equation for λ which clearly has a solution. For $n > 2$, however, elimination of both λ and r from (5.13) gives conditions on the l_i which must be satisfied for this to be possible.

For a hyperbolic system, the n characteristic equations take a particularly simple form if it should turn out that a variable r_k can be introduced corresponding to each differential form $l_i^{(k)} du_i$. Then the functions r_k can be used as new variables in place of the u_i and the characteristic equations can be written as

$$\frac{dr_k}{dt} + f_k(x, t, \mathbf{r}) = 0 \qquad \text{on } \frac{dx}{dt} = c_k(x, t, \mathbf{r}). \tag{5.14}$$

This can always be done for linear problems, and in that case the f_k are linear in $\mathbf{r}$. For nonlinear problems it can be done when $n=2$, but it may not be possible for $n>2$.

Such variables were introduced by Riemann in his work on plane waves in gas dynamics, a case with $n=2$. In that particular case (see Section 6.7), the f_k are zero so that r_1 and r_2 are constant on their respective characteristics; the functions r_1 and r_2 are then called *Riemann invariants*. In general, we might call the r_k *Riemann variables*.

5.4 Stepwise Integration Using Characteristics

Insight into the structure of solutions of hyperbolic equations, such as the correct number of boundary conditions and the domain of dependence, can be obtained by imagining a construction of the solution at successive small time increments. For simplicity, it will be assumed that the characteristic equations can be put in the form (5.14), but the qualitative features apply to the general case.

Consider the mixed initial and boundary value problem in $x>0$, $t>0$, with data prescribed on $x=0$ and $t=0$. If we take any point P in the (x,t) plane, and if Q_k is a neighboring point on the kth characteristic through P, (5.14) can be approximated to first order by

$$r_k(P) - r_k(Q_k) + f_k(Q_k)\{t(P) - t(Q_k)\} = 0, \qquad (5.15)$$

$$x(P) - x(Q_k) = c_k(Q_k)\{t(P) - t(Q_k)\}, \qquad (5.16)$$

with an obvious notation for the values of quantities at P and at Q_k. If these relations are used for all n characteristics, and if the values at the Q_k are all known, (5.15) gives n equations for the r_k at P. If some of the c_k are the same, some of the Q_k will coincide, but this does not matter provided there are the full complement of n different equations (5.15).

We now use this construction repeatedly, as shown in Fig. 5.1. The sketch is for three characteristics, with $c_1>c_2>0>c_3$. Take first the point P at the first time increment $t(P)=\Delta t$. The point is taken with sufficiently large $x(P)$ for the characteristics PQ_1, PQ_2, PQ_3 to intersect the positive x axis as shown. In all cases $t(Q_k)=0$. If all the r_k are known initially as functions of x for $t=0$, then each c_k is known as a function of x; hence for a chosen $x(P)$, (5.16) determines $x(Q_k)$ for each k. Then $r_k(Q_k), f_k(Q_k)$ are calculated from the initial values of r_k, and (5.15) determines $r_k(P)$. This can be repeated for various points P on the line $t=\Delta t$ provided that they are to the right of the point W, defined by the characteristic segment OW through the origin with the fastest velocity c_1.

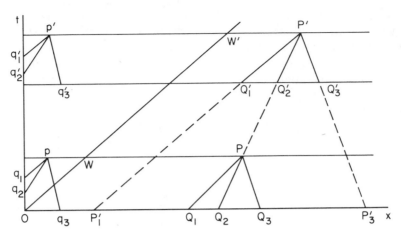

Fig. 5.1. Stepwise construction of solutions using characteristics.

These calculations can be repeated at successive time steps. For example, at P' the values of r_k are determined in terms of the values Q_1', Q_2', Q_3' which were determined at the previous step. Thus in principle the solution in the triangular region to the right of the characteristic OW can be determined from the given values of all the r_k on $t=0$, $x>0$. It is also clear that the values at P' will depend only on the data between P_1' and P_3' on the x axis, where $P'P_1'$ and $P'P_3'$ are the characteristics corresponding to the fastest and slowest speeds, respectively. The segment $P_1'P_3'$ is the *domain of dependence* of P'. The domain of dependence shows the wave character of the solution; signals propagate with speeds c_1, c_2, c_3 and waves from the points between P_1' and P_3' are the only ones that can reach P' in time.

For the full initial value problem, with r_k given on $t=0$, $-\infty < x < \infty$, it is clear that the solution can be constructed in $t>0$ and it is unique. Provided the solution is stable (a question that we do not go into here), this problem is well-posed.

But we return to the mixed problem with data given on $t=0$ only for $x>0$, and the remaining information made up by data on $x=0$, $t>0$. Consider a point p on $t=\Delta t$, but to the left of W, so that the two positive characteristics intersect the t axis at q_1, q_2. The value $r_3(p)$ is still determined from the data at q_3 on the x axis. In fact, p can be taken on the t axis and r_3 is still determined from the data on the x axis. Thus r_3 cannot be prescribed on $x=0$. To determine $r_1(p)$ and $r_2(p)$, the values of r_1, r_2, r_3 will be required at q_1 and q_2. On $x=0$, r_3 is calculated not prescribed, but clearly r_1 and r_2 must be given. At later time steps this is repeated. For example, at p' the value of r_3 is determined from q_3'; r_1, r_2 are determined from q_1', q_2', and r_1, r_2 must be given at these points. Thus the well-posed problem is

$$r_1, r_2, r_3 \quad \text{given at} \quad t = 0, x > 0,$$

$$r_1, r_2 \quad \text{given on} \quad x = 0, t > 0.$$

The data on $x = 0$ only affect the solution to the left of the "wavefront" OWW'. Of course equivalent data may be posed and, in general, any three conditions on $t = 0$ and any two conditions on $x = 0$ will be correct. The only exception is that on $x = 0$ the two conditions must not determine r_3, since this is determined by the initial data.

The results here can be generalized to n equations and other boundary values. From the stepwise construction it is clear that *the number of boundary conditions should be equal to the number of characteristics pointing into the region.* The direction along a characteristic must be defined in order to make "pointing in" unambiguous. When t is time the direction usually is chosen as t increasing; but t increasing, or x increasing, or even some function of them increasing, will all lead to well-posed problems provided the direction once chosen is used consistently.

Detailed proofs of existence and uniqueness may be based on iterations of the integral equations which express the value at a point as an integral over its characteristic triangle (e.g., $P'P_1'P_3'$ for P' in Fig. 5.1). The whole procedure is similar to the Picard iteration for ordinary differential equations. Reference may be made to Courant and Hilbert (1962, p. 476).

In nonlinear problems the characteristic speeds c_k are functions of $\mathbf{u}$; hence the number of data posed on any boundary can change with the data.

5.5 Discontinuous Derivatives

In the preceding construction of the solution it is clear that the characteristics carry information from the boundaries into the region concerned. Physically, the characteristics correspond to waves propagating with the velocities c_k. In a general way, one expects from the construction that any abrupt change in the data on a boundary will produce corresponding abrupt changes propagating on the characteristics through those boundary points. If the abrupt change is taken to be a discontinuity in some of the derivatives of $\mathbf{u}$, the vague idea becomes precise and the expectation is that discontinuities propagate along characteristics. The appropriate results can be shown directly from the equations. The arguments will be given for simple jump discontinuities in the first derivatives of the u_j. Higher derivatives and other singularities can be handled similarly.

Let $\xi(x,t)=0$ be a smooth curve separating two regions in each of which $\mathbf{u}$ is continuously differentiable. Suppose that the u_j are continuous as $\xi \to 0\pm$ and that $\partial u_j/\partial t, \partial u_j/\partial x$ have finite limits as $\xi \to 0\pm$. If $\xi(x,t)$ is a sufficiently smooth function, we can introduce a new local coordinate system $\xi(x,t)$, $\eta(x,t)$, and write (5.1) as

$$(A_{ij}\xi_t + a_{ij}\xi_x)\frac{\partial u_j}{\partial \xi} + (A_{ij}\eta_t + a_{ij}\eta_x)\frac{\partial u_j}{\partial \eta} + b_i = 0. \qquad (5.17)$$

These equations hold in each of the regions $\xi>0$ and $\xi<0$. By hypothesis,

$$u_j(0+,\eta) = u_j(0-,\eta); \qquad (5.18)$$

hence

$$\frac{\partial u_j(0+,\eta)}{\partial \eta} = \frac{\partial u_j(0-,\eta)}{\partial \eta}. \qquad (5.19)$$

This means that the tangential derivatives are continuous across $\xi=0$, and only the normal derivatives related to $\partial u_j/\partial \xi$ may jump. The limits of (5.17) are finite as $\xi \to 0\pm$, and the coefficients are all continuous. Therefore taking the difference of the limits on the two sides, we have

$$(A_{ij}\xi_t + a_{ij}\xi_x)\left[\frac{\partial u_j}{\partial \xi}\right] = 0, \qquad (5.20)$$

where

$$\left[\frac{\partial u_j}{\partial \xi}\right] = \left(\frac{\partial u_j}{\partial \xi}\right)_{\xi=0+} - \left(\frac{\partial u_j}{\partial \xi}\right)_{\xi=0-}.$$

Accordingly, the jumps $[\partial u_j/\partial \xi]$ are zero unless

$$|A_{ij}\xi_t + a_{ij}\xi_x| = 0 \qquad (5.21)$$

on $\xi=0$. If the curve $\xi(x,t)=0$ is described in the alternative form $x=X(\eta)$, $t=T(\eta)$, then

$$(\xi_t,\xi_x) \propto \{X'(\eta), -T'(\eta)\}.$$

Hence (5.21) is the same as (5.6), and we have the result that *discontinuities in the first derivatives of **u** can occur only on characteristics.*

According to this, propagating discontinuities are ruled out if the system has no characteristics, and in such a case any discontinuity in the boundary data will be immediately resolved in the solution. On the other hand, the existence of characteristics is not a sufficient guarantee that discontinuities can occur. The equations provide further restrictions on the $[\partial u_j/\partial \xi]$ and if the system is not fully hyperbolic, these may be stringent enough to require $[\partial u_j/\partial \xi]=0$. If the system is hyperbolic, however, the additional relations do not exclude discontinuities; they provide instead equations that determine the variations of the magnitudes of the discontinuities as they propagate along the characteristics.

If (5.21) is satisfied and a particular characteristic is chosen, the equations in (5.20) give a number of relations between the quantities $[\partial u_j/\partial \xi]$ on that characteristic. The number will depend on the rank of the coefficient matrix in (5.20). The simplest case is when there are $n-1$ relations, so that all the jumps $[\partial u_j/\partial \xi]$ are determined in terms of one of them, or, more symmetrically,

$$\left[\frac{\partial u_j}{\partial \xi}\right] = \sigma L_j,\tag{5.22}$$

where **L** is any nontrivial solution of

$$(A_{ij}\xi_t + a_{ij}\xi_x)L_j = 0,\tag{5.23}$$

and σ is undetermined at this stage. If the rank of the matrix is r, there will be $n-r$ independent solutions of (5.23) and a corresponding number of terms in (5.22) with $n-r$ parameters σ.

Additional information may be obtained by taking the ξ derivative of (5.17) and considering the difference of the limits as $\xi\to 0\pm$. The result takes the general form

$$(A_{ij}\xi_t + a_{ij}\xi_x)\left[\frac{\partial^2 u_j}{\partial \xi^2}\right] + E_i\left(\frac{d}{d\eta}\left[\frac{\partial u_j}{\partial \xi}\right],\left[\frac{\partial u_j}{\partial \xi}\right]\right) = 0,\tag{5.24}$$

where E_i is linear in the first argument and at most quadratic in the second. Although in the main, these equations give information on the jumps of the second derivatives $[\partial^2 u_j/\partial \xi^2]$ across $\xi=0$, the fact that the matrix

$$A_{ij}\xi_t + a_{ij}\xi_x$$

is singular means that certain linear combinations of the E_i vanish. These provide further relations to be satisfied by the $[\partial u_j / \partial \xi]$. The number of them will be $n - r$ (where r is again the rank of the matrix), and this is just the degree of arbitrariness remaining after solving (5.20). These relations, then, provide equations for the σ's introduced in (5.22).

The details become fairly complicated so we pursue them further only for the case of discontinuities at a wavefront propagating into a region of constant uniform state. This includes all the important features and is, in any case, the main application of the discontinuity analysis. We shall also take the system to be in the reduced form (5.9).

5.6 Expansion Near a Wavefront

We consider the system (5.9) in the case that it admits constant solutions $u_j = u_j^{(0)}$. Usually $\mathbf{b}$ will be independent of x, t, and we require $\mathbf{u}^{(0)}$ to satisfy

$$b_i(\mathbf{u}^{(0)}) = 0. \tag{5.25}$$

For the reduced form (5.9), the characteristics can never be in the direction of the x axis, so we may use t itself as a parameter on the wavefront and write the equation of the wavefront in the form $x = X(t)$. Instead of calculating the limits of derivatives from the equations, it is particularly convenient in the wavefront problem to use the equivalent procedure of expanding solutions in powers of

$$\xi = x - X(t).$$

If the first derivatives are discontinuous the appropriate form is

$$u_j = u_j^{(0)}, \qquad \xi > 0, \tag{5.26}$$

$$u_j = u_j^{(0)} + u_j^{(1)}(t)\xi + \frac{1}{2}u_j^{(2)}(t)\xi^2 + \cdots, \qquad \xi < 0. \tag{5.27}$$

Then

$$\left[\frac{\partial u_j}{\partial t}\right] = \dot{X}(t)u_j^{(1)}(t), \qquad \left[\frac{\partial u_j}{\partial x}\right] = -u_j^{(1)}(t), \tag{5.28}$$

and the higher derivatives are related to the other coefficients similarly.

The power series version is a convenient way of seeing the extension to include other singularities. If the mth derivatives are the first ones to be discontinuous, the power series beyond $u_j^{(0)}$ start with the terms in ξ^m, and one could also include singularities which correspond to expansions in fractional powers of $|\xi|$ or in $\log |\xi|$. The questions of convergence are not really at issue here; we are using formal power series as a device to calculate derivatives which could also be obtained by taking the appropriate limits in the equations.

The coefficients in (5.27) are obtained by substituting the series in (5.9) and equating the successive terms in powers of ξ to zero. If the a_{ij} are functions of $x, t, \mathbf{u}$, they must be expanded in powers of ξ with coefficients depending on t. That is,

$$a_{ij} = a_{ij}^{(0)} + \xi \left(\frac{\partial a_{ij}^{(0)}}{\partial u_k} u_k^{(1)} + \frac{\partial a_{ij}^{(0)}}{\partial x} \right) + \cdots, \tag{5.29}$$

where the superscript zero means that the arguments of the corresponding function are $x = X(t)$, t and $\mathbf{u} = \mathbf{u}^{(0)}$. However, in writing the resulting equations for the $u_j^{(m)}(t)$, the superscript zero will be omitted in the interests of clarity. From the substitution in (5.9), we have

$$a_{ij} u_j^{(1)} - c u_i^{(1)} = 0, \quad (5.30)$$

$$a_{ij} u_j^{(2)} - c u_i^{(2)} + \left\{ \frac{d u_i^{(1)}}{dt} + \frac{\partial a_{ij}}{\partial u_k} u_j^{(1)} u_k^{(1)} + \left(\frac{\partial a_{ij}}{\partial x} + \frac{\partial b_i}{\partial u_j} \right) u_j^{(1)} \right\} = 0, \quad (5.31)$$

and so on, where c denotes $\dot{X}$. These correspond of course to (5.20) and (5.24).

From (5.30), we deduce first that the velocity $\dot{X} = c$ must satisfy

$$|a_{ij} - c \delta_{ij}| = 0, \tag{5.32}$$

and the wavefront must be one of the characteristics. If we take the case that the rank of the matrix in (5.30) is $n - 1$, we have

$$u_j^{(1)} = \sigma L_j, \tag{5.33}$$

where L_j is any nontrivial solution of

$$(a_{ij} - c \delta_{ij}) L_j = 0. \tag{5.34}$$

There is also a nontrivial eigenvector **l** which satisfies

$$l_i(a_{ij} - c\delta_{ij}) = 0. \tag{5.35}$$

[It is this left eigenvector that arises in the characteristic form (5.10).] If this is applied to (5.31), the terms in $u_j^{(2)}$ are eliminated and we have

$$l_i\frac{du_i^{(1)}}{dt} + l_i\frac{\partial a_{ij}}{\partial u_k}u_j^{(1)}u_k^{(1)} + l_i\left(\frac{\partial a_{ij}}{\partial x} + \frac{\partial b_i}{\partial u_j}\right)u_j^{(1)} = 0. \tag{5.36}$$

Finally, substituting the expressions for $u_j^{(1)}$ from (5.33), we have an equation of the form

$$l_iL_i\frac{d\sigma}{dt} + Q\sigma^2 + P\sigma = 0, \tag{5.37}$$

where l_iL_i, Q, and P are all known functions of t.

For hyperbolic systems it may be shown that $l_iL_i \neq 0$. In other cases, however, one might have $l_iL_i = 0$, $Q = 0$, $P \neq 0$, and have to conclude that $\sigma = 0$; that is, discontinuities are impossible. For example, the sytem

$$u_t - v = 0,$$

$$v_t - u_x = 0,$$

which is equivalent to the heat equation with t and x interchanged to fit the canonical form (5.9), has the curves $x = $ constant as a double characteristic. But $\mathbf{l} = (1,0)$, $\mathbf{L} = (0,1)$, $Q = 0$, $P = -1$, so that discontinuities are impossible.

For hyperbolic systems, (5.37) may be reduced to

$$\frac{d\sigma}{dt} + q\sigma^2 + p\sigma = 0. \tag{5.38}$$

This is a Riccati equation which can be solved explicit to find the variations of σ [and hence the $u_j^{(1)}$] along the wavefront.

If the original system is linear, the a_{ij} are independent of **u** and the quadratic term is absent. The solution is

$$\sigma(t) = \sigma(0)e^{-p_1(t)}, \qquad p_1(t) = \int_0^t p(t')\,dt', \tag{5.39}$$

where $\sigma(0)$ is determined by the initial conditions. We observe, in particular, that discontinuities can only appear in the solution as a consequence of corresponding discontinuities in the boundary or initial conditions.

Moreover, once introduced, they can not disappear in a finite time.

For nonlinear systems with $q \neq 0$ in (5.38), the equation may be rewritten as

$$\frac{d}{dt}\left(\frac{1}{\sigma}\right) - \frac{p}{\sigma} - q = 0,$$

and the solution is

$$\frac{1}{\sigma} = \frac{e^{p_1(t)}}{\sigma(0)} + e^{p_1(t)} \int_0^t q(t') e^{-p_1(t')} dt'. \tag{5.40}$$

Again, discontinuities once generated cannot disappear at a finite time; they may decay to zero strength as $t \to \infty$. However, a new possibility in the nonlinear case is that $\sigma \to \infty$ for *finite* t. This occurrence will depend on the signs of $p(t), q(t)$, and on the magnitude of $\sigma(0)$. Suppose, for instance, that the equation reads

$$\frac{d\sigma}{dt} = \nu\sigma^2 - \mu\sigma, \tag{5.41}$$

where μ, ν, are positive constants and $\sigma(0) = \sigma_0 > 0$. If $\sigma_0 < \mu/\nu$, the right hand side of (5.41) is negative initially so that σ starts to decrease. But then the right hand side remains negative and so σ continues to decrease. Ultimately, $\sigma \to 0$ like $e^{-\mu t}$ as $t \to \infty$. However, if $\sigma_0 > \mu/\nu$, the reverse is true and σ continually increases. Eventually, the term $\nu\sigma^2$ dominates and leads to $\sigma \to \infty$ in *finite time*. The explicit solution is

$$\sigma(t) = \frac{\mu}{\nu} \frac{\sigma_0}{\sigma_0 - (\sigma_0 - \mu/\nu) e^{\mu t}}. \tag{5.42}$$

If $\sigma_0 - \mu/\nu > 0$, $\sigma \to \infty$ as

$$t \to \frac{1}{\mu} \log \frac{\sigma_0}{\sigma_0 - \mu/\nu}. \tag{5.43}$$

This predicts the nonlinear breaking of the wavefront, and after this time a shock wave, with discontinuities in the functions u_j themselves, must be introduced. Although this breaking discussion and a criterion such as (5.43) is limited to the special form of wave with a discontinuous derivative, it is extremely valuable because it is always possible to carry out this calculation explicitly. The functions $p(t)$, $q(t)$ appearing in (5.38) depend only on the coefficients a_{ij}, b_i. It is not necessary to find the solution in a whole region of the (x, t) plane before the information can be obtained. A continuous profile does not have to behave in precisely the same way, but one gets a rough estimate of the magnitude of the derivatives required to produce a breaking wave and an estimate of the time of breaking. In general, it may not be possible to find explicit solutions for continuous profiles to determine the criteria exactly for those cases.

5.7 An Example from River Flow

As an interesting application of the wavefront expansion, we consider the river flow equations discussed in Section 3.2. They are

$$h_t + vh_x + hv_x = 0,$$

$$v_t + vv_x + g'h_x = g'S - C_f \frac{v^2}{h}. \tag{5.44}$$

The uniform flow has constant values $h = h_0$, $v = v_0$, with $g'S = C_f v_0^2 / h_0$, and the wavefront has constant speed in this case so we take $\xi = x - ct$. Behind the wavefront the flow variables are expanded as

$$h = h_0 + \xi h_1(t) + \frac{1}{2}\xi^2 h_2(t) + \cdots,$$

$$v = v_0 + \xi v_1(t) + \frac{1}{2}\xi^2 v_2(t) + \cdots.$$

These are substituted in (5.44) and the terms in successive powers of ξ give

$$(v_0 - c)h_1 + h_0 v_1 = 0,$$

$$g'h_1 + (v_0 - c)v_1 = 0; \tag{5.45}$$

$$(v_0 - c)h_2 + h_0 v_2 + \frac{dh_1}{dt} + 2v_1 h_1 = 0,$$

$$g'h_2 + (v_0 - c)v_2 + \frac{dv_1}{dt} + v_1^2 + g'S\left(\frac{2v_1}{v_0} - \frac{h_1}{h_0}\right) = 0; \tag{5.46}$$

and so on. From the first two we have

$$(c - v_0)^2 = g'h_0, \tag{5.47}$$

so the propagation speed is

$$c = v_0 \pm \sqrt{g'h_0}\; ; \tag{5.48}$$

we also have

$$v_1 = \frac{(c - v_0) h_1}{h_0}.$$ (5.49)

The relation (5.47) allows us to eliminate h_2 and v_2 from (5.46). The resulting equation is

$$g' \frac{dh_1}{dt} + 2g' v_1 h_1 + (c - v_0) \left\{ \frac{dv_1}{dt} + v_1^2 + g' S \left(\frac{2v_1}{v_0} - \frac{h_1}{h_0} \right) \right\} = 0.$$

Finally, eliminating v_1 by use of (5.49), we have

$$\frac{dh_1}{dt} + \frac{3}{2} (c - v_0) \frac{h_1^2}{h_0} + \frac{S}{v_0} (c - v_0) \left(c - \frac{3v_0}{2} \right) \frac{h_1}{h_0} = 0.$$ (5.50)

The quantity $h_1(t)$ is the value of the derivative h_x at the wavefront.

We now consider various special cases.

Shallow Water Waves.

In the usual shallow water theory the slope and friction terms are absent from (5.44); (5.50) becomes

$$\frac{dh_1}{dt} = -\frac{3}{2} (c - v_0) \frac{h_1^2}{h_0}.$$

Downstream waves, with $c = v_0 + \sqrt{gh_0}$, break if $h_1 < 0$; upstream waves, with $c = v_0 - \sqrt{gh_0}$, break if $h_1 > 0$.

Flood Waves.

For flood waves going downstream, $c = v_0 + \sqrt{g'h_0}$ and (5.50) reads

$$\frac{dh_1}{dt} = -\frac{3}{2} \sqrt{\frac{g'}{h_0}} \, h_1^2 - \frac{g'S}{v_0} \left(1 - \frac{v_0}{2\sqrt{g'h_0}} \right) h_1.$$ (5.51)

If $v_0 / \sqrt{g'h_0} > 2$, the linear term indicates an exponential increase for h_1 of either sign. This corresponds to the instability of the steady flow under these conditions and checks with the result deduced from (3.41). If

$v_0/\sqrt{g'h_0} < 2$, the linear term shows exponential decrease corresponding to stability. However, if $h_1(0) < 0$ and

$$|h_1(0)| > \frac{2}{3} \frac{S\sqrt{g'h_0}}{v_0} \left(1 - \frac{1}{2} \frac{v_0}{\sqrt{g'h_0}}\right),$$

$dh_1/dt < 0$ and $h_1 \to -\infty$ in a finite time. This corresponds to nonlinear breaking at the wavefront, and a bore will be formed at the head of the wave. This checks with the analysis in Section 3.2 where it was shown that a sufficiently strong flood wave would be headed by a bore.

Tidal Bores.

For a wave propagating upstream, $c = v_0 - \sqrt{g'h_0}$, and (5.50) reduces to

$$\frac{dh_1}{dt} = \frac{3}{2} \sqrt{\frac{g'}{h_0}} \, h_1^2 - \frac{g'S}{v_0} \left(1 + \frac{1}{2} \frac{v_0}{\sqrt{g'h_0}}\right) h_1.$$

A wave with positive h_1 will break if

$$h_1(0) > \frac{2}{3} \frac{S\sqrt{g'h_0}}{v_0} \left(1 + \frac{1}{2} \frac{v_0}{\sqrt{g'h_0}}\right). \tag{5.52}$$

The existence of a minimum value of $h_1(0)$ which must be exceeded for bore formation is particularly interesting in view of the well-known observation that only relatively few rivers, with sufficiently high tidal variation at their mouths, develop tidal bores. The analysis here is limited to a wavefront, whereas the appropriate case for tidal bores would be an initially smooth sinusoidal variation. However, it has the usual virtues of analytic results; one can see the dependence on the various parameters explicitly, one can predict the asymptotic behavior for large values of x and t, and so on! The continuous case, which cannot be solved analytically, might require extensive numerical computations to establish clear criteria. Thus the present approach provides valuable estimates. In fact, Abbott (1956), who used this type of analysis and applied it in detail to the river Severn, found remarkably good agreement with observations. (Actually, Abbott developed his work in terms of high frequency approximations, but the two approaches are mathematically equivalent. Moreover, there is no gain in justification; the tidal variation is "low frequency" and one has to argue that only high frequency effects contribute toward breaking.)

It is important, however, to extend (5.52) to include nonuniform flow and topography in the undisturbed state of the river. The narrowing of the river upstream is particularly crucial in making the actual estimates, since it favors breaking and is needed to offset the usually overpowering damping of the frictional forces. The details can be found in Abbott's paper.

To apply the wavefront results we use the maximum rate of change in the tidal variation to determine $h_1(0)$. If the tidal variation at the mouth of the river is

$$h = h_0 + a \sin \omega t,$$

the maximum value of h_t is $a\omega$. In the discontinuity analysis the initial value of h_t at the wavefront is $\left(\sqrt{g'h_0} - v_0\right)h_1(0)$. Therefore we choose

$$h_1(0) = \frac{a\omega}{\sqrt{g'h_0} - v_0}.$$

For the uniform channel, (5.52) would predict bore formation if

$$a\omega > \frac{2S}{3v_0}\left(\sqrt{g'h_0} - v_0\right)\left(\sqrt{g'h_0} + \frac{1}{2}v_0\right).$$

The formula can be written in various forms using $g'S = C_f v_0^2 / h_0$, and the least sensitive is probably

$$a\omega > \frac{2}{3} C_f v_0 \left(1 - \frac{v_0}{\sqrt{g'h_0}}\right)\left(1 + \frac{1}{2}\frac{v_0}{\sqrt{g'h_0}}\right). \tag{5.53}$$

For rivers $v_0/\sqrt{g'h_0}$ is fairly small, so the right hand side can be approximated by the first factor. For the typical values of $v_0 = 5$ ft/sec, $C_f = 0.006$, this leads to unattainable values of the order of 100 ft for a. The effects of narrowing and other factors bring the value down considerably, although exceptionally high tides and rapidly changing topography are necessary. That is why very few rivers have bores. For the river Severn, Abbott finds that with all nonuniform effects included the required value of the tidal range $2a$ is 39.4 ft. The spring tides have an average range of 41.4 ft while the neap tides have an average range of 22.2 ft. Thus Abbott predicts that bore formation should occur for about 4 days around the times of the highest tides. This appears to be borne out by observations. His predictions of the distance upstream at which the bore forms and of its maximum height appear to be in fair agreement.

5.8 Shock Waves

The situation as regards the breaking of waves and the introduction of shock waves is very much the same as in the case of a single quasi-linear equation. Some solutions which are initially single valued, and even continuous, will develop multivalued regions: waves will break. This is again interpreted as an inadequacy of the assumptions leading to (5.1), but the appropriate saving features can be well approximated by allowing discontinuities in **u**.

We again take the view that in formulating the differential equations, there will have been an earlier stage where the equations were in integrated form

$$\frac{d}{dt}\int_{x_2}^{x_1} f_i \, dx + [\, g_i \,]_{x_2}^{x_1} + \int_{x_2}^{x_1} h_i \, dx = 0, \tag{5.54}$$

where f_i, g_i, h_i are various quantities of physical interest in the problem. For example, in problems of mechanics f_i and g_i could be the density and flux of mass, or the density and flux of momentum, or the density and flux of energy. The quantity h_i allows for a distributed source term, such as a body force in the momentum equation. Equation 5.54 is a *conservation equation* for the physical quantity concerned (mass, momentum, energy, etc.).

The densities f_i will be functions of (x,t) and of n basic variables $\mathbf{u} = (u_1,\ldots,u_n)$; in general, there will be n equations (5.54) in the statement of the appropriate physical laws. Various simplifying assumptions will then be made to relate the g_i, h_i to x, t, **u**. At the first level of approximation, the g_i and h_i will just be functions of x, t, **u**. If **u** has continuous first derivatives, (5.54) may then be written in the differentiated form

$$\frac{\partial f_i(x,t,\mathbf{u})}{\partial t} + \frac{\partial g_i(x,t,\mathbf{u})}{\partial x} + h_i(x,t,\mathbf{u}) = 0. \tag{5.55}$$

This is a differential equation *in conservation form*.

If discontinuities in **u** are to be included, the integrated form (5.54) must be used and the dependence of g_i and h_i on **u** left open at first. If a discontinuous *shock* occurs at $x = s(t)$, exactly the same argument given in Section 2.3 gives the shock conditions

$$-U[\, f_i \,] + [\, g_i \,] = 0, \qquad i = 1,\ldots,n, \tag{5.56}$$

where U is the shock velocity $\dot{s}(t)$. We then argue that in the continuous parts of the solution on the two sides of the shock, it is still a good approximation to take

$$g_i = g_i(x, t, \mathbf{u}), \qquad h_i = h_i(x, t, \mathbf{u}).$$

Therefore (5.56) is applied with the same functional dependence of g_i on $\mathbf{u}$. As in Chapter 2, a more accurate choice of the g_i will involve derivatives of $\mathbf{u}$, and the shocks will be smoothed out into thin regions of rapid change. However, the discontinuity treatment is simpler and usually is adequate.

The formal mathematical definition of *weak solutions* of (5.55), leading to the jump conditions (5.56), follows closely the discussion in Section 2.7. Evaluating the derivatives in (5.55), we see that it is a case of the system (5.1) in which

$$A_{ij} = \frac{\partial f_i}{\partial u_j}, \qquad a_{ij} = \frac{\partial g_i}{\partial u_j}. \qquad (5.57)$$

The discussion of weak solutions is applicable only to these special cases. Moreover the important warning about nonuniqueness must be emphasized. In typical cases, starting from the relevant system (5.1), it will be possible to find more than n different equations in the conservation form (5.55). Shock conditions (5.56) from a choice of any n of them will be satisfactory mathematically, but only those n equations that correspond to the original physical statements in (5.54) will give the correct solutions for the problem. A good example of this nonuniqueness occurs in gas dynamics (see Chapter 6). In view of the nonuniqueness, the connection with the physical laws is stressed here.

5.9 Systems with More Than Two Independent Variables

We comment briefly on the situation for quasi-linear equations with m independent variables where $m > 2$. The system may be written

$$a_{ij}^\nu \frac{\partial u_j}{\partial x^\nu} + b_i = 0, \qquad i = 1, \dots, n, \qquad (5.58)$$

where the dependent variables u_j are functions of the m independent variables $x^1, x^2, \dots, x^m$, and the summation extends over $\nu = 1, \dots, m$, as well as $j = 1, \dots, n$. The analogs of the characteristic curves for $m = 2$ are characteristic surfaces in the $m - 1$ dimensional $\mathbf{x}$ space. They may be introduced somewhat as before and some of their properties are similar. However, they are very much more limited in their usefulness for constructing solutions.

The limitation arises because it would be too much to expect, in general, that one could find linear combinations of the equations in (5.58) such that the directional derivatives of each u_j have the same direction. The restrictions on the system or on the solutions would make them far too special. So the analog in this case has to be that the directions involved should lie in an $m-1$ dimensional surface element. If the appropriate linear combination is

$$l_i a_{ij}^\nu \frac{\partial u_j}{\partial x^\nu} + l_i b_i = 0, \tag{5.59}$$

the directional derivative for u_j is $l_i a_{ij}^\nu$. If the surface element belongs to a surface $S(\mathbf{x}) = \text{constant}$, the normal vector is $\partial S / \partial x^\nu$, and the orthogonality requirement is

$$l_i a_{ij}^\nu \frac{\partial S}{\partial x^\nu} = 0, \qquad j = 1, \ldots, n. \tag{5.60}$$

The condition for $\mathbf{l}$ to be nontrivial is that the determinant

$$\left| a_{ij}^\nu \frac{\partial S}{\partial x^\nu} \right| = 0. \tag{5.61}$$

Surfaces with this property are *characteristic surfaces*. Again the system will be hyperbolic if n independent equations of the form (5.59) can be found with this property. Usually these will correspond to n different characteristic surfaces, but this is not necessary provided that the full complement of n vectors $\mathbf{l}$ can be found. However, these choices do not simplify the solution to the same extent as in the case $m = 2$, since we are still left with $m-1$ coupled directions within each surface.

In view of this, the main property of characteristic surfaces for wave propagation problems is that they carry singularities in the solution and, in particular, describe wavefronts. In close analogy with the approach in Section 5.5, let $S(\mathbf{x}) = 0$ be a surface across which the u_j are continuous but the $\partial u_j / \partial x^\nu$ are allowed to have simple jump discontinuities. From the continuity of the u_j, all the tangential derivatives must be continuous; hence only the normal derivatives can be discontinuous. If the surface $S = 0$ is imbedded in a family of surfaces $S = \text{constant}$, so that S can be used as a local coordinate supplemented by any choice of the other $n-1$ coordinates, the discontinuous derivatives are $\partial u_j / \partial S$. Then closely following the argument for $m = 2$, we deduce that

$$a_{ij}^\nu \frac{\partial S}{\partial x^\nu} \left[\frac{\partial u_j}{\partial S} \right] = 0. \tag{5.62}$$

Therefore discontinuities can only occur on surfaces which satisfy

$$\left| a_{ij}^{\nu} \frac{\partial S}{\partial x^{\nu}} \right| = 0. \tag{5.63}$$

This is the same as (5.61) and defines the characteristic surfaces. Equations 5.62 and 5.63 are the generalizations of (5.20) and (5.21). As before, further relations can then be derived for the jumps $[\partial u_j / \partial S]$.

5.10 Second Order Equations

Second order linear equations taking the form

$$A_{ij} \frac{\partial^2 \varphi}{\partial x_i \, \partial x_j} + B_i \frac{\partial \varphi}{\partial x_i} + C\varphi = 0 \tag{5.64}$$

appear frequently, and even in the case of two independent variables it is usually more convenient to leave them in this form than to work with a first order system. Indeed, we saw an indication in Section 5.2 that there may be some problem in finding a *satisfactory* equivalent system, unless of course (5.64) was derived from one.

There are many approaches to the classification of (5.64). In the context of wave propagation, the possibility of wavefronts carrying discontinuities in derivatives is an important question, and it provides the simplest link to show consistency with the discussion of first order systems. Obviously, in those cases where (5.64) does come from a reasonable equivalent system the definitions of hyperbolicity should agree. A detailed proof of consistency is not attempted here, but the choice of this approach shows the close connection.

Consider, then, the possibility of jump discontinuities in the second derivatives of φ. If these occur across a surface $S(x) = 0$, while φ and $\partial \varphi / \partial x_i$ remain continuous, we may introduce local coordinates based on $S(x) = 0$ as before and deduce that $\partial^2 \varphi / \partial S^2$ is discontinuous but the other second order derivatives remain continuous. Then, taking the difference of the limits of (5.64) on the two sides of $S = 0$, we have

$$A_{ij} \frac{\partial S}{\partial x_i} \frac{\partial S}{\partial x_j} \left[\frac{\partial^2 \varphi}{\partial S^2} \right] = 0. \tag{5.65}$$

A necessary condition for $[\partial^2 \varphi / \partial S^2] \neq 0$ is

$$A_{ij} \frac{\partial S}{\partial x_i} \frac{\partial S}{\partial x_j} = 0. \tag{5.66}$$

It can be shown that discontinuities in the first derivatives or even in φ itself (since the equation is linear) must also be restricted to such surfaces. But the arguments require more careful discussion, including the question of what exactly is meant by solution, and they are postponed until they are needed in Section 7.7.

The classification now depends on the quadratic form $A_{ij}\xi_i\xi_j$. At any point $\mathbf{x}$, it may be reduced by means of a linear transformation to the form

$$a_1\xi_1^2 + \cdots + a_m\xi_m^2. \tag{5.67}$$

If all the a_i are the same sign, there is clearly no solution of (5.66); the equation is *elliptic* at such a point. If some of the a_i are zero, it is *parabolic*; in the usual case, one of the a_i is zero and the remainder have the same sign. If the a_i are nonzero, but not all of the same sign, we have the *hyperbolic* case. In applications it appears that the only hyperbolic cases that arise have $m - 1$ of the a_i with the same sign and just one with the opposite sign. An explanation of this is that surfaces described by (5.66) have peculiar geometrical properties otherwise, and, for example, could not represent the simple intuitive picture of an expanding wavefront. Accordingly the term hyperbolic is restricted to this case.

To free the classification from its reliance on the discontinuity analysis, one merely notes that the linear transformation required to reduce the quadratic form $A_{ij}\xi_i\xi_j$ to (5.67) may also be used to generate a local coordinate transformation which reduces the leading term in (5.64) to the form

$$a_1\frac{\partial^2\varphi}{\partial X_1^2} + \cdots + a_m\frac{\partial^2\varphi}{\partial X_m^2}.$$

This is independent of any question of discontinuities and the classification proceeds as before on the signs of the a_i in the leading term.

The discussion here has necessarily been kept brief on those questions which will not arise directly in the later work. Further reference may be made to the many excellent texts on the general theory of partial differential equations, such as Courant and Hilbert (1962) or Petrovsky (1954).

CHAPTER 6

Gas Dynamics

As explained in Chapter 1, many of the basic ideas for hyperbolic waves and, particularly, the elucidation of shock phenomena came from gas dynamics. This chapter is a discussion of waves and shocks in gas dynamics. It provides a natural illustration of the general ideas developed in the last chapter and adds the sort of amplification and extension that can only be shown on specific problems. But gas dynamics is of course an important and interesting subject for its own sake, so this chapter is presented as a thorough introduction, not just as an illustration of the mathematical theory. Further specialized topics are taken up in later chapters, and all the material combined provides a broad coverage of gas dynamics. However, the reader interested only in the general development of wave theory may skim this chapter.

6.1 Equations of Motion

The equations of motion for a compressible fluid are derived by writing the equations of conservation of mass, momentum, and energy for an arbitrary volume of the fluid. Each of these brings in the corresponding variables to describe the balance. The description of mass flow requires two quantities: the density $\rho(\mathbf{x}, t)$ and the velocity vector $\mathbf{u}(\mathbf{x}, t)$ at any point $\mathbf{x}$ at time t. Momentum requires additional quantities to describe the forces acting on the fluid. There may be a body force, usually gravity, acting on all the fluid in any volume. This is denoted by a vector $\mathbf{F}(\mathbf{x}, t)$ per unit mass, and for gravity would be g times the unit vector in the vertical direction. There are also stresses acting across the boundary of any volume of fluid. The stress acting on any small element of area of the boundary surface is taken to be proportional to that area. In general, it depends also on the orientation of that surface element. Therefore the force per unit area will be a function of position $\mathbf{x}$, time t, and the unit vector $\mathbf{l}$ normal to the surface element. A standard argument, which will be detailed later, shows that the ith component p_i of the stress may be written as a

$$p_i = p_{ji} l_j \qquad \text{(summed over } j\text{)}, \qquad (6.1)$$

where the quantities $p_{ji}(\mathbf{x}, t)$ depend only on the position $\mathbf{x}$ and time t. Since p_i and l_i are vectors, the components p_{ji} form a tensor, and it is referred to as the stress tensor at $(\mathbf{x}, t)$. The component p_{ji} is the ith component of the force per unit area on an element of area whose normal is in the jth direction.

The energy equation introduces still further quantities. The fluid has internal energy due to the thermal agitation of the molecules. In the continuum theory it is specified as a quantity $e(\mathbf{x}, t)$ per unit mass. There is also heat flow across the boundary and this will be denoted by a vector $\mathbf{q}(\mathbf{x}, t)$ per unit surface area.

We are now in a position to write down the conservation equations, although they will not yet provide a complete system since there are more unknowns than equations. We consider a fixed arbitrary volume V of the region occupied by fluid and write down the net balance for that region, bearing in mind the transport of the fluid across the boundary surface S. For the conservation of mass the rate of change of the total mass in V,

$$\int_V \rho \, dV,$$

is balanced by the flow across S. If $\mathbf{l}$ denotes the outward normal to S, the normal component of velocity across S is $l_j u_j$. Hence

$$\frac{d}{dt} \int_V \rho \, dV + \int_S \rho l_j u_j \, dS = 0. \qquad (6.2)$$

In similar fashion, the equation for the net balance of the ith component of momentum is

$$\frac{d}{dt} \int_V \rho u_i \, dV + \int_S (\rho u_i l_j u_j - p_i) \, dS = \int_V \rho F_i \, dV. \qquad (6.3)$$

The first term is the rate of change of momentum inside V, the second term is the transport of momentum carried across the boundary by the flow, the third term is the rate of change of momentum produced by the stress p_i acting across the surface S, and the right hand side is the momentum created inside V by the body forces.

The total energy density per unit volume consists of the kinetic energy $\frac{1}{2} \rho u_i^2$ of the macroscopic motion plus the internal energy ρe of the molecular motion. For energy balance we have

$$\frac{d}{dt} \int_V \left(\frac{1}{2}\rho u_i^2 + \rho e \right) dV + \int_S \left\{ (\rho u_i^2 + \rho e) l_j u_j - p_i u_i + l_j q_j \right\} dS = \int_V \rho F_i u_i \, dV.$$

$$(6.4)$$

The first term in the surface integral is again the contribution from bodily transport across the boundary, the second term is the rate of working by the stress p_i at the boundary, and the third term is the loss or gain of heat by conduction across the boundary. The right hand side is the rate of working by the body forces.

If discontinuities are allowed in the flow quantities, these integral forms will be needed. For one dimensional problems, the volume integrals become integrals over x, from x_2 to x_1, say, and the surface integrals reduce to the differences of the integrands at x_2 and x_1; they take the form quoted in (5.54) for the treatment of shock waves. In large regions of the fluid, however, the quantities will be continuously differentiable and we may take the limit as the volume V shrinks to zero in order to obtain the corresponding differential equations. In (6.2)–(6.4) the time derivatives may be taken inside the volume integrals, since V is independent of t, and the surface integrals can be converted into volume integrals using the divergence theorem:

$$\int_S l_j v_j \, dS = \int_V \frac{\partial v_j}{\partial x_j} \, dV, \qquad (6.5)$$

for any continuously differentiable vector v_j and any reasonably smooth V. Thus (6.2) may be rewritten as

$$\int_V \left\{ \frac{\partial \rho}{\partial t} + \frac{\partial}{\partial x_j} (\rho u_j) \right\} dV = 0. \qquad (6.6)$$

Since the integrand is continuous and (6.6) is true for arbitrarily small V, we deduce that

$$\frac{\partial \rho}{\partial t} + \frac{\partial}{\partial x_j} (\rho u_j) = 0. \qquad (6.7)$$

[If it were nonzero at any point, it would have the same sign in some small volume V, by continuity, and (6.6) would be violated.] If (6.1) is accepted for the moment, (6.3) and (6.4) lead in a similar way to

$$\frac{\partial}{\partial t}(\rho u_i) + \frac{\partial}{\partial x_j}(\rho u_i u_j - p_{ji}) = \rho F_i, \qquad (6.8)$$

and

$$\frac{\partial}{\partial t}\left(\frac{1}{2}\rho u_i^2 + \rho e\right) + \frac{\partial}{\partial x_j}\left\{\left(\frac{1}{2}\rho u_i^2 + \rho e\right)u_j - p_{ji}u_i + q_j\right\} = \rho F_i u_i. \qquad (6.9)$$

The usual argument to establish (6.1) is in fact a first approximation to (6.3). If the maximum dimension of V is d, then the volume V is $O(d^3)$. Any volume integral with continuous integrand is then $O(d^3)$, by the mean value theorem. The first surface integral in (6.3) is equal to the corresponding volume integral by the divergence theorem (6.5) and hence is also $O(d^3)$. Therefore (6.3) shows that

$$\int_S p_i \, dS = O(d^3) \qquad (6.10)$$

for all S. The relation given in (6.1) is clearly sufficient, since the divergence theorem may then be used to show that (6.10) is satisfied. To show that it is also necessary, we first define quantities p_{ji} for $j = 1, 2, 3$, as the values of p_i when the area element is perpendicular to the x_1, x_2, and x_3 axes, respectively. We then apply (6.10) to the special case of a small tetrahedron with three faces normal to the three coordinates axes. If the fourth face has unit normal $\mathbf{l}$ and area ΔS, the areas of the other three faces are the projections $l_1 \Delta S, l_2 \Delta S, l_3 \Delta S$. Then (6.10) shows that

$$p_i(\mathbf{l})\Delta S = p_{1i} l_1 \Delta S + p_{2i} l_2 \Delta S + p_{3i} l_3 \Delta S + O(d^3),$$

where $p_i(\mathbf{l})$ and the p_{ji} are evaluated at appropriate mean value points in the corresponding faces. In the limit, as $d \to 0$, we have

$$p_i(\mathbf{l}) = p_{1i} l_1 + p_{2i} l_2 + p_{3i} l_3,$$

in agreement with (6.1). This is a rather inelegant proof but apparently there is no way to change it essentially.

In connection with conservation equations, it is natural to ask whether the conservation of angular momentum adds anything new. For the x_3 component of angular momentum, we would have

$$\frac{\partial}{\partial t}(x_1 \rho u_2 - x_2 \rho u_1) + \frac{\partial}{\partial x_j}\left\{(x_1 \rho u_2 - x_2 \rho u_1)u_j - (x_1 p_{j2} - x_2 p_{j1})\right\}$$

$$= x_1 \rho F_2 - x_2 \rho F_1, \qquad (6.11)$$

with similar expressions for the other components. When (6.8) is substituted in (6.11) most of the terms cancel and we are left with

$$P_{12} = P_{21}.$$

Thus angular momentum leads to the symmetry of the stress tensor:

$$P_{ji} = P_{ij}. \tag{6.12}$$

It is valuable information, but it is a subsidiary equation compared with the others.

Equations 6.7–6.9 provide five equations for the fourteen quantities $\rho, u_i, p_{ji}, q_i, e$. To complete the system various additional relations are posed between the flow variables.

6.2 The Kinetic Theory View

It adds to the understanding of the various terms in the conservation equations 6.7–6.9 to note what they represent from a molecular viewpoint. The molecules have a whole distribution of velocities, and the flow quantities are related to the distribution function $f(\mathbf{x}, \mathbf{v}, t)$, which is defined so that

$$f(\mathbf{x}, \mathbf{v}, t)\, dx_1\, dx_2\, dx_3\, dv_1\, dv_2\, dv_3$$

is the probable number of molecules in a volume element $dx_1\, dx_2\, dx_3$ centered at $\mathbf{x}$, in a velocity range $dv_1\, dv_2\, dv_3$ centered at $\mathbf{v}$. Then the density and macroscopic velocity $\mathbf{u}$ are defined by

$$\rho = \int_{-\infty}^{\infty} mf\, d\mathbf{v}, \qquad \rho u_i = \int_{-\infty}^{\infty} mv_i f\, d\mathbf{v}, \tag{6.13}$$

where m is the mass of a molecule and $\int d\mathbf{v}$ denotes the triple integral over all values of v_1, v_2, v_3. The total flux of the ith component of momentum across a surface with normal $\mathbf{l}$ is

$$\int_{-\infty}^{\infty} mv_i(l_j v_j) f\, d\mathbf{v}. \tag{6.14}$$

If we set $\mathbf{v} = \mathbf{u} + \mathbf{c}$, so that $\mathbf{c}$ measures the difference of the molecular velocity from the mean $\mathbf{u}$ defined in (6.13), this may be expanded to

$$l_j\left(u_i u_j \int_{-\infty}^{\infty} mf\, d\mathbf{c} + u_i \int_{-\infty}^{\infty} mc_j f\, d\mathbf{c} + u_j \int_{-\infty}^{\infty} mc_i f\, d\mathbf{c} + \int_{-\infty}^{\infty} mc_i c_j f\, d\mathbf{c}\right).$$

From the definition of $\mathbf{c}$ as the deviation from the mean, the middle terms vanish and we have

$$l_j\left(\rho u_i u_j + \int_{-\infty}^{\infty} mc_i c_j f\, d\mathbf{c}\right). \tag{6.15}$$

In a perfect gas where intermolecular forces are limited to relatively instantaneous "collisions" between the molecules, this is the only contribution to the surface integral in (6.3) and we see that

$$-p_i = l_j \int_{-\infty}^{\infty} mc_i c_j f\, d\mathbf{c}. \tag{6.16}$$

This agrees with the form in (6.1) and shows that the stress tensor is

$$-p_{ji} = \int_{-\infty}^{\infty} mc_i c_j f\, d\mathbf{c}; \tag{6.17}$$

the symmetry (6.12) is immediate. Thus the stress contribution in (6.3) may be interpreted as the additional momentum flux by the motion of the molecules relative to the mean.

Each molecule has translational kinetic energy $\frac{1}{2}mv_i^2$. The molecules may also have vibrational or rotational energy but, for the present, we take the case of a monatomic gas for which these additional forms of energy are absent. The total energy per unit volume is then

$$\int_{-\infty}^{\infty} \frac{1}{2} mv_i^2 f\, d\mathbf{v} = \frac{1}{2}\rho u_i^2 + \int_{-\infty}^{\infty} \frac{1}{2} mc_i^2 f\, d\mathbf{c}. \tag{6.18}$$

Therefore the internal energy term in the volume integral in (6.4) may be interpreted as the additional energy of the molecular motion relative to the mean, and we have

$$\rho e = \int_{-\infty}^{\infty} \frac{1}{2} mc_i^2 f\, d\mathbf{c}. \tag{6.19}$$

The energy flux across an element of surface with normal $\mathbf{l}$ is

$$\int_{-\infty}^{\infty} \frac{1}{2} mv_i^2 l_j v_j f\, d\mathbf{v}.$$

In terms of the quantities already defined, this can be broken down into

$$l_j\left(\rho u_i^2 u_j + \rho e u_j - p_{ji} u_i + q_j\right), \tag{6.20}$$

where

$$q_j = \int_{-\infty}^{\infty} \frac{1}{2} m c_i^2 c_j f \, d\mathbf{c}. \tag{6.21}$$

By comparing with (6.4), we see that (6.20) agrees with the energy flux in (6.4), and the heat conduction $\mathbf{q}$ is interpreted as the transfer of excess molecular energy by molecular motion.

Even for the discussion of an ideal gas, it is important to include the vibrational and rotational energy possessed by diatomic and more complicated molecules. This energy should be added to the expression in (6.19) and there will be a corresponding contribution in the heat flux vector (6.21). A basic result in statistical mechanics is that the different forms of energy reach an equilibrium value with equal contributions from each degree of freedom. This will allow us to generalize (6.19), when necessary, without going into details.

These interpretations of stress, internal energy, and heat conduction in terms of the random molecular motion show that the various quantities introduced in (6.7)–(6.9) are not just a rather ad hoc choice representing any important effects we can think of, but they follow a consistent scheme using higher and higher moments of the velocity distribution f. Indeed in kinetic theory proper, a basic equation is proposed for f and then the conservation equations (6.7)–(6.9) are *deduced* as consequences of it. The equation for f is usually taken as the Boltzmann equation or some approximation to it.

We return now to the continuum equations.

6.3 Equations Neglecting Viscosity, Heat Conduction, and Relaxation Effects

For a gas in equilibrium in the absence of body forces, we have the following:

1. The stress on any element of area is normal to the area and independent of its orientation. Hence

$$p_{ji} = -p\delta_{ji}, \tag{6.22}$$

where p is the scalar pressure.

2. The heat conduction

$$q_j = 0. \tag{6.23}$$

3. The internal energy is a definite function

$$e = e(p, \rho) \tag{6.24}$$

of the pressure and density. The form of this function is established by experiment and various thermodynamic arguments.

When the gas is nonuniform and in motion, none of these is strictly true. However, provided time and space derivatives are not too large, they are still good approximations for many purposes. With them the basic conservation equations become a complete set for the five flow quantities ρ, p, u_i. They are

$$\frac{\partial \rho}{\partial t} + \frac{\partial}{\partial x_j}(\rho u_j) = 0, \tag{6.25}$$

$$\frac{\partial(\rho u_i)}{\partial t} + \frac{\partial}{\partial x_j}(\rho u_i u_j) + \frac{\partial p}{\partial x_i} = \rho F_i, \tag{6.26}$$

$$\frac{\partial}{\partial t}\left(\frac{1}{2}\rho u_i^2 + \rho e\right) + \frac{\partial}{\partial x_j}\left\{\left(\frac{1}{2}\rho u_i^2 + \rho e + p\right)u_j\right\} = \rho F_i u_i. \tag{6.27}$$

When these equations predict shocks or other regions of high gradients the assumptions may need to be improved.

The first assumption, (6.22), corresponds to neglecting viscous effects and would be improved in the Navier-Stokes approximation by adding terms linear in the velocity gradients $\partial u_j / \partial x_i$. The second assumption, (6.23), neglects heat conduction and would be improved by taking $\mathbf{q}$ proportional to the temperature gradient. The appropriate forms for the Navier-Stokes equations are

$$p_{ji} = -p\delta_{ji} - \frac{2}{3}\mu\left(\frac{\partial u_k}{\partial x_k}\right)\delta_{ji} + \mu\left(\frac{\partial u_j}{\partial x_i} + \frac{\partial u_i}{\partial x_j}\right), \tag{6.28}$$

$$q_i = -\lambda\frac{\partial T}{\partial x_i}, \tag{6.29}$$

where μ and λ are the coefficients of viscosity and heat conduction, respectively. The temperature T is related to p and ρ by the equation of state of the gas.

The third assumption in (6.24) assumes that the gas is in local thermodynamic equilibrium. In changing flows the internal energy is always tending toward the equilibrium for the new conditions, but there is a time lag, particularly in the adjustment of the rotational and vibrational energy. This is a so-called relaxation effect and the typical time lag is referred to as the relaxation time. This is an interesting but rather special topic so the details are postponed and given as an example in Chapter 10.

6.4 Thermodynamic Relations

We could just take the view that $e(p,\rho)$ in (6.24) is some empirical function that is given to us. However, the arguments developed in thermodynamics not only provide us with formulas but suggest the important quantities to consider. It is appropriate to note here only the mathematical steps we require, and refer the reader to the numerous standard texts for motivation and a study of the deeper significance of the issues.

The differential form

$$de + p\,d\left(\frac{1}{\rho}\right) \tag{6.30}$$

plays a fundamental role. It arises first in considering the consequences when a small amount of energy is added to unit mass of the gas. If the energy is added relatively slowly so that there is no violent change in the pressure, the work done in expanding the volume $1/\rho$ by $d(1/\rho)$ is $p\,d(1/\rho)$. The rest of the energy must go into increasing the internal energy by de. In these circumstances (6.30) is equal to the amount of energy added. But in any event, for given $e(p,\rho)$ it is a differential form in two variables p,ρ. By Pfaff's theorem, this always has an integrating factor, so that there exist functions $T(p,\rho)$ and $S(p,\rho)$ such that

$$T\,dS = de + p\,d\left(\frac{1}{\rho}\right). \tag{6.31}$$

This simple mathematical statement acquires its deep significance from the fact that T is the absolute temperature and S is the entropy.

In more complicated systems, other thermodynamic variables (such as the concentration of different phases of the substance) appear besides p and ρ. Then the differential form corresponding to (6.30) involves more than two variables. On purely mathematical grounds one can no longer claim that there is always an integrating factor to relate it to a perfect differential. However, the basis of thermodynamics is that this will always be so for all real physical systems, and, moreover, the integrating factor will always be the absolute temperature.

The mathematical step in (6.31) seems to introduce T and S as subsidiary derived quantities from a given $e(p,\rho)$. But they play an equally fundamental role, and (6.31) should be viewed more as a relation between equally important quantities.

Ideal Gas.

Under normal conditions most gases obey the ideal gas law

$$p = \mathcal{R}\rho T, \tag{6.32}$$

where $\mathcal{R}$ is a constant. When that is the case, we can write (6.31) as

$$dS = \frac{de}{T} - d(\mathcal{R}\log\rho).$$

It follows that de/T must be a perfect differential and therefore e is a function of T alone:

$$e = e(T). \tag{6.33}$$

It is interesting that (6.33) can be deduced from the assumption (6.32), but actually (6.33) is more fundamental.

Equations 6.32 and 6.33 describe an ideal gas. In the equations of motion it is convenient to express e as a function of p and ρ. We see that for an ideal gas e is a function of p/ρ. The form of this function could be left open, but in fact a rather simple formula covers a wide range of phenomena in gas dynamics. It arises in considerations of the specific heats.

Specific Heats.

When heat is added slowly to unit mass of gas, it may be distributed between internal energy and volume change in various ways, provided that the sum in (6.30) is equal to the heat added. A specific heat is defined as the ratio of the heat added per unit mass to the temperature change. If the fluid is kept at a constant volume, the heat added goes entirely into internal energy; hence

$$de = c_v \, dT, \tag{6.34}$$

where c_v is the specific heat at constant volume. Alternatively, if the pressure is kept constant, and the fluid is allowed to expand, we have from (6.30)

$$d\left(e + \frac{p}{\rho}\right) = c_p \, dT, \tag{6.35}$$

where c_p is the specific heat at constant pressure. The quantity $e + p/\rho$ which appears here and also, significantly, in the flux term in (6.27), is the enthalpy

$$h = e + \frac{p}{\rho}. \tag{6.36}$$

From (6.32) and (6.33), we see that for an ideal gas e, h, c_v, and c_p are functions of the temperature alone.

Ideal Gas with Constant Specific Heats.

It is found empirically, however, that it is a good approximation to take the specific heats constant over large ranges of temperature. Hence

$$e = c_v T, \qquad h = c_p T. \tag{6.37}$$

Since the difference of these is p/ρ, the gas law (6.32) follows with

$$c_p - c_v = \mathcal{R} .$$

If the ratio of the specific heats $\gamma = c_p/c_v$ is introduced, we have

$$c_p = \gamma c_v, \qquad \mathcal{R} = (\gamma - 1) c_v,$$

$$e = \frac{1}{\gamma - 1} \frac{p}{\rho}, \qquad h = \frac{\gamma}{\gamma - 1} \frac{p}{\rho}, \qquad T = \frac{p}{\rho \mathcal{R}}. \tag{6.38}$$

From these the entropy relation (6.31) becomes

$$dS = \frac{de}{T} + \frac{p}{T} d\left(\frac{1}{\rho}\right)$$

$$= c_v \left(\frac{dp}{p} - \gamma \frac{d\rho}{\rho}\right);$$

hence

$$S = c_v \log \frac{p}{\rho^\gamma} + \text{constant},$$

or, alternatively,

$$p = \kappa \rho^\gamma e^{S/c_v}, \tag{6.39}$$

where κ is a constant.

An ideal gas with constant specific heats is sometimes referred to as a *polytropic* gas.

Kinetic Theory.

Some of these relations have a simple kinetic theory interpretation which is worth noting. First, the temperature T measures the average kinetic energy per molecule in the translational motion of the molecules. It is normalized so that this energy is $\frac{3}{2} kT$, where k is Boltzmann's constant. For an ideal monatomic gas, this is the whole of the internal energy, so that

$$e = \frac{3}{2} kTn,$$

where n is the number of molecules per unit mass. Thus the expression for e as a linear function of T is essentially an identity in this case.

In equilibrium, the expression in (6.17) for the p_{ji} must reduce to $-p\delta_{ji}$. Therefore the pressure may be related to the molecular motion by the formula

$$p = -\frac{1}{3} p_{ii} = \frac{1}{3} \int_{-\infty}^{\infty} mc_i^2 f \, d\mathbf{c}.$$

But the translational energy (6.19) involves one half of the same integral and is equal to $\frac{3}{2} kTn\rho$; hence we must have

$$p = kn\rho T. \tag{6.40}$$

This is the ideal gas law with $\mathcal{R} = kn$. The number of molecules per unit mass is Avogadro's number N divided by the molecular weight of the gas. Hence the constant $\mathcal{R}$ used here is the universal gas constant kN divided by the molecular weight of the gas.

When the molecules have other forms of internal energy, such as vibrational or rotational energy, it is a basic principle of kinetic theory that, in equilibrium, the average energy in each degree of freedom is the same. The temperature is defined so that the energy per molecule is $\frac{1}{2}kT$ for each degree of freedom. Thus it is $\frac{3}{2}kT$ for the three translational degrees of freedom. When there are α degrees of freedom the average energy per molecule is $\frac{1}{2}\alpha kT$. Therefore the average energy per unit mass is

$$e = \frac{1}{2}\alpha kTn. \tag{6.41}$$

The relation for p is unchanged, since p is related to the translational part of the energy. Combining these various results, we have

$$e = \frac{1}{2}\alpha \mathcal{R} T, \qquad h = \left(\frac{1}{2}\alpha + 1\right)\mathcal{R} T, \qquad p = \mathcal{R}\rho T, \tag{6.42}$$

and we deduce that

$$c_v = \frac{1}{2}\alpha \mathcal{R}, \qquad c_p = \left(\frac{1}{2}\alpha + 1\right)\mathcal{R}, \qquad \gamma = 1 + \frac{2}{\alpha}. \tag{6.43}$$

These agree in form with the expression obtained earlier for an ideal gas with constant specific heats, but they have the extra feature that formulas for c_v, c_p, and γ are included.

For a monatomic gas, $\alpha = 3$, $\gamma = 5/3$. For a diatomic gas including two rotational degrees of freedom, $\alpha = 5$, $\gamma = 1.4$; this is a good approximation for air.

6.5 Alternative Forms of the Equations of Motion

The conservation forms of (6.25)–(6.27) correspond to the integrated forms (6.2)–(6.4) and will be needed for the treatment of shocks. But for other purposes the equations may be simplified. It is convenient to introduce the operator

$$\frac{D}{Dt} = \frac{\partial}{\partial t} + u_j \frac{\partial}{\partial x_j}$$

for the time derivative following an individual particle. The mass equation (6.25) may be written

$$\frac{D\rho}{Dt} + \rho \frac{\partial u_j}{\partial x_j} = 0; \tag{6.44}$$

from (6.25), the ρ derivatives can be eliminated in (6.26) to give

$$\rho \frac{Du_i}{Dt} + \frac{\partial p}{\partial x_i} = \rho F_i. \qquad (6.45)$$

The energy equation (6.27) can be written in various forms. First, using the two other equations, we can reduce it to

$$\rho \frac{De}{Dt} + p \frac{\partial u_j}{\partial x_j} = 0. \qquad (6.46)$$

Then, from (6.44), an alternative form is

$$\frac{De}{Dt} - \frac{p}{\rho^2} \frac{D\rho}{Dt} = 0,$$

and, from the thermodynamic relation (6.31), this reduces to

$$T \frac{DS}{Dt} = 0. \qquad (6.47)$$

That is to say, the entropy remains constant following a particle. Flows satisfying (6.47) are usually called *adiabatic*.

It should be stressed that the arguments leading to (6.47) are purely mathematical manipulations of the conservation equations. One could, in principle, have been led to introduce an "interesting quantity $S(p,\rho)$" in this way, without any prior knowledge of thermodynamics. It is reassuring to have the results here on that basis. The discussions of (6.31) in thermodynamics refer to infinitely slow reversible changes, and we might appear to be using them unjustifiably outside that context. However, once the assumptions that $p_{ji} = -p\delta_{ji}$ and $e = e(p,\rho)$ are adopted, the rest is mathematics and consequences such as (6.47) follow without any restriction to slow flows in the thermodynamic sense.

Since the expression for S in terms of p and ρ may be solved in principle as $p = p(\rho, S)$, we may use

$$\frac{Dp}{Dt} = a^2 \frac{D\rho}{Dt}, \qquad a^2 = \left(\frac{\partial p}{\partial \rho} \right)_{S = \text{constant}} \qquad (6.48)$$

as an equivalent form of (6.47). The quantity a will subsequently be identified as the sound speed.

Unless the conservation form of the equation is particularly required, it is usual to work with (6.44), (6.45), and either (6.47) or (6.48). It will be convenient to collect them together for future reference:

$$\frac{D\rho}{Dt} + \rho\frac{\partial u_j}{\partial x_j} = 0,$$

$$\rho\frac{Du_i}{Dt} + \frac{\partial p}{\partial x_i} = \rho F_i, \tag{6.49}$$

$$\frac{DS}{Dt} = 0 \qquad \text{or} \qquad \frac{Dp}{Dt} - a^2\frac{D\rho}{Dt} = 0.$$

For a polytropic gas

$$e = \frac{1}{\gamma - 1}\frac{p}{\rho}, \qquad S = c_v\log\frac{p}{\rho^\gamma}, \qquad a^2 = \frac{\gamma p}{\rho}. \tag{6.50}$$

The entropy equation shows simply that the entropy remains constant on each particle path. In general, it may take different values on different particle paths. However, if the fluid is initially at rest with uniform entropy S_0, it follows that $S = S_0$ on each particle path and hence remains uniform in the motion. Such flows are called *isentropic*. When this is the case, p is a function of ρ alone, and the equations reduce to the first two in (6.49). For a polytropic gas

$$p = \kappa\rho^\gamma.$$

This argument requires modification when shocks or other discontinuities are present. The differential equations, and in particular the entropy equation, apply only to regions where the functions are differentiable. Across a surface of discontinuity the entropy jumps, and in general the amount of the jump will vary with time and position as the surface propagates. Thus an initially isentropic flow may not remain so after a shock passes through. This will be discussed in detail in Section 6.10.

6.6 Acoustics

The first information on wave propagation in gas dynamics is provided by the theory of acoustics, which refers to the linearized theory of small disturbances about an equilibrium state. The simplest case arises when body forces are neglected and the equilibrium state is taken to have constant values $p = p_0$, $\rho = \rho_0$, $u = 0$. If the initial disturbance also has uniform entropy, the motion remains isentropic and we may take $p = p(\rho)$.

Then for small perturbations

$$p - p_0 = a_0^2(\rho - \rho_0) \tag{6.51}$$

to first order, where

$$a_0^2 = p'(\rho_0). \tag{6.52}$$

Relation 6.51 may be viewed as the solution of the linearized form of the third equation in (6.49), and we turn to the linearization of the first two.

To first order in the small quantities $(p - p_0)/p_0$, $(\rho - \rho_0)/\rho_0$, u/a_0, and their derivatives, we have

$$\frac{\partial \rho}{\partial t} + \rho_0 \frac{\partial u_j}{\partial x_j} = 0, \tag{6.53}$$

$$\rho_0 \frac{\partial u_i}{\partial t} + \frac{\partial p}{\partial x_i} = 0. \tag{6.54}$$

(Here and elsewhere, when the unperturbed state is constant the derivatives are left in terms of original quantities to save writing or the proliferation of subscripts, but they are transposed into derivatives of the perturbations as the need arises.)

From (6.54),

$$u_i = u_i^{(0)}(\mathbf{x}) - \frac{\partial}{\partial x_i} \frac{1}{\rho_0} \int_0^t (p - p_0)\, dt,$$

where arbitrary functions $u_i^{(0)}(\mathbf{x})$ arise in the integration. It is usually appropriate in acoustics to take the functions $u_i^{(0)}(\mathbf{x})$ to be zero; this is so, for example, if $u_i = 0$ initially or if the waves move out into a region at rest. Then the vector $\mathbf{u}$ is the gradient of a scalar. If we introduce the velocity potential φ defined by $\mathbf{u} = \nabla \varphi$, we have

$$u_i = \frac{\partial \varphi}{\partial x_i}, \qquad p - p_0 = -\rho_0 \frac{\partial \varphi}{\partial t}, \qquad \rho - \rho_0 = -\frac{\rho_0}{a_0^2} \frac{\partial \varphi}{\partial t}, \tag{6.55}$$

and (6.54) is satisfied identically. The equation for φ is obtained by substitution of these expressions in (6.53); the result is the wave equation

$$\varphi_{tt} = a_0^2 \varphi_{x_i x_i} \tag{6.56}$$

for φ, with a_0 as the propagation speed. It may be noted also that all the perturbations in (6.55) satisfy the wave equation.

For one dimensional waves (6.56) may be solved immediately to give

$$\varphi = f(x - a_0 t) + g(x + a_0 t),$$

where f and g are arbitrary functions; the corresponding expressions for u and $p - p_0$ are

$$u = f'(x - a_0 t) + g'(x + a_0 t),$$

$$\frac{p - p_0}{\rho_0 a_0} = f'(x - a_0 t) - g'(x + a_0 t). \tag{6.57}$$

The functions f and g are chosen to fit initial or boundary conditions. We defer the discussion of specific examples because the full nonlinear equations for plane waves are tractable, and some of the linearized results can be seen as approximations to exact solutions. Two and three dimensional solutions of the wave equation are considered in Chapter 7.

Practically any problem of acoustics takes place in the presence of a gravitational field and, as a consequence, the unperturbed state is not uniform. In problems of propagation over large distances in the atmosphere or in the ocean, these effects may be crucially important and produce amplification and refraction of the sound waves. Even when the preceeding theory is adequate, it is not so much that the whole gravitational term ρg is negligible, but rather that its *perturbed* value may be small compared with other perturbation terms. The undisturbed pressure and density must satisfy

$$\frac{dp_0}{dz} = -\rho_0 g, \tag{6.58}$$

where z is the vertical coordinate. Since the changes of pressure and the accelerations in sound waves may be extremely small, the two terms in (6.58) may be the largest terms in the vertical momentum equation. But the point is that they balance each other, and their further effects in the perturbation equations may be negligible. We consider the case of vertical propagation of plane waves in detail. If we set $p = p_0(z) + p_1$, $\rho = \rho_0(z) + \rho_1$, $\mathbf{u} = (0, 0, w)$, in (6.49) and neglect quadratic terms and higher order terms in p_1, ρ_1, w, we have

$$\rho_{1t} + w\rho_0' + \rho_0 w_z = 0,$$

$$\rho_0 w_t + p_0' + p_{1z} = -\rho_0 g - \rho_1 g, \tag{6.59}$$

$$p_{1t} + w p_0' - a_0^2(\rho_{1t} + w\rho_0') = 0.$$

The equilibrium entropy distribution is not uniform, in general, so that entropy changes must be included and a^2 is defined as in (6.48).

From (6.58), the variations of the equilibrium quantities take place over a length scale L of the order of a_0^2/g. If $p_1 = O(\epsilon p_0)$ and λ is a typical wavelength in the perturbations,

$$p_0' = O\left(\frac{p_0}{L}\right), \qquad p_{1z} = O\left(\frac{\epsilon p_0}{\lambda}\right).$$

While λ/L may be 10^{-4}, the amplitude ϵ may easily be as low as 10^{-4} or lower, so the ambient gradients p_0' may be greater than the gradients p_{1z} produced by the sound waves. However, the terms p_0' and $-\rho_0 g$ in (6.59) cancel, and the remaining terms are all proportional to ϵ. The terms $\rho_1 g, w p_0', w \rho_0'$, have an additional factor λ/L. Therefore, unless the propagation is over distances comparable with L, the nonuniform effects would be small. Rather than make further estimates it is simplest to look at some exact solutions of (6.59).

By routine elimination of p_1 and ρ_1 in (6.59), and further use of (6.58), we find

$$w_{tt} = a_0^2 w_{zz} + \frac{(\rho_0 a_0^2)'}{\rho_0} w_z.$$

This equation is hyperbolic and the characteristic velocities are $\pm a_0(z)$. In the case of variable atmosphere, a_0 is still the sound speed in this precise sense. For a polytropic gas, $a_0^2 = \gamma p_0/\rho_0$, so that $(\rho_0 a_0^2)' = \gamma p_0' = -\gamma \rho_0 g$. Hence the equation reduces to

$$w_{tt} = a_0^2 w_{zz} - \gamma g w_z.$$

Isothermal Equilibrium.

For constant equilibrium temperature, a_0^2 is constant and (6.58) gives an exponential atmosphere

$$\rho_0(z) = \rho_0(0) e^{-z/H}, \qquad H = \frac{\mathscr{R} T_0}{g} = \frac{a_0^2}{\gamma g}.$$

The equation for w has periodic solutions

$$w = A e^{z/2H} \cos(kz - \omega t),$$

$$\omega^2 = a_0^2 k^2 + \frac{1}{4} \frac{a_0^2}{H^2}.$$

The variation of amplitude is small provided $z \ll H$, and $\omega^2 \simeq a_0^2 k^2$ provided $\lambda^2 / H^2 \ll 1$. This solution confirms the previous estimates.

Convective Equilibrium.

In convective equilibrium, the entropy is constant and $p \propto \rho^\gamma$. From (6.58),

$$\frac{1}{\rho_0} \frac{dp_0}{dz} = \frac{a_0^2}{\rho_0} \frac{d\rho_0}{dz} = \frac{2}{\gamma - 1} a_0 \frac{da_0}{dz} = -g;$$

hence

$$a_0^2(z) = a_0^2(0) - (\gamma - 1) gz.$$

Of course this distribution is realistic only below the height $a_0^2(0)/(\gamma - 1)g$. Solutions for w can be obtained in terms of Bessel functions (Lamb, 1932, p. 546), and similar conclusions about the effects of nonuniformity may be reached.

Some questions concerning the refraction of nonplanar waves will be considered in Section 7.7; other aspects can be found in Lamb (1932, pp. 547–561).

6.7 Nonlinear Plane Waves

We now consider the exact nonlinear equations for one dimensional flows in the case that body forces can be neglected. Since the interest is now in large pressure changes, gravitational effects can be wholly neglected for many applications.

The equations in (6.49) reduce to

$$\rho_t + u\rho_x + \rho u_x = 0, \tag{6.60}$$

$$\rho(u_t + uu_x) + p_x = 0, \tag{6.61}$$

$$S_t + uS_x = 0, \tag{6.62}$$

where $p(\rho, S)$ is a known function, and the last equation can also be written as

$$p_t + up_x - a^2(\rho_t + u\rho_x) = 0, \tag{6.63}$$

where

$$a^2 = \left(\frac{\partial p}{\partial \rho} \right)_{S=\text{constant}}. \qquad (6.64)$$

To study nonlinear waves the equations are manipulated into characteristic form, following the procedure described in Section 5.1. Rather than quote formulas, it is quicker to develop the characteristic equations directly from (6.60)–(6.63). We note first that (6.62) is already in characteristic form, with a characteristic velocity u. Hence

$$\frac{dS}{dt} = 0 \quad \text{on characteristics} \quad \frac{dx}{dt} = u. \qquad (6.65)$$

These characteristic curves are the particle paths, and S remains constant on each one of them.

The other two families of characteristics are most conveniently found using (6.60), (6.61), and (6.63). It is sufficient to take the linear combination: l_1 times (6.60) and l_2 times (6.61) added to (6.63). After rearrangement this combination is

$$p_t + (u + l_2)p_x + \rho l_2(u_t + uu_x) + \rho l_1 u_x + (l_1 - a^2)(\rho_t + u\rho_x) = 0.$$

The choice $l_1 = l_2 = 0$ corresponds to the characteristic form already found in (6.65). After that case is eliminated, it is clear from comparison of the terms in p and ρ that the only possible characteristic equations with $l_2 \neq 0$ require the ρ derivatives to be absent altogether: $l_1 = a^2$. Then it is observed from the p and u derivatives that we require $l_2 = l_1/l_2$. The conclusion is that $l_1 = a^2$, $l_2 = \pm a$, and the required combinations are

$$p_t + (u \pm a)p_x \pm \rho a \{ u_t + (u \pm a)u_x \} = 0. \qquad (6.66)$$

The full set of characteristic equations can be written

$$\frac{dp}{dt} + \rho a \frac{du}{dt} = 0 \quad \text{on} \quad C_+ : \frac{dx}{dt} = u + a, \qquad (6.67)$$

$$\frac{dp}{dt} - \rho a \frac{du}{dt} = 0 \quad \text{on} \quad C_- : \frac{dx}{dt} = u - a, \qquad (6.68)$$

$$\frac{dS}{dt} = 0 \quad \text{on} \quad P : \frac{dx}{dt} = u. \qquad (6.69)$$

The characteristics C_+ and C_- represent points moving with velocity $\pm a$

relative to the local velocity u of the fluid. These are the sound waves and a as defined in (6.64) is identified as the nonlinear sound speed.

In the linearized theory, these equations are approximated as

$$\frac{dp}{dt} + \rho_0 a_0 \frac{du}{dt} = 0 \quad \text{on} \quad C_+ : \frac{dx}{dt} = a_0,$$

$$\frac{dp}{dt} - \rho_0 a_0 \frac{du}{dt} = 0 \quad \text{on} \quad C_- : \frac{dx}{dt} = -a_0,$$

$$\frac{dS}{dt} = 0 \quad \text{on} \quad P : \frac{dx}{dt} = 0,$$

and may be integrated immediately to

$$(p - p_0) + \rho_0 a_0 u = F(x - a_0 t),$$

$$(p - p_0) - \rho_0 a_0 u = G(x + a_0 t), \tag{6.70}$$

$$S - S_0 = H(t).$$

If entropy changes are absent these agree with the solution in (6.57).

In the nonlinear theory the defining relations for the characteristics depend on the solution yet to be found, and the integration is not straightforward.

For isentropic flow $S = \text{constant}$ everywhere, so that (6.69) can be dropped. Moreover, $p = p(\rho)$, $a^2 = p'(\rho)$, so that the first two characteristic equations can be written

$$\int \frac{a(\rho)\,d\rho}{\rho} + u = \text{constant on } C_+ : \frac{dx}{dt} = u + a,$$

$$\int \frac{a(\rho)\,d\rho}{\rho} - u = \text{constant on } C_- : \frac{dx}{dt} = u - a.$$

These are the Riemann invariants. For a polytropic gas

$$p = \kappa \rho^\gamma, \qquad a^2 = \kappa \gamma \rho^{\gamma - 1},$$

and the Riemann invariants are

$$\frac{2}{\gamma - 1} a \pm u = \text{constant on } \frac{dx}{dt} = u \pm a. \tag{6.71}$$

6.8 Simple Waves

If, in addition to being isentropic, the flow has one of the Riemann invariants constant throughout, the solution is enormously simplified. It corresponds to wave motion in one direction only and in the linear theory would correspond to either F or $G=0$ in (6.70). As a basic model to illustrate how this type of "simple wave" may be produced, consider the waves produced by the prescribed motion of a piston in the end of a long tube. Figure 6.1 is the (x, t) diagram. Provided shocks do not appear to violate deductions from the differential equations, it may be argued that the flow must be a simple wave. For clarity, the argument is presented for a polytropic gas, but the extension to the more general form is obvious.

The gas is assumed to be at rest with a uniform state $u = 0$, $a = a_0$, $S = S_0$ in $x \geqslant 0$ at $t = 0$, and it is assumed provisionally that shocks are not formed. Since the piston is itself a particle path, it is clear that all particle paths originate on the x axis in the uniform region. From (6.69), S remains constant on any particle path P, and hence it is equal to its initial value S_0. But the initial value is the same on every particle path. Therefore

$$S = S_0 \tag{6.72}$$

throughout the flow. Since the flow is then isentropic, we may use (6.71) for the other two families of characteristics.

The C_- characteristics have a lower value of dx/dt than the particle paths, and therefore they all start on the x axis in the uniform region (see Fig. 6.1). On each of them

$$\frac{2}{\gamma-1} a - u = \frac{2a_0}{\gamma-1}, \tag{6.73}$$

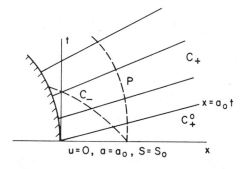

Fig. 6.1. Expansion wave produced by withdrawing piston.

because this Riemann invariant is constant on each of them, and we have inserted the initial value from the region $u=0$, $a=a_0$. Again, since the initial value is the same on every one, we have deduced that the Riemann invariant (6.73) is the same constant throughout. These are the arguments to establish that the solution is a simple wave. We now turn to the other characteristic equation in (6.71) to determine the remaining details of the solution.

For those C_+ that originate on the x axis, (6.73) also holds with the opposite sign. Hence $u=0$, $a=a_0$ in the region covered by these C_+. We deduce that the original uniform conditions apply in the region ahead of the characteristic C_+^0, which separates those C_+ meeting the x axis from those meeting the piston. Since we are assuming the flow is continuous with no shocks, we have $u=0$, $a=a_0$ ahead of and on C_+^0. Hence C_+^0 is given by

$$x = a_0 t.$$

For those C_+ that meet the piston, we use (6.71) with the positive signs:

$$\frac{2a}{\gamma-1} + u = \text{constant on each } C_+ : \frac{dx}{dt} = u + a.$$

In view of (6.73), which holds everywhere, this may be reduced to

$$u = \text{constant on each } C_+ : \frac{dx}{dt} = a_0 + \frac{\gamma+1}{2}u. \tag{6.74}$$

The value of u on each one will be different, depending on where it meets the piston, but we see in general that the family of positive characteristics will be a family of straight lines, each having a slope $a_0 + \{(\gamma+1)/2\}u$ corresponding to the value of u that it carries.

The boundary condition on the piston is that the fluid velocity is equal to the piston velocity. Therefore if the piston path is given by $x = X(t)$, the boundary condition is

$$u = \dot{X}(t) \qquad \text{on} \quad x = X(t). \tag{6.75}$$

For a C_+ that intersects the piston path at time τ, it follows that $u = \dot{X}(\tau)$ along it, and then from (6.74) that its equation is

$$x = X(\tau) + \left\{ a_0 + \frac{\gamma+1}{2}\dot{X}(\tau) \right\}(t-\tau). \tag{6.76}$$

Thus the solution is

$$u = \dot{X}(\tau), \qquad a = a_0 + \frac{\gamma - 1}{2} \dot{X}(\tau), \qquad S = S_0, \tag{6.77}$$

with $\tau(x,t)$ determined implicitly from (6.76).

Since the C_+ are straight lines with slope dx/dt increasing with u, it is clear that the characteristics will overlap if u ever increases on the piston; that is, $\ddot{X}(\tau) > 0$ for any τ. This is the typical nonlinear breaking described in Fig. 2.1 and it shows that shocks will be formed in such cases. If u increases, so do a, p, and ρ, so that breaking occurs and shocks appear in *compressive* parts of the disturbance. When shocks appear, we shall need to reexamine the arguments leading to (6.72) and (6.73) as well as discuss appropriate shock conditions.

For expansion waves the solution in (6.76) and (6.77) is complete. The limiting case in which the piston is suddenly withdrawn with velocity $-V$ is of special interest. There is a uniform region with

$$u = -V, \qquad a = a_0 - \frac{\gamma - 1}{2} V \tag{6.78}$$

next to the piston, and the adjustment to this from the initial undisturbed region is through a centered fan of characteristics as shown in Fig. 6.2. In the fan, since all the characteristics meet the piston at $x = t = 0$, they are given by

$$x = \left(a_0 + \frac{\gamma + 1}{2} u \right) t, \qquad -V \leqslant u \leqslant 0.$$

All the values of u between 0 and $-V$ are taken instantaneously at the origin, but each value leads to a different member of the fan. Solving this

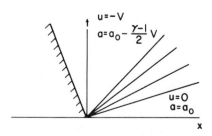

Fig. 6.2. Centered expansion fan.

relation for u and adding the expression for a from (6.73), we have

$$
\left.
\begin{array}{l}
u = \dfrac{2a_0}{\gamma+1}\left(\dfrac{x}{a_0 t} - 1\right), \\[3mm]
\\
a = a_0\left(\dfrac{\gamma-1}{\gamma+1}\dfrac{x}{a_0 t} + \dfrac{2}{\gamma+1}\right),
\end{array}
\right\}
\qquad 1 - \dfrac{\gamma+1}{2}\dfrac{V}{a_0} \leqslant \dfrac{x}{a_0 t} \leqslant 1.
\qquad (6.79)
$$

If the piston is pushed forward with positive velocity V, the fan is reversed and forms a multivalued region, which can be envisaged as a fold in the (x,t) plane (compare Fig. 2.3). It corresponds of course to immediate breaking and must be replaced by a shock.

For other problems the simple wave argument applies in general near the front of any disturbance propagating into a uniform state. There is a region in which particle paths and one set of characteristics originate in the uniform state ahead, so that it is isentropic and one Riemann invariant is constant. The other family of characteristics "carries" the disturbance; on each of them, flow quantities remain constant and each of them is a straight line. The simple wave region extends as far back as the first particle path from the nonuniform region. Figure 6.1 applies with the piston path replaced by this bounding characteristic. The occurrence of such simple wave regions adjacent to uniform regions is well illustrated in the initial value problem of Section 6.12.

6.9 Simple Waves as Kinematic Waves

In Section 2.2 we saw that the continuity equation

$$
\rho_t + q_x = 0 \tag{6.80}
$$

together with a functional relation $q = Q(\rho)$ leads to the simplest nonlinear waves. The simple waves discussed here can be viewed in this light. Whereas in gas dynamics there are three basic equations for the three quantities p, ρ, u, the special simple wave argument uses the two integrals

$$
S = S_0, \qquad \frac{2a}{\gamma-1} - u = \frac{2a_0}{\gamma-1}.
$$

This means that two of the equations may be eliminated and any two of p,

ρ, u can be expressed as functions of the third. We are then reduced to one equation, which may be taken in conservation form, and we have a functional relation between the flux and the density.

For example, if we choose to express all quantities in terms of ρ, the condition for isentropic flow provides the relation

$$p = p_0 \left(\frac{\rho}{\rho_0} \right)^\gamma,$$

and the Riemann invariant provides the relation

$$u = V(\rho) = \frac{2a}{\gamma - 1} - \frac{2a_0}{\gamma - 1}$$

$$= \frac{2a_0}{\gamma - 1} \left\{ \left(\frac{\rho}{\rho_0} \right)^{(\gamma - 1)/2} - 1 \right\}. \qquad (6.81)$$

We may then take the mass conservation as the final equation to determine ρ. It is in the form (6.80) with $q = \rho u$. Therefore the kinematic wave formulation is to take (6.80) and

$$q = Q(\rho) = \rho V(\rho), \qquad (6.82)$$

with the $V(\rho)$ given in (6.81). In this case the function $Q(\rho)$ is obtained from the other two differential equations rather than being proposed as part of the original formulation of the problem. But the analysis may be continued as in the kinematic theory. The equation for ρ is

$$\rho_t + c(\rho)\rho_x = 0, \qquad c(\rho) = Q'(\rho),$$

and we verify that

$$c(\rho) = Q'(\rho) = V(\rho) + \rho V'(\rho)$$

$$= V(\rho) + a(\rho)$$

from the above relations. The conclusion, in agreement with the first derivation, is that flow quantities remain constant on characteristics and the characteristic velocity is $u + a$.

This is essentially the approach used by Earnshaw (1858) in one of the earliest derivations of the solution. He assumed isentropic flow from the outset and wrote the equations as

$$\rho_t + u\rho_x + \rho u_x = 0,$$

$$u_t + u u_x + \frac{a^2(\rho)}{\rho} \rho_x = 0.$$

Then, based on the observation that a linearized acoustic wave moving to the right has $u = a_0(\rho - \rho_0)/\rho_0$, he considered the possibility of exact solutions with $u = V(\rho)$. The equations reduce to

$$\rho_t + (V + \rho V')\rho_x = 0,$$

$$(\rho_t + V\rho_x)V' + \frac{a^2}{\rho}\rho_x = 0.$$

Therefore, for consistency,

$$V' = \pm \frac{a}{\rho},$$

and the common form of the equation is

$$\rho_t + (V \pm a)\rho_x = 0.$$

The choice of the upper sign leads to

$$V(\rho) = \int_{\rho_0}^{\rho} \frac{a(\rho)}{\rho}\, d\rho = \frac{2}{\gamma - 1}\{a(\rho) - a_0\}$$

and agrees with the above. Riemann (1858) gave the deeper argument presented in the last section.

The choice of ρ as the working variable brings the description closest to the description in Chapter 2, but the formulas are simpler in terms of u. We can transfer freely between the two by means of (6.81). In terms of u the basic equation is

$$u_t + \left(a_0 + \frac{\gamma + 1}{2}u\right)u_x = 0, \tag{6.83}$$

and we have

$$a = a_0 + \frac{\gamma - 1}{2}u, \qquad S = S_0,$$

to determine p and ρ. The propagation speed is

$$c(u) = a_0 + \frac{\gamma + 1}{2}u.$$

The solution of (6.83) is then found as before. For the piston problem, in particular, the solution is given by (6.76) and (6.77).

In the simple wave solution, at least, waves break in exactly the way described in Chapter 2, and the solution has to be completed by the introduction of shock waves.

6.10 Shock Waves

When waves break, realistic solutions are recovered by fitting in discontinuities, and we do this with the same general point of view developed in Chapter 2. In the region where breaking occurs, derivatives become large and, strictly speaking, the assumptions in (6.22)–(6.24) become inadequate there. But the real behavior usually can be closely approximated by introducing discontinuities satisfying the appropriate shock conditions and retaining (6.22)–(6.24) in continuous parts of the flow. Subsequently the detailed shock structure can be examined by including the effects of viscosity and heat conductivity.

As noted earlier, the simple wave arguments leading to the integrals

$$S = S_0, \qquad \frac{2a}{\gamma - 1} - u = \frac{2a_0}{\gamma - 1}$$

have to be reexamined once shocks are formed, and we have to return to the complete system of three equations for the discussion of discontinuities.

The shock conditions are derived by the arguments presented in Section 5.8. It is crucial that we work with the equations in conservation form, and that we choose the three equations that are known to remain valid in the integrated form. To make the correct choice it is necessary to go back to the original integral formulation in (6.2)–(6.4). Specialized to one dimension and omitting the body force (although it would not affect the jump conditions), they are

$$\frac{d}{dt} \int_{x_2}^{x_1} \rho \, dx + [\rho u]_{x_2}^{x_1} = 0, \qquad (6.84)$$

$$\frac{d}{dt} \int_{x_2}^{x_1} \rho u \, dx + [\rho u^2 - p_{11}]_{x_2}^{x_1} = 0, \qquad (6.85)$$

$$\frac{d}{dt} \int_{x_2}^{x_1} \left(\frac{1}{2} \rho u^2 + \rho e \right) dx + \left[\left(\frac{1}{2} \rho u^2 + \rho e \right) u - p_{11} u + q_1 \right]_{x_2}^{x_1} = 0. \qquad (6.86)$$

Each of these is in the form (5.54) and the corresponding jump condition is

(5.56). *After deriving the jump conditions*, the assumptions $p_{11} = -p$, $q_1 = 0$, $e = e(p, \rho)$ are again used for the continuous flow on the two sides and therefore may be inserted in the jump conditions. Then they become

$$- U[\rho] + [\rho u] = 0, \tag{6.87}$$

$$- U[\rho u] + [\rho u^2 + p] = 0, \tag{6.88}$$

$$- U\left[\frac{1}{2}\rho u^2 + \rho e\right] + \left[\left(\frac{1}{2}\rho u^2 + \rho e\right)u + pu\right] = 0. \tag{6.89}$$

The corresponding differential equations in conservation form are

$$\rho_t + (\rho u)_x = 0,$$

$$(\rho u)_t + (\rho u^2 + p)_x = 0, \tag{6.90}$$

$$\left(\frac{1}{2}\rho u^2 + \rho e\right)_t + \left\{\left(\frac{1}{2}\rho u^2 + \rho e\right)u + pu\right\}_x = 0.$$

These are equivalent to the set (6.60)–(6.62).

Another conservation equation can be derived from this set:

$$(\rho S)_t + (\rho u S)_x = 0, \tag{6.91}$$

which follows immediately from (6.60) and (6.62). But this does *not* apply more generally in integrated form. In fact, from (6.7)–(6.9) and (6.31), we have

$$(\rho S)_t + (\rho u S)_x = \frac{(p_{11} + p)u_x - q_{1x}}{T}. \tag{6.92}$$

Hence

$$\frac{d}{dt}\int_{x_2}^{x_1} \rho S \, dx + [\rho u S]_{x_2}^{x_1} = \int_{x_2}^{x_1} \frac{(p_{11} + p)u_x - q_{1x}}{T} \, dx. \tag{6.93}$$

The term on the right in (6.93) is crucially different from the source term h_i in (5.54), since it involves *derivatives* of the flow quantities and there is no way to integrate them out. Therefore the argument leading to (5.56) does not apply. (It should be remembered that the assumptions $p_{11} = -p$, $q_1 = 0$ are introduced only *after* the appropriate limits have been taken.) Thus the jump condition corresponding formally to (6.91) cannot be derived. In fact, we shall show later from (6.87)–(6.89) that

$$- U[\rho S] + [\rho u S] \neq 0. \tag{6.94}$$

In the discussion of shock structure the actual contribution of the right hand side of (6.93) will be studied in more detail.

It is interesting that the four equations in (6.90) and (6.91) are the only conservation equations that can be found for the set (6.60)–(6.62). To prove this result consider

$$\frac{\partial f}{\partial t} + \frac{\partial g}{\partial x} + h = 0,$$

where f, g, h are functions of p, ρ, u. If this equation is expanded in terms of derivatives of p, ρ, u, and if (6.60), (6.61), and (6.63) are used to replace the t derivatives, we have

$$(g_\rho - u f_\rho) \rho_x + \left(g_p - u f_p - \frac{1}{\rho} f_u\right) p_x + \left(g_u - u f_u - \rho f_\rho - \rho a^2 f_p\right) u_x + h = 0.$$

Since this is to be an identity, the coefficients of the derivatives must vanish separately and h must be equal to zero. The three equations for f and g can be solved to show that the most general solution for f is a linear combination of ρ, ρu, $\frac{1}{2}\rho u^2 + \rho e$, ρS. Thus the only independent conservation equations are those already noted. Any three of the four can be used to generate a "weak solution," but only the choice (6.90) with jump conditions (6.87)–(6.89) corresponds to the real physical situation.

Useful Forms of the Shock Conditions.

First it is convenient to write the shock conditions (6.87)–(6.89) in terms of the relative velocities $v = U - u$. With this substitution for u, they become

$$[\rho v] = 0,$$

$$[p + \rho v^2 - \rho v U] = 0,$$

$$\left[\rho v \left(h + \frac{1}{2}v^2\right) - (p + \rho v^2) U + \frac{1}{2}\rho v U^2\right] = 0,$$

where h is the enthalpy $e + p/\rho$. By taking linear combinations, they may be reduced to

$$[\rho v] = 0, \qquad [p + \rho v^2] = 0, \qquad \left[\rho v\left(h + \frac{1}{2}v^2\right)\right] = 0.$$

These are the shock conditions for steady flow in a frame of reference in which the shock is at rest. If $\rho_1 v_1 = \rho_2 v_2 \neq 0$, the constant factor ρv may be dropped in the third condition and we have

$$\rho_2 v_2 = \rho_1 v_1, \tag{6.95}$$

$$p_2 + \rho_2 v_2^2 = p_1 + \rho_1 v_1^2, \tag{6.96}$$

$$h_2 + \frac{1}{2}v_2^2 = h_1 + \frac{1}{2}v_1^2. \tag{6.97}$$

The usual situation is that the flow ahead of the shock is known and the shock conditions are used either to determine the flow behind in terms of the shock velocity or to determine the shock velocity and the remaining flow quantities in terms of one of the flow quantities behind. We note explicit formulas in the case of a polytropic gas. It will be useful to include the expressions for the sound speed even though they follow from those for p and ρ. For a polytropic gas

$$e = \frac{1}{\gamma - 1}\frac{p}{\rho}, \qquad h = \frac{\gamma}{\gamma - 1}\frac{p}{\rho}, \qquad a^2 = \gamma\frac{p}{\rho}, \tag{6.98}$$

and the required formulas are derived by straightforward manipulation of (6.95)–(6.97).

When the flow quantities are expressed in terms of U it is convenient to use the parameter

$$M = \frac{U - u_1}{a_1},$$

which is the Mach number of the shock relative to the flow ahead. Then

$$\frac{u_2 - u_1}{a_1} = \frac{2(M^2 - 1)}{(\gamma + 1)M}, \tag{6.99}$$

$$\frac{\rho_2}{\rho_1} = \frac{(\gamma + 1)M^2}{(\gamma - 1)M^2 + 2}, \tag{6.100}$$

$$\frac{p_2 - p_1}{p_1} = \frac{2\gamma(M^2 - 1)}{\gamma + 1},$$ (6.101)

$$\frac{a_2}{a_1} = \frac{\{2\gamma M^2 - (\gamma - 1)\}^{1/2}\{(\gamma - 1)M^2 + 2\}^{1/2}}{(\gamma + 1)M}.$$ (6.102)

When p_2 is taken as known, it is convenient to introduce the shock strength

$$z = \frac{p_2 - p_1}{p_1}$$

and solve the shock relations in the form

$$M = \frac{U - u_1}{a_1} = \left(1 + \frac{\gamma + 1}{2\gamma}z\right)^{1/2}$$ (6.103)

$$\frac{u_2 - u_1}{a_1} = \frac{z}{\gamma\left(1 + \frac{\gamma + 1}{2\gamma}z\right)^{1/2}}$$ (6.104)

$$\frac{\rho_2}{\rho_1} = \frac{1 + \frac{\gamma + 1}{2\gamma}z}{1 + \frac{\gamma - 1}{2\gamma}z},$$ (6.105)

$$\frac{a_2}{a_1} = \left\{\frac{(1 + z)\left(1 + \frac{\gamma - 1}{2\gamma}z\right)}{1 + \frac{\gamma + 1}{2\gamma}z}\right\}^{1/2}$$ (6.106)

Properties of Shocks.

Certain important properties of shocks will be noted from these formulas for polytropic gases, but the qualitative results are true in general. First we check the condition (6.94). For a polytropic gas, $S = c_v \log p / \rho^\gamma$;

hence from (6.105) and the definition of z, we have

$$\frac{S_2 - S_1}{c_v} = \log \frac{(1+z)\left(1 + \frac{\gamma - 1}{2\gamma} z\right)^{\gamma}}{\left(1 + \frac{\gamma + 1}{2\gamma} z\right)^{\gamma}}. \tag{6.107}$$

This is not zero when $z \neq 0$; the entropy does indeed jump at the shock.

According to the second law of thermodynamics, the entropy can only increase following a particle. Therefore if a particle passes from side 1 to side 2, we require $S_2 > S_1$. It may be shown from (6.107) that $d(S_2 - S_1)/dz$ > 0 for $\gamma > 1$, $z > -1$, and these always hold. Hence $S_2 - S_1 > 0$ implies $z > 0$. Thus a shock should always be compressive with $p_2 > p_1$, and it then follows from the other relations in (6.103)–(6.106) that

$$p_2 > p_1, \qquad \rho_2 > \rho_1, \qquad a_2 > a_1, \qquad u_2 > u_1, \qquad M > 1. \tag{6.108}$$

Another approach to this question of the sign of the jumps is to ask when shock waves are required by the breaking of waves. In the simple waves of Section 6.8, the propagation speed is $u + a$, so that the shock formation condition discussed in Section 2.6 becomes

$$u_2 + a_2 > U > u_1 + a_1. \tag{6.109}$$

That is, the shock is supersonic viewed from ahead and subsonic viewed from behind. From (6.103), it is clear that $U > u_1 + a_1$ requires $z > 0$, and the other inequality then follows from (6.103), (6.104), and (6.106).

The jump in the entropy at a shock, (6.107), depends on the strength of the shock. As a consequence, the flow behind a shock of changing strength cannot be isentropic. This has an important bearing on the simple wave argument in Section 6.8. At the same time, we should consider the possibility of a jump in the Riemann invariant,

$$s = \frac{2a}{\gamma - 1} - u,$$

which was used in (6.73). It turns out that there is a jump in this quantity at the shock. On both counts, the simple wave solution would not be valid, strictly speaking, when shock waves occur. However, the jumps are surprisingly small when the shock strength is small or even moderate.

Weak Shocks.

For weak shocks, all the expressions in (6.103)–(6.107) may be expanded in powers of z. The first few terms are

$$\frac{U - u_1}{a_1} = 1 + \frac{\gamma + 1}{4\gamma} z - \frac{(\gamma + 1)^2}{32\gamma^2} z^2 + O(z^3),$$

$$\frac{u_2 - u_1}{a_1} = \frac{z}{\gamma} - \frac{\gamma + 1}{4\gamma^2} z^2 + \frac{3(\gamma + 1)^2}{32\gamma^3} z^3 + O(z^4),$$

$$\frac{\rho_2 - \rho_1}{\rho_1} = \frac{z}{\gamma} - \frac{\gamma - 1}{2\gamma^2} z^2 + O(z^3),$$

$$\frac{a_2 - a_1}{a_1} = \frac{\gamma - 1}{2\gamma} z - \frac{\gamma^2 - 1}{8\gamma^2} z^2 + \frac{(\gamma - 1)(\gamma + 1)^2}{16\gamma^3} z^3 + O(z^4),$$

$$\frac{S_2 - S_1}{c_v} = \frac{\gamma^2 - 1}{12\gamma^2} z^3 + O(z^4),$$

$$\frac{s_2 - s_1}{a_1} = \frac{2}{\gamma - 1} \frac{a_2 - a_1}{a_1} - \frac{u_2 - u_1}{a_1} = \frac{1}{32} \frac{(\gamma + 1)^2}{\gamma^3} z^3 + O(z^4).$$

In general the jumps in flow quantities are proportional to z, but the jumps in S and s are only $O(z^3)$. Even for moderate shocks they remain surprisingly small; typical values for $\gamma = 1.4$ are given below.

z	$\dfrac{S_2 - S_1}{c_v}$	$\dfrac{\gamma - 1}{2} \cdot \dfrac{s_2 - s_1}{a_1}$
0.5	0.003	0.001
1.0	0.013	0.005
2.0	0.052	0.019
5.0	0.215	0.085
10.0	0.478	0.209

For shocks of weak or moderate strength, it is a reasonable approximation to neglect changes in the entropy and the Riemann invariant. With these approximations, the simple wave solution of Section 6.8 can be retained and used even when weak shocks are included.

Strong Shocks.

At the other extreme of very strong shocks, the asymptotic behavior as $z \to \infty$ may be used. The most convenient form is obtained from (6.99)–(6.101). Usually when these are required $U \gg u_1$, so we assume that $M \sim U/a_1$ and $M \gg 1$. Then (6.99)–(6.101) may be approximated by

$$u_2 \sim \frac{2}{\gamma + 1} U, \qquad \frac{p_2}{p_1} \sim \frac{\gamma + 1}{\gamma - 1}, \qquad p_2 \sim \frac{2}{\gamma + 1} \rho_1 U^2, \qquad (6.110)$$

where we have used $a_1^2 = \gamma p_1 / \rho_1$ to eliminate both p_1 and a_1. The expressions involve only the parameter ρ_1 from the flow ahead; u_1, p_1, a_1 are now small compared with values behind the shock and so are negligible in this approximation. It is interesting that there is only a finite compression ρ_2 / ρ_1, however strong the shock may be.

6.11 Weak Shocks in Simple Waves

As noted previously, it will be a good approximation when the only shocks occurring are weak or of moderate strength to retain the relations

$$S = S_0, \qquad a = a_0 + \frac{\gamma - 1}{2} u \qquad (6.111)$$

in a simple wave. Then, as we saw in (6.83), the remaining equation may be written

$$u_t + c(u) u_x = 0, \qquad c = a + u = a_0 + \frac{\gamma + 1}{2} u. \qquad (6.112)$$

In addition to eliminating two equations, the relations in (6.111) already satisfy two of the shock conditions approximately and there is just one left to accompany (6.112). Since this treatment is approximate, there are many choices that agree to lowest order. As we saw in Chapter 2, the most convenient one to use with an equation like (6.112) is

$$U = \frac{c_1 + c_2}{2}, \qquad (6.113)$$

which reduces to

$$U = \frac{1}{2}(a_1 + u_1 + a_2 + u_2) = a_0 + \frac{\gamma + 1}{4}(u_1 + u_2) \qquad (6.114)$$

in this case. From the weak shock conditions,

$$\frac{U-u_1}{a_1} - 1 = \frac{\gamma+1}{4\gamma} z - \frac{(\gamma+1)^2}{32\gamma^2} z^2 + O(z^3),$$

$$\frac{1}{2}\left(\frac{a_2-a_1}{a_1} + \frac{u_2-u_1}{a_1}\right) = \frac{\gamma+1}{4\gamma} z - \frac{(\gamma+1)^2}{16\gamma^2} z^2 + O(z^3),$$

so that (6.114) is true to first order in z, but not to second order. This assumption is not as accurate as (6.111). One could relate U to u_1 and u_2 by an expression that is accurate to second order in z, but the extra complication in the shock fitting is usually not worth it.

The problem of shock fitting is essentially the same as in Chapter 2 and, with the simple form (6.113), follows closely the account in Section 2.8.

We could still consider the initial value problem for (6.112) but this is only a special case of the full initial value problem in gas dynamics; it applies only if the prescribed values satisfy (6.111) so that the flow produced is a simple wave. It is more natural here to consider again the piston problem with

$$u = g(t) = \dot{X}(t) \qquad \text{on} \quad x = X(t),$$

and complete the discussion of Sections 6.8 and 6.9 in the case that shocks are formed.* The solution is

$$u = g(\tau),\tag{6.115}$$

$$x = X(\tau) + \left\{a_0 + \frac{\gamma+1}{2} g(\tau)\right\}(t-\tau),\tag{6.116}$$

where the characteristic variable τ is chosen so that $t=\tau$ on the piston path. Shocks will be required on characteristics with $\dot{g}(\tau)>0$. We could relate the problem to the initial value problem by extending the characteristics back to the x axis as in Fig. 6.3 and taking

$$\xi = X(\tau) - \left\{a_0 + \frac{\gamma+1}{2} g(\tau)\right\}\tau, \qquad F(\xi) = a_0 + \frac{\gamma+1}{2} g(\tau).$$

Then the results of Section 2.8 could be applied. It is perhaps clearer, however, to proceed directly.

---

*The formulas will be given in a form that covers also the more general case in which the prescribed $g(t)$ is not equal to $\dot{X}(t)$.

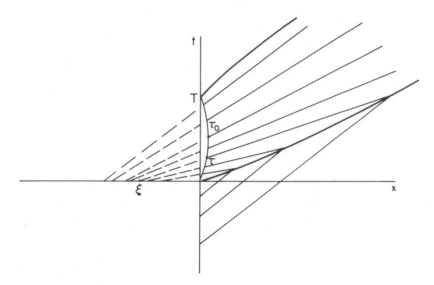

Fig. 6.3. Shocks produced by a piston motion.

We treat the case of a front shock propagating into an undisturbed region with $u = 0$ ahead (Fig. 6.3). If the shock path is $x = s(t)$, the shock condition (6.114) gives

$$\frac{ds}{dt} = a_0 + \frac{\gamma+1}{4} g(\tau),$$ (6.117)

and the characteristic equation (6.116) gives

$$s(t) = X(\tau) + \left\{ a_0 + \frac{\gamma+1}{2} g(\tau) \right\} (t - \tau).$$ (6.118)

These two equations are solved to find s and t as functions of the parameter τ and thus determine the shock. Since $\dot{X}(\tau)/a_0 \ll 1$, $X(\tau)/a_0 \ll \tau$, and $t \gg \tau$ at the shock, it is sufficient in the shock calculation to approximate (6.118) by

$$s(t) = \left\{ a_0 + \frac{\gamma+1}{2} g(\tau) \right\} t - a_0 \tau.$$ (6.119)

Then we have

$$\frac{ds}{dt} = a_0 + \frac{\gamma+1}{2} g(\tau) + \left\{ \frac{\gamma+1}{2} \dot{g}(\tau) t - a_0 \right\} \frac{d\tau}{dt}.$$

Comparing this with (6.117), we have

$$\frac{\gamma+1}{4a_0}g(\tau)+\frac{\gamma+1}{2a_0}\dot{g}(\tau)t\frac{d\tau}{dt} = \frac{d\tau}{dt}.$$

This may be integrated to

$$\frac{\gamma+1}{4a_0}g^2(\tau)t = \int_0^\tau g(\tau')d\tau'. \tag{6.120}$$

Relations (6.119) and (6.120) determine the shock.

If $g(0)>0$, the shock starts immediately at the origin. The relation between τ and t is

$$\frac{\gamma+1}{4a_0}g(0)t\sim\tau,$$

so that

$$s\sim\left\{a_0+\frac{\gamma+1}{4}g(0)\right\}t.$$

The shock starts with velocity $a_0+\{(\gamma+1)/4\}\,g(0)$; this checks with (6.114).

If the piston comes to rest so that $g(\tau)\to0$ as $\tau\to\tau_0$, the asymptotic behavior is

$$\frac{\gamma+1}{4a_0}g^2(\tau)t\sim\int_0^{\tau_0}g(\tau)d\tau$$

$$s\sim a_0t+\left\{(\gamma+1)a_0\int_0^{\tau_0}g(\tau)d\tau\right\}^{1/2}t^{1/2}-a_0\tau_0,$$

$$u=g(\tau)\sim\left\{\frac{4a_0}{\gamma+1}\int_0^{\tau_0}g(\tau)d\tau\right\}^{1/2}t^{-1/2}.$$

The shock path is roughly parabolic in the (x,t) diagram and the shock strength decays like $t^{-1/2}$. Between the shock and the limiting characteristic $\tau=\tau_0$, we have

$$x\sim\left\{a_0+\frac{\gamma+1}{2}g(\tau)\right\}t-a_0\tau_0,$$

so that

$$u = g(\tau) \sim \frac{2}{\gamma+1} \frac{x - a_0(t - \tau_0)}{t}.$$

Further shocks can be handled similarly. If the piston returns to its original position at $t = T$ and remains there, the asymptotic behavior is a balanced N wave with shocks at

$$x = a_0(t - \tau_0) \pm \left\{ (\gamma+1) a_0 \int_0^{\tau_0} g(\tau) d\tau \right\}^{1/2} t^{1/2},$$

and

$$u \sim \frac{2}{\gamma+1} \frac{x - a_0(t - \tau_0)}{t}$$

between them.

Other special cases and limiting results can be derived following the arguments in Section 2.8.

6.12 Initial Value Problem; Wave Interaction

The simple wave is a disturbance propagating on one family of characteristics. In general there will be waves on all three families. The complete initial value problem will require all three, unless the values prescribed already satisfy the simple wave relations. If the initial values are uniform with $u = 0$, $a = a_0$, $S = S_0$ except in the range $a \leqslant x \leqslant b$, the (x, t) diagram is as shown in Fig. 6.4. There is an interaction region $adceb$, but then, provided shocks do not occur, the disturbance separates into three simple waves as shown. Their existence is established by noting that in each of them two of the characteristics originate in the initial uniform region.

Between the simple waves, all three characteristics originate in one or the other of the uniform regions; hence they are uniform states with $u = 0$, $a = a_0$, $S = S_0$. Once the interaction is solved, the simple waves can be described analytically (similar to Section 6.8) with boundary conditions provided by the interaction region.

If the entropy is initially uniform everywhere, the flow is isentropic. The P wave in Fig. 6.4 drops out and the interaction region is limited to abc.

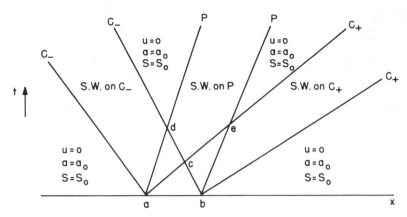

Fig. 6.4. Initial value problem in gas dynamics.

　　In practice, the interaction region is best solved numerically. However, some analytic simplification can be made in the isentropic case, and a related problem was solved completely by Riemann. For isentropic flow, a may be used in place of ρ and the equations may be written

$$a_t + ua_x + \frac{\gamma-1}{2} au_x = 0,$$

$$u_t + uu_x + \frac{2}{\gamma-1} aa_x = 0. \tag{6.121}$$

Equations such as these can be transformed into a linear set of equations by interchanging the roles of the dependent and independent variables: the so-called "hodograph" transformation. In this transformation,

$$x_u = \frac{-a_t}{J}, \qquad x_a = \frac{u_t}{J}, \qquad t_u = \frac{a_x}{J}, \qquad t_a = -\frac{u_x}{J},$$

where J is the Jacobian:

$$J = u_t a_x - a_t u_x.$$

Since the equations (6.121) have one and only one derivative in each term, the Jacobian cancels through and the equations become

$$x_u - ut_u + \frac{\gamma-1}{2} at_a = 0,$$

$$x_a - ut_a + \frac{2}{\gamma-1} at_u = 0.$$

These are now linear equations for $x(u,a)$ and $t(u,a)$. Notice it is crucial that J cancel through; otherwise the equations would still be highly nonlinear.

A single equation for $t(u,a)$ may be obtained if x is eliminated by cross-differentiation. Then we have

$$\left(\frac{\gamma-1}{2}\right)^2 t_{aa} + \frac{\gamma^2-1}{4}\frac{1}{a}t_a = t_{uu}.$$

With $b = 2a/(\gamma-1)$, $n = (\gamma+1)/2(\gamma-1)$, the equation becomes

$$t_{bb} + \frac{2n}{b}t_b = t_{uu}.$$

For $n=1$ this is the wave equation for spherical symmetry, which has a relatively simple general solution. Indeed, when n is any integer, the general solution is not too difficult and may be written

$$t = \left(\frac{\partial^2}{\partial u^2} - \frac{\partial^2}{\partial b^2}\right)^{n-1}\left\{\frac{F(u+b) + G(u-b)}{b}\right\}, \qquad (6.122)$$

where F and G are arbitrary functions. It is fortunate that the two interesting cases of $\gamma = \frac{5}{3}$ and $\gamma = \frac{7}{5}$ are included as the cases $n=2$ and $n=3$, respectively. The Riemann invariants are $u \pm b$, so the basic characteristic structure is apparent in this expression.

In the linearized approximation to (6.121), that is,

$$a_t + \frac{\gamma-1}{2}a_0 u_x = 0, \qquad u_t + \frac{2}{\gamma-1}a_0 a_x = 0,$$

the general solution is

$$u = f(x - a_0 t) + g(x + a_0 t),$$

$$b - b_0 = \frac{2}{\gamma-1}(a - a_0) = f(x - a_0 t) - g(x + a_0 t).$$

The nonlinear solution takes a form relating to the inverse of this.

For the initial value problem, a and u are prescribed functions

$$a = \mathcal{Q}(x), \qquad u = \mathcal{U}(x) \qquad \text{at } t = 0.$$

In principle, by elimination of x, this defines a certain curve $\mathcal{I}$ in the (b,u) plane, and we have $t = 0$, x given, on $\mathcal{I}$. These two boundary conditions are sufficient to determine the solution in the appropriate characteristic triangle. In practice, however, it is simpler to use a numerical solution in the (x, t) plane, as noted previously.

Simple waves have either $u + b$ or $u - b$ constant, so the mapping into the (u, b) plane is singular; the entire simple wave region in the (x, t) plane is mapped into a line in the (u, b) plane. The interaction of two simple waves is then formulated in the (u, b) plane as requiring the solution with t prescribed on characteristics

$$u + b = b_1, \qquad u - b = b_2.$$

The determination of F, G is easier in this case. The solution can be found also for general γ; the details are given in Courant and Friedrichs (1948). This analysis is limited to shock-free flows and seems to be mainly of academic interest. In view of this, the details are not documented here.

6.13 Shock Tube Problem

One special initial value problem, which includes a shock, can be solved exactly and very simply. It is also important since it refers to the main device for producing shocks in experimental studies. A shock tube is a long tube with an end section partitioned off by a thin diaphragm. The gas held behind the diaphragm is pumped up to a high pressure. The initial state is two uniform regions with

$$u = 0, \qquad p = p_1, \qquad \rho = \rho_1 \qquad \text{in } x > 0$$

and

$$u = 0, \qquad p = p_4 > p_1, \qquad \rho = \rho_4 \qquad \text{in } x < 0.$$

The diaphragm separating the two uniform initial states is burst to produce a shock propagating down the tube. If the viscous effect of the side walls of the tube are ignored so that this may be treated as a plane wave problem, and if the solution is limited to times before the waves are reflected from the ends of the tube, the exact solution may be obtained analytically.

The (x, t) diagram is shown in Fig. 6.5. The interface separating the two gases moves down the tube, there is a compressive shock moving into the gas on the low pressure side, and an expansion wave moves into the

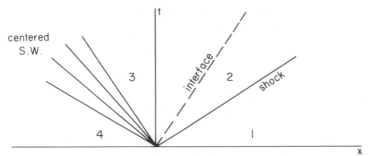

Fig. 6.5. The (x,t) diagram for a shock tube.

high pressure gas. Since the parameters provided by the initial conditions do not specify a basic length or time, and the lengths of the tube sections are irrelevant in the early stages before the disturbances reach the ends, dimensional arguments show that the solution must be constant on the lines $x/a_1 t = $ constant in the (x,t) plane. Therefore the velocities of the shock and the interface must be constant, and the expansion wave must be a fan centered at the origin. There are uniform regions 1, 2, 3, 4 as shown; regions 1 and 4 are still in the original uniform states. The problem may be viewed as the combination of two piston problems, using the interface as an effective piston. The fluid velocities on the two sides of the interface must be the same as the velocity of the interface, so it is like a solid wall as far as the flow on either side is concerned. However, its motion must be determined as part of the problem and cannot be prescribed in advance. If the velocity of the interface is V, the shock conditions (6.103)–(6.105) may be used with $u_2 = V$, $u_1 = 0$, to determine p_2, ρ_2, U in terms of V. In particular, the shock strength $z = (p_2 - p_1)/p_1$ is determined from (6.104):

$$\frac{V}{a_1} = \frac{z}{\gamma_1 \left(1 + \dfrac{\gamma_1 + 1}{2\gamma_1} z\right)^{1/2}}. \qquad (6.123)$$

The gases on the two sides of the diaphragm may have different values of γ. We use γ_1 for the gas ahead and γ_4 for the gas behind. The expansion wave between regions 4 and 3 is a simple wave with

$$S_3 = S_4, \qquad \frac{2a_3}{\gamma_4 - 1} + V = \frac{2a_4}{\gamma_4 - 1}.$$

For a polytropic gas, $S = c_v \log p/\rho^{\gamma_4}$, $a^2 = \gamma_4 p/\rho$, so that these two relations can be solved to find also p_3, ρ_3. In particular, p_3 is determined from

$$\frac{V}{a_1} = \frac{2}{\gamma_4 - 1} \frac{a_4}{a_1} \left\{ 1 - \left(\frac{p_3}{p_4} \right)^{(\gamma_4 - 1)/2\gamma_4} \right\}. \qquad (6.124)$$

If required, the details of the flow in the centered simple wave can be calculated; the solution is similar to (6.79) but on the other family of characteristics.

At this stage, the solution is completely determined in terms of V. The final relation is to enforce the condition $p_2 = p_3$, since the interface has no mass and cannot sustain a net force. These pressures are determined from (6.123) and (6.124), and they provide the equation for V. The result of most interest, however, is the shock strength z. If we substitute $p_3 = p_2 = p_1(1 + z)$ in (6.124) and equate the two expressions for V/a_1, we have

$$\frac{z}{\gamma_1 \left(1 + \dfrac{\gamma_1 + 1}{2\gamma_1} z \right)^{1/2}} = \frac{2}{\gamma_4 - 1} \frac{a_4}{a_1} \left\{ 1 - \left[\frac{p_1}{p_4} (1 + z) \right]^{(\gamma_4 - 1)/2\gamma_4} \right\}.$$

This is the equation for z in terms of the known quantities p_4/p_1, a_4/a_1.

6.14 Shock Reflection

The reflection of the shock from the end wall can also be solved exactly. The normal reflection of a plane shock from a plane wall can be analyzed from the shock conditions. Let subscripts 1 and 2 refer to the states ahead of and behind the incident shock, and subscript 3 refer to the state behind the reflected shock. If the incident shock strength is denoted by $z_i = (p_2 - p_1)/p_1$, the state 2 is determined by (6.104)–(6.106) with $z = z_i$. The reflected shock has state 2 ahead and state 3 behind, so that if the reflected shock strength is $z_r = (p_3 - p_2)/p_2$, then from (6.104), with suitable changes in subscripts and change in sign of the velocities because the reflected shock travels in the opposite direction, we have

$$\frac{u_2 - u_3}{a_2} = \frac{z_r}{\gamma \left(1 + \dfrac{\gamma + 1}{2\gamma} z_r \right)^{1/2}}.$$

Next to the wall the gas must be at rest; hence $u_3 = 0$. But we also know u_2

and a_2 in terms of z_i, so this is a relation for z_r in terms of z_i:

$$\frac{z_i}{\gamma\left\{(1+z_i)\left(1+\dfrac{\gamma-1}{2\gamma}z_i\right)\right\}^{1/2}} = \frac{z_r}{\gamma\left(1+\dfrac{\gamma+1}{2\gamma}z_r\right)^{1/2}}.$$

This leads to a quadratic for z_r and the relevant solution is

$$z_r = \frac{z_i}{1+\dfrac{\gamma-1}{2\gamma}z_i}. \tag{6.125}$$

For weak shocks, $z_i \to 0$ and (6.125) is approximately $z_r \sim z_i$; hence

$$p_3 - p_1 \sim 2(p_2 - p_1),$$

and the pressure increase at the wall is approximately twice that in the incident shock. For strong shocks, $z_i \to \infty$, we have

$$z_r \sim \frac{2\gamma}{\gamma-1}, \qquad \frac{p_3}{p_2} \sim \frac{3\gamma-1}{\gamma-1} = 8 \qquad \text{for } \gamma = 1.4.$$

6.15 Shock Structure

In accordance with the general point of view that evolved in Chapter 2, a shock is interpreted as a thin region in which rapid changes of the flow quantities occur. It is a discontinuity in one level of description, but it is replaced by a thin region in a more accurate level of description. As a check on this, particularly regarding the correct choice of conserved quantities, and in order to estimate the shock thickness if necessary, we consider the shock structure in the special case of a transition between two uniform states.

For one dimensional flow, the equations for conservation of mass, momentum, and energy are

$$\rho_t + (\rho u)_x = 0,$$

$$(\rho u)_t + (\rho u^2 - p_{11})_x = 0, \tag{6.126}$$

$$\left(\frac{1}{2}\rho u^2 + \rho e\right)_t + \left\{\left(\frac{1}{2}u^2 + e\right)\rho u - p_{11}u + q_1\right\}_x = 0.$$

An improved description over the one used so far is to take the Navier-Stokes relations for the stress p_{11} and the heat flux q_1, while retaining the assumption of local thermodynamic equilibrium. The assumptions are that p_{11} depends linearly on the velocity gradient and q_1 depends linearly on the temperature gradient. They are noted in general in equations (6.28) and (6.29); for one dimensional flow they reduce to

$$p_{11} = -p + \frac{4}{3}\mu u_x, \qquad q_1 = -\lambda T_x. \tag{6.127}$$

The thermodynamic relations

$$e = \frac{1}{\gamma - 1}\frac{p}{\rho}, \qquad p = \mathcal{R}\rho T \tag{6.128}$$

complete the system for a polytropic gas.

For the shock structure solution the flow is steady relative to the shock. Hence all flow quantities are functions of $X = x - Ut$ alone. For such functions

$$\frac{\partial}{\partial t} = -U\frac{d}{dX}, \qquad \frac{\partial}{\partial x} = \frac{d}{dX}$$

and the equations (6.126), being in conservation form, may be integrated to

$$-U\rho + \rho u = A,$$

$$-U(\rho u) + \left(\rho u^2 + p - \frac{4}{3}\mu u_x\right) = B, \tag{6.129}$$

$$-U\left(\frac{1}{2}\rho u^2 + \rho e\right) + \left\{\left(\frac{1}{2}u^2 + e\right)\rho u + pu - \frac{4}{3}\mu u u_x - \lambda T_x\right\} = C,$$

where A, B, and C are constants of integration. As $X \to +\infty$, the flow tends to a uniform state denoted by subscript 1. The constants A, B, and C are determined then in terms of U, u_1, ρ_1, p_1. If the flow also tends to a uniform state u_2, ρ_2, p_2 as $X \to -\infty$, it is clear that the states at $\pm\infty$ are related by the shock conditions (6.87)–(6.89).

The relations (6.127) may also be used to explore further the equations for the entropy change. We may now write (6.92) explicit as

$$\frac{d}{dX}\{\rho(U - u)S\} = -\frac{(4/3)\mu u_x^2 + (\lambda T_x)_x}{T},$$

or, better still,

$$\frac{d}{dX}\left\{\rho(U-u)S+\frac{\lambda T_X}{T}\right\}=-\frac{(4/3)\mu Tu_X^2+\lambda T_X^2}{T^2}.$$

Hence

$$[\rho(U-u)S]_1^2=\int_{-\infty}^{\infty}\frac{(4/3)\mu Tu_X^2+\lambda T_X^2}{T^2}dX>0.$$

The entropy change across the shock is clearly seen to be a consequence of the dissipation of energy by viscosity and heat conduction, and the sign of the entropy change is automatically predicted.

The details of the shock profile between the limiting values at $\pm\infty$ are provided by the ordinary differential equations in (6.129). With $v=U-u$ and new constants related to A, B, and C, they may be put in the form

$$\rho v=Q,$$

$$\rho v^2+p+\frac{4}{3}\mu v_X=P,$$

$$\left(h+\frac{1}{2}v^2\right)\rho v+\frac{4}{3}\mu vv_X+\lambda T_X=E.$$

These are the steady flow equations in a frame of reference moving with the shock, but with v measured positive in the negative X direction. The shock conditions relating the uniform states at $\pm\infty$ are now in the corresponding form (6.95)–(6.97).

The continuity equation $\rho v=Q$ and the relations in (6.128) may be used to reduce the system to two equations for v and T. For a polytropic gas,

$$h=\frac{\gamma}{\gamma-1}\mathcal{R}T=c_pT,$$

and it is a little more convenient to work with v and h. The equations are

$$\frac{4}{3}\mu v_X=P-Q\left(v+\frac{\gamma-1}{\gamma}\frac{h}{v}\right), \qquad (6.130)$$

$$\frac{\lambda}{c_p}h_X+\frac{4}{3}\mu vv_X=E-Q\left(h+\frac{1}{2}v^2\right). \qquad (6.131)$$

It is a simple matter now to show qualitatively that a solution of the required form exists. In the special case $\lambda/c_p = \frac{4}{3}\mu$, which is a good approximation for air, an integral can be found and the solution exhibited explicitly. (The quantity $\mu c_p/\lambda$ is the Prandtl number, and it is 0.71 for air at normal temperatures.) For this value of λ/c_p, (6.131) can be written

$$\frac{4}{3}\mu\left(h + \frac{1}{2}v^2\right)_X = E - Q\left(h + \frac{1}{2}v^2\right).$$

The right hand side vanishes as $X \to \infty$, that is, $h_1 + \frac{1}{2}v_1^2 = E/Q$. Therefore the only solution that is bounded as $X \to -\infty$ is

$$h + \frac{1}{2}v^2 = \frac{E}{Q}$$

throughout. In this case, $h + \frac{1}{2}v^2$ is not only the same on the two sides of the shock; it remains constant throughout the shock. Equation 6.130 may then be written

$$\frac{4}{3}\mu v_X = P - Q\left(\frac{\gamma+1}{2\gamma}v + \frac{\gamma-1}{\gamma}\frac{E}{Q}\frac{1}{v}\right).$$

Since the constants *must* be such that the right hand side vanishes for both $v = v_1$ and $v = v_2$, it must be possible to write it

$$\frac{4}{3}\mu v_X = \frac{\gamma+1}{2\gamma}Q\frac{(v_1-v)(v-v_2)}{v}.$$

This is easily integrated to give

$$\frac{3Q}{4\mu}\frac{\gamma+1}{2\gamma}X = \frac{v_2}{v_1-v_2}\log(v-v_2) - \frac{v_1}{v_1-v_2}\log(v_1-v).$$

We have $Q = \rho_1 v_1$, so the shock thickness is proportional to

$$\frac{4\mu}{3\rho_1}\frac{2\gamma}{\gamma+1}\frac{1}{v_1-v_2}.$$

As expected it becomes thinner as μ decreases for fixed strength and also as the strength increases for fixed μ.

6.16 Similarity Solutions

The simple wave solution is limited to plane waves moving into a uniform region. The problems of cylindrical or spherical symmetry and of plane waves moving into a nonuniform region are more complicated. A fairly general approximate theory for weak waves can be obtained (and this is developed in Chapter 9), but there are also some special exact solutions, which fit more conveniently into the account in this chapter.

We consider first cylindrical or spherical wave motion. The equations (6.49) reduce to

$$\rho_t + u\rho_r + \rho\left(u_r + \frac{ju}{r}\right) = 0, \tag{6.132}$$

$$u_t + uu_r + \frac{1}{\rho}p_r = 0, \tag{6.133}$$

$$p_t + up_r - a^2(\rho_t + u\rho_r) = 0, \tag{6.134}$$

where r is the distance from the center and $j = 1, 2$ for cylindrical and spherical waves, respectively.

The characteristic equations are nearly the same as in Section 6.7; the extra term $j\rho u/r$ does not involve derivatives and therefore does not change the appropriate choice of the linear combinations. The characteristic equations become

$$\frac{dp}{dt} \pm \rho a\frac{du}{dt} + j\frac{\rho a^2 u}{r} = 0 \qquad \text{on } \frac{dr}{dt} = u \pm a, \tag{6.135}$$

$$\frac{dp}{dt} - a^2\frac{d\rho}{dt} = 0 \qquad \text{on } \frac{dr}{dt} = u. \tag{6.136}$$

The innocuous-looking extra term in (6.135) invalidates the simple wave argument. In isentropic flow, the C_- characteristic equation becomes

$$\frac{d}{dt}\left(\frac{2}{\gamma-1}a - u\right) + j\frac{au}{r} = 0, \qquad \frac{dr}{dt} = u - a. \tag{6.137}$$

This can no longer be integrated once and for all to give the simple relation between a and u. As a consequence there are no exact solutions corresponding to the simple waves of plane flow. One can proceed by means of certain approximate methods to get analogous solutions, but they are limited to weak disturbances. These approximate theories are taken up in Chapter 9.

However, turning to another approach, we can find a different class of exact solutions which are surprisingly useful. The set of equations (6.132)–(6.134) has special similarity solutions in which all the flow quantities take the form $t^m f(r/t^n)$. These have the simplifying feature that the partial differential equations reduce to ordinary differential equations with independent variable r/t^n.

Point Blast Explosion.

One of the best known similarity solutions describes the blast wave produced by an intense explosion. It was found independently by Taylor, Sedov, and von Neumann in connection with atomic bomb research. Its form can be argued on dimensional grounds. It is supposed first that the explosion can be idealized as the sudden release of an amount of energy E concentrated at a point, and that this is the only dimensional parameter introduced by the explosion. Second it is supposed that the resulting disturbance will be so strong that the initial pressure and sound speed of the ambient air are negligible compared with the pressures and velocities produced in the disturbed flow. Then the only dimensional parameter relating to the ambient gas is the density ρ_0. In particular, the strong shock relations (6.110) apply; that is,

$$u = \frac{2}{\gamma+1} U, \qquad \rho = \frac{\gamma+1}{\gamma-1} \rho_0, \qquad p = \frac{2}{\gamma+1} \rho_0 U^2, \qquad (6.138)$$

behind the shock moving with velocity U.

The dimensional argument is based on the fact that the only parameters in the problem are E, with dimensions ML^2/T^2, and ρ_0, with dimensions M/L^3. The only parameter involving dimensions of length and time is E/ρ_0 with dimensions L^5/T^2 or some function of it. Let us now consider various quantities that arise in the solution. The flow is headed by a shock at $r = R(t)$. Since $R(t)$ is a length, the only possible form for its dependence on t is

$$R(t) = k \left(\frac{E}{\rho_0} \right)^{1/5} t^{2/5}, \qquad (6.139)$$

where k is a dimensionless number. It then follows from the shock conditions (6.138) that the pressure and velocity immediately behind the shock are

$$p = \frac{8}{25} \frac{k^2 \rho_0}{\gamma + 1} \left(\frac{E}{\rho_0} \right)^{2/5} t^{-6/5}, \qquad u = \frac{4}{5} \frac{k}{\gamma + 1} \left(\frac{E}{\rho_0} \right)^{1/5} t^{-3/5},$$

or, equivalently,

$$p = \frac{8}{25} \frac{k^5}{\gamma + 1} E R^{-3}, \qquad u = \frac{4}{5} \frac{k^{5/2}}{\gamma + 1} \left(\frac{E}{\rho_0} \right)^{1/2} R^{-3/2}. \qquad (6.140)$$

It is striking, as usual, that such valuable information can be obtained at the outset from a simple dimensional argument.

The argument can be taken further to show the functional form of u, ρ, p in the entire flow field. Since there are no independent length and time scales provided by the parameters in the problem, only the combination E/ρ_0 with dimensions L^5/T^2, any dimensionless function of r and t can depend only on the combination $\zeta = Et^2/\rho_0 r^5$ or some function of it. We shall in fact use

$$\xi = \frac{r}{R(t)},$$

which is proportional to $\zeta^{-1/5}$ from (6.139). Then, for example, ut/R, ρ/ρ_0, $pt^2/\rho_0 R^2$ are all dimensionless and must be functions of ξ alone. Following Taylor (1950) we take

$$u = \frac{2}{5} \frac{R}{t} \varphi(\xi), \qquad \rho = \rho_0 \psi(\xi), \qquad p = \left(\frac{2}{5} \frac{R}{t} \right)^2 \frac{\rho_0}{\gamma} f(\xi); \qquad (6.141)$$

the factor of $\frac{2}{5}$ is included because $2R/5t$ is the shock velocity. There are other equivalent forms and the choice

$$u = \frac{2}{5} \frac{r}{t} V(\xi), \qquad \rho = \rho_0 \Omega(\xi), \qquad p = \left(\frac{2}{5} \frac{r}{t} \right)^2 \rho_0 P(\xi)$$

fits a general scheme discussed below. Clearly the connection is

$$\varphi = \xi V, \qquad \psi = \Omega, \qquad f = \gamma \xi^2 P.$$

The shock is at $\xi = 1$ and the shock velocity U is $\dot{R} = 2R/5t$, so the shock conditions are

$$\varphi(1) = \frac{2}{\gamma+1}, \qquad \psi(1) = \frac{\gamma+1}{\gamma-1}, \qquad f(1) = \frac{2\gamma}{\gamma+1}. \qquad (6.142)$$

When the expressions (6.141) are substituted in the equations of motion, three first order *ordinary* differential equations are found for the functions $\varphi(\xi), \psi(\xi), f(\xi)$. These have to be integrated from $\xi = 1$ to $\xi = 0$, with initial conditions (6.142). The parameter k does not appear in the equations. It is fixed from the definition of E as the total energy in the flow. That is, we require

$$E = \int_0^{R(t)} \left(\frac{p}{\gamma-1} + \frac{1}{2}\rho u^2 \right) 4\pi r^2 \, dr,$$

which leads to

$$1 = 4\pi k^5 \left(\frac{2}{5} \right)^2 \int_0^1 \left\{ \frac{f}{\gamma(\gamma-1)} + \frac{1}{2}\psi\varphi^2 \right\} \xi^2 \, d\xi.$$

The functions φ, ψ, f in Fig. 6.6 are taken from Taylor's numerical integration of the equations in the case $\gamma = 1.4$. It was shown by Sedov, using different variables, that the equations could in fact be solved analytically; reference should be made to Sedov's book on similarity solutions. (1959, Chapter 4).

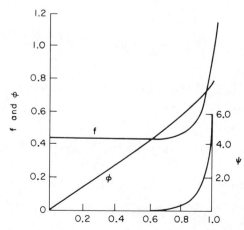

Fig. 6.6. The normalized velocity φ, density ψ, and pressure f in the point blast explosion (Taylor).

Similarity Equations.

The point blast similarity solution is one of a family in which

$$u = n\frac{r}{t} V(\xi), \qquad p = n^2 \rho_0 \frac{r^2}{t^2} P(\xi),$$

$$\rho = \rho_0 \Omega(\xi), \qquad \xi = \frac{r}{(Ct)^n}. \tag{6.143}$$

If these expressions are substituted in (6.132)–(6.134), and the forms

$$\frac{\partial}{\partial t} f(\xi) = \xi_t f'(\xi) = -n\frac{r}{t} f'(\xi),$$

$$\frac{\partial}{\partial r} f(\xi) = \xi_r f'(\xi) = \frac{\xi}{r} f'(\xi),$$

are substituted for derivatives of functions of ξ, the factors in r and t cancel through and leave ordinary differential equations in ξ alone. They are most conveniently written in terms of V, $A = (\gamma P / \Omega)^{1/2}$ and Ω; the sound speed is

$$a = n\frac{r}{t} A(\xi).$$

The equations are

$$\{(V-1)^2 - A^2\}\xi \frac{dV}{d\xi} = \left\{ (j+1)V - \frac{2(1-n)}{n\gamma} \right\} A^2 - V(V-1)\left(V - \frac{1}{n}\right), \tag{6.144}$$

$$\{(V-1)^2 - A^2\}\frac{\xi}{A}\frac{dA}{d\xi} = \left\{ 1 - \frac{1-n}{\gamma n}(V-1)^{-1} \right\} A^2 + \frac{\gamma-1}{2} V\left(V - \frac{1}{n}\right)$$

$$- \frac{\gamma-1}{2}(j+1)V(V-1) - (V-1)\left(V - \frac{1}{n}\right), \tag{6.145}$$

$$\{(V-1)^2 - A^2\}\frac{\xi}{\Omega}\frac{d\Omega}{d\xi} = 2\left\{ (j+1)U - \frac{1-n}{n\gamma} \right\}(V-1)^{-1}A^2$$

$$- V\left(V - \frac{1}{n}\right) - (j+1)V(V-1). \tag{6.146}$$

There is a possible singularity if

$$(V-1)^2 - A^2 = 0, \qquad (6.147)$$

and this would be a singularity on a curve $\xi = $ constant. The original equations are hyperbolic so we know that singularities can occur only on the characteristics

$$\frac{dr}{dt} = u \pm a = \frac{nr}{t}(V \pm A). \qquad (6.148)$$

On a curve $\xi = $ constant, we have

$$\frac{dr}{dt} = \frac{nr}{t}.$$

Therefore a curve $\xi = $ constant which is also a characteristic must have $V \pm A = 1$. This agrees with (6.147). Thus, a singularity can occur only on the characteristic through the origin, and this limiting characteristic is one of the family $\xi = $ constant.

In the point blast problem, $n = 2/5$, the limiting characteristic does not appear in the flow behind the shock. Its position would be in the region ahead of the shock, as shown in Fig. 6.7, but the flow is taken to be uniform there, with $V = A = 0$, and there is no singularity. It represents the edge of the fold in the (r, t) plane in the multivalued solution which is replaced by the shock.

Guderley's Implosion Problem.

The limiting characteristic plays a crucial role in the problem of an incoming spherical or cylindrical shock collapsing to the center. In this case there is no dimensional argument to establish that the solution must

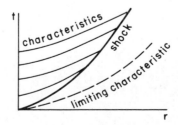

Fig. 6.7. The (x, t) diagram for the point blast explosion.

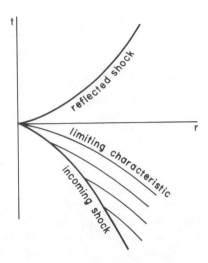

Fig. 6.8. The (x,t) diagram for Guderley's implosion solution.

be a similarity solution. However, Guderley (1942) proposed that it be of the above form (6.143) for some exponent n to be determined. The origin of t is taken to be the instant at which the shock reaches the center, so that $t \leqslant 0$ and $C < 0$ in (6.143). One might argue for a similarity solution on the grounds that there must be a singularity at the center with a shock coming in on some curve in the (r,t) plane and being reflected back along another curve. This perhaps suggests that the solution is related to a family of curves coming into the origin, and the family $r/(-t)^n = \text{constant}$ is the simplest to try. In any event, it works! The equations are then as derived above (6.144)–(6.146) for some n to be determined. The incoming shock may be normalized to be at

$$\xi = \frac{r}{(Ct)^n} = 1,$$

since C is an adjustable parameter.

In this geometry, it is clear from an outline of the (r,t) plane (see Fig. 6.8) that the limiting characteristic through the origin is in the flow region. And the question of a singularity on it becomes most important. In integrating (6.144)–(6.146) for $V(\xi)$, $P(\xi)$, $\Omega(\xi)$ from their values at $\xi = 1$, the curve $(V-1)^2 - A^2 = 0$ will be reached. There would then be a singularity in the solution except in the special case when the right sides of (6.144)–(6.146) also vanish. Obviously a nonsingular solution is required, and the

TABLE 6.1

	Cylindrical $j = 1$		Spherical $j = 2$	
γ	n	$(1-n)/n$	n	$(1-n)/n$
5/3	0.815625	0.226054	0.688377	0.452692
7/5	0.835217	0.197294	0.717173	0.394364
6/5	0.861163	0.161220	0.757142	0.320756

exponent n has not yet been specified. Guderley put the two together and proposed that the value of n should be the one that makes the right hand sides of (6.144)–(6.146) vanish at the curve (6.147) and allows the solution to be continued smoothly across the limiting characteristic.

The numerical integration has been repeated with great accuracy by Butler (1954) and the values found for n are presented in Table 6.1. The pressure at the shock and the shock velocity are given by

$$p \propto r^{-2(1-n)/n}, \qquad U \propto r^{-(1-n)/n};$$

the exponent $(1-n)/n$ is also presented. It is intriguing that the exponent $(1-n)/n$ for $j = 2$ is nearly double the value for $j = 1$. One is tempted to try arguments to prove that this should be exactly true. An approximate theory, described later (Chapter 8), gives this result automatically, but the proposed result does not appear to be *exactly* true.

Guderley also showed that the reflected shock can be fitted into the same similarity solution. Of prime interest is the increase of strength on reflection. At the center both the incoming and outgoing shocks have infinite strength and velocity in this idealized model, but the ratio of the strengths remains finite. Guderley shows that for $\gamma = 7/5$ the ratio of the pressure behind the reflected shock to the pressure behind the incident shock is about 26 for spherical waves ($j = 2$) and about 17 for cylindrical waves ($j = 1$). These compare with the ratio 8 for plane waves (see Section 6.14).

The infinity at the center would be modified by viscous effects in any case, but a more important question concerns stability. It will be shown later that an approximate theory predicts that small, unsymmetrical derivations in the shape of the shock would grow and the shock would not focus perfectly at the center. However, it seems that this instability affects the behavior in only a small neighborhood of the center, and over much of the motion Guderley's solution applies.

Other Similarity Solutions.

On dimensional grounds one can argue that the flow produced by a uniformly expanding sphere must be a similarity solution with u, p, ρ functions only of r/Ct, where C is the velocity of the sphere. The equations are (6.143)–(6.146) with $n = 1$, and in this case the similarity solution is not limited to strong shocks. This was proposed and solved by Taylor (1946).

In some ways, problems of plane shocks propagating through a nonuniform density distribution $\rho_0(x)$ are analogous to spherical or cylindrical waves. It turns out that there are some corresponding similarity solutions. Sakurai (1960) investigated cases in which $\rho_0(x) \propto x^m$ and found that they could be analyzed like the implosion problem. In particular, the exponent in the similarity variable is found by suppressing a possible singularity on the limiting characteristic through $x = 0$. (See Section 8.2.)

Other cases, including an exponential density stratification, have been studied by Sedov, (1959, Chapter 4), Zeldovich and Raizer (1966), and Hayes (1968).

6.17 Steady Supersonic Flow

Problems of steady supersonic flow can also be treated by the methods developed for unsteady waves. There are, in fact, close analogies between problems in the two fields. Two dimensional steady flow corresponds to unsteady plane waves, axisymmetric steady flow to cylindrical waves.

If body forces are neglected, the equations for steady flow may be written

$$\nabla \cdot (\rho \mathbf{q}) = 0, \tag{6.149}$$

$$\nabla \left(\frac{1}{2} \mathbf{q}^2 \right) + \omega \times \mathbf{q} = -\frac{1}{\rho} \nabla p, \tag{6.150}$$

$$\mathbf{q} \cdot \nabla S = 0. \tag{6.151}$$

These are taken from (6.49) with minor changes in form. It is convenient to use vector notation, to use $\mathbf{q}$ for the velocity vector (since later in two dimensional flow the components will be denoted by u and v) and to replace the original expression $(\mathbf{q} \cdot \nabla)\mathbf{q}$ on the left of (6.150) by the equivalent expression shown, where $\omega = \text{curl } \mathbf{q}$ is the vorticity.

Now the thermodynamic relation (6.31) can be taken to be

$$T\,dS = dh - \frac{1}{\rho}\,dp$$

and used with (6.150) and (6.151) to deduce the equations

$$\nabla\left(\frac{1}{2}q^2 + h\right) + \omega \times \mathbf{q} = T\nabla S, \qquad (6.152)$$

$$\mathbf{q}\cdot\nabla\left(\frac{1}{2}q^2 + h\right) = 0. \qquad (6.153)$$

Therefore, from (6.151) and (6.153), the entropy S and the "total enthalpy" $h + \frac{1}{2}q^2$ remain constant along the streamlines in any continuous part of the flow. If a continuous flow comes from a uniform state $S = S_0$, $h = h_0$, $q = U$, at infinity we have

$$h + \frac{1}{2}q^2 = h_0 + \frac{1}{2}U^2, \qquad (6.154)$$

$$S = S_0 \qquad (6.155)$$

throughout the flow. These relations have to be reexamined if shocks occur, since the quantities may jump discontinuously as the streamlines cross a shock and the jumps may be different on different streamlines. But we accept them for the present. Then (6.152) reduces to

$$\omega \times \mathbf{q} = 0. \qquad (6.156)$$

We are interested only in two dimensional or axisymmetrical flows; for these, ω and $\mathbf{q}$ are orthogonal so the conclusion is $\omega = 0$, that is, the flow is *irrotational*. In general, there are special flows—the so-called Beltrami flows—which satisfy (6.156) with $\omega \neq 0$.

For a polytropic gas, $h = a^2/(\gamma - 1)$ and Bernoulli's equation (6.154) may be used in the form

$$a^2 = a_0^2 - \frac{\gamma - 1}{2}(q^2 - U^2) \qquad (6.157)$$

to express a in terms of q. If the flow is also isentropic p and ρ can be determined from a and therefore from q.

We now consider a continuous two dimensional flow with uniform conditions upstream. The flow is taken to be in the (x,y) plane with

$\mathbf{q} = (u, v)$. Since all the thermodynamic quantities are known in terms of q, via (6.155) and (6.157), we need two equations for u and v. We may choose (6.149) and the irrotational condition $\omega = 0$. The remaining equations are then automatically satisfied by the various integrals used to arrive at this point. We have

$$(\rho u)_x + (\rho v)_y = 0,$$

$$v_x - u_y = 0,$$

where ρ is to be expressed in terms of u and v by (6.157). The relation between a and ρ is $a^2 \propto \rho^{\gamma-1}$; hence

$$\frac{d\rho}{\rho} = \frac{1}{\gamma-1} \frac{d(a^2)}{a^2} = -\frac{u\,du + v\,dv}{a^2}.$$

The equations may therefore be transformed into

$$(u^2 - a^2) u_x + 2uv u_y + (v^2 - a^2) v_y = 0, \tag{6.158}$$

$$v_x - u_y = 0, \tag{6.159}$$

where a^2 is given by (6.157).

Characteristic Equations.

The next step is to investigate the characteristic forms, following the procedures of Chapter 5. The manipulation, although elementary, is a little complicated and a variety of alternative tricks may be used to keep it as short as possible. We choose to start in a straightforward way and consider the linear combination

$$(u^2 - a^2) u_x + (2uv - l) u_y + l v_x + (v^2 - a^2) v_y = 0.$$

This is in characteristic form

$$(u^2 - a^2)(u_x + m u_y) + l(v_x + m v_y) = 0,$$

provided that

$$(u^2 - a^2)m = 2uv - l \quad \text{and} \quad lm = v^2 - a^2.$$

The requirement on m is

$$(u^2 - a^2)m^2 - 2uvm + (v^2 - a^2) = 0. \tag{6.160}$$

It has two real roots if $u^2 + v^2 > a^2$. Therefore the system is hyperbolic in regions where the flow is supersonic. The corresponding characteristic equations may be written as

$$(u^2 - a^2)m\frac{du}{dx} + (v^2 - a^2)\frac{dv}{dx} = 0 \qquad \text{on} \qquad \frac{dy}{dx} = m. \tag{6.161}$$

Since only two variables are involved, the differential form

$$(u^2 - a^2)m\,du + (v^2 - a^2)\,dv \tag{6.162}$$

can be integrated for each choice of m, and two Riemann variables can be obtained. The procedure is clear, but it is at this point that a little ingenuity (combined with knowing the answer!) can be used. The guide here is symmetry.

Since m is the slope of the characteristic, (6.160) can be expressed as a relation between the differentials dx, dy, on the characteristic and taken as

$$(u^2 - a^2)dy^2 - 2uv\,dx\,dy + (v^2 - a^2)dx^2 = 0,$$

or, better still,

$$(u\,dy - v\,dx)^2 = a^2(dx^2 + dy^2).$$

If χ is the inclination of the characteristic to the x axis, and θ is the inclination of the streamline, we have

$$dx = \cos\chi\,ds, \qquad dy = \sin\chi\,ds,$$

$$u = q\cos\theta, \qquad v = q\sin\theta.$$

Then the differential relation reduces to

$$q^2\sin^2(\chi - \theta) = a^2. \tag{6.163}$$

But a is a function of q, so if we introduce a variable μ defined by

$$\sin\mu = \frac{a}{q}, \qquad 0 < \mu < \frac{\pi}{2}, \tag{6.164}$$

the characteristic condition (6.162) becomes simply

$$\chi = \theta \pm \mu. \tag{6.165}$$

The characteristics make angles $\pm\mu$ with the stream direction. The quantity μ is called the *Mach angle* and is related to q by (6.164) and (6.157). In view of their significant roles, we now work towards the two dependent variables μ, θ, in place of either q, θ or u, v.

The remaining problem is to transpose the differential relation (6.162) for the Riemann variables. On a characteristic, (6.162) vanishes; therefore (6.160) can also be used to relate du and dv. The relation is

$$(v\,dv + u\,du)^2 = a^2 (du^2 + dv^2).$$

In terms of q and θ, it becomes

$$q^2\,dq^2 = a^2 (dq^2 + q^2 d\theta^2),$$

that is,

$$d\theta \pm \left(\frac{q^2}{a^2} - 1 \right)^{1/2} \frac{dq}{q} = 0.$$

Therefore the Riemann variables are

$$\theta \pm P(\mu),$$

where

$$P(\mu) = \int \left(\frac{q^2}{a^2} - 1 \right)^{1/2} \frac{dq}{q}$$

$$= \int \frac{\cos^2\mu}{\sin^2\mu + (\gamma-1)/2}\,d\mu$$

$$= \sqrt{\frac{\gamma+1}{\gamma-1}}\ \tan^{-1}\left(\sqrt{\frac{\gamma+1}{\gamma-1}}\ \tan\mu \right) - \mu. \qquad (6.166)$$

Finally, the characteristic equations are

$$\theta + P(\mu) = \text{constant} \qquad \text{on} \quad C_+ : \frac{dy}{dx} = \tan(\theta+\mu),$$

$$\qquad\qquad\qquad\qquad\qquad\qquad\qquad\qquad\qquad (6.167)$$

$$\theta - P(\mu) = \text{constant} \qquad \text{on} \quad C_- : \frac{dy}{dx} = \tan(\theta-\mu).$$

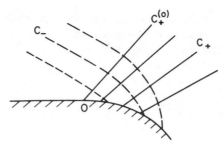

Fig. 6.9. Expanding supersonic flow around a continuous corner.

Simple Waves.

The special solution in which one of the Riemann variables is constant throughout the flow will again be called a simple wave. It arises in the study of flow around an expansive corner as indicated in Fig. 6.9. Upstream of the corner the flow is uniform, with $\mu = \mu_0$, $\theta = 0$, say. The C_- characteristics all start in this uniform region and therefore

$$P(\mu) - \theta = P(\mu_0) \tag{6.168}$$

on each of them. As a consequence, this Riemann variable is the same constant throughout the flow. Then, from the equations for the C_+, μ and θ must be individually constant on each C_+, and each C_+ is a straight line with slope $\tan(\theta + \mu)$. Since $\theta = \theta_w$ is given at the wall in terms of the wall shape $y = Y_w(x)$, the solution may be written

$$\theta = \theta_w(\xi), \qquad P(\mu_w) = P(\mu_0) + \theta_w,$$
$$y = Y_w(\xi) + (x - \xi)\tan(\theta_w + \mu_w). \tag{6.169}$$

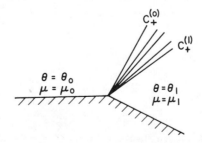

Fig. 6.10. The centered Prandtl-Meyer fan for supersonic flow around a sharp corner.

There is a close analogy with the piston problem, both with the solution (6.76)–(6.77) and its derivation; the correspondence is

$$y \leftrightarrow x, \qquad x \leftrightarrow t, \qquad \xi \leftrightarrow \tau.$$

All quantities of interest may be calculated from the expressions for μ and θ. Of particular interest is the pressure at the wall and this requires only the Riemann invariant (6.168); the value of μ_w is determined in terms of θ_w and the pressure is related with μ by the relation

$$\frac{p}{p_0} = \left(\frac{a}{a_0}\right)^{2\gamma/(\gamma-1)} = \left\{ \frac{1+(\gamma-1)/2\sin^2 \mu_0}{1+(\gamma-1)/2\sin^2 \mu} \right\}^{\gamma/(\gamma-1)}. \tag{6.170}$$

In the limiting case of a sharp corner, shown in Fig. 6.10, the simple wave becomes a fan (the Prandtl-Meyer fan) and the solution in the fan is given by

$$P(\mu) - \theta = P(\mu_0),$$

$$\tan(\theta + \mu) = \frac{y}{x}. \tag{6.171}$$

When the corner is compressive, multivalued regions develop due to the convergence of characteristics (as shown in Fig. 6.11) and a discontinuous shock must be fitted in. The extreme case of a sharp compressive corner is shown in Fig. 6.12. In general, once a shock appears, the various integrals along the streamlines and along the C_- are no longer strictly valid, since the quantities concerned jump discontinuously at the shock. However, this situation is closely analogous to the situation for the corres-

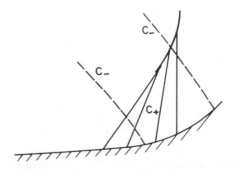

Fig. 6.11. Shock formation in supersonic flow.

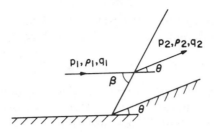

Fig. 6.12. Shock formation in supersonic flow past a wedge.

ponding unsteady problems treated earlier. For weak shocks the various integrated relations are true to first order and the simple wave solution is an approximation valid to first order in the strength of the disturbance. It is completed by fitting in shocks appropriately.

Oblique Shock Relations.

The shock conditions will be required for the oblique shock shown in Fig. 6.12. These are easily obtained from the normal shock relations in (6.95)–(6.97). If the flow in Fig. 6.12 is viewed by an observer moving with speed $q_1 \cos \beta$ along the shock, the flow on side 1 will appear to be normal to the shock. Then (6.95)–(6.97) give the normal jump relations with $v_1 = q_1 \sin \beta$, $v_2 = q_2 \sin(\beta - \theta)$. Moreover, in this moving frame, the flow is one dimensional; therefore the flow on side 2 must also be normal to the shock. Hence $q_1 \cos \beta = q_2 \cos(\beta - \theta)$. The complete set of shock conditions may be written (in a slightly more general form to allow for an inclination θ_1 of the upstream flow to some arbitrary reference direction) as

$$\rho_2 q_2 \sin(\beta - \theta_2) = \rho_1 q_1 \sin(\beta - \theta_1),$$

$$p_2 + \rho_2 q_2^2 \sin^2(\beta - \theta_2) = p_1 + \rho_1 q_1^2 \sin^2(\beta - \theta_1),$$

$$q_2 \cos(\beta - \theta_2) = q_1 \cos(\beta - \theta_1), \tag{6.172}$$

$$h_2 + \frac{1}{2} q_2^2 \sin^2(\beta - \theta_2) = h_1 + \frac{1}{2} q_1^2 \sin^2(\beta - \theta_1).$$

The shock conditions can also be derived directly from the steady flow equations in conservation form. Then they are related to conservation of mass, normal momentum, tangential momentum, and energy, respectively.

It might be noted that $h + \frac{1}{2}q^2$ is in fact continuous, so that the argument leading to (6.154) is still valid. However, the entropy and the Riemann invariant do jump.

The shock conditions determine p_2, ρ_2, q_2, β in terms p_1, ρ_1, q_1, θ, and they can be used to solve the problem shown in Fig. 6.12 exactly. If the wall curves around after the sharp corner, as for a supersonic airfoil, we may use the simple wave solution as an approximation when all the values of θ are small. Three of the shock conditions are satisfied to first order by the simple wave relations; the remaining one is used to fit in the shock. It may be shown from (6.172) that

$$\beta = \mu_1 + \frac{\mu_2 - \mu_1 + \theta}{2} + O(\theta^2).$$

Therefore, to first order in θ and $\mu_2 - \mu_1$, the shock bisects the angle between the characteristics on the two sides. This corresponds to (6.113) and the shock fitting follows very closely the similar steps in the unsteady case. The procedure will be seen later in the discussion of the more interesting case of the shock produced at the nose of an axisymmetric body.

Oblique Shock Reflection.

Finally, we note for later use that if an oblique shock is reflected from a plane wall, a possible local flow pattern is as shown in Fig. 6.13. If the initial uniform state and β are known, the shock conditions between regions 1 and 2 can be used to determine all the flow quantities in 2. Then in region 3, θ is known, since it must take its original value parallel to the wall, so the shock conditions applied to jumps from 2 to 3 are again sufficient to determine the remaining flow quantities in 3, together with the reflected shock angle β_r.

An important result of this analysis is that the proposed solution covers only a certain range of cases. If the shock is weak enough or at a

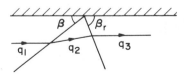

Fig. 6.13. Regular shock reflection.

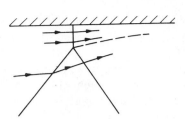

Fig. 6.14. Mach reflection.

sufficiently glancing angle, there is no solution and what happens is that the range of reflected shocks available can not turn the flow in region 3 back parallel to the wall. The whole pattern gets pushed away from the wall and assumes the pattern with three shocks shown in Fig. 6.14. It is known as "Mach reflection" in honor of Ernst Mach, who first observed it experimentally. The analysis of it is still incomplete with some theoretical results in apparent disagreement with observations.

CHAPTER 7

The Wave Equation

The equation

$$\frac{\partial^2 \varphi}{\partial t^2} = c^2 \nabla^2 \varphi, \qquad c = \text{constant}, \qquad (7.1)$$

has become known as the wave equation even though the majority of waves are not governed by it! However, it does occur in many problems and it is the simplest equation for starting the discussion of two and three dimensional waves. There is an enormous number of possible topics and we must make some choice. Following the general theme of this book, we restrict the discussion to basic results which contribute to an understanding of waves and play a role in extensions to nonlinear theory. We make no attempt to give even an introduction to the vast amount of special and intricate analysis developed for the various boundary value problems of diffraction theory. The elementary aspects of interference and diffraction patterns are well documented in a variety of books and the more advanced theory rapidly becomes a matter of skillful use of "mathematical methods" rather than one of understanding the nature of waves more deeply.

On the other hand, the approximate theory of geometrical optics involves valuable general ideas which can be extended to other contexts both for linear and nonlinear problems. It is developed here for the wave equation and the extensions for nonuniform media and for anisotropic waves are noted. These extensions go beyond the strict discussion of (7.1) but the material fits in conveniently here. Other aspects of geometrical optics and the development of similar ideas in nonlinear problems are considered in later chapters.

7.1 Occurrence of the Wave Equation

The wave equation (7.1) occurs primarily in three fields: acoustics, elasticity, and electromagnetism.

Acoustics.

The equations for acoustics have been given in Section 6.6. The expressions are noted here for easy reference. The gas dynamic equations are linearized for small perturbations about a constant state in which

$$u = 0, \qquad \rho = \rho_0, \qquad p = p_0 = p(\rho_0).$$

The propagation speed is

$$a_0^2 = p'(\rho_0), \tag{7.2}$$

and in terms of a velocity potential φ the perturbations are given by

$$u = \nabla \varphi, \tag{7.3}$$

$$p - p_0 = -\rho_0 \varphi_t, \tag{7.4}$$

$$\rho - \rho_0 = -\frac{\rho_0}{a_0^2} \varphi_t. \tag{7.5}$$

Substitution in the linearized continuity equation leads to the equation for φ:

$$\varphi_{tt} = a_0^2 \nabla^2 \varphi. \tag{7.6}$$

Linearized Supersonic Flow.

The acoustic formulation may be used when the disturbance is caused by a moving solid body. If the disturbance is to remain small, the motion of the body must be very small (this applies to the cone of a loud speaker, for example) or the body must be very slender. The former is a typical source for sound waves and the equation must be solved subject to appropriate boundary conditions. The case of a slender body moving with arbitrary constant velocity relates acoustics to aerodynamics. If the body moves with constant velocity, there is an obvious advantage in transforming to a moving frame of reference fixed in the body. Let (x_1, x_2, x_3) refer to the original frame in which the motion of the gas is small and described by (7.3)–(7.6). If the body moves with speed U in the negative x_1 direction, and (x, y, z) refer to coordinates fixed with respect to the body, the transformation of coordinates is given by

$$x = x_1 + Ut, \qquad y = x_2, \qquad z = x_3.$$

The velocity components in the new frame are $(U + u_1, u_2, u_3)$ where $u_i = \partial \varphi / \partial x_i$. Moreover, the flow appears steady in the new frame so that

$$\varphi(x_1, x_2, x_3, t) = \Phi(x, y, z)$$

$$= \Phi(x_1 + Ut, x_2, x_3).$$

Therefore (7.6) becomes

$$(M^2 - 1)\Phi_{xx} = \Phi_{yy} + \Phi_{zz}, \qquad M = \frac{U}{a_0}; \tag{7.7}$$

(7.4) becomes

$$p - p_0 = -\rho_0 U \Phi_x; \tag{7.8}$$

and the velocity components relative to the body are

$$(U + \Phi_x, \Phi_y, \Phi_z). \tag{7.9}$$

For supersonic flow, $M > 1$, we recover the wave equation in a reduced number of variables with x playing the role of time. This is a linearized version of the analogy noted in Section 6.16.

Elasticity.

The derivation of the wave equation in the elementary treatments of transverse vibrations in strings and membranes, and of longitudinal and torsional waves in bars, is taken to be well known. Here we note only how the wave equation arises in the full three dimensional theory.

The motion of an elastic solid can be described in terms of the displacement $\xi(\mathbf{x}, t)$ of a point from its position $\mathbf{x}$ in the unstrained configuration. It will also be convenient to introduce $\mathbf{X}(\mathbf{x}, t) = \mathbf{x} + \xi(\mathbf{x}, t)$ for the new position at time t. The forces acting across a surface in the deformed body can be described in terms of a stress tensor p_{ji}, just as in the case of a fluid (Section 6.1). If we temporarily consider the stress as a function of the current variable $\mathbf{X}$ in the deformed configuration, the stresses produce a net force $\partial p_{ji} / \partial X_j$ per unit volume. This follows from the divergence theorem, exactly as in the case of a fluid. However, the preceding

"Lagrangian" description of the displacements (which is usually more convenient in elasticity) relates all quantities back to the original unstrained configuration. Accordingly, the net force *per unit volume of the unstrained configuration* is

$$J\frac{\partial x_k}{\partial X_j}\frac{\partial p_{ji}}{\partial x_k},\qquad(7.10)$$

where J is the Jacobian

$$J=\frac{\partial(X_1,X_2,X_3)}{\partial(x_1,x_2,x_3)}.\qquad(7.11)$$

Moreover, $\partial x_k/\partial X_j$ is J_{jk}/J, where J_{jk} is the cofactor of the element $\partial X_j/\partial x_k$ in the determinant (7.11). Therefore the net force per unit unstrained volume, (7.10), is

$$J_{jk}\frac{\partial p_{ji}}{\partial x_k},$$

and the equations of motion are

$$\rho_0\frac{\partial^2 X_i}{\partial t^2}=J_{jk}\frac{\partial p_{ji}}{\partial x_k}.\qquad(7.12)$$

The extension of any line element from the unstrained to the strained position is obtained from

$$dX_i^2-dx_j^2=\frac{\partial X_i}{\partial x_j}\frac{\partial X_i}{\partial x_k}dx_j\,dx_k-dx_j^2$$

$$=2\epsilon_{jk}\,dx_j\,dx_k,$$

where

$$\epsilon_{jk}=\frac{1}{2}\left(\frac{\partial X_i}{\partial x_j}\frac{\partial X_i}{\partial x_k}-\delta_{jk}\right)$$

$$=\frac{1}{2}\left(\frac{\partial \xi_k}{\partial x_j}+\frac{\partial \xi_j}{\partial x_k}+\frac{\partial \xi_i}{\partial x_j}\frac{\partial \xi_i}{\partial x_k}\right).\qquad(7.13)$$

In general, the stresses p_{ji} depend on the strains ϵ_{jk} and the temperature. In the linear theory of elasticity for small displacements $\boldsymbol{\xi}$ from an unstrained configuration, the equations are linearized as follows. Since $J_{jk} = \delta_{jk} + O(\nabla\xi)$, (7.12) is linearized to

$$\rho_0 \frac{\partial^2 \xi_i}{\partial t^2} = \frac{\partial p_{ji}}{\partial x_j} ; \tag{7.14}$$

the strains (7.13) are linearized to

$$\epsilon_{jk} = \frac{1}{2} \left(\frac{\partial \xi_k}{\partial x_j} + \frac{\partial \xi_j}{\partial x_k} \right); \tag{7.15}$$

the stress-strain relations are taken to be

$$p_{ji} = 2\mu\epsilon_{ji} + \lambda\epsilon_{kk}\delta_{ji}, \tag{7.16}$$

where λ, μ are the Lamé constants. Strictly speaking, different constants λ, μ should appear, for example, for isentropic and isothermal motions, but the difference is small for most materials. (A good elementary discussion of the thermodynamics is given by Landau and Lifshitz, 1959, p. 8.)

From (7.14)–(7.16), the three equations for the displacements ξ_i are

$$\rho_0 \frac{\partial^2 \xi_i}{\partial t^2} = \mu\nabla^2\xi_i + (\lambda+\mu)\frac{\partial}{\partial x_i}(\nabla\cdot\xi). \tag{7.17}$$

The divergence and curl of (7.17) lead to

$$\frac{\partial^2}{\partial t^2}(\nabla\cdot\xi) = \frac{\lambda+2\mu}{\rho_0}\nabla^2(\nabla\cdot\xi), \tag{7.18}$$

$$\frac{\partial^2}{\partial t^2}(\nabla\times\xi) = \frac{\mu}{\rho_0}\nabla^2(\nabla\times\xi), \tag{7.19}$$

respectively. Thus there are two modes, each satisfying a wave equation; (7.18) describes compression waves propagating with velocity $\{(\lambda+2\mu)/\rho_0\}^{1/2}$ while (7.19) describes shear waves which have velocity $\{\mu/\rho_0\}^{1/2}$. The two modes are coupled through the boundary and initial conditions placed on the ξ_i or the p_{ji}, and the full solutions to problems involve considerably more than just solving the wave equation.

Electromagnetic Waves.

Maxwell's equations for a nonconducting medium with permeability μ

and permittivity ϵ may be written

$$\frac{\partial \mathbf{B}}{\partial t} + \nabla \times \mathbf{E} = 0, \qquad \epsilon \frac{\partial \mathbf{E}}{\partial t} = \frac{1}{\mu} \nabla \times \mathbf{B},$$

$$\nabla \cdot \mathbf{B} = 0, \qquad \nabla \cdot \mathbf{E} = 0,$$

where $\mathbf{B}$ is the magnetic induction and $\mathbf{E}$ is the electric field. Therefore

$$\frac{\partial^2 \mathbf{B}}{\partial t^2} = -\frac{1}{\epsilon \mu} \nabla \times (\nabla \times \mathbf{B}) = \frac{1}{\epsilon \mu} \nabla^2 \mathbf{B},$$

and $\mathbf{E}$ satisfies the same equation. All the components of $\mathbf{E}$ and $\mathbf{B}$ satisfy the wave equation with the propagation speed $c = (\epsilon \mu)^{-1/2}$. However, choosing the components to satisfy $\nabla \cdot \mathbf{E} = 0$, $\nabla \cdot \mathbf{B} = 0$ couples the components, as do the boundary and initial conditions, so the solution to problems is again more than just the solution of the scalar wave equation.

7.2 Plane Waves

For one space dimension x, the wave equation is

$$\varphi_{tt} = c^2 \varphi_{xx}.$$

If characteristic coordinates $\alpha = x - ct$, $\beta = x + ct$ are introduced, it reduces to

$$\frac{\partial^2 \varphi}{\partial \alpha \, \partial \beta} = 0,$$

and the general solution is

$$\varphi = f(\alpha) + g(\beta)$$

$$= f(x - ct) + g(x + ct),$$

where f and g are arbitrary functions. The arbitrary functions are readily determined to fit prescribed initial or boundary conditions. For the signaling problem of outgoing waves with*

$$\varphi_x = Q(t), \qquad x = 0,$$

---

*The boundary condition is taken with prescribed φ_x rather than φ to compare with the source solutions in spherical and cylindrical waves.

the solution is

$$\varphi = -cQ_1\left(t - \frac{x}{c}\right) \tag{7.20}$$

where $Q_1(t)$ is the integral of $Q(t)$. For the initial value problem,

$$\varphi = \varphi_0(x), \qquad \varphi_t = \varphi_1(x), \qquad t = 0, \qquad -\infty < x < \infty,$$

the solution is

$$\varphi = \frac{1}{2}\{\varphi_0(x - ct) + \varphi_0(x + ct)\} + \frac{1}{2c}\int_{x-ct}^{x+ct}\varphi_1(\xi)\,d\xi. \tag{7.21}$$

7.3 Spherical Waves

For waves symmetric about the origin $\varphi = \varphi(R, t)$, where R is the distance from the origin. The wave equation reduces to

$$\frac{1}{c^2}\frac{\partial^2\varphi}{\partial t^2} = \frac{\partial^2\varphi}{\partial R^2} + \frac{2}{R}\frac{\partial\varphi}{\partial R}.$$

Surprisingly enough, this may also be written

$$\frac{1}{c^2}\frac{\partial^2(R\varphi)}{\partial t^2} = \frac{\partial^2(R\varphi)}{\partial R^2},$$

which is the one dimensional wave equation. Thus the general solution takes the simple form

$$\varphi = \frac{f(R - ct)}{R} + \frac{g(R + ct)}{R}. \tag{7.22}$$

For a source producing only outgoing waves, the solution is

$$\varphi = \frac{f(R - ct)}{R}$$

and f is determined from the properties of the source. A convenient standard form is to prescribe

$$Q(t) = \lim_{R \to 0} 4\pi R^2 \frac{\partial\varphi}{\partial R}; \tag{7.23}$$

this gives

$$Q(t) = -4\pi f(-ct)$$

and

$$\varphi = -\frac{1}{4\pi} \frac{Q(t - R/c)}{R}. \tag{7.24}$$

In acoustics, $\partial\varphi/\partial R$ is the radial velocity and $Q(t)$ is the volume flux of fluid.

For an initial value problem, although it is merely a matter of determining the functions f and g in (7.22), the solution is more interesting than might have been expected. Consider the "balloon problem" in acoustics: the pressure inside a region of radius R_0 is $p_0 + P$ while the pressure outside is p_0, the gas is initially at rest, and the balloon is burst at $t = 0$. From (7.3) and (7.4), the initial conditions may be formulated as

$$\varphi = 0, \qquad \varphi_t = \begin{cases} -\dfrac{P}{p_0}, & R < R_0, \\[2mm] 0, & R > R_0. \end{cases}$$

Therefore the solution

$$\varphi = \frac{f(R - a_0 t)}{R} + \frac{g(R + a_0 t)}{R} \tag{7.25}$$

must have

$$f(R) + g(R) = 0, \qquad\qquad 0 < R < \infty,$$

$$f'(R) - g'(R) = \begin{cases} \dfrac{P}{p_0 a_0} R, & 0 < R < R_0, \\[2mm] 0, & R_0 < R < \infty. \end{cases} \tag{7.26}$$

These conditions determine f and g for positive values of their arguments. However, in the solution (7.25), the values of f are also required for negative argument. The remaining condition comes from the behavior of the solution at the origin. Since there is no source at the origin, we require

$$\lim_{R \to 0} R^2 \frac{\partial\varphi}{\partial R} = 0;$$

hence,

$$f(-a_0 t) + g(a_0 t) = 0, \qquad 0 < t < \infty. \tag{7.27}$$

This condition determines f for negative argument in terms of the values of g for positive argument.

Solving (7.26) and (7.27), we have

$$f(\xi) = \begin{cases} \dfrac{1}{4}\dfrac{P}{\rho_0 a_0}(\xi^2 - R_0^2), & -R_0 < \xi < R_0, \\[2mm] 0, & R_0 < |\xi|, \end{cases}$$

$$g(\xi) = \begin{cases} -\dfrac{1}{4}\dfrac{P}{\rho_0 a_0}(\xi^2 - R_0^2), & 0 < \xi < R_0, \\[2mm] 0, & R_0 < \xi. \end{cases}$$

Finally, the solution for the pressure disturbance is

$$p - p_0 = \frac{P}{2R}\left\{(R - a_0 t)F + (R + a_0 t)G\right\},$$

where

$$F = \begin{cases} 1, & -R_0 < R - a_0 t < R_0, \\ 0, & \text{otherwise,} \end{cases}$$

$$G = \begin{cases} 1, & 0 < R + a_0 t < R_0, \\ 0, & \text{otherwise.} \end{cases}$$

The variation of pressure with time is shown in Fig. 7.1. For a point $R > R_0$, a discontinuous pressure increase equal to $PR_0/2R$ arrives at time $t = (R - R_0)/a_0$; the excess pressure then decreases linearly in time to a value $-PR_0/2R$ at $t = (R + R_0)/a_0$ and it then returns discontinuously to zero. Even at $R = R_0$, the discontinuity at the front of the wave is only $P/2$; the remaining $P/2$ to make up the initial discontinuity P is taken by the incoming expansion wave.

For interior points $R < R_0$, a discontinuous decrease in pressure, reducing the initial value P to $P(1 - R_0/2R)$, arrives at time $t = (R_0 - R)/a_0$; the excess pressure then decreases linearly with time to $-PR_0/2R$ at $t = (R_0 + R)/a_0$ and then returns discontinuously to zero. Notice at the center $R = 0$ these changes are infinite but the whole disturbance lasts for zero time!

It is interesting that an entirely positive initial distribution of pressure leads to an outgoing wave with equal positive and negative phases. In fact this N wave profile is typical in two and three dimensional waves. The

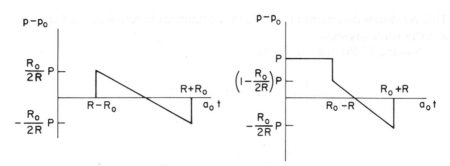

Fig. 7.1. Pressure signatures in the balloon problem.

reasons for it can be understood as follows. In an outgoing wave the pressure and radial velocity are given by

$$p - p_0 = \frac{\rho_0 a_0 f'(R - a_0 t)}{R},$$

$$u = \frac{f'(R - a_0 t)}{R} - \frac{f(R - a_0 t)}{R^2}.$$

The first point we can make is that in any wave that returns *both* $p - p_0$ and u to zero after the whole wave passes, *both* f' and f must return to zero. Hence, f' has to have both positive and negative parts if the total integral, which is f, has to return to zero. A second point concerns the total volume flow at large distances. For large R, the volume flow across a sphere of radius R is

$$4\pi R^2 u \sim 4\pi R f'(R - a_0 t).$$

This is large in R. If f', which is proportional to the pressure, were always positive, there would be an infinitely large outward flow as $R \to \infty$. An N wave, however, has a large outward flow immediately followed by a balancing large inward flow, so the net outflow is finite.

For plane waves neither of these effects arises and a positive source provides a wave with wholly positive $p - p_0$ and u.

7.4 Cylindrical Waves

Hadamard pointed out in his classic studies of the wave equation that the general character of the solution is different in even and odd numbers of space dimensions. Precise instances will appear below, but we might express the result crudely by saying that odd dimensions are easier than even. For this reason the three dimensional spherical case was considered first and the cylindrical wave solution will be deduced from the spherical wave solution. Here only the solution for outgoing waves will be obtained.

We start from the solution (7.24) for a point source. Suppose such sources are distributed uniformly on the z axis with a uniform strength $q(t)$ per unit length (see Fig. 7.2). The total disturbance from this distribution is clearly a function only of the distance r from the z axis and the time t; it is the cylindrical wave produced by a line source. The total disturbance is

$$\varphi(r,t) = -\frac{1}{4\pi}\int_{-\infty}^{\infty}\frac{q(t-R/c)\,dz}{R} = -\frac{1}{2\pi}\int_{0}^{\infty}\frac{q(t-R/c)\,dz}{R},$$

where $R = \sqrt{r^2 + z^2}$.

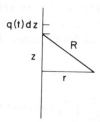

Fig. 7.2. Construction detail for a line source.

Various forms of this solution are valuable. If $z = r\sinh\zeta$, $R = r\cosh\zeta$ are substituted in the integral, we have

$$\varphi = -\frac{1}{2\pi}\int_{0}^{\infty}q\!\left(t-\frac{r}{c}\cosh\zeta\right)d\zeta; \qquad (7.28)$$

alternatively, if

$$t - \frac{R}{c} = \eta, \qquad z = c\sqrt{(t-\eta)^2 - \frac{r^2}{c^2}},$$

we have

$$\varphi(r,t) = -\frac{1}{2\pi} \int_{-\infty}^{t-r/c} \frac{q(\eta)\,d\eta}{\sqrt{(t-\eta)^2 - r^2/c^2}}. \tag{7.29}$$

Formula 7.28 is particularly useful for calculating the derivatives of φ and therefore for verifying directly that the wave equation is satisfied. It is easily shown that

$$c^2\left(\varphi_{rr} + \frac{1}{r}\varphi_r\right) - \varphi_{tt} = \frac{1}{2\pi} \int_0^\infty \frac{d}{d\zeta}\left\{\frac{c}{r}\sinh\zeta\, q'\left(t - \frac{r}{c}\cosh\zeta\right)\right\}d\zeta$$

$$= \lim_{\zeta\to\infty}\left\{\frac{c}{2\pi r}\sinh\zeta\, q'\left(t - \frac{r}{c}\cosh\zeta\right)\right\}.$$

If $q'(t)\to 0$ sufficiently fast as $t\to -\infty$, for example if q is identically zero until $t=0$, this is zero.

For a periodic source $q(t) = e^{-i\omega t}$, we satisfy the condition $q(t)\to 0$ as $t\to -\infty$ by allowing ω to have a small positive imaginary part, which makes insignificant changes at finite times. From (7.28), the solution for the periodic source is

$$\varphi = -\frac{1}{2\pi} \int_0^\infty e^{i(\omega r/c)\cosh\zeta} d\zeta\, e^{-i\omega t}.$$

It is just

$$\varphi = \frac{1}{4i} H_0^{(1)}\left(\frac{\omega r}{c}\right) e^{-i\omega t},$$

since the integral is one of the representations of the Hankel function. This solution would be obtained more simply from the equation by separation of variables. The Fourier superposition of such elementary solutions provides another derivation of (7.28).

The first derivatives of φ describe important physical quantities (pressure and velocity in acoustics); from (7.28), followed by the substitution

$\cosh \zeta = c(t - \eta)/r,$

$$\varphi_t = -\frac{1}{2\pi} \int_0^\infty q'\left(t - \frac{r}{c}\cosh\zeta\right) d\zeta = -\frac{1}{2\pi} \int_{-\infty}^{t-r/c} \frac{q'(\eta)\,d\eta}{\sqrt{(t-\eta)^2 - r^2/c^2}},$$

$$\varphi_r = \frac{1}{2\pi c} \int_0^\infty \cosh\zeta \, q'\left(t - \frac{r}{c}\cosh\zeta\right) d\zeta$$

$$= \frac{1}{2\pi} \int_{-\infty}^{t-r/c} \frac{t-\eta}{r} \frac{q'(\eta)\,d\eta}{\sqrt{(t-\eta)^2 - r^2/c^2}}.$$

$$(7.30)$$

These formulas can also be derived directly from (7.29) if judicious integration by parts is used to avoid the divergent integrals or if Hadamard's "finite part" of an integral is used. The latter is incorporated into the version using the theory of generalized functions.

Behavior Near the Origin.

From the final integral in (7.30) it is easy to see that

$$\varphi_r \sim \frac{1}{2\pi r} \int_{-\infty}^t q'(\eta)\,d\eta = \frac{1}{2\pi r} q(t) \qquad \text{as } r \to 0.$$

Hence the flux per unit length of the line source is

$$\lim_{r \to 0} 2\pi r \varphi_r = q(t),$$

which checks our definition of $q(t)$. This gives

$$\varphi \sim \frac{q(t)}{2\pi} \log r,$$

but often the next term in the expansion is required. The expression in (7.29) may be integrated by parts to give

$$\varphi = -\frac{1}{2\pi} \int_{-\infty}^{t-r/c} q'(\eta) \log\left\{ \frac{(t-\eta) + \sqrt{(t-\eta)^2 - r^2/c^2}}{r/c} \right\} d\eta.$$

If we now approximate this for small r by

$$\varphi \sim -\frac{1}{2\pi} \int_{-\infty}^{t} q'(\eta) \log \frac{2c(t-\eta)}{r} \, d\eta, \qquad \frac{r}{ct} \to 0, \qquad (7.31)$$

it may be shown by careful estimation that the errors are proportional to r. The expression for φ_t is the same as (7.31) with $q'(\eta)$ replaced by $q''(\eta)$.

Behavior Near the Wavefront and at Large Distances.

If $q(t) = 0$ for $t < 0$, the lower limit in (7.29) may be taken to be zero, and the solution is nonzero in $t - r/c > 0$. The first signal arrives with the wavefront $t - r/c = 0$. If we introduce

$$\tau = t - \frac{r}{c}$$

to measure the time elapsed after arrival of the wavefront, (7.29) may be written

$$\varphi = -\frac{1}{2\pi} \int_{0}^{\tau} \frac{q(\eta) \, d\eta}{\sqrt{(\tau-\eta)(\tau-\eta+2r/c)}}, \qquad \tau = t - \frac{r}{c} > 0.$$

Since η ranges from 0 to τ, the second factor under the radical may be approximated by $2r/c$ when $c\tau/r \ll 1$. Hence

$$\varphi \sim -\frac{1}{2\pi} \int_{0}^{\tau} \frac{q(\eta) \, d\eta}{\sqrt{\tau-\eta}} \left(\frac{c}{2r} \right)^{1/2}, \qquad \frac{c\tau}{r} \ll 1.$$

That is,

$$\varphi \sim -\left(\frac{c}{2r} \right)^{1/2} Q(t-r/c), \qquad \frac{ct-r}{r} \ll 1, \qquad (7.32)$$

where

$$Q(\tau) = \frac{1}{2\pi} \int_{0}^{\tau} \frac{q(\eta) \, d\eta}{\sqrt{\tau-\eta}}. \qquad (7.33)$$

The expression (7.32) may be compared with (7.20) and (7.24). In all three cases we have an amplitude falling off like $r^{-(n-1)/2}$, where n is the dimension of the space. In the present case, however, the formula is not exact, and Q is not simply the source strength. A simple interpretation of the dependence of the amplitude will be given in Section 7.7, equation 7.70.

The expansion in (7.32) can be continued by noting that

$$\varphi = -\frac{1}{2\pi} \int_0^\tau \frac{q(\eta)}{(\tau-\eta)^{1/2}} \left(\frac{c}{2r}\right)^{1/2} \left\{1 + c\frac{(\tau-\eta)}{2r}\right\}^{-1/2} d\eta$$

$$= -\frac{1}{2\pi} \int_0^\tau \frac{q(\eta)}{(\tau-\eta)^{1/2}} \left(\frac{c}{2r}\right)^{1/2} \sum_{m=0}^\infty \binom{-1/2}{m} \left\{\frac{c(\tau-\eta)}{2r}\right\}^m d\eta$$

$$= -\sum_{m=0}^\infty \frac{(m-1/2)!}{(-m-1/2)!} \frac{Q_m(\tau)}{m!} \left(\frac{c}{2r}\right)^{m+1/2}, \qquad \frac{c\tau}{2r} < 1, \qquad (7.34)$$

where

$$Q_m(\tau) = \frac{(-1/2)!}{(m-1/2)!} \frac{1}{2\pi} \int_0^\tau q(\eta)(\tau-\eta)^{m-1/2} d\eta. \qquad (7.35)$$

It is interesting that if $q(0+) > 0$, so the source switches on with a finite strength,

$$Q_m(\tau) \propto \tau^{m+1/2}, \qquad \tau \to 0. \qquad (7.36)$$

Thus the expansion in time after the wavefront passes is in one half powers.

Tail of the Cylindrical Wave.

One of the important differences between odd and even dimensions, noted by Hadamard, is in the behavior of the solution for a source which lasts for only a finite time. Suppose $q(t)$ is zero except for the time interval $0 < t < T$. For plane or spherical waves, we see from (7.20)* and (7.24) that the disturbance is confined to

$$\frac{x}{c} < t < \frac{x}{c} + T \quad \text{and} \quad \frac{R}{c} < t < \frac{R}{c} + T,$$

---

*For (7.20), φ_x was prescribed, so by "disturbance" we mean the quantities φ_x and φ_t. The fact that φ may be a constant not equal to zero in $t > x/c + T$ is not counted as a disturbance.

respectively. The first signal arrives with the wavefront which left the source at $t = 0$; this must be true in all cases. The interesting point is that the disturbance ceases with the signal which left the source at the final time $t = T$. For the cylindrical wave, (7.29), an integral over the source strength $q(t)$ is involved and the disturbance continues after $t = r/c + T$. We have

$$\varphi = -\frac{1}{2\pi} \int_0^T \frac{q(\eta)\, d\eta}{\sqrt{(t-\eta)^2 - r^2/c^2}}, \qquad t > \frac{r}{c} + T.$$

For fixed r,

$$\varphi \sim -\left\{ \frac{1}{2\pi} \int_0^T q(\eta)\, d\eta \right\} \frac{1}{t}, \qquad t \to \infty. \tag{7.37}$$

The disturbance decays to zero only asymptotically as $t \to \infty$.

7.5 Supersonic Flow Past a Body of Revolution

The most interesting use of the cylindrical wave solution is probably in supersonic aerodynamics. As noted in (7.7), the perturbation velocity potential satisfies the two dimensional wave equation with

$$x \leftrightarrow t, \qquad M^2 - 1 \leftrightarrow \frac{1}{c^2}.$$

For a body of revolution, (7.7) becomes

$$B^2 \Phi_{xx} = \Phi_{rr} + \frac{1}{r} \Phi_r, \qquad B = \sqrt{M^2 - 1}\,,$$

where r is distance from the flight path and x is distance from the nose of the body. The solution is zero for $x < Br$ and

$$\Phi = -\frac{1}{2\pi} \int_0^{x - Br} \frac{q(\eta)}{\sqrt{(x-\eta)^2 - B^2 r^2}}\, d\eta, \qquad x > Br. \tag{7.38}$$

The source strength $q(\eta)$ is related to the shape of the body. The boundary condition on the body is that the velocity normal to it is zero. Hence if the body shape is given by $r = R(x)$,

$$\Phi_r = R'(x)(U + \Phi_x) \qquad \text{on} \quad r = R(x).$$

For the linearization in the equations, the body must be slender, that is, $R'(x)$ is small, and Φ_x and Φ_r are both small. Accordingly, the boundary condition is linearized to

$$\Phi_r = UR'(x) \qquad \text{on} \quad r = R(x). \tag{7.39}$$

But $\Phi_r \sim q(x)/2\pi r$ as $r \to 0$, therefore (7.39) gives

$$q(x) = 2\pi UR(x)R'(x) = US'(x),$$

where $S(x) = \pi R^2(x)$ is the cross-sectional area of the body at a distance x from the nose. Intuitively one can see that $US'(x)$ is the rate at which the increasing cross-sectional area is pushing fluid out, and this is the source strength. The solution for the given body therefore is,

$$\Phi = -\frac{U}{2\pi} \int_0^{x-Br} \frac{S'(\eta)\,d\eta}{\sqrt{(x-\eta)^2 - B^2 r^2}}, \qquad x - Br > 0. \tag{7.40}$$

The components of the velocity perturbation are obtained by suitable modification of (7.30) as

$$\Phi_x = -\frac{U}{2\pi} \int_0^{x-Br} \frac{S''(\eta)\,d\eta}{\sqrt{(x-\eta)^2 - B^2 r^2}}, \tag{7.41}$$

$$\Phi_r = \frac{U}{2\pi r} \int_0^{x-Br} \frac{(x-\eta)S''(\eta)\,d\eta}{\sqrt{(x-\eta)^2 - B^2 r^2}}. \tag{7.42}$$

In linear theory, the pressure is given by (7.8). However, an interesting question arises here about the consistency of the linear theory, particularly with regard to the pressure. The exact expression for the pressure in potential flow is given by Bernoulli's equation [see (6.157)] as

$$\frac{p}{p_0} = \left(\frac{a}{a_0}\right)^{2\gamma/(\gamma-1)} = \left\{1 - \frac{\gamma-1}{a_0^2}\left(U\Phi_x + \frac{1}{2}\Phi_x^2 + \frac{1}{2}\Phi_r^2\right)\right\}^{\gamma/(\gamma-1)}.$$

Therefore, since $a_0^2 = \gamma p_0/\rho_0$,

$$\frac{p-p_0}{\rho_0} = -\left(U\Phi_x + \frac{1}{2}\Phi_x^2 + \frac{1}{2}\Phi_r^2\right) + \cdots.$$

When r is not small compared with the length of the body, Φ_x and Φ_r are comparable small quantities being of order δ^2 in the thickness ratio of the body δ (defined as maximum diameter divided by length). Then the linearization in which Φ_x^2, Φ_r^2, are neglected is correct. However, on the body $r = R(x) = O(\delta)$, and for small r,

$$\Phi_r \sim \frac{US'(x)}{2\pi r}, \qquad \Phi_x \sim \frac{US''(x)}{2\pi} \log r.$$

Therefore on the body

$$\Phi_r = O(\delta), \qquad \Phi_x = O\left(\delta^2 \log \frac{1}{\delta}\right).$$

Apart from the $\log(1/\delta)$, which in practical situations in not really large, the term $\frac{1}{2}\Phi_r^2$ is as important as the term Φ_x. It appears then that one should take

$$\frac{p - p_0}{\rho_0} = -U\Phi_x - \frac{1}{2}\Phi_r^2, \tag{7.43}$$

rather than (7.8), for a good approximation to the pressure. Lighthill (1945) and Broderick (1949) showed by careful consideration of higher approximations that (7.43) is correct with an error $O(\delta^4 \log^2 1/\delta)$. At the same time the consistency of the linear theory must be questioned, since the boundary condition is applied in the region where Φ_x and Φ_r are not of the same order. It is shown in the references cited that (7.41) and (7.42) are the correct lowest order terms, and the adoption of the nonlinear relation (7.43) is the only essential change.

Drag.

The drag on the surface of the body due to the perturbed pressure is

$$D = \int_0^l (p - p_0) S'(x)\, dx,$$

where the integration extends over the length l of the body. The expression for Φ near the body is

$$\Phi \sim -\frac{U}{2\pi} \int_0^x S''(\eta) \log \frac{2(x-\eta)}{Br}\, d\eta,$$

[see (7.31)], and the pressure is given by (7.43). Hence the drag can be written

$$\frac{2\pi D}{\rho_0 U^2} = \int_0^l S'(x) \left\{ -S''(x)\log R(x) \right.$$

$$\left. + \frac{\partial}{\partial x} \int_0^x S''(\eta)\log\frac{2(x-\eta)}{B}\,d\eta - \frac{1}{4\pi}\frac{S'^2(x)}{R^2(x)} \right\} dx.$$

The first and third terms combine into

$$-\int_0^l \left\{ S'(x)S''(x)\log R(x) + \frac{S'^2(x)}{2R(x)}R'(x) \right\} dx,$$

since $S' = 2\pi R R'$, and this is

$$-\int_0^l \frac{d}{dx}\left\{ \frac{1}{2}S'^2(x)\log R(x) \right\} dx = 0,$$

for a body with $S'(0) = S'(l) = 0$. After integration by parts, the second term gives

$$D = \frac{\rho_0 U^2}{2\pi} \int_0^l S''(x) \int_0^x S''(\eta)\log\frac{1}{(x-\eta)}\,d\eta\,dx$$

$$= \frac{\rho_0 U^2}{4\pi} \int_0^l \int_0^l S''(x)S''(\eta)\log\frac{1}{|x-\eta|}\,dx\,d\eta. \qquad (7.44)$$

[The term in $\log(2/B)$ integrates to zero.] This is the famous supersonic drag formula first obtained by von Karman and Moore in 1932.

Behavior Near the Mach Cone and at Large Distances.

The wavefront is $x - Br = 0$; this is the Mach cone making an angle $\sin^{-1} 1/M$ with the x axis. When $(x - Br)/Br \ll 1$, we have from (7.32) and (7.33), suitably transcribed to supersonic flow, that

$$\Phi \sim -\frac{U}{2\pi\sqrt{2Br}} \int_0^\xi \frac{S'(\eta)}{\sqrt{\xi-\eta}}\,d\eta, \qquad \xi = x - Br.$$

Hence the velocity components are

$$\Phi_x \sim -\frac{UF(x-Br)}{\sqrt{2Br}}, \qquad \Phi_r \sim UB\frac{F(x-Br)}{\sqrt{2Br}}, \qquad \frac{x-Br}{Br} \ll 1, \quad (7.45)$$

where

$$F(\xi) = \frac{1}{2\pi} \int_0^\xi \frac{S''(\eta)}{\sqrt{\xi-\eta}} \, d\eta. \qquad (7.46)$$

In this limit, Φ_x and Φ_r are the same order and the pressure is given to the first order by

$$\frac{p-p_0}{\rho_0 U^2} \sim \frac{F(x-Br)}{\sqrt{2Br}}. \qquad (7.47)$$

It should again be noted that the behavior near the wavefront and at large distances can be combined in one expression.

If the body is sharp nosed with $R'(0) = \epsilon$, then $S(x) \sim \pi\epsilon^2 x^2$ for small x, and we have

$$F(\xi) \sim 2\epsilon^2 \xi^{1/2} \qquad \text{as} \quad \xi \to 0. \qquad (7.48)$$

Figure 7.3 is a typical $F(\xi)$ curve. The appearance of negative phase is typical, even for a shell shaped body for which the source strength $US'(x)$ is always positive. In fact it is easily shown that

$$\int_0^\infty F(\xi) \, d\xi = 0,$$

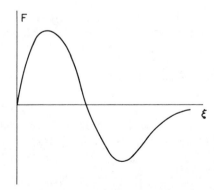

Fig. 7.3. Typical F curve for axisymmetric supersonic flow past a body.

and the physical explanation in terms of mass flow is similar to that given for spherical waves at the end of Section 7.3.

It may be remarked that the velocity components and pressure are continuous at the Mach cone according to this linear theory. In reality a shock wave is produced, and we have the important phenomenon of the sonic boom. This is missed because it is a nonlinear effect. The theory for it is developed in detail in Chapter 9.

7.6 Initial Value Problem in Two and Three Dimensions

One of the many "Poisson integrals" appearing in the theory of partial differential equations provides the solution of the wave equation for initial conditions

$$\varphi = \varphi_0(\mathbf{x}), \qquad \varphi_t = \varphi_1(\mathbf{x}), \qquad t = 0.$$

According to Hadamard's ideas, the three dimensional problem will be easier than the two dimensional, and we start with it.

We know from the spherical wave solution discussed in Section 7.3 that

$$\psi(\mathbf{x}, t) = \frac{f(|\mathbf{x} - \boldsymbol{\xi}| - ct)}{|\mathbf{x} - \boldsymbol{\xi}|}$$

is a solution for arbitrary $\boldsymbol{\xi}$. We now argue intuitively that the initial prescribed disturbance at any point $\boldsymbol{\xi}$ gives rise to such a spherical wave and propose that the solution should be something like a superposition of all the spherical waves. That is, we consider

$$\psi(\mathbf{x}, t) = \int_{-\infty}^{\infty} \Psi(\boldsymbol{\xi}) \frac{f(|\mathbf{x} - \boldsymbol{\xi}| - ct)}{|\mathbf{x} - \boldsymbol{\xi}|} d\boldsymbol{\xi}. \qquad (7.49)$$

The arbitrary function $\Psi(\boldsymbol{\xi})$ is inserted because, depending on the initial conditions, the spherical waves from different points $\boldsymbol{\xi}$ will have different strengths. The form of (7.49) suggests the introduction of spherical polars (R, θ, λ) based on the point $\mathbf{x}$. It then becomes

$$\psi(\mathbf{x}, t) = \int_0^{\infty} \int_0^{\pi} \int_0^{2\pi} \Psi(\mathbf{x} + R\mathbf{l}) f(R - ct) R \sin\theta \, dR \, d\theta \, d\lambda,$$

where $\mathbf{l}$ is the unit vector from $\mathbf{x}$ to ξ and its Cartesian components may be written

$$\mathbf{l} = (\sin\theta\cos\lambda, \sin\theta\sin\lambda, \cos\theta).$$

With the idea that the initial source strength determining f acts only for an instant, we specialize this formula to the case $f(R - ct) = \delta(R - ct)$. Then

$$\psi(\mathbf{x}, t) = ct \int_0^\pi \int_0^{2\pi} \Psi(\mathbf{x} + ct\mathbf{l})\sin\theta \, d\theta \, d\lambda. \tag{7.50}$$

Formally, this is a solution for arbitrary Ψ. It may also be written as a surface integral

$$\psi(\mathbf{x}, t) = \frac{1}{ct} \int_{S(t)} \Psi \, dS,$$

where $S(t)$ is the surface of the sphere with origin at $\mathbf{x}$ and radius ct. For a continuously differentiable function Ψ, we see from (7.50) that

$$\psi \to 0, \qquad \psi_t \to 4\pi c\Psi(\mathbf{x}) \qquad \text{as} \quad t \to 0. \tag{7.51}$$

If we choose $\Psi(\mathbf{x}) = \varphi_1(\mathbf{x})/4\pi c$, we shall have solved the special initial value problem

$$\psi \to 0, \qquad \psi_t \to \varphi_1(\mathbf{x}) \qquad \text{as} \quad t \to 0. \tag{7.52}$$

The solution is

$$\psi(\mathbf{x}, t) = \frac{1}{4\pi c^2 t} \int_{S(t)} \varphi_1 \, dS. \tag{7.53}$$

It represents the total contribution of the instantaneous sources which send spherical waves to the point $\mathbf{x}$ in time t; they are all exactly a distance ct away and their contributions traveling with speed c arrive at $\mathbf{x}$ just at the time t (see Fig. 7.4). Notice that all points inside $S(t)$ could still in principle have been contributing. But there is no "tail" for spherical waves: the sources act for an infinitesimal time and each contribution lasts only for an infinitesimal time. This will not be so in two dimensions. In any event, (7.53) is formally the solution for the initial values (7.52). It may also be written

$$\psi(\mathbf{x}, t) = t M[\varphi_1],$$

where

$$M[\varphi_1] = \frac{1}{4\pi c^2 t^2} \int_{S(t)} \varphi_1 \, dS$$

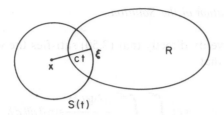

Fig. 7.4. Construction detail in Poisson's solution of the initial value problem. R is the region of the initial disturbance.

stands for the mean value of φ_1 over the sphere $S(t)$.

To furnish the other half of the initial condition we could specialize f in (7.49) to δ' and pursue the consequences similarly to the above. However, it is quicker to use a trick that is often useful: if ψ is a solution of a partial differential equation with constant coefficients, then so is any t or x derivative. In this case, we note that

$$\chi(\mathbf{x},t) = \frac{\partial \psi(\mathbf{x},t)}{\partial t}$$

is a solution of the wave equation, where $\psi(\mathbf{x},t)$ is given by (7.50). Moreover, as $t \to 0$, we see from (7.51) that

$$\chi = \psi_t \to 4\pi c \Psi(\mathbf{x}),$$

$$\chi_t = \psi_{tt} = c^2 \nabla^2 \psi \to 0.$$

To give $\chi \to \varphi_0(\mathbf{x})$, $\chi_t \to 0$ as $t \to 0$, we must now choose $\Psi(\mathbf{x}) = \varphi_0(\mathbf{x})/4\pi c$, and take

$$\chi(\mathbf{x},t) = \frac{\partial}{\partial t} \left\{ \frac{1}{4\pi c^2 t} \int_{S(t)} \varphi_0 \, dS \right\}.$$

The complete solution for general initial values therefore is

$$\varphi(\mathbf{x},t) = \frac{\partial}{\partial t} \left\{ \frac{1}{4\pi c^2 t} \int_{S(t)} \varphi_0 \, dS \right\} + \frac{1}{4\pi c^2 t} \int_{S(t)} \varphi_1 \, dS$$

$$= \frac{\partial}{\partial t} \{ t M[\varphi_0] \} + t M[\varphi_1]. \tag{7.54}$$

Direct Verification of the Solution.

It remains to verify directly that (7.50) satisfies the wave equation. We have immediately that

$$\psi_{x_i x_i} = ct \int_0^\pi \int_0^{2\pi} \frac{\partial^2 \Psi(\xi)}{\partial \xi_i^2} \sin\theta \, d\theta \, d\lambda$$

$$= \frac{1}{ct} \int_{S(t)} \frac{\partial^2 \Psi}{\partial \xi_i^2} dS,$$

where $\xi = x + ct\mathbf{l}$. The t derivatives require a little more manipulation. First,

$$\psi_t = \frac{\psi}{t} + c^2 t \int_0^\pi \int_0^{2\pi} l_i \frac{\partial \Psi}{\partial \xi_i} \sin\theta \, d\theta \, d\lambda$$

$$= \frac{\psi}{t} + \frac{1}{t} \int_{S(t)} l_i \frac{\partial \Psi}{\partial \xi_i} dS$$

$$= \frac{\psi}{t} + \frac{1}{t} \int_{V(t)} \frac{\partial^2 \Psi}{\partial \xi_i^2} dV,$$

where $V(t)$ is the volume inside the sphere $S(t)$. Then

$$\psi_{tt} = -\frac{\psi}{t^2} + \frac{\psi_t}{t} - \frac{1}{t^2} \int_{V(t)} \frac{\partial^2 \Psi}{\partial \xi_i^2} dV + \frac{c}{t} \int_{S(t)} \frac{\partial^2 \Psi}{\partial \xi_i^2} dS,$$

which reduces to

$$\psi_{tt} = \frac{c}{t} \int_{S(t)} \frac{\partial^2 \Psi}{\partial \xi_i^2} dS$$

in view of the expression for ψ_t. We see that

$$\psi_{tt} = c^2 \psi_{x_i x_i},$$

as required.

These arguments assume that Ψ is twice continuously differentiable. The solution (7.54) requires only that φ_0 and φ_1 are integrable in order for it to be meaningful. We might extend the meaning of solution to include all cases in which (7.54) is defined. In particular, if φ_0 and φ_1 are piecewise continuous, (7.54) is defined and $\varphi \to \varphi_0(\mathbf{x})$, $\varphi_t \to \varphi_1(\mathbf{x})$ at points of continuity.

For the balloon problem in Section 7.3, $\varphi_0 = 0$, $\varphi_1 = -P/\rho_0$ in an initial sphere. This is an example of piecewise continuous data. It is interesting to construct the solution using Poisson's integral not only for the spherically symmetric case but for an arbitrary shape of initial pressure region. This is left for the reader.

Wavefront.

If the nonzero values of $\varphi_0(\mathbf{x})$ and $\varphi_1(\mathbf{x})$ are confined to a finite region R as shown in Fig. 7.4, the solution at any point outside R is zero until the time when $S(t)$ first intersects points of R. It is clear that this occurs when ct is equal to the least distance from $\mathbf{x}$ to the boundary of R. This least distance is on the normal from R to $\mathbf{x}$. The wavefront at time t can be determined by turning this argument around. Construct all the normals to the boundary surface of R. Measure a distance ct out along each normal. The surface formed by these points is the wavefront. Notice that where the surface of R is concave, the wavefront will be folded over itself after a while (see Fig. 7.5). This construction will be studied further in the discussion of geometrical optics.

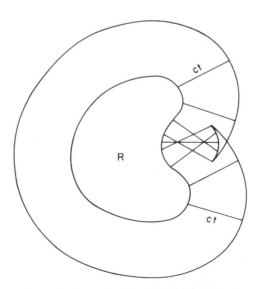

Fig. 7.5. Wavefront construction for a disturbance initially confined to the region R.

The disturbance at any point $\mathbf{x}$ outside R ceases when $S(t)$ becomes so large that R is entirely within it. Thus in three dimensions an initial disturbance of finite extent produces disturbances which last only for a finite time. There is no "tail."

Two Dimensional Problem.

The solution for a two dimensional distribution of initial values may be treated as the special case in which $\varphi_0(\mathbf{x})$ and $\varphi_1(\mathbf{x})$ are independent of x_3. Suppose nonzero values of $\varphi_0(x_1, x_2)$, $\varphi_1(x_1, x_2)$ are confined to a finite region R_0 of the (x_1, x_2) plane. From the three dimensional point of view, they occupy the cylinder R with generators parallel to the x_3 axis based on the cross section R_0. The initial disturbance is no longer finite in extent. For a point $\mathbf{x}$ outside the cylinder R, the wavefront construction is as before, but the spheres $S(t)$ centered at $\mathbf{x}$ will intersect R at all times after the first time of intersection. This accounts for the "tail" in two dimensional waves, and it vividly shows the difference between two and three dimensions.

In the solution (7.54), the value of $\varphi(\mathbf{x}, t)$ must be independent of x_3. The integrals can be reduced to two dimensional form to show this explicitly. We consider the value of

$$M[\varphi_0] = \frac{1}{4\pi c^2 t^2} \int_{S(t)} \varphi_0 \, dS$$

at a point $(x_1, x_2, 0)$.

At a point (ξ_1, ξ_2, ξ_3) of $S(t)$ (see Fig. 7.6), the value of φ_0 is $\varphi_0(\xi_1, \xi_2)$. The outward normal has a direction cosine l_3 with the x_3 axis equal to

$$\frac{\xi_3}{ct} = \pm \frac{\sqrt{c^2 t^2 - (x_1 - \xi_1)^2 - (x_2 - \xi_2)^2}}{ct}.$$

The surface element dS is equal to $d\xi_1 d\xi_2 / |l_3|$ where $d\xi_1 d\xi_2$ is its projection in the (x_1, x_2) plane. Therefore, remembering the two equal contributions from above and below the plane, we may write

$$M[\varphi_0] = \frac{1}{2\pi ct} \int\int_{\sigma(t)} \frac{\varphi_0(\xi_1, \xi_2) d\xi_1 d\xi_2}{\sqrt{c^2 t^2 - (x_1 - \xi_1)^2 - (x_2 - \xi_2)^2}},$$

where $\sigma(t)$ is the interior of the projection of $S(t)$ onto the (x_1, x_2) plane:

$$\sigma(t): \quad (x_1 - \xi_1)^2 + (x_2 - \xi_2)^2 \leqslant c^2 t^2.$$

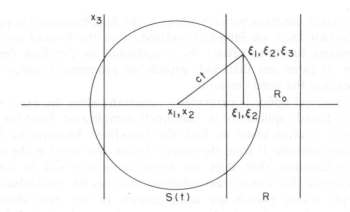

Fig. 7.6. Geometry involved in descending from three dimensions to two in the initial value problem.

The full solution reduces to

$$\varphi(x_1, x_2, t) = \frac{\partial}{\partial t} \frac{1}{2\pi c} \int\int\limits_{\sigma(t)} \frac{\varphi_0(\xi_1, \xi_2) \, d\xi_1 \, d\xi_2}{\sqrt{c^2 t^2 - (x_1 - \xi_1)^2 - (x_2 - \xi_2)^2}}$$

$$+ \frac{1}{2\pi c} \int\int\limits_{\sigma(t)} \frac{\varphi_1(\xi_1, \xi_2) \, d\xi_1 \, d\xi_2}{\sqrt{c^2 t^2 - (x_1 - \xi_1)^2 - (x_2 - \xi_2)^2}}. \qquad (7.55)$$

One notes the similarity with (7.29).

Since the integration is over the whole of the inside of the circle $(x_1 - \xi_1)^2 + (x_2 - \xi_2)^2 = c^2 t^2$, not just its boundary, the disturbance continues even after this circle completely surrounds the initial region R_0.

7.7 Geometrical Optics

In the discussion of one dimensional problems in Sections 5.5 and 5.6, the role of the characteristics as carriers of discontinuities was developed. It was also shown that the variation in the magnitude of the discontinuity could be found directly from the equations without finding the complete solution. The same is true in more dimensions, and the theory of discontinuities for linear equations is one version of geometrical optics. The second version concerns periodic waves in the high frequency approximation. The two cases are intimately related, since the Fourier analysis of discontinuous functions relates the singularities to the high frequency behavior. It turns out that both aspects of geometrical optics can be combined into the same derivation.

Geometrical optics is particularly important when the exact solution cannot be found explicitly or is exceedingly complicated. Even for simpler problems it is often easier to find the wavefront behavior in this way rather than untangle it from the exact solution. We develop the ideas on the wave equation, then show the applications to waves in nonhomogeneous media (for which exact solutions may not be obtainable) and to anisotropic waves (which are complicated). In the next chapter, an approximate theory for shockwave propagation is developed from the ideas of geometrical optics. Due to nonlinearity and number of dimensions, such problems are extremely difficult to tackle in other ways.

The main use of the discontinuity theory is to determine the behavior of a wavefront spreading into an undisturbed region. We suppose that the wavefront is specified by the equation $S(\mathbf{x}, t) = 0$, and that φ is identically zero in $S(\mathbf{x}, t) < 0$. The surface $S = 0$ and the behavior of the discontinuity in φ or its derivatives are to be determined. If the mth derivatives of φ are the first ones to be discontinuous at the wavefront, we assume that φ may be expanded in the form

$$\varphi = \begin{cases} \Phi_0(\mathbf{x})\dfrac{S^m}{m!} + \Phi_1(\mathbf{x})\dfrac{S^{m+1}}{(m+1)!} + \cdots, & S > 0 \\ \\ 0, & S < 0 \end{cases}.$$

The coefficient $\Phi_0(\mathbf{x})$ determines the variation in the magnitude of the discontinuity. As we saw in the case of cylindrical waves, the singularity at the wavefront may involve fractional powers, so we allow m to be noninteger. The idea then is to substitute this series into the equation for φ, equate the coefficients of successive powers of S to zero, and thus obtain equations for S and the Φ_n. In doing this, however, it turns out that all we require is that φ should take the form

$$\varphi = \sum_{n=0}^{\infty} \Phi_n(\mathbf{x}) f_n(S), \tag{7.56}$$

where the $f_n(S)$ have the property

$$f_n'(S) = f_{n-1}(S). \tag{7.57}$$

In calculating the derivatives of φ, derivatives of the f_n will appear but they can all be replaced in terms of earlier members of the set by making use of (7.57). Then the equation may be satisfied by equating the coefficients of the successive f_n to zero. This allows a number of important extensions, and the preceding case is included because

$$f_n(S) = \begin{cases} \dfrac{S^{n+m}}{(n+m)!}, & S > 0, \\ 0, & S < 0, \end{cases} \tag{7.58}$$

satisfies (7.57).

We will carry out the substitution in the form (7.56), but for the present one should think of this as notation for (7.58). Then we will come back to consider the extensions. Also for the first run through we will consider $m > 2$. Strictly speaking, the meaning of "solution" is extended if the derivatives appearing in the equation are not continuous. For the wave equation this requires continuous second derivatives and $m > 2$. As we shall see, there is a real point to this, not just ultra caution!

In the substitution, first and second derivatives of the $f_n(S)$ will appear, but they are replaced according to (7.57) by

$$f_n'(S) = f_{n-1}(S), \qquad f_n''(S) = f_{n-2}(S).$$

For $n = 0, 1$, these will introduce $f_{-1}(S)$ and $f_{-2}(S)$ which do not appear in the original set. For (7.58) with $m > 2$, they are defined by the same formula; in other cases their definitions will have to be included in defining the $f_n(S)$. After substitution in the wave equation, we have

$$\left[S_{x_i}^2 - c^{-2} S_t^2 \right] \Phi_0 f_{-2}$$

$$+ \left\{ \left[S_{x_i}^2 - c^{-2} S_t^2 \right] \Phi_1 + 2 S_{x_i} \Phi_{0x_i} + \left(S_{x_i x_i} - c^{-2} S_{tt} \right) \Phi_0 \right\} f_{-1}$$

$$+ \sum_{n=0}^{\infty} E_n f_n(S) = 0. \tag{7.59}$$

The expression for E_n will not be needed beyond noting that it contains the leading term

$$\left[S_{x_i}^2 - c^{-2} S_t^2 \right] \Phi_{n+2}$$

followed by further terms in $\Phi_{n+1}, \ldots, \Phi_0$ and derivatives of S.

The wave equation will be satisfied if the coefficients of $f_{-2}, f_{-1}, \ldots$, are individually zero. Thus

$$S_{x_i}^2 - c^{-2} S_t^2 = 0, \tag{7.60}$$

$$2 S_{x_i} \frac{\partial \Phi_0}{\partial x_i} + \left(S_{x_i x_i} - c^{-2} S_{tt} \right) \Phi_0 = 0, \tag{7.61}$$

and the subsequent equations determine $\Phi_1, \Phi_2, \ldots$, successively. The equations for S and Φ_0 are the ones of main interest. Before discussing their solution, we return to the question of extending the application by suitable choice of the f_n.

Discontinuities in φ or its First Derivatives.

If $m < 2$ in (7.58), the question of the definition of f_{-2} and f_{-1} arises, and this is bound up with the extended meaning of the solution. For $m = 2$, that is, for discontinuities in second derivatives, the definition of f_{-2} and f_{-1} by (7.58) is still straightforward and the extended meaning of the solution is merely that the equation is satisfied separately on the two sides of the wavefront $S = 0$.

If φ itself is discontinuous, $m = 0$, the series behind the wavefront is

$$\varphi = \Phi_0(x) + \Phi_1(x) S + \frac{1}{2} \Phi_2(x) S^2 + \cdots.$$

If this alone is substituted into the wave equation, the first two terms in (7.59) do not appear, and (7.60), (7.61) are lost. If, however, we take

$$\varphi = \Phi_0(x) H(S) + \Phi_1(x) H_1(S) + \cdots,$$

where $H(S)$ is the Heaviside function

$$H(S) = \begin{cases} 1, & S > 0, \\ 0, & S < 0, \end{cases}$$

and the $H_n(S)$ are its integrals defined by

$$H_n(S) = \begin{cases} \dfrac{1}{n!} S^n, & S > 0, \\ 0, & S < 0, \end{cases}$$

and if, in addition, the derivatives are taken in the generalized sense, we have

$$\varphi_t = \Phi_0 S_t \delta(S) + \Phi_1 S_t H(S) + \cdots,$$

$$\varphi_{tt} = \Phi_0 S_t^2 \delta'(S) + (\Phi_1 S_t^2 + \Phi_0 S_{tt})\delta(S) + \cdots,$$

and so on. Then (7.59) is obtained with

$$f_0 = H(S), \qquad f_{-1} = \delta(S), \qquad f_{-2} = \delta'(S),$$

and the information on S and Φ_0 is not lost. In the case $m \geqslant 2$, this difference does not arise. The explanation is that for $m < 2$, we really are extending the discussion into the realm of "weak solutions," and the extended definition does include information on the possible discontinuities, by whatever means the extension is carried out. For linear problems, the required extension is immediately obtained by allowing generalized functions such as the delta function and interpreting the derivatives in that sense. It is equivalent to the method indicated in Section 2.7.

If φ is continuous, but first derivatives are discontinuous, the appropriate series would be

$$\varphi = \Phi_0(\mathbf{x}) H_1(S) + \Phi_1(\mathbf{x}) H_2(S) + \cdots,$$

and in this case

$$f_0(S) = H_1(S), \qquad f_{-1}(S) = H(S), \qquad f_{-2}(S) = \delta(S).$$

Wavefront Expansion and Behavior at Large Distances.

If (7.56) is viewed as an approximation for the behavior of the solution near the wavefront, rather than just as a device for studying the discontinuities in derivatives, the validity can be extended by allowing the $f_n(S)$ to be more general even than powers or step functions. For instance, in cylindrical waves, the wavefront expansion (7.34) takes the form (7.56) if we introduce

$$f_n(S) = \frac{(-1/2)!}{(n-1/2)!} \frac{1}{2\pi} \int_0^S q(\eta)(S-\eta)^{n-1/2} d\eta, \qquad S = t - \frac{r}{c},$$

$$\Phi_n(r) = \frac{(n-1/2)!}{n!(-n-1/2)!} \left(\frac{c}{2r}\right)^{n+1/2}$$

This expansion was found to be valid for

$$\frac{cS}{2r} < 1.$$

Thus S need not be small, provided r is large enough. The functions $f_n(S)$ satisfy the crucial relation (7.57) and so this expansion is included in the development here.

We expect in general that the expansion (7.56) with appropriate $f_n(S)$ will give the behavior in some extended region behind the wavefront. Of course, the precise forms of these more general $f_n(S)$ can be known only from more complete solutions; they are not determined from the substitution of (7.56) into the equation. But the determination of S and Φ_0 from (7.60) and (7.61) still gives valuable information. Typically, this extension gives the behavior for large distances in the sense that $cS/|\mathbf{x}|$ is small. In the first approximation $f_0(S)$ gives the wave profile and $\Phi_0(\mathbf{x})$ gives the amplitude decay as $\mathbf{x} \to \infty$.

High Frequencies.

In wave propagation there is often interest in solutions periodic in time with a given frequency ω. If the equation for φ is taken more generally now in the form

$$L\Phi = \varphi_{tt},$$

where L is some linear operator independent of t, periodic solutions may be written

$$\varphi = \Phi(\mathbf{x})e^{-i\omega t}$$

where

$$L\Phi + \frac{\omega^2}{c^2}\Phi = 0.$$

For large values of ω/c (normalized by some suitable length in the problem, possibly x itself), a standard method of finding the asymptotic solutions is to take

$$\Phi \sim e^{i\omega\sigma(\mathbf{x})} \sum_{n=0}^{\infty} \Phi_n(\mathbf{x})(-i\omega)^{-n},$$

where the functions $\sigma(\mathbf{x})$ and $\Phi_n(\mathbf{x})$ are to be determined. In terms of φ this is

$$\varphi \sim e^{-i\omega(t-\sigma(\mathbf{x}))} \sum_{n=0}^{\infty} \Phi_n(\mathbf{x})(-i\omega)^{-n}. \tag{7.62}$$

It may be rewritten as

$$\varphi \sim \sum_{n=0}^{\infty} \Phi_n(\mathbf{x})f_n(S),$$

with

$$S = t - \sigma(\mathbf{x}), \qquad f_n(S) = \frac{e^{-i\omega S}}{(-i\omega)^n} .$$

Moreover, with this definition of $f_n(S)$, (7.57) is satisfied. Therefore the equations for S and Φ_n are exactly the same as in the wavefront expansion; there is no need to rederive them, and we also see why results for the two cases are the same.

In this application, the surfaces $S = $ constant are surfaces of constant phase (e.g., crests and troughs), while $\Phi_0(\mathbf{x})$ determines the amplitude of the oscillations at $\mathbf{x}$.

Determination of S and Φ_0.

We continue now with the discussion of the equations for S and Φ_0 in the case of the wave equation. The description will be given in the language of wavefront propagation, but the high frequency interpretation is obvious.

Equation 7.60 for S is often called the *eikonal equation*. It defines how the surface $S = 0$ moves in $\mathbf{x}$ space. The unit normal to the surface is given by the vector $\mathbf{l}$ with components

$$l_i = \frac{-S_{x_i}}{|\nabla S|} .$$

The normal velocity can be calculated by noting that neighboring points $(\mathbf{x}_0, t_0)$ and $(\mathbf{x}_0 + \mathbf{l}\delta s, t_0 + \delta t)$ are on the surface at neighboring times provided

$$S(\mathbf{x}_0, t_0) = 0, \qquad S(\mathbf{x}_0 + \mathbf{l}\delta s, t_0 + \delta t) = 0.$$

Hence to first order in δs and δt,

$$l_i S_{x_i} \delta s + S_t \delta t = 0,$$

and the normal velocity is

$$\lim_{\delta t \to 0} \frac{\delta s}{\delta t} = \frac{-S_t}{l_i S_{x_i}} = \frac{S_t}{|\nabla S|} . \tag{7.63}$$

Thus the eikonal equation states simply that the wavefront has normal velocity $\pm c$.

In the further construction of the solution, it is convenient to specify the wavefront in the form

$$S(\mathbf{x},t) \equiv t - \sigma(\mathbf{x}) = 0. \tag{7.64}$$

The family of surfaces $\sigma(\mathbf{x}) = $ constant gives the successive positions of the wavefront in $\mathbf{x}$ space. Equations 7.60 and 7.61 become

$$\sigma_{x_i}^2 = \frac{1}{c^2}, \tag{7.65}$$

$$2\sigma_{x_i}\frac{\partial\,\Phi_0}{\partial x_i} + \sigma_{x_ix_i}\Phi_0 = 0. \tag{7.66}$$

The nonlinear equation for σ may be solved according to the method of Section 2.13 by integration along its characteristic curves. If we introduce $p_i = \sigma_{x_i}$ and write the equation as

$$H \equiv \frac{1}{2}cp_i^2 - \frac{1}{2}c^{-1} = 0,$$

the characteristics defined by (2.86) are curves in the $\mathbf{x}$ space with direction

$$\frac{dx_i}{ds} = cp_i.$$

Normalized in this way the parameter s is the distance along the characteristic, because reuse of the equation shows that $c^2p_i^2 = 1$. The full set of characteristic equations (2.86)–(2.88) is

$$\cdot\,\frac{dx_i}{ds} = cp_i, \qquad \frac{dp_i}{ds} = 0, \qquad \frac{d\sigma}{ds} = \frac{1}{c}. \tag{7.67}$$

These can also be derived directly from (7.65), using

$$\frac{d}{ds} = cp_j\frac{\partial}{\partial x_j} = c\sigma_{x_j}\frac{\partial}{\partial x_j},$$

without quoting the general formulas. Since the vector $p_i = \sigma_{x_i}$ is normal to the wavefront $\sigma = $ constant, the first of the equations (7.67) shows that the rays are also normal; they are the orthogonal trajectories of the wavefronts $\sigma = $ constant. The second equation shows that $\mathbf{p}$ is constant on the ray; hence so is the ray direction $c\mathbf{p}$, and the rays must be straight lines. The rays can be constructed, then, by drawing the family of straight lines normal to the initial wavefront. The third equation in (7.67) may be integrated to

$$\sigma = \frac{s}{c},$$

where s is measured from the initial wavefront. At any time $t > 0$, the wavefront $t = \sigma = s/c$ is a distance ct out along the rays. This is exactly the construction deduced for Poisson's exact solution and represented in Fig. 7.5. Formally, if $\mathbf{x}_0$ is a point on the initial wavefront and $\mathbf{l}(\mathbf{x}_0)$ is the unit normal at that point, the solution of (7.67) is

$$\mathbf{x} = \mathbf{x}_0 + \mathbf{l}(\mathbf{x}_0)s, \qquad \mathbf{p} = c^{-1}\mathbf{l}(\mathbf{x}_0), \qquad \sigma = \frac{s}{c}.$$

This is an implicit form for $\sigma(\mathbf{x})$; given $\mathbf{x}$, the initial point $\mathbf{x}_0$ and distance s are determined, in principle, from the first of these, then $\sigma = s/c$.

These results are special to the wave equation. In general the rays, defined as the characteristics of the eikonal equation, are neither straight (nonhomogeneous medium) nor orthogonal to the wavefronts (anisotropic).

Equation 7.66 is a linear equation for Φ_0, and its characteristics are the same rays already introduced for the eikonal equation. It may be written in characteristic form as

$$\frac{1}{\Phi_0} \frac{d\Phi_0}{ds} = -\frac{1}{2} c\sigma_{x_i x_i}; \tag{7.68}$$

the integration is straightforward, in principle, once $\sigma(\mathbf{x})$ has been determined. It shows that Φ_0 is to be obtained by integration along the rays and its variation is somehow related to the divergence of the rays as measured by $\sigma_{x_i x_i}$. But, due to the implicit form for $\sigma(\mathbf{x})$, it is more illuminating to proceed a little differently.

First we note that (7.66) takes the divergence form

$$\frac{\partial}{\partial x_i}\left(\sigma_{x_i}\Phi_0^2\right) = 0, \tag{7.69}$$

which suggests that something is conserved, and perhaps also suggests use of the divergence theorem. We consider a tube formed by rays, going from the initial wavefront S_0: $\sigma = 0$ to the wavefront S: $\sigma = t$ at time t, as shown in Fig. 7.7. We integrate (7.69) over the volume inside the ray tube and use the divergence theorem to write the result as

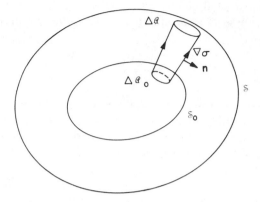

Fig. 7.7. Wavefronts and ray tube in geometrical optics.

$$\int n_i \sigma_{x_i} \Phi_0^2 \, dS = 0,$$

where **n** is the outward normal and the surface integral is over the sides Σ and ends $\mathcal{S}_0, \mathcal{S}_1$ of the ray tube. In this case the rays are orthogonal to the wavefronts $\sigma = $ constant. Therefore

$$n_i \sigma_{x_i} = 0 \qquad \text{on} \quad \Sigma,$$

and that contribution drops out. On $\mathcal{S}$, **n** and $\nabla \sigma$ are in the same direction, so

$$n_i \sigma_{x_i} = |\nabla \sigma| \qquad \text{on} \quad \mathcal{S},$$

and from the eikonal equation (7.65), $|\nabla \sigma| = c^{-1}$. On $\mathcal{S}_0$, **n** and $\nabla \sigma$ are in the opposite direction, so

$$n_i \sigma_{x_i} = -|\nabla \sigma| = -c^{-1} \qquad \text{on} \quad \mathcal{S}_0.$$

In this case c is constant, and we have

$$\int_{\mathcal{S}} \Phi_0^2 \, dS = \int_{\mathcal{S}_0} \Phi_0^2 \, dS.$$

If the ray tube is taken to be very narrow with small cross-sectional areas $\Delta \mathcal{C}_0$ on $\mathcal{S}_0$ and $\Delta \mathcal{C}$ on $\mathcal{S}$, this may be written to first order as

$$\Phi_0^2(\mathbf{x})\Delta \mathcal{C} = \Phi_0^2(\mathbf{x}_0)\Delta \mathcal{C}_0,$$

or, in the limit $\Delta \mathcal{C}_0$, $\Delta \mathcal{C} \rightarrow 0$, as

$$\frac{\Phi_0(\mathbf{x})}{\Phi_0(\mathbf{x}_0)} = \left(\frac{d\mathcal{C}}{d\mathcal{C}_0} \right)^{-1/2}. \tag{7.70}$$

It is usual to interpret (7.70) in terms of the flow of energy down the ray tube, particularly when the context is the high frequency approximation for periodic waves noted in (7.62). There is an average flow of energy across any section of the ray tube, and without performing a detailed calculation it is clear that the flux is proportional to $\Phi_0^2 \Delta \mathcal{C}$. Hence (7.70) is equivalent to a "law" of constant energy flux along the ray tube. For nonuniform media (as will be indicated below) there are extra factors, depending on the medium, multiplying $\Phi_0^2 \Delta \mathcal{C}$, but the law of constant energy flux remains true. It is in fact a general result of geometrical optics for nondispersive waves and it is often used directly to determine the amplitude variation without going through the detailed calculation each time. Recent work on dispersive waves has provided general arguments on this kind of question but has also changed the point of view. The more general concept appears to be conservation of "wave action," which in the simplest linear cases is energy flux divided by an appropriate frequency. Here the frequency is constant and so the two are the same. These general questions are discussed in Part II.

For plane, cylindrical, and spherical waves, the ray tubes are straight channels, wedges, and cones, respectively. Therefore we have

$$\frac{d\mathcal{C}}{d\mathcal{C}_0} = 1, \qquad \Phi_0 = \text{constant},$$

$$\frac{d\mathcal{C}}{d\mathcal{C}_0} = r, \qquad \Phi_0 \propto r^{-1/2}$$

$$\frac{d\mathcal{C}}{d\mathcal{C}_0} = R^2, \qquad \Phi_0 \propto R^{-1},$$

in these cases. They check with the earlier results, obtained from the exact solutions, for the behavior near the front or at large distances. In two dimensions, without cylindrical symmetry,

$$\Delta \mathcal{C}_0 = R_1 \Delta \theta, \qquad \Delta \mathcal{C} = (R_1 + s)\Delta \theta,$$

where R_1 is the radius of curvature of the initial wavefront, and $\Delta \theta$ is the angle subtended at the center of curvature (see Fig. 7.8). [The ray construction shows that the radius of curvature of the wavefront at a distance s along the ray is $(R_1 + s)$ and the angle subtended remains as $\Delta \theta$.] Hence

$$\frac{\Phi_0(\mathbf{x})}{\Phi_0(\mathbf{x}_0)} = \left(\frac{R_1}{R_1 + s} \right)^{1/2}.$$

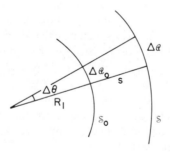

Fig. 7.8. Geometry of wavefronts and rays.

In three dimensions, a little differential geometry shows that the surface elements are proportional to the Gaussian curvature:

$$\Delta \mathcal{C}_0 \propto R_1 R_2, \qquad \Delta \mathcal{C} \propto (R_1 + s)(R_2 + s),$$

where R_1, R_2 are the principle radii of curvature of the initial wavefront. Hence

$$\frac{\Phi_0(\mathbf{x})}{\Phi_0(\mathbf{x}_0)} = \left\{ \frac{R_1 R_2}{(R_1 + s)(R_2 + s)} \right\}^{1/2}.$$

Caustics.

At points where the initial wavefront is concave outward, the rays form an envelope as shown in Fig. 7.9. This is typically cusp shaped, and between the two arms the region is triply covered by rays; it is like a fold in the sheet. The envelope is called a *caustic*. At the caustic, since neighboring rays touch each other there, $d\,\mathcal{Q}\,/\,d\,\mathcal{Q}_0\to 0$. According to (7.70) this means that $\Phi_0\to\infty$ there. For the wavefront problem this result is correct and can also be established from the exact solution of the wave equation. Whether the wave equation still applies is another question. In acoustics, for example, it is obtained only after linearization; it would be invalid due to nonlinear effects when $\Phi_0\to\infty$. In Chapter 8 the nonlinear behavior of shocks is discussed, and it will be argued (except possibly for extremely weak shocks which are affected more by viscous effects) that as a concave shock focuses it also speeds up and avoids overlapping itself.

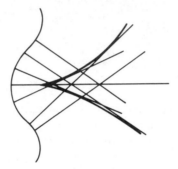

Fig. 7.9. Formation of a caustic.

For high frequency waves, the geometrical optics approximation is invalid at the caustic, even as an approximation to the wave equation. The singular behavior of Φ_0 makes the expansion (7.62) nonuniformly valid in the neighborhood of the caustic. The correct behavior was first investigated by Airy and more recent work has been carried out by Keller and his co-workers (see, e.g., Kay and Keller, 1954). The correct result is that the amplitude remains finite but large in ω, typically being proportional to a fractional power of ω. This topic is a little more special than the rest of this chapter and the reader is referred to the original papers.

7.8 Nonhomogeneous Media

In nonhomogeneous media the governing equations have coefficients which are functions of **x**. The expansion method still goes through, but the equations for σ and Φ_0 are modified. In isotropic cases the only change to the eikonal equation is that $c = c(\mathbf{x})$, so we can usefully discuss the consequences before looking at specific examples. The dependence on $\mathbf{x}$ modifies the characteristic form for the rays. With $p_i = \sigma_{x_i}$, as before, and

$$H \equiv \frac{1}{2} c(\mathbf{x}) p_i^2 - \frac{1}{2} c^{-1}(\mathbf{x}) = 0,$$

the characteristic equations are

$$\frac{dx_i}{ds} = \frac{\partial H}{\partial p_i} = c p_i,$$

$$\frac{dp_i}{ds} = -p_i \frac{\partial H}{\partial \sigma} - \frac{\partial H}{\partial x_i} = -\frac{c_{x_i}}{c^2}, \qquad (7.71)$$

$$\frac{d\sigma}{ds} = p_i \frac{\partial H}{\partial p_i} = \frac{1}{c}.$$

Since $\mathbf{p}$ is normal to the wavefront $\sigma = $ constant, the first equation shows that the rays are still orthogonal to the wavefronts. However, the second equation shows that $c p_i$ is no longer constant on a ray, and so the rays bend around in response to the gradient in c. The rays have to be found by solving the first two equations simultaneously for $x_i(s), p_i(s)$, then the third equation gives

$$\sigma = \int_{\text{ray}} \frac{ds}{c(\mathbf{x})}.$$

Of course σ is the time of travel of the wavefront along the ray.

Fermat's principle states that this time is stationary compared with neighboring paths between the two points. We can check this. Let an arbitrary path between the two points $\mathbf{x} = \mathbf{a}$ and $\mathbf{x} = \mathbf{b}$ be specified parametrically by

$$\mathbf{x} = \mathbf{x}(\mu), \qquad 0 < \mu < 1, \qquad \mathbf{x}(0) = \mathbf{a}, \qquad \mathbf{x}(1) = \mathbf{b}.$$

(To apply the usual methods of the calculus of variations, it is convenient to normalize the parameter so the integration is over the same fixed range for all paths. Thus s is not a convenient parameter for this argument.) The time on each path is

$$T = \int_0^1 \frac{1}{c\{\mathbf{x}(\mu)\}} \sqrt{\left(\frac{dx_j}{d\mu}\right)^2} \, d\mu. \qquad (7.72)$$

From a standard argument in the calculus of variations, the $x_i(\mu)$ for a stationary value of any integral

$$T = \int_0^1 \mathcal{F}\{\mathbf{x}(\mu), \dot{\mathbf{x}}(\mu)\} \, d\mu, \qquad \dot{\mathbf{x}}(\mu) = \frac{d\mathbf{x}}{d\mu},$$

satisfies

$$\frac{d}{d\mu}\left(\frac{\partial \mathcal{F}}{\partial \dot{x}_i}\right) - \frac{\partial \mathcal{F}}{\partial x_i} = 0.$$

In our case, this reduces to

$$\frac{d}{d\mu}\left(\frac{1}{c} \frac{1}{\sqrt{\dot{x}_j^2}} \frac{dx_i}{d\mu}\right) + \frac{1}{c^2} \sqrt{\dot{x}_j^2} \, c_{x_i} = 0.$$

If, *on this stationary path*, we revert to s as parameter, by means of $\sqrt{\dot{x}_j^2} \, d\mu = ds$, we have

$$\frac{d}{ds}\left(\frac{1}{c} \frac{dx_i}{ds}\right) + \frac{1}{c^2} c_{x_i} = 0.$$

This agrees with (7.71) and we have proved Fermat's principle in this case. Fermat's principle gives an immediate and illuminating picture of why the rays are straight when c is constant.

Stratified Media.

For a stratified medium in which c depends only on the vertical coordinate, y say, we can simplify the ray calculation further. First, any ray remains in the same vertical plane in which it starts. Thus it is sufficient to discuss the two dimensional case with x horizontal and y vertical. The speed is $c = c(y)$, and (7.71) reduce to

$$\frac{dx}{ds} = cp_1, \qquad \frac{dy}{ds} = cp_2,$$

$$\frac{dp_1}{ds} = 0, \qquad \frac{dp_2}{ds} = -\frac{c'(y)}{c^2}, \qquad \frac{d\sigma}{ds} = \frac{1}{c}.$$

Since p_1 is constant and $dx/ds = cp_1$, the angle θ of the ray to the horizontal is given by $\cos\theta = p_1 c(y)$, and if subscript zero refers to some initial point on the ray, we have

$$p_1 = \frac{\cos\theta_0}{c_0}, \qquad \frac{\cos\theta}{\cos\theta_0} = \frac{c(y)}{c_0}. \tag{7.73}$$

This is just Snell's law in optics.

The component p_2 can be found by solving the equations for y and p_2, or better still by noting that $p_1^2 + p_2^2 = 1/c^2$ (the eikonal equation itself) gives

$$p_2 = \sqrt{\frac{1}{c^2(y)} - \frac{\cos^2\theta_0}{c_0^2}}.$$

The ray equations may be combined into

$$\frac{dx}{dy} = \frac{p_1}{p_2} = \frac{c(y)\cos\theta_0/c_0}{\sqrt{1 - c^2(y)\cos^2\theta_0/c_0^2}}. \tag{7.74}$$

Therefore the ray with initial angle θ_0 at (x_0, y_0) is given by

$$x - x_0 = \int_{y_0}^{y} \frac{c(y)\cos\theta_0/c_0}{\sqrt{1 - c^2(y)\cos^2\theta_0/c_0^2}} \, dy. \tag{7.75}$$

The time of arrival of the wavefront is

$$\sigma = \int_0^s \frac{ds}{c} = \int_{y_0}^{y} \frac{dy}{c^2 p_2} = \int_{y_0}^{y} \frac{dy}{c(y)\sqrt{1 - c^2(y)\cos^2\theta_0/c_0^2}}. \tag{7.76}$$

It should be noted that all we have really used in deriving these results is Snell's law; we viewed it from the more general characteristic theory.

Harmless appearing distributions for $c(y)$ give some remarkable effects for the rays. We note two examples.

Ocean Waveguide.

Suppose $c(y)$ is as shown in Fig. 7.10 with variations in c confined to a layer $|y| < Y$, $c = $ constant c_1 outside the layer, $c < c_1$ inside the layer, with a minimum c_0 at $y = 0$. We consider the rays from a point source at $x = y = 0$.

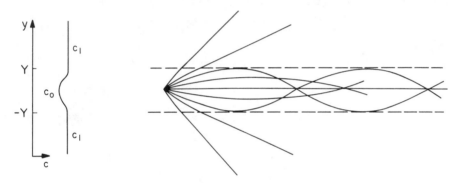

Fig. 7.10. Rays in an ocean wave guide.

As c increases along a ray, $\cos\theta = c\cos\theta_0 / c_0$ increases, θ decreases; the ray bends toward the horizontal. If $\theta_0 > \cos^{-1}(c_0/c_1)$, the ray penetrates into the region $c = c_1$ and remains straight thereafter. However, if $\theta_0 < \cos^{-1}(c_0/c_1)$, $\cos\theta$ increases to 1 and θ decreases to zero at the value of y for which

$$c(y) = \frac{c_0}{\cos\theta_0}.$$

At this point the ray turns around, crosses the minimum in c again, and repeats the pattern symmetrically about the x axis. These rays, then, oscillate about the x axis as shown in Fig. 7.10.

The channel $|y| < Y$ forms a sort of waveguide, and at points inside it sufficiently far from the source there may be a number of overlapping rays. Thus geometrical optics predicts a succession of signals. Moreover, it can be shown from (7.76) that the signals off the center line may arrive faster than those on the center line. They have farther to go but benefit from the faster propagation speed. In this situation the amplitude predictions of geometrical optics are not valid and one must turn to more exact treatments (see Cohen and Blum, 1971, for some recent work).

Shadow Zones.

For a source below a maximum of c, as shown in Fig. 7.11, one can similarly deduce that a shadow may be formed, into which the rays do not penetrate. The rays are sketched in Fig. 7.11.

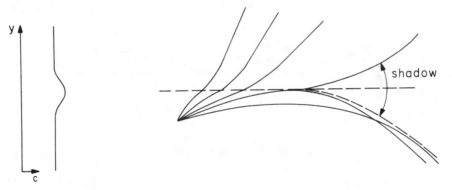

Fig. 7.11. Formation of a shadow zone.

Energy Propagation.

The modifications to (7.66) depend on the particular problem and on which particular physical quantity is denoted by φ. The remarkable thing is that the result is always constant energy flux down the ray tubes. We may verify this for acoustics in a nonuniform medium. To keep the analysis simple, we consider a fluid initially at rest with uniform pressure and no body forces, but with arbitrary density distribution $\rho(\mathbf{x})$. We might imagine a heated layer of fluid in which the gravitational effects are of smaller order and may be neglected. The linearized equations* for the perturbation pressure P, perturbation density R, and perturbation velocity $\mathbf{V}$ are

$$R_t + \nabla \cdot (\rho \mathbf{V}) = 0,$$

$$\rho \mathbf{V}_t + \nabla P = 0, \tag{7.77}$$

$$P_t - c^2(R_t + \mathbf{V} \cdot \nabla \rho) = 0,$$

where $c^2(\mathbf{x})$ is the sound speed. In terms of P alone, the equation is

$$P_{tt} = c^2 \left(\nabla^2 P - \frac{\nabla \rho}{\rho} \cdot \nabla P \right).$$

From the first two terms of the series

$$P = \sum_0^\infty P_n(\mathbf{x}) f_n \{ t - \sigma(\mathbf{x}) \},$$

we find

$$\sigma_{x_i}^2 = \frac{1}{c^2}, \qquad 2\sigma_{x_i} \frac{\partial P_0}{\partial x_i} + \left(\sigma_{x_i x_i} - \sigma_{x_i} \frac{\rho_{x_i}}{\rho} \right) P_0 = 0.$$

If $\mathbf{V}$ and R are expanded in similar series, the coefficients $\mathbf{V}_n$ and R_n may be determined in terms of P_n and σ by going back to the original three first order equations in (7.77). In particular,

$$\mathbf{V}_0 = \frac{P_0 \nabla \sigma}{\rho}, \qquad R_0 = \frac{P_0}{c^2}. \tag{7.78}$$

The equation for $P_0(\mathbf{x})$ may be written in divergence form as

$$\frac{\partial}{\partial x_i} \left(\frac{P_0^2 \sigma_{x_i}}{\rho} \right) = 0.$$

Integration over a narrow ray tube as in the argument preceding (7.70)

---

*The notation is changed from Section 6.6 to avoid clashing here with the use of subscripts to denote the successive terms in the wavefront expansion.

gives

$$\frac{P_0^2}{\rho c} \Delta \mathscr{Q} = \text{constant}. \tag{7.79}$$

The extra factor ρc, which is a function of $\mathbf{x}$, modifies the pressure amplitude in addition to the modifications of ray divergence.

The interpretation of (7.79) as energy flux is best given in the high frequency application of geometrical optics. The average rate of working of the pressure on the fluid crossing the element of area $\Delta \mathscr{Q}$ is

$$P_0 V_{0n} \Delta \mathscr{Q}, \tag{7.80}$$

where V_{0n} is the component of $\mathbf{V}$ normal to $\Delta \mathscr{Q}$. From (7.78),

$$V_{0n} = \frac{P_0}{\rho} \mathbf{n} \cdot \nabla \sigma = \frac{P_0}{\rho c}.$$

Hence

$$P_0 V_{0n} \Delta \mathscr{Q} = \frac{P_0^2}{\rho c} \Delta \mathscr{Q}.$$

Thus (7.79) shows that the energy flux remains constant along the ray tube.

7.9 Anisotropic Waves

When there are preferred directions in the medium, the eikonal equation may not be symmetric in the σ_{x_i}. As a consequence, the rays defined as the characteristics of the eikonal equation are no longer orthogonal to the wavefront. If we assume that the medium is homogeneous, so that $\mathbf{x}$ does not appear explicitly in the eikonal equation, we may write the eikonal equation as

$$H(p_1, p_2, p_3) = 0, \qquad p_i = \sigma_{x_i}.$$

The characteristic equations reduce in this case to

$$\frac{dx_i}{d\lambda} = H_{p_i}, \qquad \frac{dp_i}{d\lambda} = 0, \qquad \frac{d\sigma}{d\lambda} = p_i H_{p_i}. \tag{7.81}$$

The p_i are constant on these rays; hence the ray direction H_{p_i} is constant and the rays are straight lines. However, the ray direction is parallel to the wavefront normal if and only if

$$H_{p_i} \propto p_i.$$

This is true if and only if H is a function of $p_1^2 + p_2^2 + p_3^2$, that is, if the propagation is isotropic.

A simple example of anisotropy is provided by the wave equation in a moving medium. If the medium has velocity U in the x_1 direction, and if a_0 is the propagation speed when the medium is at rest,

$$\nabla^2 \varphi = \frac{1}{a_0^2} \left(\frac{\partial}{\partial t} + U \frac{\partial}{\partial x_1} \right)^2 \varphi.$$

(The symbol c is reserved for the normal speed of the wavefront which is not a_0.) The eikonal equation is

$$\sigma_{x_i}^2 = \frac{1}{a_0^2} \left(1 - U\sigma_{x_1} \right)^2, \tag{7.82}$$

and we take

$$H = \frac{1}{2} \left\{ p_i^2 - a_0^{-2} (1 - Up_1)^2 \right\}.$$

The ray direction has components

$$\frac{dx_1}{d\lambda} = \frac{U}{a_0^2} + \left(1 - \frac{U^2}{a_0^2} \right) p_1, \qquad \frac{dx_2}{d\lambda} = p_2, \qquad \frac{dx_3}{d\lambda} = p_3.$$

This clearly is not in the direction of the wavefront normal $\mathbf{p}$. For a point source the wavefront is just a sphere of radius $a_0 t$ swept downstream a distance Ut. The rays are shown in Fig. 7.12.

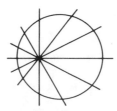

Fig. 7.12. Wavefront and rays for an acoustic pulse in a wind.

The result that the rays are not orthogonal to the wavefronts is surely surprising at first encounter. The intuitive feeling that the wavefront is moving out along its normals is natural and one might expect as a consequence that the orthogonal trajectories play a basic role in the geometry. But it just is not so. The point is that the rays are concerned with energy propagation and neither the speed nor the direction need coincide with the normal velocity of the wavefront. This is the first appearance, in a restricted form, of the important difference between phase velocity and group velocity. It will be discussed in the general form in Chapter 11, and more detailed examination of the underlying concepts is postponed until then.

We return to the general case in (7.81). Since the p_i are constant, the integration of the equations is immediate. We consider the case of a point source at the origin. In terms of the distance s from the source, the solution of (7.81) is

$$x_i = l_i s, \qquad \sigma = p_i l_i s, \tag{7.83}$$

where the unit ray vector $\mathbf{l}$ is defined by

$$l_i = \frac{H_{p_i}}{\sqrt{H_{p_j}^2}}. \tag{7.84}$$

The family of rays is obtained by varying the p_i over all values satisfying

$$H(p_1, p_2, p_3) = 0: \tag{7.85}$$

Each choice of the p_i determines a ray, and at time t the wavefront $\sigma = t$ is a distance

$$s = \frac{t}{p_i l_i} \tag{7.86}$$

along it. The coordinates of the point on the wavefront are

$$x_i = \frac{l_i}{p_j l_j} t; \tag{7.87}$$

varying the parameters p_1, p_2, p_3, subject to (7.85), gives the whole wavefront.

From (7.63), with the reduced form $S = t - \sigma(\mathbf{x})$ which is used here, the normal velocity of the wavefront has magnitude

$$c = \frac{1}{|\nabla \sigma|} = \frac{1}{p}.$$

Therefore the unit normal to the wavefront is given by

$$n_i = cp_i,$$

and the angle μ between the ray vector and the normal vector is given by

$$\cos \mu = l_j n_j = c l_j p_j.$$

Therefore (7.86) may be written as

$$s = \frac{c}{\cos \mu} t. \tag{7.88}$$

The wavefront moves along the ray with increased speed $c/\cos \mu$ so that its speed along the normal is c.

It is sometimes convenient to use $c = 1/p$ and the unit normal $n_i = cp_i$ as parameters in place of p_1, p_2, p_3. Then the eikonal equation (7.85) determines c as a function of the direction $\mathbf{n}$. This function $c(\mathbf{n})$ specifies the anisotropic medium and the geometry of the rays and the wavefront can be expressed completely in terms of it. This description is particularly convenient for two dimensional and axisymmetric problems. We describe the two dimensional case in (x_1, x_2) space, but note that the axisymmetric case is exactly the same with x_2 interpreted as distance from the axis of symmetry.

Two Dimensional or Axisymmetric Problems.

If the normal $\mathbf{n}$ makes an angle ψ with the x_1 axis, the eikonal equation provides the function $c(\psi)$ for the propagation speed. To express the rays and wavefront in terms of $c(\psi)$, we shall need to find the direction of the ray vector $\mathbf{l}$ in terms of $c(\psi)$. It turns out that it is most useful to do this by finding the angle μ between the ray vector $\mathbf{l}$ and the normal $\mathbf{n}$ (see Fig. 7.13). Since $\mathbf{l}$ has the direction $\partial H/\partial \mathbf{p}$ and $\mathbf{n}$ has the same direction as $\mathbf{p}$, this angle can be found from an argument in $\mathbf{p}$ space. The vector H_{p_i} in $\mathbf{p}$ space is first written in terms of polar coordinates p and ψ. It has a component $\partial H/\partial p$ in the direction of $\mathbf{p}$ and a component $\partial H/p\partial \psi$ perpendicular to $\mathbf{p}$. Hence

$$\tan \mu = \frac{1}{p} \frac{\partial H/\partial \psi}{\partial H/\partial p}. \tag{7.89}$$

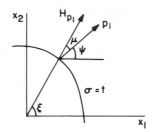

Fig. 7.13. Geometry of wavefront, rays, and normals in an anisotropic medium.

In terms of the function $c(\psi)$, an equivalent eikonal equation for $H(p,\psi)$ can be written:

$$H \equiv pc(\psi) - 1 = 0.$$

Therefore

$$\tan\mu = \frac{c'(\psi)}{c(\psi)}. \tag{7.90}$$

This is the crucial expression relating the directions of $\mathbf{l}$ and $\mathbf{n}$.
   The rays are given by

$$x_1 = s\cos(\mu+\psi)t, \qquad x_2 = s\sin(\mu+\psi)t.$$

The wavefront (7.88) is at a distance

$$s = \frac{c(\psi)t}{\cos\mu} = t\sqrt{c'^2 + c^2} \tag{7.91}$$

along the ray with parameter ψ. Therefore in Cartesians the wavefront is

$$x_1 = \frac{c(\psi)t}{\cos\mu}\cos(\mu+\psi), \qquad x_2 = \frac{c(\psi)t}{\cos\mu}\sin(\mu+\psi).$$

If these are expanded and μ eliminated from (7.90), we obtain

$$x_1 = \{c(\psi)\cos\psi - c'(\psi)\sin\psi\}t, \tag{7.92}$$

$$x_2 = \{c(\psi)\sin\psi + c'(\psi)\cos\psi\}t. \tag{7.93}$$

The geometry is shown in Fig. 7.13.

Another derivation of (7.92) and (7.93) is obtained by arguing that the wavefront is the envelope of elementary plane waves

$$x_1 \cos\psi + x_2 \sin\psi = c(\psi)t.$$

The envelope is found by solving this simultaneously with its ψ derivative:

$$-x_1 \sin\psi + x_2 \cos\psi = c'(\psi)t.$$

The solution is (7.92), (7.93). This derivation is simpler, but it is limited to homogeneous media and it does not bring out the ray properties. We have preferred to unify all cases by one method, the method of characteristics applied to the eikonal equation.

Source in a Moving Medium.

To illustrate the results, let us apply them to (7.82). With $p_1 = \cos\psi/c$, $p_2 = \sin\psi/c$, the eikonal equation gives

$$\frac{1}{c^2} = \frac{1}{a_0^2}\left(1 - \frac{U\cos\psi}{c}\right)^2.$$

Therefore

$$c(\psi) = U\cos\psi + a_0$$

for an outgoing wave. The wavefront (7.92)–(7.93) is

$$x_1 = (U + a_0\cos\psi)t,$$

$$x_2 = (a_0\sin\psi)t.$$

This is a circle of radius $a_0 t$ centered at a point Ut downstream as required.

Magnetogasdynamics.

Again rather harmless appearing problems lead to surprisingly complicated geometry. An interesting example of this arises in magnetogas-dynamics. In an infinitely conducting medium with uniform magnetic field in the x_1 direction, perturbations satisfy

$$\frac{\partial^4 \varphi}{\partial t^4} - (a^2 + b^2)\frac{\partial^2}{\partial t^2}\nabla^2\varphi + a^2 b^2 \frac{\partial^2}{\partial x_1^2}\nabla^2\varphi = 0. \tag{7.94}$$

The eikonal equation is

$$1 - (a^2 + b^2)p^2 + a^2 b^2 p_1^2 p^2 = 0.$$

With $p_1 = \cos\psi/c$, $p = 1/c$, we have

$$c^4 - (a^2 + b^2)c^2 + a^2 b^2 \cos^2\psi = 0. \tag{7.95}$$

There are two outgoing wavefronts (the fast and slow waves), corresponding to the increased order of (7.94).

In this case, it is convenient to work with the polar form (7.91), where the wavefront is distance

$$s = t\sqrt{c'^2 + c^2} \tag{7.96}$$

in the direction

$$\xi(\psi) = \mu(\psi) + \psi, \qquad \tan\mu = \frac{c'(\psi)}{c(\psi)}. \tag{7.97}$$

From (7.95), the derivative $c'(\psi)$ satisfies

$$\{2c^3 - (a^2 + b^2)c\}c' - a^2 b^2 \sin\psi\cos\psi = 0. \tag{7.98}$$

Consider now the range $0 \leqslant \psi \leqslant \pi/2$ for the parameter ψ, and for definiteness suppose $a > b$.

From (7.95) to (7.98), we have the following values as $\psi \to 0$ and $\psi \to \pi/2$:

$\psi \to 0$:

$$c \to a, \quad b,$$
$$c' \to 0, \quad 0,$$
$$\mu \to 0, \quad 0,$$
$$\xi \to 0, \quad 0,$$
$$s \to at, \quad bt,$$

$\psi \to \pi/2$:

$$c \to \sqrt{a^2+b^2}, \quad 0,$$

$$c' \to 0, \quad\quad -\frac{ab}{\sqrt{a^2+b^2}},$$

$$\mu \to 0, \quad\quad -\frac{\pi}{2},$$

$$\xi \to \frac{\pi}{2}, \quad\quad 0,$$

$$s \to \sqrt{a^2+b^2}\, t, \quad \frac{abt}{\sqrt{a^2+b^2}},$$

The first solution gives points A, B on the x_1 and x_2 axes and suggests a distorted but reasonable outgoing wavefront S_1 as shown in Fig. 7.14. The second solution is surprising. In both limits $\psi \to 0, \pi/2$ we have $\xi \to 0$ and s finite. Thus we have points P, Q on the x axis. At P, $\psi = \pi/2$ so the wavefront is tangential to the axis; at Q, $\psi = 0$ so the wavefront is perpendicular to the axis. Between $\psi = 0$ and $\pi/2$ there must be a maximum or minimum of ξ. Since $\xi = \mu + \psi$, this occurs at the value of ψ given by

$$\frac{d\mu}{d\psi} + 1 = 0;$$

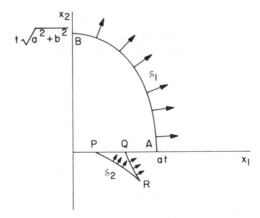

Fig. 7.14. Wavefronts in magnetogasdynamics.

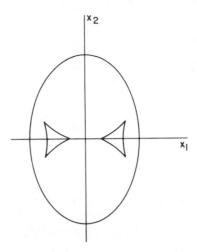

Fig. 7.15. Wavefronts in magnetogasdynamics.

from $\tan\mu = c'/c$, the condition may be written

$$c''(\psi) + c(\psi) = 0.$$

It may be shown that the wavefront has a cusp at this point. It may be shown also that ξ is negative. Thus the second front has the remarkable shape $\mathcal{S}_2$ shown in Fig. 7.14. Even though $\psi > 0$ and the wavefront is locally moving with a component in the positive x_2 direction, the energy propagation has a component in the negative x_2 direction, and as a consequence this wavefront appears *below* the x_1 axis. This is a striking example of the difference between the wavefront velocity and the ray velocity, between the phase velocity and the group velocity.

The complete wavefronts are symmetrical in the x_1 and x_2 axes and the complete picture is presented in Fig. 7.15.

CHAPTER 8

Shock Dynamics

The discussion of the wave equation has brought out most of the main ideas in the linear theory of hyperbolic waves in two or three dimensions, and we turn now to nonlinear effects. For plane waves in uniform media, it was possible to give a thorough treatment of the nonlinear theory. However, it required sophisticated ideas and methods in contrast to the linear theory which was almost trivial. In more dimensions or in nonuniform media, where even the linear theory becomes complicated, we should expect considerable difficulties in the analysis. Cylindrical and spherical waves still involve only two independent variables, but some complication arises because their equations have nonconstant coefficients. Plane waves in nonuniform media are similar. In the general case of two or three dimensional propagation, we have to deal with more independent variables and the geometry becomes even more involved. It is not surprising therefore that we have to resort to approximate methods. Indeed, the only exact analytic solutions are similarity solutions for special problems, and these usually require numerical integration of the reduced equations. The similarity solutions for cylindrical and spherical waves and for waves in nonuniform media have been discussed in Section 6.16; others will be referred to below. Apart from similarity solutions, one must use approximate theories or numerical methods. This chapter and Chapter 9 are devoted to some of the approximate theories that have been developed for shock propagation in these circumstances. The description is for shock waves in gases, but the ideas and mathematical procedures may be used for analogous problems in other fields.

The most obvious type of approximation is for those problems that can be treated as small perturbations to simpler problems with known solution. For example, a plane shock propagating through a slightly nonuniform medium or along a slightly corrugated wall can be analyzed as a perturbation of the uniform case. A problem like this will be analyzed in detail below, since it is needed in another connection, and others will be indicated. But the perturbation procedures of this type generally are

263

obvious, and the resulting mathematical problems, although often difficult, no longer involve new concepts about the behavior of waves. We again draw the line at developing purely mathematical methods and stress, rather, approximations that are intimately related with the wave propagation.

In an intuitive way, we can say that the difficulty in these problems is due to the combination of two effects: the shock is adjusting to changes in the geometry (or in the medium) at the same time that it is coping with a complicated nonlinear interaction with the flow behind it. Nonlinear plane waves are free of the first, and linear nonplanar waves are free of the second. In the more general case, if one of the effects can be dealt with fairly simply so that emphasis can be placed on the other, there is hope for an approximate theory.

This chapter concerns problems where the nonlinear geometrical effects play the biggest role and the interactions with the flow behind are not responsible for the major changes in the shock motion. "Shock dynamics" seems to be a convenient name since the motion of the shock is stressed over the dynamics of the whole fluid flow. In the next chapter we consider problems where the opposite emphasis is appropriate. They are problems of weak shocks and the idea is that for weak shocks the geometrical effects, although important, can be taken unchanged from the linear theory. Then the nonlinear analysis consists of introducing, within that geometrical framework, the crucial effects of the nonlinear interaction with the flow.

In both cases, the approximations become intuitive and are based on incorporating known effects into a mathematical description. The "justification" comes from checks on particular cases that can be handled precisely and from comparison with observations. The problems are too hard for the more routine approximation procedures.

In the discussion of shock dynamics, the description will be based on the picture provided by geometrical optics. There the geometry is in terms of wavefronts propagating down ray tubes and, for an *isotropic* medium, the rays are the orthogonal trajectories of the successive positions of the wavefront. For a shock moving into a gas at rest the medium is isotropic. Therefore, by analogy, we shall introduce rays orthogonal to the successive positions of the shock, and study how an element of the shock propagates down a ray tube. However, there is a crucial difference between a shock and a linear wavefront. The shock velocity at any point depends on its strength, so the geometry cannot be mapped out in advance, independently of the determination of the strength of the wave. The two are coupled together and the ray tube geometry itself has to be determined at the same time as the shock strength is determined in terms of the ray tube area. The equations corresponding to (7.60) and (7.61) are coupled together. It is as if c in (7.60) depends on Φ_0. But we have to analyze the whole thing again.

As a building block in this theory, we shall need to study propagation down a *given* tube of arbitrary cross section. This is interesting for its own sake, and of course propagation in a wedge-shaped channel is identical with cylindrical waves, propagation in a cone is identical with spherical waves, so that we have an opportunity for further discussion of those problems. Propagation of a plane wave in a nonuniform medium is similar, and some details of that are also included.

8.1 Shock Propagation Down a Nonuniform Tube

We consider the one dimensional (hydraulic) formulation for flow in a tube of given cross-sectional area $A(x)$. Even in a uniform tube, the shock can be changing in a complicated way due to interactions with the flow behind it, as described in the piston problem of Sections 6.8 and 6.11. But we are concerned with isolating as much as possible the effects due to a nonuniform $A(x)$ and, effectively, want to take the very simplest piston problem. That is, we want to formulate the problem in such a way that the shock would continue with constant speed in the case $A(x) = $ constant. To do this, assume that

$$A(x) = A_0 = \text{constant in } x < 0,$$

and that the shock is initially moving in this section with constant Mach number M_0. We may imagine the shock to be produced by a piston moving with appropriate *constant* speed far back in the uniform section. The piston is still providing the thrust to keep the shock moving, but there are no *changes* due to this; the changes are due entirely to the cross-sectional area. The problem then is to determine how the Mach number of the transmitted shock depends on $A(x)$ in $x > 0$.

The flow is not strictly one dimensional but if the cross section $A(x)$ does not vary too rapidly, the equations obtained by averaging across the tube will provide a good approximation. They are

$$\rho_t + u\rho_x + \rho u_x + \rho u \frac{A'(x)}{A(x)} = 0, \tag{8.1}$$

$$u_t + uu_x + \frac{1}{\rho}p_x = 0, \tag{8.2}$$

$$p_t + up_x - a^2(\rho_t + u\rho_x) = 0. \tag{8.3}$$

The area change appears only in the continuity equation (8.1), and this equation follows immediately from conservation of mass in the form

$$(\rho A)_t + (\rho u A)_x = 0. \tag{8.4}$$

We note that for propagation into a wedge with apex at x_0,

$$A(x) \propto (x_0 - x), \qquad \frac{A'(x)}{A(x)} = \frac{-1}{x_0 - x}, \tag{8.5}$$

and for a cone

$$A(x) \propto (x_0 - x)^2, \qquad \frac{A'(x)}{A(x)} = \frac{-2}{(x_0 - x)}. \tag{8.6}$$

With $r = (x_0 - x)$ and the sign of u reversed (to be measured positive for increasing r), the equations are then identical with those for cylindrical and spherical waves in (6.132)–(6.134), *and they are exact*. The fact that the equations are exact in this case indicates that the true criterion for the one dimensional formulation is really that the curvature of the walls in the x direction should be small. But the question of precise validity seems to be one which has never been completely investigated.

Figure 8.1 is the (x,t) diagram for the problem, with the origin of t taken as the time when the incident shock arrives at $x = 0$. For $t < 0$ the

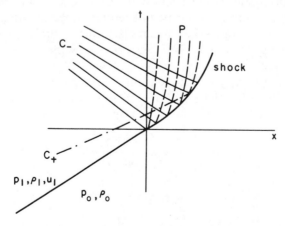

Fig. 8.1. The (x,t) diagram for a shock entering a nonuniform region.

flow consists of uniform regions separated by the moving shock. We take $u=0$, $p=p_0$, $\rho=\rho_0$ in the undisturbed state ahead of the shock, and we take $u=u_1$, $p=p_1$, $\rho=\rho_1$ in the initial uniform state behind it. The quantities u_1, p_1, ρ_1 are determined in terms of p_0, ρ_0, M_0 by the shock conditions. When the shock reaches $x=0$, disturbances to this state propagate on the particle paths P and the negative characteristics C_-. The C_- may have positive or negative slope depending on whether $u_1 > a_1$ or $u_1 < a_1$; the latter case is shown in Fig. 8.1. The problem is to determine these disturbances from (8.1)–(8.3), together with the modifications of the shock position and strength. The shock conditions are

$$u = a_0 \frac{2}{\gamma+1}\left(M - \frac{1}{M}\right),\tag{8.7}$$

$$p = \rho_0 a_0^2 \left\{ \frac{2}{\gamma+1} M^2 - \frac{\gamma-1}{\gamma(\gamma+1)} \right\},\tag{8.8}$$

$$\rho = \rho_0 \frac{(\gamma+1)M^2}{(\gamma-1)M^2+2},\tag{8.9}$$

and, when required, the sound speed is given by $a^2 = \gamma p/\rho$.

The Small Perturbation Case.

The one dimensional formulation is not limited to small changes in $A(x)$ itself, since large changes may be attained over large enough distances, even though the derivatives of $A(x)$ are small. However, in the case when the change of $A(x)$ from A_0 remains small,

$$\frac{A(x)-A_0}{A_0} \ll 1,$$

we may assume that the disturbances to the state u_1, p_1, ρ_1 behind the shock and the change in shock Mach number are correspondingly small. We may then solve the problem as a perturbation on the solution for the uniform tube. The equations (8.1)–(8.3) and the shock conditions (8.7)–(8.9) are linearized about the state u_1, p_1, ρ_1. However, it should be noted that there is no assumption that $p_1-p_0,\ldots$, are small; the shock is of arbitrary strength.

The linearized equations are

$$\rho_t + u_1\rho_x + \rho_1 u_x + \frac{\rho_1 u_1 A'(x)}{A_0} = 0,$$

$$u_t + u_1 u_x + \frac{1}{\rho_1} p_x = 0, \qquad (8.10)$$

$$p_t + u_1 p_x - a_1^2(\rho_t + u_1\rho_x) = 0,$$

where, to save writing, we leave it to be understood that ρ_t is interpreted as $(\rho - \rho_1)_t$, $A'(x)$ as $\{A(x) - A_0\}'$, and so on. The general solution is readily obtained since the equations are linear with constant coefficients; the most significant derivation is via the characteristic form of the equations. The characteristic equations for (8.10) are

$$C_+ : \left\{ \frac{\partial}{\partial t} + (u_1 + a_1)\frac{\partial}{\partial x} \right\}(p + \rho_1 a_1 u) + \rho_1 a_1^2 u_1 \frac{A'(x)}{A_0} = 0, \quad (8.11)$$

$$C_- : \left\{ \frac{\partial}{\partial t} + (u_1 - a_1)\frac{\partial}{\partial x} \right\}(p - \rho_1 a_1 u) + \rho_1 a_1^2 u_1 \frac{A'(x)}{A_0} = 0, \quad (8.12)$$

$$P : \left\{ \frac{\partial}{\partial t} + u_1 \frac{\partial}{\partial x} \right\}(p - a_1^2 \rho) = 0, \quad (8.13)$$

and the general solution taking each in turn is

$$(p - p_1) + \rho_1 a_1(u - u_1) = -\frac{\rho_1 a_1^2 u_1}{u_1 + a_1} \cdot \frac{A(x) - A_0}{A_0} + F\{x - (u_1 + a_1)t\}, \quad (8.14)$$

$$(p - p_1) - \rho_1 a_1(u - u_1) = -\frac{\rho_1 a_1^2 u_1}{u_1 - a_1} \cdot \frac{A(x) - A_0}{A_0} + G\{x - (u_1 - a_1)t\}, \quad (8.15)$$

$$(p - p_1) - a_1^2(\rho - \rho_1) = H(x - u_1 t), \quad (8.16)$$

where F, G, and H are arbitrary functions. In the linearized form, because of the constant coefficients, we have been able to carry out the integration on the three families of characteristics explicitly, and the characteristics have been approximated by the straight lines, $x - (u_1 \pm a_1)t = $ constant, $x - u_1 t = $ constant. The three arbitrary functions are to be determined from the initial conditions of the problem and the boundary conditions at the shock.

First, and most decisive, F must be identically zero. This is because the C_+ characteristics behind the shock, that is, the lines $x-(u_1+a_1)t<0$, all originate in the uniform region where $u=u_1, p=p_1, \rho=\rho_1, A=A_0$ (see Fig. 8.1); hence from (8.14), $F=0$. It is in this crucial step that modifying disturbances overtaking the shock are excluded. It should also be stressed that in this perturbation analysis the conclusion $F=0$ is a strict deduction from the formulation of the initial conditions, not an intuitive argument.

The other two functions G and H are not zero: they describe the disturbances on the C_- characteristics and the particle paths P shown in Fig. 8.1. These originate at the perturbed shock and the *three* shock conditions are sufficient to determine G, H, and the change in shock Mach number (from which the change in shock position may also be deduced). The functions G and H are of subsidiary interest. The main result we want is the change in shock Mach number, and this can be determined without involving G and H. The shock conditions give the perturbations $p-p_1$, $u-u_1$, at the shock in terms of the change in Mach number $M-M_0$. From (8.7) and (8.8), they are

$$p-p_1 = \frac{4\rho_0 a_0^2}{\gamma+1} M_0(M-M_0), \qquad u-u_1 = \frac{2}{\gamma+1} a_0 \left(1+\frac{1}{M_0^2}\right)(M-M_0).$$

$$(8.17)$$

When these are substituted in (8.14) with $F=0$, we have

$$\left\{ \frac{4}{\gamma+1} M_0 + \frac{2}{\gamma+1}\left(1+\frac{1}{M_0^2}\right)\frac{\rho_1 a_1}{\rho_0 a_0} \right\}(M-M_0) = -\frac{\rho_1 a_1^2}{\rho_0 a_0^2}\frac{u_1}{u_1+a_1}\frac{A-A_0}{A_0}.$$

$$(8.18)$$

The expressions for u_1, ρ_1, a_1 in terms of M_0 are given by (8.7)–(8.9) with $M=M_0$. Then, after some algebraic manipulation, (8.18) becomes

$$\frac{A-A_0}{A_0} = -g(M_0)(M-M_0), \qquad (8.19)$$

where

$$g(M) = \frac{M}{M^2-1}\left(1+\frac{2}{\gamma+1}\frac{1-\mu^2}{\mu}\right)\left(1+2\mu+\frac{1}{M^2}\right), \quad (8.20)$$

$$\mu^2 = \frac{(\gamma-1)M^2+2}{2\gamma M^2-(\gamma-1)}. \qquad (8.21)$$

The quantity μ is in fact the Mach number of the shock relative to the flow behind it.

When needed, the expressions for G and H may be found by applying (8.15) and (8.16) at the shock; in the small perturbation theory, it is consistent to apply these conditions at the unperturbed shock position $x = a_0 M_0 t$, since the errors would be second order. The details may be found in the paper by Chester (1954), where small perturbation results were first derived. It is interesting that the area term in (8.15) changes sign at $u_1 = a_1$, and in fact is singular at $u_1 = a_1$. Yet (8.19) shows neither a change in sign nor any singularity. Friedman (1960) has investigated the case $u_1 = a_1$, which is the case of exactly sonic flow behind the shock. He shows that small nonlinear effects must be incorporated into the disturbances on the C_- in order to obtain a uniformly valid solution, but that (8.19) is unaltered.

Before discussing (8.19) in detail, we go on to an important extension.

The Finite Area Changes; The Characteristic Rule.

For a tube which varies slowly but which accumulates large changes in $A(x)$ over a sufficiently large length, we might break down the problem into successive small lengths in each of which the change in A is small. In each such small length of tube, it would be admissible to linearize about the local conditions and develop a small perturbation theory as in (8.14)–(8.16). But it would no longer be strictly valid to take $F = 0$, because the entry conditions into each of these subsections would not be a uniform state. After a number of successive sections the errors might accumulate. However, *if we neglected this*, (8.19) would apply to each subsection with A_0 and M_0 taken to be the area and Mach number at the entry to the subsection. But then the theory is extremely simple. We are saying, in effect, that (8.19) is the differential form of a functional relation $M = M(A)$:

$$\frac{1}{A} \frac{dA}{dM} = -g(M), \tag{8.22}$$

and we do not have to discuss the subdivision into small subsections explicitly at all! Moreover, (8.19) was itself merely a substitution of the shock conditions into the characteristic relation on the C_+ characteristics. So the whole derivation can be put into the following characteristic rule:

Write down the exact nonlinear differential relation for the C_+ characteristics. Substitute the expressions for p, ρ, u, a in terms of M from the *shock conditions. The resulting differential equation gives the variation of M wih x.*

Although we have all the ingredients already, let us follow this prescription to emphasize its simplicity. The basic equations governing this particular problem are (8.1)–(8.3). The characteristic equation for the C_+ characteristic is

$$\frac{dp}{dx} + \rho a \frac{du}{dx} + \frac{\rho a^2 u}{u+a} \frac{1}{A} \frac{dA}{dx} = 0. \tag{8.23}$$

The shock conditions are given in (8.7)–(8.9). On substitution we have

$$g(M) \frac{dM}{dx} + \frac{1}{A} \frac{dA}{dx} = 0, \tag{8.24}$$

where $g(M)$ is given by (8.20). It is convenient to write this as

$$\frac{M}{M^2-1} \lambda(M) \frac{dM}{dx} + \frac{1}{A} \frac{dA}{dx} = 0, \tag{8.25}$$

where

$$\lambda(M) = \left(1 + \frac{2}{\gamma+1} \frac{1-\mu^2}{\mu}\right)\left(1 + 2\mu + \frac{1}{M^2}\right), \tag{8.26}$$

$$\mu^2 = \frac{(\gamma-1)M^2+2}{2\gamma M^2-(\gamma-1)}. \tag{8.27}$$

The reason for this choice is that $\lambda(M)$ varies little over the range of Mach numbers. The limits are

$$M \to 1, \qquad \lambda \to 4, \tag{8.28}$$

$$M \to \infty, \qquad \lambda \to n = 1 + \frac{2}{\gamma} + \sqrt{\frac{2\gamma}{\gamma-1}} = 5.0743 \text{ for } \gamma = 1.4. \tag{8.29}$$

Formula 8.25 was first obtained by Chisnell (1957), who used a different approach. The area distribution $A(x)$ was approximated by a sequence of discontinuous steps and the solution was built up from analysis of the elementary interaction of the shock at each discontinuity. At each interaction the shock is transmitted with a modified strength, and disturbances are reflected (the C_- and P disturbances in our analysis). But the reflected disturbances are themselves re-reflected when they travel back through the earlier steps; the re-reflected waves overtake the shock

and contribute to the later interaction. If all re-reflected waves are neglected, (8.25) follows. Chisnell analyzed the effects of all singly re-reflected
disturbances and found that their *total* modification to (8.25) was much
smaller than their individual contributions. Earlier Moeckel (1952) had
applied similar ideas to steady oblique shocks in nonuniform supersonic
streams. The nonuniform stream was replaced by layers, in each of which
the flow quantities were constant, with surfaces of discontinuity between
the layers. The solution was built up from the elementary interactions at
the interfaces.

   Although the Moeckel-Chisnell approach offers in principle the
possibility of successive improvement by including more and more
multireflections, it does not seem practicable to go beyond the first
re-reflections. It is also difficult to assess to what degree of approximation
the equations (8.1)–(8.3) have been solved. However, the relatively small
modifications from the first re-reflections indicate that (8.25) may be
unexpectedly good. This is indeed the case, as we shall see below.

   When the quick derivation of (8.25) by the characteristic rule occurred
to me, I hoped also that a full analysis of the approximation could be
based directly on (8.1)–(8.3). So far this has not been completed! To see
what is involved, note that the characteristic equation (8.23) may be
written

$$\frac{p_t}{u+a}+p_x+\rho a\left(\frac{u_t}{u+a}+u_x\right)+\frac{\rho a^2 u}{u+a}\frac{A'(x)}{A(x)}=0. \qquad (8.30)$$

This is exact and holds throughout the flow since it is just a combination
of the basic equations (8.1)–(8.3). If (8.23) is applied to a shock moving
with velocity U, we are claiming that

$$\frac{p_t}{U}+p_x+\rho a\left(\frac{u_t}{U}+u_x\right)+\frac{\rho a^2 u}{u+a}\frac{A'(x)}{A(x)}=0 \qquad (8.31)$$

is a good approximation at the shock. Taking (8.30) and (8.31) together we
see that the approximation is based on the assumption that

$$\left(\frac{1}{U}-\frac{1}{u+a}\right)(p_t+\rho a u_t) \qquad (8.32)$$

is relatively small at the shock; that is, this expression is small compared
with p_t/U. The smallness of the first factor would correspond to the
thought that the C_+ characteristic in Fig. 8.1 is fairly close to the shock, so
that we are merely transferring the relation that holds on a C_+ to the

shock. However, whereas $(u + a - U)/U$ is zero for $M = 1$, it tends to 0.274 (for $\gamma = 1.4$) as $M \to \infty$. The consequences of the characteristic rule are sometimes a hundred times more accurate than this! In the case of a cylindrical or spherical implosion noted below the relative error is about 0.003. Thus although the first factor may contribute a little to the accuracy, the rule works well because

$$\frac{p_t + \rho a u_t}{p_t} \tag{8.33}$$

is extremely small at the shock. Although some further discussion was presented in the original paper (Whitham, 1958), no really satisfactory explanation of this was found. Of course we know that the result is correct in the small perturbation theory, and we check from (8.14) with $F = 0$ that

$$p_t + \rho_1 a_1 u_t = 0,$$

in that theory.

With only the partial justification provided by the small perturbation case and the Moeckel-Chisnell analysis, the accuracy of (8.25) has been confirmed by comparison with known solutions. First for weak shocks, $M \simeq 1$, we have $\lambda \simeq 4$; hence

$$M - 1 \propto A^{-1/2}. \tag{8.34}$$

This is the correct result of geometrical acoustics for weak pulses, as $M - 1$ is proportional to the strength of the pulse. Secondly, we may apply (8.25) to converging cylindrical or spherical shocks by taking $A \propto x_0 - x$ or $(x_0 - x)^2$, respectively, and compare the results with Guderley's exact similarity solutions described in Section 6.16. For infinitely strong shocks, λ tends to the value n given in (8.29), and (8.25) becomes

$$\frac{n}{M} \frac{dM}{dx} + \frac{1}{A} \frac{dA}{dx} = 0, \qquad M \propto A^{-1/n}. \tag{8.35}$$

Therefore the rule gives

$$M \propto r^{-1/n} \qquad \text{for cylindrical shocks,}$$
$$M \propto r^{-2/n} \qquad \text{for spherical shocks.} \tag{8.36}$$

A comparison with the exponents from the exact similarity solution is given in Table 8.1. The accuracy is amazing in view of the simplicity of the approximate theory. Among other things, it shows that converging shocks are reacting primarily to the changing geometry as assumed in the approximate theory, and are very little affected by the further disturbances from the source of the motion; the strength of the initial shock enters only through the constants of proportionality in (8.36). This would not be true for outgoing shocks. They would slow down due to the expanding geometry and the continuing interaction with the flow behind over large distances would be important; this approximate theory is *not* appropriate for such problems.

TABLE 8.1

	Cylindrical		Spherical	
γ	Approximate	Exact	Approximate	Exact
6/5	0.163112	0.161220	0.326223	0.320752
7/5	0.197070	0.197294	0.394142	0.394364
5/3	0.225425	0.226054	0.450850	0.452692

Another point of interest is that according to the approximate theory the exponent for the spherical case is just double the exponent in the cylindrical case. This is not true, however, in the exact similarity solution, although it is very nearly so.

With the partial justification mentioned earlier and these independent checks, we conclude that the characteristic rule gives a good simple approximation to problems of this type and it may be used with confidence in a wide range of problems.

For general M, the solution of (8.25) may be written

$$\frac{A}{A_0} = \frac{f(M)}{f(M_0)}, \qquad f(M) = \exp\left\{ - \int \frac{M\lambda(M)}{M^2 - 1} \, dM \right\}. \qquad (8.37)$$

In particular this formula may be used to extend the results for converging and spherical shocks to include shocks of intermediate strength. Of course, as the center is approached, $A \to 0$ and $M \to \infty$. Table 8.3 presents values of $f(M)$ for $\gamma = 1.4$. (See p. 288.)

In the next section the characteristic rule will be used for the problem of a shock propagating through a nonuniform density layer, and further examples may be found in the original paper (Whitham, 1958). It will also be the basis for the geometrical treatment of two and three dimensional shock propagation in Section 8.3.

8.2 Shock Propagation through a Stratified Layer

Here the method is applied to the one dimensional problem of a plane shock moving in the x direction through a given equilibrium distribution $u = 0$, $p = p_0(x)$, $\rho = \rho_0(x)$. If $p_0(x)$ is not constant, there must be a body force in the problem to maintain the pressure gradients and we include this in the equations. The one dimensional equations are

$$\rho_t + u\rho_x + \rho u_x = 0,$$

$$u_t + uu_x + \frac{1}{\rho}p_x = \mathcal{F},$$ (8.38)

$$p_t + up_x - a^2(\rho_t + u\rho_x) = 0,$$

where $\mathcal{F}$ is the body force per unit mass. In the atmosphere or for propagation in the outer layers of a star, $\mathcal{F}$ would be the gravitational acceleration. In equilibrium, the distribution of $\rho_0(x)$ and $p_0(x)$ must satisfy

$$\frac{1}{\rho_0}\frac{dp_0}{dx} = \mathcal{F},$$ (8.39)

and the entropy distribution must be given to complete the determination of $\rho_0(x), p_0(x)$. In the atmosphere $\mathcal{F} = -g$ and we have, for example,

$$\rho_0(x) = \rho_0(0)e^{-gx/\mathcal{R}T_0} \qquad \text{(isothermal)},$$

$$\frac{\gamma\kappa}{\gamma-1}\rho_0^{\gamma-1}(x) = c - gx \qquad \text{(isentropic)},$$

as discussed in Section 6.6.

We now apply the characteristic rule to the propagation of a shock through such a layer, remembering that the theory applies only for the local effects of the layer and should be used only when additional effects would be small. The appropriate characteristic relation may be written in differential form as

$$dp + \rho a\, du - \frac{\rho a}{u+a}\mathcal{F}\, dx = 0 \qquad \text{on} \qquad \frac{dx}{dt} = u + a.$$

But we apply it along the shock. That is, we use the differential equation

$$\frac{dp}{dx} + \rho a \frac{du}{dx} - \frac{\rho a}{u+a} \mathcal{F} = 0, \tag{8.40}$$

with u, p, ρ, a expressed in terms of $p_0(x)$, $\rho_0(x)$, and the shock Mach number $M(x)$. In general, numerical integration will be required, but the result for strong shocks may be obtained analytically. For strong shocks [see (6.110)], the shock conditions simplify to

$$u = \frac{2}{\gamma+1} U, \qquad p = \frac{2}{\gamma+1} \rho_0 U^2, \qquad \rho = \frac{\gamma+1}{\gamma-1} \rho_0, \qquad a^2 = \frac{\gamma p}{\rho} = \frac{2\gamma(\gamma-1)}{(\gamma+1)^2} U^2,$$

where U is the shock velocity. In this limit U is relatively large and the third term in (8.40) is negligibly small compared with the other two; the body force $\mathcal{F}$ enters indirectly through its control of $\rho_0(x)$. Equation 8.40 reduces to

$$\frac{1}{U} \frac{dU}{dx} + \beta \frac{1}{\rho_0} \frac{d\rho_0}{dx} = 0,$$

where

$$\beta = \left(2 + \sqrt{\frac{2\gamma}{\gamma-1}}\right)^{-1} \tag{8.41}$$

Hence

$$U \propto \rho_0^{-\beta}, \qquad p \propto \rho_0^{1-2\beta}. \tag{8.42}$$

For $\gamma = 1.4$, $\beta = 0.21525$.

These results allow a further check to be made against exact solutions. Sakurai (1960) investigated similarity solutions to this problem in the case where $\rho_0 \propto x^\alpha$. He found $U \propto x^{-\lambda}$ and determined the value of λ for different values of α. His values of λ/α are given in Table 8.2, and they are compared with β. Although not quite as good as for the implosion problem, the approximation is still remarkably close.

The limitation to problems where *local* modification of the shock is intense must be borne in mind. For an exponential fall off in density, similarity solutions can also be found and compared with this approximation. The comparison has been made by Hayes (1968) and the difference in the exponents is as much as 15%. We attribute this to the fact that an exponential change in density does not have the strong local effect of a power law with $\rho_0 \to 0$ at finite x.

TABLE 8.2

γ	$\alpha = 2$	$\alpha = 1$	$\alpha = 1/2$	β
5/3	0.21779	0.22335	0.22820	0.23608
7/5	0.19667	0.20214	0.20704	0.21525
6/5	0.16545	0.17040	0.17498	0.18301

Chisnell (1955) originally studied the problem of this section from the successive interaction approach, in the case $p_0 =$ constant, $\mathscr{F} = 0$, and found the small corrections due to re-reflections. As before there is some beneficial cancellation.

8.3 Geometrical Shock Dynamics

We now turn to the development of the approximate geometrical theory for shock propagation in two or three dimensional problems when there is no special symmetry (Whitham, 1957, 1959b). We consider a shock propagating into a uniform gas at rest and, based on the experience with geometrical optics for linear problems, we introduce "rays" defined as the orthogonal trajectories of the successive positions of the shock. As a specific example, consider the case of shock diffraction around a corner in Fig. 8.2. The shock positions are shown by full line curves and the rays by broken lines. The idea is to treat the propagation of each element of the shock down each elementary ray tube as a problem of shock propagation in a tube with solid walls. The equivalence would be exactly valid if the rays were particle paths, since solid walls are particle paths in inviscid flow. However, this is only approximately true. The shock conditions require the induced flow immediately behind the shock to be normal to it, but as the distance from the shock increases the particle paths will deviate from the rays in general. So a definite approximation is involved, and it is one that may be quite severe. However, it is only by this step, or a similar one, that the geometrical effects can be extricated from the whole complicated flow. In the problem of diffraction around a corner (Fig. 8.2) the wall itself has to be both a ray and a particle path along its entire length, so there is some additional resistraint on the deviation between the rays and particle paths further back.

The accuracy of this type of approximatiom is hard to assess in advance and higher approximations are virtually impossible. As justifica-

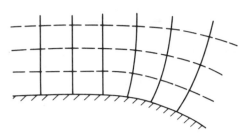

Fig. 8.2. Shock positions (solid lines) and rays (dashed lines) in the diffraction of a shock around a continuous corner.

tion we shall see that the theory does reduce precisely to geometrical optics for linear problems and the nonlinear results will be compared both with other theoretical results for special cases and with experiments. It is perhaps appropriate to comment that approximations which are easy to assess usually involve small effects. Here we are concerned with large effects in extremely difficult problems.

The ray tube approximation is independent of how the propagation in each ray tube is handled. However, we assume that the local Mach number will be a function of the ray tube area, and in the absence of any other explicit formula we adopt the results established in Section 8.1 and use the relation in (8.37).

It is convenient to specify the shock position at time t in the form

$$\alpha(\mathbf{x}) = a_0 t, \tag{8.43}$$

where a_0 is the undisturbed sound speed. The successive shock positions are then given by the family of surfaces $\alpha(\mathbf{x}) = \text{constant}$. It is clear that in principle we have a procedure to determine the function $\alpha(\mathbf{x})$. First of all, the shock Mach number at any point can be determined in terms of $\alpha(\mathbf{x})$ from (8.43). Secondly, it must be possible to determine all the geometry of the rays from the function $\alpha(\mathbf{x})$, since it specifies the family of shock positions: this determination gives the ray tube area. The A-M relation then provides the bridge to derive an equation for $\alpha(\mathbf{x})$.

The normal velocity of any moving surface $S(t, \mathbf{x}) = 0$ has been noted in (7.63). If this is applied to $S = a_0 t - \alpha(\mathbf{x})$, the shock velocity is found to be $U = a_0 / |\nabla \alpha|$. Therefore

$$M = \frac{1}{|\nabla \alpha|}. \tag{8.44}$$

To study the geometry of the ray tubes it is convenient to introduce a unit vector $\mathbf{l}$ for the ray direction at any point and a function A related to the ray tube area. The definition of $\mathbf{l}$ is clear and it is given in terms of α by

$$\mathbf{l} = \frac{\nabla \alpha}{|\nabla \alpha|}, \tag{8.45}$$

since the rays are normal to the surfaces $\alpha = \text{constant}$. The definition of A may need a little amplification. We want to introduce a finite function of position that can be used to measure the area of arbitrary infinitesimal ray tubes. To do this we consider any particular ray and construct a narrow ray tube around it consisting of a bundle of neighboring rays. We may then introduce the ratio of the cross-sectional area at any location along the ray tube to the area at a standard reference section. In the limit as the maximum diameter of the ray tube tends to zero, this ratio approaches a finite limit and the limit function is taken as the function A along that ray. It is defined in similar fashion along each ray and so becomes a function of position. For any infinitesimal ray tube, A is now proportional to the ray tube area rather than being the area itself. However, in (8.37) only the ratio of areas appears so that this quantity A is still related to the local Mach number by

$$\frac{A}{A_0} = \frac{f(M)}{f(M_0)}. \tag{8.46}$$

For the same reason the original reference point for the ratio of cross-sectional areas along the ray tube drops out and is replaced by the initial condition $A = A_0$ for $M = M_0$, which is incorporated in (8.46). In fact, A can be taken to be any finite function proportional to the infinitesimal ray tube area along the ray. Different constants of proportionality on different rays would be compensated for by different values of A_0.

The relation of A to the function $\alpha(\mathbf{x})$ defining the shock positions comes essentially from the fact that increases in A along a ray are related to the divergence of the ray vector $\mathbf{l}$ in (8.45). In fact, we now show that

$$\nabla \cdot \left(\frac{\mathbf{l}}{A} \right) = 0, \tag{8.47}$$

and this can also be written

$$\frac{1}{A} \frac{dA}{ds} = \frac{\mathbf{l} \cdot \nabla A}{A} = \nabla \cdot \mathbf{l}. \tag{8.48}$$

To prove (8.47), the divergence theorem is applied to the volume V in a

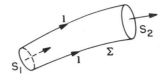

Fig. 8.3. Ray tube geometry.

narrow ray tube between the two successive shock positions as in Fig. 8.3. We have

$$\int_V \nabla \cdot \left(\frac{1}{A} \right) dV = \int_{S_1 + \Sigma + S_2} \frac{\mathbf{l} \cdot \mathbf{\nu}}{A} \, dS, \qquad (8.49)$$

where Σ refers to the sides of the tube, S_1 and S_2 refer to the ends, and $\mathbf{\nu}$ is the outward normal. On the side walls $\mathbf{l} \cdot \mathbf{\nu} = 0$ by definition of $\mathbf{l}$ so the contribution from Σ is zero. On S_2, $\mathbf{l} \cdot \mathbf{\nu} = +1$, while on S_1, $\mathbf{l} \cdot \mathbf{\nu} = -1$. Hence the right hand side of (8.49) reduces to

$$\int_{S_2} \frac{dS}{A} - \int_{S_1} \frac{dS}{A}.$$

From the definition of A, both integrals tend to the same value so the difference tends to zero as the diameter of the tube shrinks to zero. Therefore the integral in (8.49) is zero. Since the choice of V is arbitrary, it follows that (8.47) holds everywhere.

Equations 8.44–8.47 provide a partial differential equation for $\alpha(\mathbf{x})$. Collecting the results we have

$$M = \frac{1}{|\nabla \alpha|}, \qquad (8.50)$$

$$\nabla \cdot \left(\frac{M}{A} \nabla \alpha \right) = 0, \qquad (8.51)$$

$$\frac{A}{A_0} = \frac{f(M)}{f(M_0)}, \qquad (8.52)$$

where $\mathbf{l} = \nabla \alpha / |\nabla \alpha| = M \nabla \alpha$ has been used in obtaining (8.51) from (8.47).

This is a convenient form for comparison with the results of geometrical optics in linear theory. The linear limit corresponds to $M \rightarrow 1$ and the linear theory replaces (8.50)–(8.52), respectively, by

$$|\nabla \alpha| = 1, \tag{8.53}$$

$$\nabla \cdot \left(\frac{1}{A} \nabla \alpha \right) = 0, \tag{8.54}$$

$$\frac{A}{A_0} = \left(\frac{M_0 - 1}{M - 1} \right)^2. \tag{8.55}$$

Note that M is replaced identically by unity in the first two, where the geometry is concerned, but $M - 1$ appears as a measure of the strength (which is small) in (8.55). In this way the geometry is uncoupled from the determination of the strength of the wave. The flow quantities such as $z = (p - p_0)/p_0$ are proportional to $(M - 1)$ and the linear theory would refer to z rather than $M - 1$. From (8.54) and (8.55) we have

$$\nabla \cdot (z^2 \nabla \alpha) = 0. \tag{8.56}$$

Equation 8.53 is the same as the eikonal equation (7.65), bearing in mind the normalization of α by the sound speed, and (8.56) is the same as the transport equation (7.66) with $z \propto \Phi_0$. In the linear theory (8.56) was obtained first and then interpreted as $z \propto A^{-1/2}$, which corresponds to (8.55). Our arguments here have led directly to what was the "interpretation." The main point, however, is that the theory developed here does reduce to the linear theory in the appropriate limits. The crucial difference in the nonlinear theory is the coupling of the strength z with M. Even for weak shocks, with $M - 1 \ll 1$, this coupling can make important qualitative differences.

8.4 Two Dimensional Problems

In two dimensions the shock positions and rays form an orthogonal coordinate system as shown in Fig. 8.4, and for some purposes it is convenient to formulate the equations in terms of these intrinsic coordinates. The successive shock positions are already described by the family of curves $\alpha = $ constant, and we introduce a function $\beta(\mathbf{x})$ to describe the rays as the family $\beta = $ constant. The required equations, using α and β as independent coordinates, can be derived by direct transformation of equations (8.50) to (8.52), but it is instructive in bringing out further points in the geometry to give an independent derivation. (This was, in fact, the first derivation of the theory.)

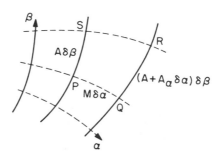

Fig. 8.4. Line elements in shock dynamics.

In the description based on the net of curves $\alpha = $ constant, β $= $ constant, the geometry is closely tied to the line elements for increments $d\alpha, d\beta$ in the coordinates, and the ray tube geometry is introduced via the coefficients for the line elements. The line element for an increment $d\beta$ will be $A(\alpha, \beta)d\beta$ for some function A. This function A is clearly proportional to the width of the ray channel between rays β and $\beta + d\beta$. In the two dimensional problem ray tubes are of constant depth in the third dimension; hence A is proportional to the ray tube area and may be used as in the first formulation. An increment $d\alpha$ corresponds to a change of shock position in time $dt = d\alpha / a_0$. Therefore the distance traveled is $U\,dt = M\,d\alpha$. This shows that the line element for the increment $d\alpha$ is $M\,d\alpha$. The general line element for neighboring points is given by the metric

$$\cdot\, ds^2 = M^2 d\alpha^2 + A^2 d\beta^2. \tag{8.57}$$

The functions M and A in such orthogonal coordinates are not arbitrary functions of (α, β). They satisfy a differential equation which follows because we know that the 2-space described by (8.57) is, in fact, flat. The curvature calculated in terms of M and A must vanish. Another way of saying this is that M and A must be such that (8.57) can be transformed into

$$ds^2 = dx^2 + dy^2.$$

The appropriate condition, which will be derived below, is

$$\frac{\partial}{\partial\alpha}\left(\frac{1}{M}\frac{\partial A}{\partial\alpha}\right) + \frac{\partial}{\partial\beta}\left(\frac{1}{A}\frac{\partial M}{\partial\beta}\right) = 0. \tag{8.58}$$

When the A-M relation is added we have a complete set of equations to determine $A(\alpha,\beta)$, $M(\alpha,\beta)$. From these the rays and shock positions can be determined as functions of x and y.

To establish (8.58), consider the curvilinear quadrilateral $PQRS$ in Fig. 8.4 with vertices (α,β), $(\alpha+\delta\alpha,\beta)$, $(\alpha,\beta+\delta\beta)$, $(\alpha+\delta\alpha,\beta+\delta\beta)$. Let $\theta(\alpha,\beta)$ be the angle between the ray and a fixed direction, the x axis say. Since the sides PS and QR are of length $A\delta\beta$ and $(A+A_\alpha\delta\alpha)\delta\beta$, respectively, and the distance between them is $M\delta\alpha$, the change in ray inclination from P to S is

$$\delta\theta = \frac{QR - PS}{PQ} = \frac{1}{M}\frac{\partial A}{\partial\alpha}\delta\beta.$$

Hence

$$\frac{\partial\theta}{\partial\beta} = \frac{1}{M}\frac{\partial A}{\partial\alpha}. \tag{8.59}$$

Since the inclination of the β curves is $\theta + \frac{1}{2}\pi$, a similar argument shows that

$$\frac{\partial\theta}{\partial\alpha} = -\frac{1}{A}\frac{\partial M}{\partial\beta}. \tag{8.60}$$

Equation 8.58 follows by elimination of θ, but it will be more convenient to work with the pair of equations (8.59) and (8.60). The system is completed by the A-M relation

$$A = A_0\frac{f(M)}{f(M_0)}. \tag{8.61}$$

The equivalence of (8.59)–(8.60) with (8.50)–(8.51) is easily established using

$$\alpha_x = \frac{\cos\theta}{M}, \qquad \alpha_y = \frac{\sin\theta}{M},$$
$$\beta_x = -\frac{\sin\theta}{A}, \qquad \beta_y = \frac{\cos\theta}{A}, \tag{8.62}$$

Once θ, M, A have been found as functions of α and β, the shock positions may be obtained by integration along rays. On a ray

$$\frac{\partial x}{M \, \partial \alpha} = \cos \theta, \qquad \frac{\partial y}{M \, \partial \alpha} = \sin \theta.$$

Therefore

$$x = x_0(\beta) + \int_0^\alpha M \cos \theta \, d\alpha,$$

$$\hspace{8cm} (8.63)$$

$$y = y_0(\beta) + \int_0^\alpha M \sin \theta \, d\alpha,$$

give the position of the shock at time $t = \alpha/a_0$ in terms of the position $x = x_0(\beta)$, $y = y_0(\beta)$ at $t = 0$.

In general, the coefficient $A_0/f(M_0)$ in (8.61) may be a function of β, since both A_0 and M_0 may vary along the initial shock position $\alpha = 0$. But we may define a new variable β and a new A to eliminate this. The invariant quantity is the line element $A \, d\beta$. If $A = k(\beta)\overline{A}$, then

$$A \, d\beta = k(\beta)\overline{A} \, d\beta = \overline{A} \, d\overline{\beta},$$

where

$$\overline{\beta} = \int k(\beta) d\beta.$$

Thus any unwanted factor $k(\beta)$ can be absorbed into a new $\overline{\beta}$. We shall assume this has been done, unless otherwise stated, and we take $A = A(M)$. In the diffraction problem of Fig. 8.2, the initial shock $\alpha = 0$ is plane and $M_0 = $ constant. We choose β to be the distance from the wall in this uniform region. Hence $A_0 = 1$ and

$$A = \frac{f(M)}{f(M_0)}. \hspace{6cm} (8.64)$$

8.5 Wave Propagation on the Shock

It is interesting that (8.59)–(8.61) turn out to be hyperbolic and represent a wave motion for disturbances propagating on the shock. A little thought shows that this should have been expected. The flow in the region behind a deforming shock involves two dimensional waves propa-

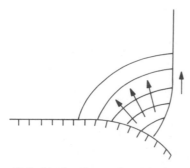

Fig. 8.5. Cylindrical waves produced in the diffraction of a shock.

gating with the local sound speed relative to the local flow as sketched in Fig. 8.5. Our approximate equations describe in some way the trace of these cylindrical waves where they intersect the shock.

The wave propagation on the shock is brought out by studying the characteristic form of the equations. When $A = A(M)$ is substituted in (8.59) and (8.60), they become

$$\frac{\partial \theta}{\partial \beta} - \frac{A'(M)}{M} \frac{\partial M}{\partial \alpha} = 0,$$

$$\frac{\partial \theta}{\partial \alpha} + \frac{1}{A(M)} \frac{\partial M}{\partial \beta} = 0. \tag{8.65}$$

The characteristic form is

$$\left(\frac{\partial}{\partial \alpha} \pm c \frac{\partial}{\partial \beta} \right) \left(\theta \pm \int \frac{dM}{Ac} \right) = 0, \tag{8.66}$$

where c is the function of M given by

$$c(M) = \sqrt{\frac{-M}{AA'}}. \tag{8.67}$$

Since $A'(M) < 0$, the characteristics are real and we have nonlinear waves propagating with velocities

$$\frac{d\beta}{d\alpha} = \pm c$$

relative to the (α, β) mesh. These waves carry the changes of the shock shape and shock strength along the shock. The Riemann invariants are given by (8.66). We have

$$\theta + \int \frac{dM}{Ac} = \text{constant on} \frac{d\beta}{d\alpha} = c, \qquad (8.68)$$

$$\theta - \int \frac{dM}{Ac} = \text{constant on} \frac{d\beta}{d\alpha} = -c. \qquad (8.69)$$

The wave motion is analogous in every way to the prototype of one dimensional nonlinear gas dynamics discussed in Chapter 6, and the ideas and techniques established there may be taken over to these waves propagating along the shock.

The $A(M)$ relation is derived from (8.25), which may be written

$$\frac{1}{A} \frac{dA}{dM} = - \frac{M}{M^2 - 1} \lambda(M). \qquad (8.70)$$

Hence

$$Ac = \left\{ \frac{M^2 - 1}{\lambda(M)} \right\}^{1/2}, \qquad (8.71)$$

and the integral in the Riemann invariants is

$$\omega(M) = \int_1^M \frac{dM}{Ac} = \int_1^M \left\{ \frac{\lambda(M)}{M^2 - 1} \right\}^{1/2} dM. \qquad (8.72)$$

The explicit formulas for weak shocks, $M - 1 \ll 1$, and strong shocks, $M \gg 1$, will be useful. They are

$$\left. \begin{array}{ll} \lambda \sim 4, & \dfrac{A}{A_0} \sim \dfrac{(M_0 - 1)^2}{(M - 1)^2} \\[2mm] Ac \sim \left(\dfrac{M-1}{2} \right)^{1/2}, & \omega(M) \sim 2^{3/2} (M - 1)^{1/2} \end{array} \right\} \quad \text{as } M \to 1, \quad (8.73)$$

and

$$\left. \begin{array}{ll} \lambda \sim n = 5.0743 \text{ for } \gamma = 1.4, & \dfrac{A}{A_0} \sim \left(\dfrac{M_0}{M} \right)^n \\[2mm] Ac \sim n^{-1/2} M, & \omega(M) \sim n^{1/2} \log M \end{array} \right\} \quad \text{as } M \to \infty. \quad (8.74)$$

The characteristic relations are most easily obtained in the (α, β) coordinates, but in applications to specific boundary value problems the description in Cartesian coordinates (x,y) is sometimes preferable. It is a simple matter to transform (8.68) and (8.69) to this form. We note that

$$\frac{dy}{dx} = \frac{y_\alpha + y_\beta \, d\beta / d\alpha}{x_\alpha + x_\beta \, d\beta / d\alpha}, \tag{8.75}$$

where $x_\alpha = M \cos\theta$, $y_\alpha = M \sin\theta$, $x_\beta = -A \sin\theta$, $y_\beta = A \cos\theta$. Therefore the characteristics $d\beta / d\alpha = \pm c$ become

$$\frac{dy}{dx} = \tan(\theta \pm m), \tag{8.76}$$

where

$$\tan m = \frac{Ac}{M} = \left(-\frac{A}{MA'} \right)^{1/2}, \tag{8.77}$$

and the characteristic equations are

$$\theta \pm \omega(M) = \text{constant} \qquad \text{on} \quad C_\pm : \frac{dy}{dx} = \tan(\theta \pm m). \tag{8.78}$$

Values of $m(M)$ and $\omega(M)$ are given in Table 8.3, which is taken from a paper by Bryson and Gross (1961).

These equations could also be deduced directly from the two dimensional form of (8.50), (8.51). With $\alpha_x = \cos\theta / M$, $\alpha_y = \sin\theta / M$, the equivalent set of equations in θ and M is

$$\frac{\partial}{\partial x} \left(\frac{\sin\theta}{M} \right) - \frac{\partial}{\partial y} \left(\frac{\cos\theta}{M} \right) = 0,$$
$$\frac{\partial}{\partial x} \left(\frac{\cos\theta}{A} \right) + \frac{\partial}{\partial y} \left(\frac{\sin\theta}{A} \right) = 0. \tag{8.79}$$

It is straightforward, but longer, to show that the characteristic equations are the above.

TABLE 8.3

Mach no. M	Ray area A $A \times 10^{-N}$	N	Characteristic angle m (degrees)	The integral ω
1	∞		0	0
1·000001	3·668749	$+10$	—	0·003
1·00001	3·668672	$+8$	—	0·009
1·0001	3·667902	$+6$	0·403	0·028
1·001	3·660213	$+4$	1·280	0·089
1·01	3·584696	$+2$	4·002	0·283
1·05	1·310728	$+1$	8·544	0·633
1·10	2·946288	$+0$	11·474	0·896
1·15	1·184152	$+0$	13·142	1·097
1·20	6·053638	-1	14·843	1·266
1·25	3·536658	-1	15·958	1·414
1·30	2·250720	-1	16·859	1·547
1·35	1·520662	-1	17·604	1·669
1·40	1·074028	-1	18·231	1·728
1·45	7·850741	-2	18·766	1·887
1·50	5·898186	-2	19·228	1·984
1·55	4·531934	-2	19·630	2·077
1·60	3·548150	-2	19·983	2·165
1·65	2·822580	-2	20·295	2·249
1·70	2·276434	-2	20·572	2·330
1·75	1·858064	-2	20·820	2·406
1·80	1·532637	-2	21·042	2·480
1·85	1·276079	-2	21·242	2·551
1·90	1·071389	-2	21·423	2·619
1·95	9·063299	-3	21·587	2·685
2·00	7·719471	-3	21·736	2·749
2·05	6·615861	-3	21·872	2·811
2·10	5·702352	-3	21·997	2·871
2·15	4·940726	-3	22·111	2·929
2·20	4·301517	-3	22·216	2·985
2·25	3·761766	-3	22·312	3·040
2·30	3·303423	-3	22·401	3·094
2·40	2·576553	-3	22·560	3·203
2·50	2·037086	-3	22·696	3·302
2·60	1·630023	-3	22·814	3·388
2·70	1·318343	-3	22·916	3·477
2·80	1·076566	-3	23·006	3·563
2·90	8·868121	-4	23·085	3·645
3·00	7·363072	-4	23·154	3·724
3·20	5·184216	-4	23·271	3·875
3·40	3·740925	-4	23·364	4·015
3·60	2·757067	-4	23·439	4·148
3·80	2·069662	-4	23·501	4·272
4·00	1·578970	-4	23·552	4·398
4·50	8·519558	-5	23·647	4·660
5·00	4·926060	-5	23·710	4·900
6·00	1·921342	-5	23·788	5·314
7·00	8·705958	-6	23·832	5·672
8·00	4·395269	-6	23·859	5·966
9·00	2·408270	-6	23·876	6·232
10·00	1·407051	-6	23·889	6·470
15·00	1·786391	-7	23·917	7·385
20·00	4·141420	-8	23·926	8·033
100·00	1·172427	-11	23·937	11·67
∞	0		23·938	∞

8.6 Shock-Shocks

The function $c(M)$ is an increasing function of M. Therefore waves moving in the positive direction and carrying an increase of M and θ will break in typical nonlinear fashion. From the earlier experience in analogous problems, we assume that a discontinuous jump in M and θ will be required; that is, the shock develops a corner as shown in Fig. 8.6. Within this approximate theory, we follow the usual philosophy of such "shock" discontinuities and derive jump conditions from the conservation form of the basic equations. These "shocks" in the waves on the original gas dynamic shock will be referred to as "shock-shocks." The waves on the shock are interpreted as the trace of roughly cylindrical waves which are spreading out in the flow behind the shock. A shock-shock is the trace of a true gas dynamic shock in the flow behind the main shock. Thus it corresponds to the three-shock Mach reflection (described at the end of Chapter 6) which has been studied and investigated directly. We shall discuss the relation with Mach reflection in more detail later.

The differential equations (8.59) and (8.60) of this theory are derived from the metric (8.57) of the shock-ray network. The corresponding finite form is required to deduce jump conditions. Consider the neighborhood of the discontinuity in two successive positions of the shock as shown in Fig. 8.6. Let the difference in the α coordinates for the two shock positions be $\Delta\alpha$, and let the difference in the β coordinates of the rays be $\Delta\beta$. Let subscripts 1 and 2 refer to values ahead of and behind the discontinuity. Then, in Fig. 8.6, $PQ = M_2\Delta\alpha$, $QR = A_2\Delta\beta$, $SR = M_1\Delta\alpha$, $PS = A_1\Delta\beta$.

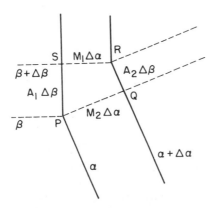

Fig. 8.6. Line elements for a shock-shock.

Expressing the distance PR in two alternative ways, we have

$$(M_1\Delta\alpha)^2 + (A_1\Delta\beta)^2 = (M_2\Delta\alpha)^2 + (A_2\Delta\beta)^2.$$

But the ratio $\Delta\beta/\Delta\alpha$ is the shock-shock velocity C in the (α,β) coordinates; hence

$$C^2 = -\frac{M_2^2 - M_1^2}{A_2^2 - A_1^2}. \tag{8.80}$$

The corresponding jump in θ is deduced from

$$\cot(\theta_2 - \theta_1) = \tan(RPQ + RPS)$$

$$= \frac{A_2 C/M_2 + M_1/A_1 C}{1 - A_2 M_1/M_2 A_1}.$$

Substituting for C from (8.80), we have

$$\tan(\theta_2 - \theta_1) = \frac{(M_2^2 - M_1^2)^{1/2}(A_1^2 - A_2^2)^{1/2}}{A_2 M_2 + A_1 M_1}. \tag{8.81}$$

For the description in Cartesian coordinates (x,y), (8.80) is transformed to the equivalent forms

$$\tan(\chi - \theta_i) = \frac{A_i}{M_i}\left(\frac{M_2^2 - M_1^2}{A_1^2 - A_2^2}\right)^{1/2}, \qquad i = 1 \text{ or } 2, \tag{8.82}$$

where χ is the angle of the shock-shock line with the x axis.

It is assumed that the functional relation (8.61) between A and M still applies even for the sharp change in channel section at a shock-shock, the velocity C is determined by (8.80) in terms of M_1 and M_2, and the jump in θ is determined by (8.81). Then for weak shock-shocks, $M_2 - M_1 \to 0$, it is easily checked that (8.80) reduces correctly to the characteristic velocity (8.67), and (8.81) reduces correctly to the Riemann invariant relation. However, for sufficiently strong shock-shocks the dependence of A_2 on M_2 will not be given accurately by (8.61) since the relation was derived on the supposition that the channel section varies slowly. Nor is it just a question of establishing the correct formulas relating M and A at an abrupt change in channel section. In fact, these formulas have been found by Laporte (1954). In reality, as is known from the traditional discussion of Mach reflection, there is a third shock and a vortex sheet behind the main shock; in principle, then, one should take the additional relations from the analysis of the three-shock configurations. This would be a complication which does not seem to be worth pursuing in detail in view of the approximate nature of the theory. We might note, however, that if it were done, the A-M relation for the waves following the shock-shock would take the form

$$A = k(\beta)f(M),$$

where $k(\beta) = A_2/f(M_2)$ has to be found from the three-shock relations; it would be incorrect to continue the A-M relation through the shock-shock back to the initial position and take $k = A_0/f(M_0)$. The whole thing is like the question of entropy in ordinary gas dynamics, where first of all one assumes that $p = p(\rho)$ and this leads to simple waves. But then, since compression waves break, shocks have to be considered and they involve entropy changes so that p is no longer a function of ρ alone; behind the shock the entropy is constant on each particle path. We have the analogous situation with A and M similar to p and ρ, and k playing a role similar to the entropy. The simpler theory of shock-shocks with $A = A(M)$ in (8.80) and (8.81) is rather like neglecting entropy changes at gas dynamic shocks. This is known to give accurate results if the discontinuity is not too strong, and it is expected that the same will be true here. A comparison between these simpler shock-shock conditions and the three shock results for Mach reflection will be given in Fig. 8.11. It substantiates the view that the more elaborate treatment would not be worthwhile in the context of this approximate theory.

8.7 Diffraction of Plane Shocks

We now consider specific applications of the general theory and start with the problem of the diffraction of a plane shock as it moves along a curved wall. The geometry for a convex curve is shown in Fig. 8.2. The wall is a ray and the shape of the wall provides a given boundary value $\theta = \theta_w$ on the wall. If we use the (α, β) coordinates, the wall can be taken to be the ray $\beta = 0$. In the first instance, θ_w is known as a function of distance s along the wall. However, if we pose $\theta = \theta_w(\alpha)$ on the wall, we can determine the relation between α and s from the final solution. For simple shapes, such as sharp corners, this relation is not needed since θ_w just takes constant values on the two sides of the corner, and the (α, β) description is simpler. For more general shapes, for which this implicit relation would be a nuisance, it is usually better to work in the (x,y) description and use the equivalent set of equations (8.78).

The wall is initially straight with $\theta_w = 0$ and the shock is uniform with $\theta = 0$, $M = M_0$. We choose β as the distance from the wall in the uniform region so that $A_0 = 1$ and (8.64) applies. We choose $\alpha = 0$ as the initial position at the start of the corner. The complete problem then is to solve the equations (8.65) for initial and boundary values given by

$$\theta = 0, \quad M = M_0 \qquad \text{for } \alpha = 0, \quad 0 < \beta < \infty,$$

$$\theta = \theta_w(\alpha) \qquad \text{for } \beta = 0, \quad 0 < \alpha < \infty.$$

The propagation of the waves along the shock is analogous to one dimensional waves in gas dynamics. The displacement of the wall corresponds to a piston motion, and we may think of the relative displacement of the wall pulling or pushing the foot of the shock and sending out waves along it. A convex corner corresponds to the piston being withdrawn and sending out expansion waves, whereas a concave corner corresponds to pushing the piston in to produce compression waves. In either case, until shock-shocks are produced by breaking, the solution is diagnosed as a simple wave by exactly the same type of argument used in Section 6.8. The C_- invariant in (8.69) is constant everywhere since all the C_- originate in the uniform region ahead of the waves in which $\theta = 0$, $M = M_0$. Therefore

$$\theta - \omega(M) = -\omega(M_0) \qquad (8.83)$$

throughout, where $\omega(M)$ is given by (8.72). In particular, the Mach number M_w at the wall, which is perhaps the most important result, is given in terms of θ_w without further calculation of the solution. We have

$$\theta_w = \omega(M_w) - \omega(M_0). \qquad (8.84)$$

For any particular θ_w, the corresponding value of M_w is obtained by solving this relation using Table 8.3.

In the simple wave, the values at the wall propagate out and remain constant on the C_+ characteristics shown in Fig. 8.7. If a characteristic variable τ is specified as the value of α at the point of intersection of the characteristic with the wall $\beta = 0$, the simple wave solution is

$$\theta = \theta_w(\tau), \qquad M = M_w(\theta_w), \qquad \beta = (\alpha - \tau)c(M_w). \qquad (8.85)$$

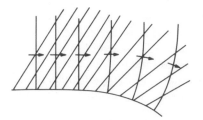

Fig. 8.7. The characteristics in the diffraction of a shock.

The corresponding equations in the (x,y) description are

$$\theta = \theta_w = \tan^{-1} y_w'(\xi), \qquad M = M_w(\theta_w),$$

$$y = y_w(\xi) + (x - \xi)\tan(\theta_w + m_w). \tag{8.86}$$

Expansion Around a Sharp Corner.

For a sharp convex corner, θ jumps from 0 to a negative value θ_w and remains at that value. The corresponding Mach number at the wall jumps from M_0 to the value M_w given by (8.84). The disturbance is a centered simple wave (see Fig. 8.8), and the solution for M in the wave is found from

$$\frac{\beta}{\alpha} = c(M), \qquad c(M_w) < \frac{\beta}{\alpha} < c(M_0). \tag{8.87}$$

The corresponding θ is given by (8.83). Along the shock

$$\frac{\partial x}{A\,\partial\beta} = -\sin\theta, \qquad \frac{\partial y}{A\,\partial\beta} = \cos\theta;$$

therefore at time $t = \alpha/a_0$ the shock is given in terms of the parameter β by

$$x = \alpha M_w \cos\theta_w - \int_0^\beta \frac{f(M)}{f(M_0)}\sin\theta\,d\beta,$$

$$y = \alpha M_w \sin\theta_w + \int_0^\beta \frac{f(M)}{f(M_0)}\cos\theta\,d\beta. \tag{8.88}$$

Since M and θ are functions of the single variable β/α, it follows that x/α and y/α are also functions of this single variable. Hence the shock pattern expands uniformly in time. This could have been deduced directly in advance by dimensional arguments. There is no fundamental length or time in the problem so that all flow quantities must be functions of x/a_0t and y/a_0t. This is equally true in the exact formulation and in the approximate geometrical theory.

The first disturbance propagates out on the characteristic $\beta = \alpha c(M_0)$. Since β measures distance from the wall on the initial undisturbed shock and $\alpha = a_0t$, the velocity in physical space is $a_0c(M_0)$. One of the few quantities that can be determined in the exact formulation of this problem is the speed of the first signal. According to the theory of sound, the first possible disturbance from the corner travels out into the flow behind the shock with the local sound speed a relative to the local flow velocity u. Therefore the disturbance travels along the shock with speed

$$\left\{ a^2 - (U-u)^2 \right\}^{1/2}, \tag{8.89}$$

where U is the shock velocity. The quantities U, a, u all can be expressed in terms of M_0, and it is found that (8.89) is a_0c^* where

$$c^* = \left\{ \frac{(M_0^2-1)[(\gamma-1)M_0^2+2]}{(\gamma+1)M_0^2} \right\}^{1/2}.$$

This is to be compared with

$$c_0 = c(M_0) = \left\{ \frac{M_0^2-1}{\lambda(M_0)} \right\}^{1/2},$$

where $\lambda(M)$ is given by (8.26). For weak shocks,

$$c_0 \sim \left\{ \frac{1}{2}(M_0-1) \right\}^{1/2}, \qquad c^* \sim \left\{ 2(M_0-1) \right\}^{1/2}, \qquad M_0 \to 1; \tag{8.90}$$

for strong shocks, taking $\gamma = 1.4$,

$$c_0 \sim 0.4439 M_0, \qquad c^* \sim 0.4082 M_0, \qquad M_0 \to \infty. \tag{8.91}$$

Thus the dependence on M_0 is the same, and in fact there is reasonable numerical agreement for $M_0 > 2$. For weak shocks, $c_0 = \frac{1}{2}c^*$. One could claim that c^* gives only the speed of the first signal and the main disturbance could in fact come later. But on the whole the evidence seems to be that for weak shocks the true disturbance is distributed over the whole sonic circle and the approximate theory concentrates the disturbance roughly halfway out. We shall see below that the prediction of the total magnitude of the disturbance, as evidenced by the value of M_w, is very good and the concentration of the disturbance in this theory is unavoidable. For stronger shocks the disturbance is more concentrated and the approximate theory represents the behavior very well. One might add that the theory emphasizes local behavior near the shock, and this is obviously better for stronger shocks. Fortunately, the problems of weak shocks are less interesting and can be handled by linear acoustics in any case.

Even though the exact formulation for diffraction around a sharp corner can be reduced to a similarity solution in $x/a_0 t, y/a_0 t$, very little can be done with it in general. However, for small angles θ_w, the flow behind the shock can be linearized and the solution carried through. This was done by Lighthill (1949). We may compare our results with Lighthill's results in this special case. For small θ_w, (8.84) for the Mach number M_w at the wall may be approximated by

$$M_w - M_0 = c(M_0)\theta_w$$

$$= \left\{ \frac{M_0^2 - 1}{\lambda(M_0)} \right\}^{1/2} \theta_w, \tag{8.92}$$

We compare this with Lighthill's results in the two extreme cases $M_0 \to 1$ and $M_0 \to \infty$. For weak shocks

$$M_w - M_0 \sim \left\{ \frac{1}{2}(M_0 - 1) \right\}^{1/2} \theta_w,$$

whereas Lighthill has $8/3\pi$ times this. For strong shocks

$$M_w - M_0 \sim 0.4439 M_0 \theta_w;$$

Lighthill's value has to be taken from a graph but the numerical factor appears to be about 0.5. Lighthill's theory shows that the disturbance is spread out over the entire sonic circle for weak shocks, but for stronger shocks it concentrates more and, in fact, the curvature tends to infinity as $M_0 \to \infty$.

In view of the relative simplicity of this approximate theory, the results are remarkably good. Lighthill's analysis, limited to a sharp corner and to a small angle, is already quite heavy by comparison, and the approximate theory may be applied to the enormous variety of problems for which other analytic solutions have not been found. The results should be good except for very weak shocks and even there the prediction of total Mach number change should be good. Experimental tests showing this agreement will be noted later.

The solution for any initial Mach number and any corner angle is given in (8.87) and (8.88). The formulas simplify in the limit of strong shocks, $M_0 \to \infty$, and the form of the solution becomes clearer. For strong shocks, the appropriate expression for M_w becomes

$$M_w = M_0 \exp\left(\frac{\theta_w}{\sqrt{n}}\right), \tag{8.93}$$

and in the fan

$$\frac{M}{M_0} = \left(\frac{\beta\sqrt{n}}{M_0\alpha}\right)^{1/(n+1)},$$

$$\theta = \frac{\sqrt{n}}{n+1} \log \frac{\beta\sqrt{n}}{M_0\alpha}. \tag{8.94}$$

The equation of the shock at time $t = \alpha/a_0$ is found from (8.88) with $f(M) = M^{-n}$. It is easiest to use θ as a parameter instead of β, and to fix the constants of integration from $x = M_0 a_0 t$, $y = M_0 a_0 t / \sqrt{n}$, when $\theta = 0$. Then we have,

$$\left.\begin{aligned}
\frac{x}{M_0 a_0 t} &= \left(\frac{n+1}{n}\right)^{1/2} e^{\theta/\sqrt{n}} \sin(\eta - \theta), \\
\frac{y}{M_0 a_0 t} &= \left(\frac{n+1}{n}\right)^{1/2} e^{\theta/\sqrt{n}} \cos(\eta - \theta),
\end{aligned}\right\} \quad \theta_w < \theta < 0, \tag{8.95}$$

where $\tan\eta = \sqrt{n}$. The shape of the shock is plotted in Fig. 8.8 for $\theta_w = -\pi/2$. The similarity form of the solution has already been noted, and we add that for strong shocks the solution scales with M_0 also.

The solution for various M_0 has been compared with experimental results by Skews (1967). The agreement is found to be reasonably good and typical results are reproduced in Fig. 8.9.

For strong shocks there is no limit on the magnitude of θ_w for which a solution can be found. For sufficiently weak shocks there is a limit, since M_w cannot decrease below unity. Hence if θ_w decreases below the value

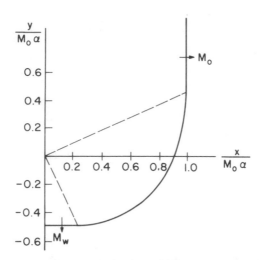

Fig. 8.8. Theoretical shock shape in diffraction around a 90° corner.

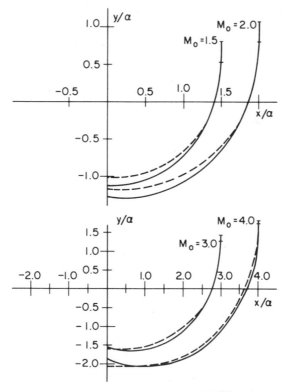

Fig. 8.9. Shock diffraction: comparison of experimental results (solid lines) with theory (dashed lines). (Skews)

$\theta_{\lim}$ given by

$$\theta_{\lim} = -\int_1^{M_0} \frac{dM}{Ac},$$

there is no solution. Presumably this corresponds to strong separation or other effects at the corner, but at present the interpretation is unclear.

Diffraction by a Wedge.

For a concave corner, the waves on the shock break and a shock-shock must be introduced into the solution (8.85), using the jump conditions established in Section 8.6. We consider in detail only the solution for a sharp concave corner, which is equivalent to the problem of diffraction of a plane shock by a wedge. This problem, has received considerable attention in the literature (see Courant and Friedrichs, 1948 p. 338). In the approximate theory the solution is simple. The solution is a shock-shock separating two regions in which M and θ are constant, as in Fig. 8.10. From (8.81), the Mach number at the wall is obtained by solving

$$\tan\theta_w = \frac{\left(M_w^2 - M_0^2\right)^{1/2}\left(A_0^2 - A_w^2\right)^{1/2}}{A_w M_w + A_0 M_0}, \qquad \frac{A_w}{A_0} = \frac{f(M_w)}{f(M_0)}. \qquad (8.96)$$

The angle χ for the line of the shock-shock shown in Fig. 8.10 is expressed from (8.82) as

$$\tan(\chi - \theta_w) = \frac{A_w}{A_0}\left\{\frac{1 - (M_0/M_w)^2}{1 - (A_w/A_0)^2}\right\}^{1/2}. \qquad (8.97)$$

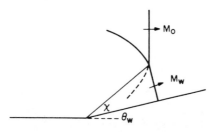

Fig. 8.10.   Mach reflection at a wedge.

For strong shocks

$$\frac{A_w}{A_0} = \left(\frac{M_0}{M_w}\right)^n ,$$

and χ becomes a function of θ_w alone. It is plotted in Fig. 8.11.

In reality, the true configuration is the Mach reflection with a third reflected shock and a vortex sheet, as indicated in Fig. 8.10. Moreover, the "Mach stem," the part of the shock near the wall, is slightly curved. The gas dynamic shock relations for the three shocks provide relations between the angles of the flow and the shocks at the triple point. If we assume that the Mach stem is straight, these allow an alternative determination of χ as a function of θ_w. This is the broken line in Fig. 8.11. The difference for small θ_w is about as expected, being of the same order as the discrepancy in (8.91). Then, fortuitously, the curves come closer together and cross. In the three-shock theory there is an upper limit on θ_w at which Mach reflection goes over into regular reflection (see Section 6.17), while the simplified shock-shock relations continue to predict a very tiny Mach stem. However, for θ greater than about 70° the Mach stem is so small that we have virtually the same picture as regular reflection. Again we conclude that the theory is remarkably good.

Diffraction by a Circular Cylinder.

Perhaps the most severe test of this theory is the application to diffraction by a circular cylinder which was carried out by Bryson and Gross (1961) and then compared with their experimental results. There is a

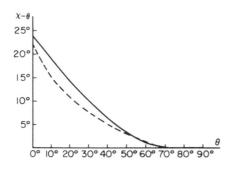

Fig. 8.11. The Mach reflection angle χ-θ versus wedge angle θ. The solid line refers to the present theory; the broken line refers to the three-shock theory.

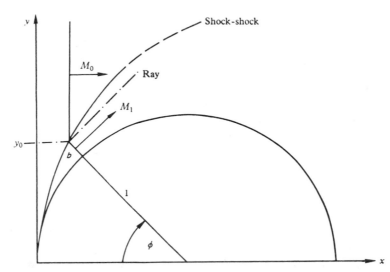

Fig. 8.12. Diffraction of a shock by a sphere or a cylinder.

difficulty in starting the solution at the nose, but Bryson and Gross
proposed a satisfactory way to handle this. First of all, the shock suffers
regular reflection until an angle of about 45° around from the nose is
reached, when a Mach stem is formed and subsequently grows. As noted
in Fig. 8.11, the approximate theory predicts a Mach stem for all θ_w up to
$\pi/2$. Bryson and Gross adopted the view that this is virtually regular
reflection if the Mach stem is extremely small. There is still a difficulty in
getting the calculation started at the nose, however, since the behavior
there is singular. They adopted the following procedure. It is assumed in
the early stages that the small Mach stem is straight and radial, as in Fig.
8.12. If its length is b at an angle φ around from the nose, and the radius of
the cylinder is normalized to unity, the undisturbed rays contained in a
stream tube of area $A_0 = (1 + b)\sin\varphi$ pass through the area $A_1 = b$. Hence

$$\frac{b}{(1+b)\sin\varphi} = \frac{f(M_1)}{f(M_0)}.\tag{8.98}$$

Since α is continuous at the shock-shock and is given by x/M_0 in the
undisturbed part of the shock, we have $\alpha = \{1 - (1 + b)\cos\varphi\}/M_0$. The
Mach number M is given by

$$\frac{1}{M} = |\nabla\alpha| = \frac{\partial\alpha}{R\partial\varphi}$$

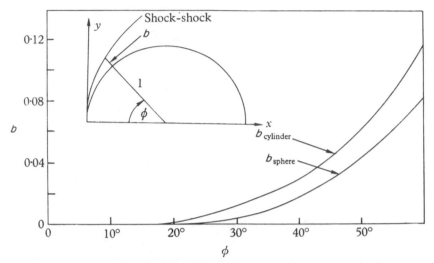

Fig. 8.13. The shock-shock standoff distance.

at radius R. Putting these two together and taking the mean position $R = 1 + \frac{1}{2}b$ for M_1, we have

$$\frac{M_0}{M_1} = \frac{1}{1 + b/2} \frac{d}{d\varphi} \{1 - (1 + b)\cos\varphi\}. \tag{8.99}$$

Equations 8.98 and 8.99 provide a differential equation for $b(\varphi)$ to be solved subject to the initial condition $b = 0, \varphi = 0$. For strong shocks, $M_0 \gg 1$, it is

$$\frac{db}{d\varphi} = (1 + b)\tan\varphi - \frac{1 + b/2}{\cos\varphi} \left[\frac{b}{(1 + b)\sin\varphi} \right]^{1/n}. \tag{8.100}$$

For small φ,

$$b = \sin^{n+1}\varphi, \qquad \varphi \ll 1.$$

The solution of (8.100) is plotted in Fig. 8.13. Bryson and Gross use this solution up to $\varphi = 45°$ and then continue with the detailed characteristics solution. When the two Mach stems meet behind the cylinder a second shock-shock is formed. The results are shown in Fig. 8.14 and are compared with their experimental observations. The theoretical shock positions

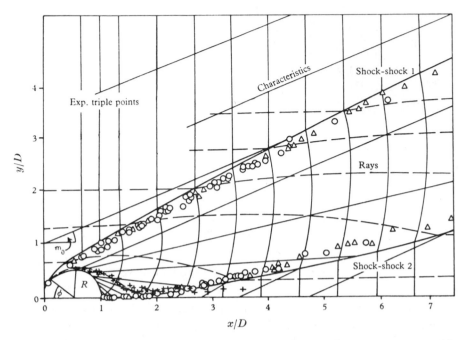

Fig. 8.14. Diffraction by a cylinder. $M_0 = 2.81$: circle = Re 7.79×10^4; triangle = Re 0.87×10^4, cross = vortex locus. (Bryson and Gross, 1961.)

and the two shock-shocks are the full lines, the rays are the broken lines. The circles and triangles are experimental points for the shock-shock positions at Reynolds numbers Re = 7.79×10^4, Re = 0.87×10^4, respectively. In the experiments a vortex is formed near the front and its locus is shown by the crosses; it is, of course, not included in the simple theory. Schlieren photographs of the flow pattern are presented in Figs. 8.15a,b,c.

Diffraction by a Cone or a Sphere.

For three dimensional problems, the formulation in (8.50)–(8.52) is used. For axially symmetric problems, the first two equations are

$$\alpha_x^2 + \alpha_r^2 = \frac{1}{M^2},$$

$$\frac{\partial}{\partial x}\left(\frac{rM}{A}\alpha_x\right) + \frac{\partial}{\partial r}\left(\frac{rM}{A}\alpha_r\right) = 0,$$

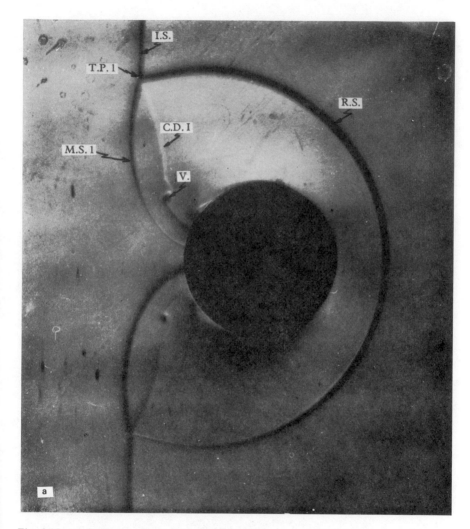

Fig. 8.15. (a) Schlieren photograph of shock diffraction on a cylinder of $\frac{1}{2}$ in. diameter. $M_0 = 2.82$. Note the boundary-layer separation starting. Notation: I.S., indicent shock; M.S., mach shock; R.S., reflected shock; C.D., contact discontinuity; T.P., triple point; V., vortex. (Bryson and Gross 1961.)

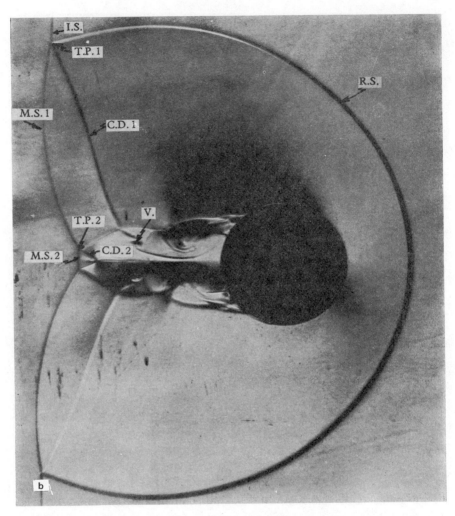

Fig. 8.15 (b) Schlieren photograph of shock diffraction on a cylinder of $\frac{1}{2}$ in. diameter. $M_0 = 2.81$. Notation: I.S., incident shock; M.S., mach shock; R.S., reflected shock; C.D., contact discontinuity; T.P., triple point; V., vortex. (Bryson and Gross 1961.)

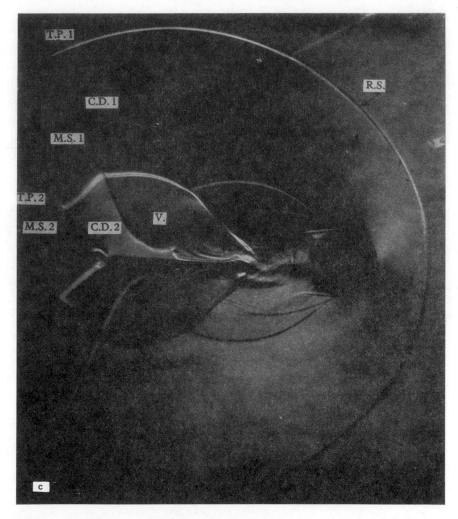

Fig. 8.15 (c) Schlieren photograph of shock diffraction on a cylinder of $\frac{1}{2}$ in. diameter. $M_0 = 2.84$. Notation: M.S. mach shock; R.S. reflected shock; C.D., contact discontinuity; T.P., triple point; V., vortex. (Bryson and Gross 1961.)

305

where x is distance along the axis and r is the radial distance. It is again convenient to introduce the ray angle θ by

$$\alpha_x = \frac{\cos\theta}{M}, \qquad \alpha_r = \frac{\sin\theta}{M},$$

and work with the set

$$\frac{\partial}{\partial x}\left(\frac{\sin\theta}{M}\right) - \frac{\partial}{\partial r}\left(\frac{\cos\theta}{M}\right) = 0,$$

$$\frac{\partial}{\partial x}\left(\frac{r\cos\theta}{A}\right) + \frac{\partial}{\partial r}\left(\frac{r\sin\theta}{A}\right) = 0, \qquad (8.101)$$

$$\frac{A}{A_0} = \frac{f(M)}{f(M_0)}.$$

The boundary condition on a solid wall $r = r_w(x)$ is $\tan\theta = r_w'(x)$.

For diffraction by a cone, the solution is a similarity solution in which all quantities are functions of r/x. Equations 8.101 can be reduced to ordinary differential equations which have to be solved subject to the

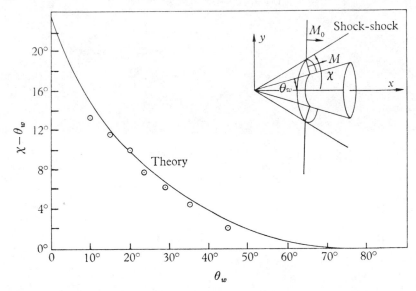

Fig. 8.16. Comparison of theoretical and experimental results for the shock-shock angle in diffraction by a cone. (Bryson and Gross, 1961.)

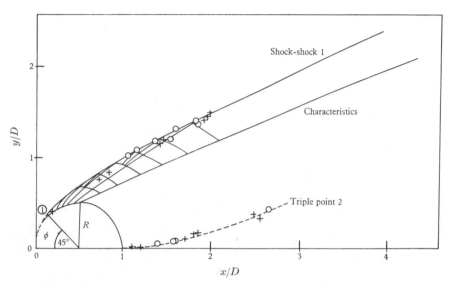

Fig. 8.17. Diffraction by a sphere. Circles represent $M_0 = 2.85$; crosses represent $M_0 = 4.41$. (Bryson and Gross, 1961.)

conditions on the wall and at the shock-shock. The details are given in the original paper (Whitham, 1959b). Bryson and Gross extended the calculations and compared the results with experiments. Figure 8.16 compares shock-shock angle χ with wall angle θ_w for $M_0 = 3.68$.

For a sphere, Bryson and Gross performed a characteristics calculation for (8.101), starting the calculation from an approximate treatment of the nose region analogous to their method for the cylinder. Their results are not as detailed as for the cylinder, but the agreement between theory and experiment shown in Fig. 8.17 is equally good.

8.8 Stability of Shocks

The theory puts into quantitative terms one of the arguments that has always been used to explain the stability of plane shocks. Suppose that for some reason a portion of the shock has developed a bulge as shown in Fig. 8.18. The delayed part is now concave forward and so will strengthen as it propagates. As it strengthens, it speeds up and thus tends to reduce the bulge. Similarly, any section of the shock ahead of the rest weakens and slows down,. The overall effect is one of stability. The arguments for changes of strength depending on the curvature are put into quantitative terms in the A-M relation.

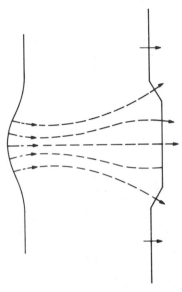

Fig. 8.18. Sketch of a shock positions (solid lines) and rays (dashed lines) for nonlinear resolution of a caustic.

In linear geometrical optics, a concave portion of a wavefront would produce a caustic, since the linear rays would be normal to the initial wave front and form an envelope (refer back to p. 247). As the wavefront propagates down the converging ray tubes it strengthens, and its strength tends to infinity as it reaches the caustic. But in the linear theory the speed is unchanged, and hence the rays remain the same. In the nonlinear theory developed here, the shock speeds up as it strengthens. This pushes the rays apart and there is no overlap and no caustic. The shock overshoots as shown in Fig. 8.18 and the disturbance evens out as it spreads along the shock.

In detail the problem would be formulated as an initial value problem with M and θ prescribed on the initial shock position. In two dimensions the problem would be exactly analogous to the problem discussed in Section 6.12. There would be an initial interaction region and then the disturbance would separate into two simple waves moving in the positive and negative directions along the shock. In each the total change in θ and M would be zero, so that they would ultimately take the N wave form with shock-shocks at the front and at the back and a linear decrease of θ between. The shape would be that of Fig. 8.18. Detailed calculations will not be given here. According to the general results established earlier, the shock-shocks decay like $t^{-1/2}$. For a uniformly distributed disturbance such as an initial sinusoidal shape, the disturbance eventually decays like $1/t$ (see Section 2.8).

Stability of Converging Cylindrical Shocks.

An interesting and important question arises concerning the stability of converging cylindrical and spherical shocks. The expected intense pressure at the center would be considerably reduced by imperfect focusing. Experiments by Perry and Kantrowitz (1951) showed very symmetrical shapes for weak and moderate shocks, and some indication of instability for strong shocks, although the conclusions do not seem to be clear-cut. It is interesting to analyze the question using this theory. Local corrugations on the shock will have the tendencies described for plane shocks, but these have to be superimposed on an overall convergence and strengthening of the whole shock. Delayed parts will have the tendency to strengthen but the other parts are already strengthening due to the general convergence and are nearer the center. So the delayed parts could continue to lag behind and could possibly be left further and further behind. While the radius is sufficiently large it seems clear that the behavior would be close to that of plane shocks and the propagation would be stable. The question then concerns the behavior as the strength becomes large close to the center.

The problem for strong cylindrical shocks was analyzed by Butler (1955) using a small perturbation treatment which implicitly included the approximations of the ray tube theory. With the general formulation developed here, it can be handled more easily and without making small perturbation assumptions. For strong shocks, the two dimensional equations (8.59)–(8.61) may be written

$$\frac{\partial \theta}{\partial \beta} + \frac{nM_0^n}{M^{n+2}} \frac{\partial M}{\partial \alpha} = 0, \tag{8.102}$$

$$\frac{\partial \theta}{\partial \alpha} + \frac{M^n}{M_0^n} \frac{\partial M}{\partial \beta} = 0. \tag{8.103}$$

The symmetrical solution for a shock with initial radius R_0 is

$$\theta = -\frac{\beta}{R_0}, \qquad M = M_0\left(-\frac{n+1}{n} \cdot \frac{M_0\alpha}{R_0}\right)^{-1/(n+1)}, \qquad \alpha < 0. \tag{8.104}$$

This is, of course, Guderley's solution.

To study perturbations to this solution, we use a hodograph transformation of (8.102)–(8.103), and interchange the roles of dependent and independent variables. This produces linear equations without any approximation as to the size of the perturbations. First we introduce new variables

$$q = \left(\frac{M}{M_0}\right)^{n+1}, \qquad \Theta = \frac{(n+1)\theta}{\sqrt{n}}, \qquad s = \frac{M_0 \alpha}{\sqrt{n}},$$

and (8.102)–(8.103) become

$$\frac{\partial q}{\partial s} + q^2 \frac{\partial \Theta}{\partial \beta} = 0,$$

$$\frac{\partial \Theta}{\partial s} + \frac{\partial q}{\partial \beta} = 0. \tag{8.105}$$

In these variables the symmetrical solution is $q \propto 1/s$, $\Theta \propto \beta$. In the hodograph transformation, β, s are treated as functions of q, Θ. The transformation formulas are $\Theta_\beta = J s_q$, $\Theta_s = -J \beta_q$, $q_\beta = -J s_\Theta$, $q_s = J \beta_\Theta$, where J is the Jacobian $\partial(q, \Theta)/\partial(s, \beta)$. The equations in (8.105) become

$$\beta_\Theta + q^2 s_q = 0, \qquad \beta_q + s_\Theta = 0.$$

When β is eliminated we have the single equation

$$q^2 s_{qq} + 2q s_q = s_{\Theta\Theta}. \tag{8.106}$$

Solutions to this (by separation of variables) are

$$s = q^\mu e^{im\Theta}, \qquad \mu = -\frac{1}{2} \pm \left(\frac{1}{4} - m^2\right)^{1/2}. \tag{8.107}$$

If $m = 0$, $\mu = -1$ gives the symmetrical solution. If $m \geq 1$, $\Re \mu = -\frac{1}{2}$. Therefore when $q \to \infty$ as the shock contracts to the center, the harmonics dominate the symmetrical mode. Hence the shock is unstable.

The imaginary part of μ shows that the disturbance consists of waves traveling around the shock. When the disturbance becomes large it is possible for J to vanish. This means that the mapping from the (q, Θ) to the (s, β) plane is no longer single valued, and it corresponds to the appearance of shock-shocks. When this stage is reached, further calculations would have to be carried out numerically in the (s, β) plane.

8.9 Shock Propagation in a Moving Medium

For propagation into a moving stream, the linear theory shows con-
clusively that rays are not orthogonal to the wavefronts (see Section 7.9
and Fig. 7.12). Correspondingly, we cannot expect the rays to be orthogonal to the shocks in the nonlinear theory. At first sight this poses a
problem since the nonlinear formulation relied on the orthogonality to
compare propagation in a ray tube with propagation down a given channel. However, a way out is to consider propagation in a *uniform* stream as
a test case. In a frame of reference moving with the stream the original
formulation applies. It remains only to make a Galilean transformation to
another moving frame to see the correct formulation for moving media.
Then, as expected, the rays are not orthogonal to the shocks. The results have been given (Whitham, 1968) and applied by Huppert and
Miles (1968). It would be interesting to pursue this investigation to see how
the theory for a moving medium might have been formulated directly. It
indicates that the closeness of rays to streamlines is not as important as
was originally thought, and this might lead to more novel points of view.

CHAPTER 9

The Propagation of Weak Shocks

As indicated at the beginning of Chapter 8, we can pursue a different approach and cover a different class of problems when the waves are moderately weak. The geometrical effects are accepted unchanged from linear theory and we are then able to cope with more general nonlinear interactions within the wave profile. The approximation procedures will be developed for unsteady waves, using spherical and cylindrical waves as prime examples, and then applied more specifically to the sonic boom problem, which is perhaps the most interesting situation where weak shocks have to be studied. The unpleasant aspects of sonic booms loom large, but they are in fact extremely weak shocks and, naturally, the aim is to make them weaker. The maximum overpressure at the ground for current and contemplated supersonic transports is about 2 lb/ft^2; this corresponds to a shock strength of about 10^{-3}. The basic problem for uniform velocity and flight path may be treated as one in steady supersonic flow, so this work also continues the development in Section 6.17.

9.1 The Nonlinearization Technique

Geometrical effects arise in their simplest form for spherical waves. If the linearized theory is governed by the wave equation, the solution for an outgoing wave may be written*

$$\varphi = \frac{f(t - r/c_0)}{r}, \tag{9.1}$$

where c_0 is the propagation speed. The amplitude decays like $1/r$ as the energy spreads out across a surface area increasing like r^2. The wave

---

*It was convenient in Section 7.3 to use different symbols R and r for the radial distances in spherical and cylindrical geometry, since the solution for a point source was used to generate the solution for a line source. It is no longer necessary and we use r in each case.

312

profile is reduced by the factor $1/r$ but is otherwise undistorted. If this is the linearized solution to a nonlinear problem, we know that the nonlinear distortion of the profile will be crucial for a correct treatment of wave breaking and shock propagation. Suppose that the correct nonlinear propagation speed determined from the characteristic equations is in fact $c(\varphi)$, the linearized speed c_0 being its value at $\varphi=0$. Then the nonlinear distortion may be introduced by modifying (9.1) to

$$\varphi = \frac{f(\tau)}{r},\tag{9.2}$$

where $\tau(t,r)$ is to be determined so that the curves $\tau=\text{constant}$ satisfy the *exact* characteristic condition. That is, we require

$$\frac{dr}{dt}=c(\varphi)\qquad\text{on }\tau=\text{constant.}\tag{9.3}$$

Since φ is expressed as a function of τ and r, this provides a differential equation to determine τ, and the inverted form can be integrated immediately. We have

$$\frac{dt}{dr}=\frac{1}{c\{f(\tau)/r\}}\qquad\text{on }\tau=\text{constant;}\tag{9.4}$$

hence

$$t=\int\frac{1}{c\{f(\tau)/r\}}\,dr+T(\tau),\tag{9.5}$$

where the integration is carried out holding τ constant and $T(\tau)$ is an arbitrary function allowed by the integration with respect to r. Equation 9.5 determines $\tau(t,r)$ implicitly and the combined equations 9.2 and 9.5 provide us with a "nonlinearized" solution. This nonlinearization technique was proposed first by Landau (1945) and independently by the author in the context of the sonic boom problem (Whitham, 1950, 1952).

The function $T(\tau)$ corresponds to the arbitrariness in the choice of the characteristic variable. Once $T(\tau)$ has been chosen, $f(\tau)$ is determined from an appropriate boundary condition at the source. Different choices of $T(\tau)$ are compensated for in the resulting $f(\tau)$. For the general development one may take $T(\tau)=\tau$ for simplicity but the extra flexibility is sometimes useful in specific problems. It should be noted that the functions f in the linear result (9.1) and in the nonlinearized version (9.2) will be the same only if $T(\tau)$ is chosen so that $\tau=t-r/c_0$ (to a sufficient approximation) where the boundary condition is applied.

Of course this nonlinearized solution usually does not satisfy the relevant nonlinear equations exactly, nor has it been deduced at this point as a formal approximation. However, it appears to include the important nonlinear effects for small φ. In the simpler case discussed in Section 2.10 and in the case of plane waves, we saw that the linearization of the characteristics was the source of the nonuniform validity. Nonlinearity will also modify the amplitude factor $1/r$, but one would expect this second modification to be uniformly small in φ. This view will be substantiated later when we look more carefully at the justification of the procedure. First, the further consequences and extensions of the procedure are mapped out to see its full scope.

Since φ takes a particularly simple form for spherical waves, the integral in (9.5) is tractable and can be simplified by a change of variable to $f(\tau)/r$. In other problems, however, the corresponding expressions are less amenable and it is valuable to streamline the analysis. In using the linearized result as a starting point, we have already assumed that φ is small so it is consistent to expand $c(\varphi)$ as $c_0 + c_1\varphi + O(\varphi^2)$, say, and to use (9.3) in the approximate form

$$\frac{dt}{dr} = \frac{1}{c_0} - \frac{c_1}{c_0^2}\varphi. \tag{9.6}$$

The expansion is taken for dt/dr rather than dr/dt in view of the subsequent integration with respect to r. Then corresponding to (9.4) we have

$$\frac{dt}{dr} = \frac{1}{c_0} - \frac{c_1}{c_0^2}\frac{f(\tau)}{r} \qquad \text{on } \tau = \text{constant}, \tag{9.7}$$

and the characteristics are

$$t = \frac{r}{c_0} - \frac{c_1}{c_0^2}f(\tau)\log r + T(\tau). \tag{9.8}$$

The linear theory would take $\tau = t - r/c_0$ or a function of it, and we see that it is not uniformly valid because the additional term *tends to infinity* as $r \to \infty$. The term in $\log r$ is small compared with r, but it should be compared rather with $c_0 t - r$, which measures distance from the head of the wave. The view that the correction of the propagation speed would be crucial is confirmed, and there is a close analogy with the situation discussed in Section 2.10.

The approximation of (9.4) by (9.6) is not only a simplification. In most problems it would actually be inconsistent to keep higher order terms in φ, since (9.2) is itself only a first order approximation to φ.

The singular behavior of $\log r$ as $r \to 0$ is only a minor nuisance since the correction term is not important near the origin and we could revert to linear theory there. However, to use (9.8) as it stands we must exclude the origin and apply the solution outside some sphere $r = r_0(t)$ on which boundary data are given. (For example, a fluid source could be represented by an expanding sphere pushing out the fluid.) We may then choose $T(\tau)$ so that (9.8) becomes

$$t = \frac{r}{c_0} - \frac{c_1}{c_0^2} f(\tau) \log \frac{r}{r_0(\tau)} + \tau. \tag{9.9}$$

With this choice, the nonlinear τ agrees with $t - r/c_0$ on the boundary curve and the function f is the same as in the linear theory.

Waves described by (9.2) and (9.8) will break whenever the characteristics form an envelope and the solution becomes multivalued. If $c_1 > 0$, a wave carrying an increase of φ breaks. Assuming that $T'(\tau) > 0$, this means that breaking occurs when $f'(\tau) > 0$. On the envelope,

$$\frac{c_1}{c_0^2} f'(\tau) \log r - T'(\tau) = 0.$$

For (9.9), breaking first occurs at a distance given by

$$\log \frac{r}{r_0(\tau_m)} \simeq \frac{c_0^2}{c_1 f'(\tau_m)}, \tag{9.10}$$

where τ_m corresponds to the maximum of $f'(\tau)$. Then a shock must be fitted in. The techniques of shock fitting follow closely the earlier treatment and we defer the discussion for the present.

We now consider extensions of the procedure. First, the linear solution may not be as simple as (9.1). For example, in cylindrical waves the solution (7.29) is

$$\varphi = -\frac{1}{2\pi} \int_0^{t - r/c_0} \frac{q(\eta)\, d\eta}{\sqrt{(t - \eta)^2 - r^2/c_0^2}}. \tag{9.11}$$

The characteristic variable $t - r/c_0$ is significantly visible in the upper limit, but both t and r appear also in the integrand. However, the evidence from plane and spherical waves is that the nonlinear effects become important at large distances. And at large distances, we saw in (7.32) that

$$\varphi \sim \frac{f(t - r/c_0)}{r^{1/2}}.$$ \qquad (9.12)

Therefore the nonlinearization for large distances can follow closely the spherical case. If the correct propagation speed is $c(\varphi) = c_0 + c_1\varphi + O(\varphi^2)$, we take

$$\varphi = \frac{f(\tau)}{r^{1/2}},$$ \qquad (9.13)

$$\frac{dt}{dr} = \frac{1}{c_0} - \frac{c_1}{c_0^2} \frac{f(\tau)}{r^{1/2}},$$

$$t = \frac{r}{c_0} - \frac{2c_1}{c_0^2} f(\tau) r^{1/2} + \tau.$$ \qquad (9.14)

Here the function $T(\tau)$ arising in the integration with respect to r has been taken to be τ; the correction term remains small as $r \to 0$ and there is no need for more elaborate choices of $T(\tau)$. Again, the linear theory, which takes $\tau = t - r/c_0$, is not uniformly valid as $r \to \infty$. Moreover, although the emphasis so far has been on the behavior at large distances, the deviation of the characteristics from the linear ones depends on

$$\frac{f(\tau) r^{1/2}}{t - r/c_0};$$ \qquad (9.15)

this is large near $t - r/c_0 = 0$ as well as for large r. So the nonlinear correction will be equally important near the wavefront $t - r/c_0 = 0$ *at all distances*. Significantly, (9.12) is valid for $(c_0 t - r)/r \ll 1$, so that it covers both situations; the corresponding nonlinearized solution given by (9.13) and (9.14) is also valid near the front of the wave for all r. This is of tremendous importance since the most interesting problems will have a shock at the head of the wave and the nonlinearized solution derived from (9.13) can be used to study it in its entirety.

We may nonlinearize the whole solution by taking

$$\varphi = -\frac{1}{2\pi} \int_0^\tau \frac{q(\eta)\, d\eta}{\sqrt{(\tau - \eta)(\tau - \eta + 2r/c_0)}}$$ \qquad (9.16)

and determining the nonlinear τ from this. In this more complete form, the characteristic equation corresponding to (9.14) becomes quite complicated. But the extra terms remain small compared with $f(\tau)r^{1/2}$. It is therefore sufficient to take (9.16) combined with (9.14) as a nonlinearized solution over the field. In any case, the nonlinearization is of prime importance in the region $c_0\tau/r \ll 1$, where (9.16) may be approximated by (9.13) with

$$f(\tau) = -\frac{1}{2\pi}\left(\frac{c_0}{2}\right)^{1/2}\int_0^\tau \frac{q(\eta)d\eta}{\sqrt{\tau-\eta}}.$$

It should be noted, however, that appropriate boundary conditions are normally prescribed outside the region $c_0\tau/r \ll 1$, so that either (9.16) or the full linear solution to which it reduces is required to determine the function $f(\tau)$ that appears in (9.13)–(9.14).

The basic role of geometrical optics now becomes apparent, for (9.12) is the geometrical optics approximation to cylindrical waves. In general, geometrical optics provides a ray geometry and (for uniform media) we have

$$\varphi \simeq \Phi(s)f\left(t - \frac{s}{c_0}\right) \tag{9.17}$$

along each ray, where s is the distance along the ray, $\Phi(s)$ is the amplitude, and $f(t - s/c_0)$ describes the wave profile. This is the natural form for nonlinearization and is applicable just where the nonlinear effects are most important—near the head of the wave and at large distances. The nonlinearization follows by taking

$$\varphi = \Phi(s)f(\tau),$$

$$t = \frac{s}{c_0} - \frac{c_1}{c_0^2}f(\tau)\int_0^s \Phi(s')\,ds' + T(\tau). \tag{9.18}$$

Combining the results for each ray we have a complete nonlinearized solution. It should be noted that $c(\varphi)$ refers here to the velocity on the ray and this is not the same as the normal wavefront velocity for anisotropic media. For nonuniform media, c and c_0 may also depend on s. The s/c_0 is replaced by $\int ds/c_0$ and any dependence of c_1/c_0^2 on s must be included under the integral sign in (9.18).

At this point, the procedure can be compared with the one in the last chapter. Roughly speaking, the $f(\tau)$ for those problems was a step function, so the nonlinear interaction took a mild form and strong nonlinear effects on the ray geometry could be included. Here the ray geometry and its effect on the amplitude $\Phi(s)$ is accepted unchanged from linear theory, but more general profiles $f(\tau)$ can be handled. Presumably a combination of both approaches would be needed in still more general problems, but the analysis looks forbidding.

A second extension of the techniques is required because the non-linear propagation speed is often a function of the derivatives, φ_t and φ_s, rather than of φ itself. However, the procedure goes through. The expressions for φ_t and φ_s are each written in the form corresponding to (9.17) and the revised characteristics are determined from the appropriate expansion

$$\frac{dt}{ds} = \frac{1}{c} = \frac{1}{c_0} - \alpha_1 \varphi_t - \alpha_2 \varphi_s \qquad \text{on } \tau = \text{constant.}$$

From (9.17) the corresponding first terms for φ_t, φ_s are

$$\varphi_t = \Phi(s) f'(\tau), \qquad \varphi_s = -\frac{1}{c_0} \Phi(s) f'(\tau),$$

and the characteristic relation becomes

$$\frac{\partial t}{\partial s} = \frac{1}{c_0} - k f'(\tau) \Phi(s), \qquad k = \alpha_1 - \alpha_2 c_0^{-1}.$$

The characteristics are given by

$$t = \frac{s}{c_0} - k f'(\tau) \int_0^s \Phi(s') \, ds' + T(\tau). \qquad (9.19)$$

A specific example of this is provided by spherical waves in gas dynamics. The linear theory is the acoustic theory and φ_t, φ_r are related to the pressure and velocity perturbations. From (7.3)–(7.4) we have

$$\frac{p - p_0}{p_0} = -\frac{\gamma}{a_0^2} \varphi_t = \frac{\gamma F(t - r/a_0)}{r}$$

$$\frac{u}{a_0} = \frac{1}{a_0} \varphi_r = \frac{F(t - r/a_0)}{r} - \frac{f(t - r/a_0)}{a_0 r^2},$$

where $F(\tau) = -f'(\tau)/a_0^2$. The perturbation in the sound speed a will also be needed, and it is given by

$$\frac{a-a_0}{a_0} = \frac{\gamma-1}{2\gamma}\frac{p-p_0}{p_0} = \frac{\gamma-1}{2}\frac{F(t-r/a_0)}{r}.$$

The nonlinearized solution is

$$\frac{p-p_0}{p_0} = \frac{\gamma F(\tau)}{r}, \qquad \frac{a-a_0}{a_0} = \frac{\gamma-1}{2}\frac{F(\tau)}{r}, \qquad (9.20)$$

$$\frac{u}{a_0} = \frac{F(\tau)}{r} - \frac{f(\tau)}{a_0 r^2}, \qquad (9.21)$$

where $\tau(t,r)$ is to be determined from the improved characteristics. The exact characteristic equations were given in (6.135). The outgoing characteristics have velocity $a+u$. Therefore the first order correction to the characteristics requires

$$\frac{dt}{dr} = \frac{1}{a+u} \simeq \frac{1}{a_0} - \frac{a+u-a_0}{a_0^2}. \qquad (9.22)$$

From (9.20) and (9.21), this means that

$$\frac{dt}{dr} = \frac{1}{a_0} - \frac{\gamma+1}{2a_0}\frac{F(\tau)}{r} + \frac{1}{a_0^2}\frac{f(\tau)}{r^2},$$
$$\qquad\qquad (9.23)$$
$$t = \frac{r}{a_0} - \frac{\gamma+1}{2a_0}F(\tau)\log r - \frac{1}{a_0^2}\frac{f(\tau)}{r} + T(\tau).$$

[The relation between $F(\tau)$ and $f(\tau)$ is modified to $f'(\tau) = -a_0^2 F(\tau)T'(\tau)$ if $T'(\tau) \neq 1$.] Since our interest is in the region $a_0\tau/r \ll 1$, and the term $f(\tau)/r$ is always relatively small in this region, it is sufficient to use

$$\frac{p-p_0}{p_0} = \frac{\gamma F(\tau)}{r}, \qquad \frac{a-a_0}{a_0} = \frac{\gamma-1}{2}\frac{F(\tau)}{r}, \qquad u = \frac{F(\tau)}{r},$$
$$\qquad\qquad (9.24)$$
$$t = \frac{r}{a_0} - \frac{\gamma+1}{2a_0}F(\tau)\log r + T(\tau).$$

This is a rather trivial example of retaining only the geometrical acoustics approximation to (9.21) and (9.23). Cylindrical and other waves in gas dynamics are handled similarly and the geometrical acoustics approximation provides a greater simplification similar to the step from (9.11) to (9.12).

When derivatives appear in the expression for c it is more convenient to take them as new dependent variables. Then, in all cases, the geometrical optics approximation leads to expressions for the dependent variables which are proportional to

$$\Phi(s)F(\tau),$$

where $\Phi(s)$ is an amplitude function and $F(\tau)$ describes the wave profile. The corrected propagation speed, *using this approximation*, takes the form

$$c \simeq c_0 + c_0^2 k \Phi(s)F(\tau), \tag{9.25}$$

where the coefficient k is a constant determined by the particular relation of c to the dependent variables. The improved characteristics satisfy

$$\frac{\partial t}{\partial s} = \frac{1}{c_0} - k\Phi(s)F(\tau) \tag{9.26}$$

and are given by

$$t = \frac{s}{c_0} - kF(\tau) \int_0^s \Phi(s')\, ds' + T(\tau). \tag{9.27}$$

Shock Determination.

Shocks, when required, are fitted in using the weak shock condition

$$U = \frac{1}{2}(c_1 + c_2),$$

where U is the shock velocity and c_1, c_2 now denote the propagation speeds on the two sides. In the present context, it is convenient to describe curves in the (s,t) plane by giving t as a function of s, so the shock condition is used in the form

$$\left(\frac{dt}{ds}\right)_{\text{shock}} = \frac{1}{2}\left\{\left(\frac{dt}{ds}\right)_{c_1} + \left(\frac{dt}{ds}\right)_{c_2}\right\}, \tag{9.28}$$

which is equivalent to first order in the deviation of velocities from c_0. If the shock is specified by

$$t = \frac{s}{c_0} - G(s),$$

we have

$$G'(s) = \frac{1}{2}k\{F(\tau_1) + F(\tau_2)\}\Phi(s),$$

$$G(s) = kF(\tau_1)\int_0^s \Phi(s')\,ds' - T(\tau_1),$$

$$G(s) = kF(\tau_2)\int_0^s \Phi(s')\,ds' - T(\tau_2).$$

We then deduce the typical "equal area" relation

$$\frac{1}{2}\{F(\tau_1) + F(\tau_2)\}\{T(\tau_2) - T(\tau_1)\} = \int_{\tau_1}^{\tau_2} F(\tau)\,dT(\tau). \quad (9.29)$$

For a head shock moving into the undisturbed region, the shock is determined by (9.27) with τ related to s by

$$\frac{1}{2}kF^2(\tau)\int_0^s \Phi(s')\,ds' = \int_0^\tau F(\tau')\,dT(\tau'). \quad (9.30)$$

As $s \to \infty$, the equation of the shock asymptotes to

$$t = \frac{s}{c_0} - K\left\{\int_0^s \Phi(s')\,ds'\right\}^{1/2} + T(\tau_0), \quad (9.31)$$

where

$$K = \left\{2k\int_0^{\tau_0} F(\tau)\,dT(\tau)\right\}^{1/2}, \qquad F(\tau_0) = 0. \quad (9.32)$$

At the shock the flow quantities are proportional to

$$K\Phi(s)\left\{\int_0^s \Phi(s')\,ds'\right\}^{-1/2}. \quad (9.33)$$

The typical asymptotic wave form is the N wave, with balanced shock waves, and between the shocks the linear decrease in time is proportional to

$$\Phi(s)\left\{\int_0^s \Phi(s')\,ds'\right\}^{-1}. \quad (9.34)$$

For spherical waves $\Phi(s) = 1/s$ and the shock strength (9.33) decays like $s^{-1}(\log s)^{-1/2}$, only slightly faster than the decay for linear pulses. For cylindrical waves $\Phi = s^{-1/2}$ and the shock strength decays like $s^{-3/4}$. Of course plane waves are also covered by these formulas; Φ is constant and the decay law is $s^{-1/2}$, in agreement with earlier results. The asymptotic decay laws for cylindrical and spherical waves were obtained independently by various writers and the first was probably Landau (1945).

For more general two and three dimensional waves in a uniform medium

$$\Phi(s) \propto A^{-1/2}(s),$$

where $A(s)$ is the ray tube area. Further details and applications may be found in an earlier paper (Whitham, 1956). For nonuniform media s/c_0 is replaced by $\int ds/c_0$ and all the dependence on s in (9.26) must be included in $\Phi(s)$.

9.2 Justification of the Technique

There are several approaches by which the nonlinearization technique can be examined mathematically on specific systems and each one throws light on different aspects of the approximations.

First, suppose the nonlinear equation for φ is

$$\varphi_t + (c_0 + c_1\varphi)\varphi_x + \frac{\beta c_0}{x}\varphi = 0. \tag{9.35}$$

This is proposed as a model in the first instance, but we shall see a tie-in with other cases later. The linearized equation is

$$\varphi_t + c_0\varphi_x + \frac{\beta c_0}{x}\varphi = 0, \tag{9.36}$$

and its solution is

$$\varphi = \frac{f(t - x/c_0)}{x^\beta}. \tag{9.37}$$

This is analogous to spherical waves for $\beta = 1$ and cylindrical waves for $\beta = \frac{1}{2}$. The characteristic form of (9.35) is

$$(c_0 + c_1\varphi)\frac{d\varphi}{dx} = -\frac{\beta c_0}{x}\varphi, \tag{9.38}$$

$$\frac{dt}{dx} = \frac{1}{c_0 + c_1\varphi}. \tag{9.39}$$

The exact solution of (9.38) is

$$\varphi e^{c_1 \varphi / c_0} = \frac{f(\tau)}{x^{\beta}},$$
(9.40)

where τ is the characteristic variable to be determined from (9.39). It is clear that

$$\varphi = \frac{f(\tau)}{x^{\beta}}$$
(9.41)

is a uniformly valid approximation to (9.40) for small φ. This confirms the main step in the technique. The determination of τ can be examined using the expansions of (9.39) and (9.40) in powers of φ, and the expansions are convergent for $|\varphi| < c_0 / c_1$. We obtain

$$\frac{dt}{dx} = \frac{1}{c_0} + \frac{\gamma_1 f(\tau)}{x^{\beta}} + \frac{\gamma_2 f^2(\tau)}{x^{2\beta}} + \cdots,$$

with coefficients γ_n related to c_0, c_1; in particular $\gamma_1 = -c_1 / c_0^2$. Hence

$$t = T(\tau) + \frac{x}{c_0} + \frac{\gamma_1 f(\tau)}{1 - \beta} x^{1-\beta} + \frac{\gamma_2 f^2(\tau)}{1 - 2\beta} x^{1-2\beta} + \cdots.$$
(9.42)

(Logarithms replace the corresponding powers when $\beta = 1, \frac{1}{2}$, etc.) The first uniformly valid approximation is

$$t = T(\tau) + \frac{x}{c_0} + \frac{\gamma_1 f(\tau)}{1 - \beta} x^{1-\beta},$$
(9.43)

and it agrees with the result obtained from (9.41) and

$$\frac{dt}{dx} = \frac{1}{c_0} - \frac{c_1}{c_0^2} \varphi.$$
(9.44)

Therefore the uniformly valid approximation given by (9.41) and (9.43) is exactly the one that would be obtained by the nonlinearization procedure. Notice that it would, in fact, be inconsistent to use further terms in (9.44) without further terms in (9.41).

The remaining approaches are illustrated on the equation for spherical waves in gas dynamics to keep the algebra as simple as possible, but it seems clear that they would go through (with possibly minor alterations) in other cases. It will be sufficient, again for simplicity, to give the details for isentropic flow; the methods are not limited by this and, even when shocks are present, the entropy changes for weak waves do not affect the lowest order terms. The full equations are given in (6.132)–(6.134), and for isentropic flow they may be reduced to the following pair of equations in sound speed a and radial velocity u:

$$a_t + ua_r + \frac{\gamma - 1}{2} a \left(u_r + \frac{2u}{r} \right) = 0, \tag{9.45}$$

$$u_t + uu_r + \frac{2}{\gamma - 1} aa_r = 0. \tag{9.46}$$

Small Parameter Expansions.

One obvious approach is to continue the naive expansions in small amplitude beyond linear theory, see what goes wrong, and correct it. It may now be helpful to display a small parameter ϵ explicitly; ϵ would be taken, for example, as the maximum value of u/a_0 on some initial surface. The naive expansions would then be

$$u = \epsilon u_1(r,t) + \epsilon^2 u_2(r,t) + \cdots,$$

$$a = a_0 + \epsilon a_1(r,t) + \epsilon^2 a_2(r,t) + \cdots.$$

These are substituted in (9.45) and (9.46). The coefficients of successive powers of ϵ are equated to zero to obtain a hierarchy of successive pairs of equations for (u_1, a_1), (u_2, a_2),.... Of course u_1 and a_1 are found to be the linear expressions given earlier, the main terms being proportional to $F(t - r/a_0)/r$. The expressions for u_2 and a_2 are then found to contain terms in $r^{-1}\log r$, $r^{-2}\log r$, and r^{-2}. The first of these is responsible for the nonuniformity, since it makes the ratios u_2/u_1 and a_2/a_1 tend to infinity as $r \to \infty$; the others are harmless. The expressions are

$$\frac{u}{a_0} = \epsilon \left\{ \frac{F(\tau^*)}{r} - \frac{f(\tau^*)}{a_0 r^2} \right\} + \epsilon^2 \left\{ \frac{\gamma + 1}{2a_0} \frac{F(\tau^*)F'(\tau^*)}{r} \log r + \bar{u}_2 \right\} + \cdots,$$

$$\frac{2}{\gamma - 1} \frac{a - a_0}{a_0} = \epsilon \frac{F(\tau^*)}{r} + \epsilon^2 \left\{ \frac{\gamma + 1}{2a_0} \frac{F(\tau^*)F'(\tau^*)}{r} \log r + \bar{a}_2 \right\} + \cdots,$$

where τ^* is the linearized characteristic $t - r/a_0$. Here $\bar{u}_2$, $\bar{a}_2$, denote terms uniformly bounded with respect to u_1 and a_1; $F(\tau^*) = -f'(\tau^*)/a_0^2$ as before, and ϵF now replaces the function F in (9.20) and (9.21). We see immediately that the appearance of the nonuniform terms may be interpreted as the consequence of an injudicious use of Taylor's expansion on the expressions

$$\frac{u}{a_0} = \epsilon \left\{ \frac{F(\tau)}{r} - \frac{f(\tau)}{a_0 r^2} \right\},$$

$$\frac{2}{\gamma - 1} \frac{a - a_0}{a_0} = \epsilon \frac{F(\tau)}{r},$$

with

$$\tau = \tau^* + \frac{\gamma + 1}{2a_0} \epsilon F(\tau) \log r.$$

But these expressions would just be the proposed nonlinear solution (9.20), (9.21), *with τ determined as in* (9.24) *and with $T(\tau) = \tau$.* The situation is closely similar to the one discussed in Section 2.10. The use of Taylor's theorem in reverse is a familiar one in perturbation theory. The arbitrary function $T(\tau)$ allows more freedom in the choice of the characteristic variable τ; changes in the choice of τ are absorbed by changes in the determination of $F(\tau)$ from boundary data so the final solution is unique.

The foregoing investigation shows that to avoid nonuniformities one should start with the expansions

$$u = \epsilon u_1(r, \tau) + \epsilon^2 u_2(r, \tau) + \cdots,$$
$$a = a_0 + \epsilon a_1(r, \tau) + \epsilon^2 a_2(r, \tau) + \cdots, \tag{9.47}$$

where $\tau(t, r, \epsilon)$ is to be suitably chosen. Better still, one should add the expansion

$$t = t_0(r, \tau) + \epsilon t_1(r, \tau) + \cdots \tag{9.48}$$

to (9.47) and determine rather the function $t(r, \tau, \epsilon)$ by choosing $t_1(r, \tau)$, $t_2(r, \tau), \ldots,$ to avoid nonuniformities in the validity. For wave problems we expect the latter to stem from the requirement that curves $\tau = $ constant be characteristic curves. (Cases other than wave problems were proposed using this "strained coordinate" technique by Lighthill, 1949.) Assuming in advance that τ will turn out to be the characteristic variable, it is clearly preferable to start from the equations (9.45)–(9.46) written with τ and r as independent variables and $t(\tau, r)$ defined by

$$\frac{\partial t}{\partial r} = \frac{1}{a + u}. \tag{9.49}$$

The equations may then be written

$$\frac{2}{\gamma-1}a_r + u_r + \frac{2au}{a+u}\frac{1}{r} = 0,$$ (9.50)

$$\frac{2}{\gamma-1}a_\tau - u_\tau - \left(\frac{2}{\gamma-1}aa_r + uu_r\right)\frac{a+u}{a}t_\tau = 0.$$ (9.51)

The form is unsymmetrical because of the mixed use of one characteristic variable τ and the radial distance r. However, (9.50) can be recognized as the characteristic equation for variations along the characteristics $\tau = $ constant.

Equations 9.49–9.51 are now solved by the expansions 9.47–9.48. To lowest order in (9.49) we have

$$\frac{\partial t_0}{\partial r} = \frac{1}{a_0};$$

hence

$$t_0 = \frac{r}{a_0} + T(\tau).$$

The first order terms in (9.50)–(9.51) are then

$$\frac{2}{\gamma-1}a_{1r} + u_{1r} + \frac{2u_1}{r} = 0,$$

$$\frac{2}{\gamma-1}a_{1\tau} - u_{1\tau} - \frac{2}{\gamma-1}a_0 a_{1r} T'(\tau) = 0.$$

To solve these equations it should be remembered that they must be the linearized equations in disguise. It is easily verified that the solution is

$$\frac{u_1}{a_0} = \frac{F(\tau)}{r} - \frac{f(\tau)}{a_0 r^2},$$

$$\frac{2}{\gamma-1}\frac{a_1}{a_0} = \frac{F(\tau)}{r},$$

where $f'(\tau) = -a_0^2 F(\tau)T'(\tau)$. In the next order, (9.49) gives

$$\frac{\partial t_1}{\partial r} = -\frac{u_1 + a_1}{a_0^2}.$$

Hence

$$t_1 = -\frac{\gamma+1}{2a_0} F(\tau)\log r - \frac{1}{a_0^2}\frac{f(\tau)}{r}.$$

These lowest terms are precisely the nonlinearized solution (9.20)–(9.23), thus justifying the technique and providing a consistent scheme for higher approximations.

Expansions at Large Distances.

A variant of this approach is to use expansions of $u(r,\tau)$, $a(r,\tau)$, $t(r,\tau)$, not in powers of a small amplitude parameter ϵ but in inverse powers of r (supplemented by logarithmic terms when necessary) whose coefficients are functions of τ. This is essentially the approach used in the author's earlier papers (Whitham, 1950a, 1950b).

Wavefront Expansion.

Another approach, which is not as strongly based on expansions in ϵ, proceeds from a close analogy with the simple waves of the plane case. The full characteristic forms for (9.45)–(9.46) are

$$\left\{ \frac{\partial}{\partial t} + (u-a)\frac{\partial}{\partial r} \right\}\left(\frac{2}{\gamma-1}a - u \right) + \frac{2au}{r} = 0, \qquad (9.52)$$

$$\left\{ \frac{\partial}{\partial t} + (u+a)\frac{\partial}{\partial r} \right\}\left(\frac{2}{\gamma-1}a + u \right) + \frac{2au}{r} = 0. \qquad (9.53)$$

In the plane case the terms in $2au/r$ are absent, and for a simple wave (9.52) is dispensed with quickly by arguing that

$$\frac{2a}{\gamma-1} - u$$

is constant everywhere. This conclusion can no longer be reached exactly. However, the change in this Riemann variable will depend on the integral

$$\int_{C_-} 2au\frac{dt}{r}$$

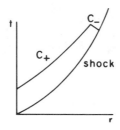

Fig. 9.1. Characteristics and shock in discussion of spherical waves.

taken along the C_-. Near the head of the wave the contribution will be small, since the range of integration is small (see Fig. 9.1). The relative change in the Riemann variable due to the integral will in fact be of order $a_0\tau/r$, since τ provides an estimate of the time change from the head of the wave. Furthermore, the arguments so far have indicated the region $a_0\tau/r$ $\ll 1$ as the one of interest. This suggests the constancy of the Riemann invariant as a good first approximation for this region. If we take

$$\frac{2a}{\gamma - 1} - u = \frac{2a_0}{\gamma - 1} \tag{9.54}$$

as an approximation to (9.52); the second equation (9.53) provides a single first order equation for u. Its solution requires integration along the C_+ and the range of integration on the C_+ is *not* small. The equation for u is

$$\frac{\partial u}{\partial t} + \left(a_0 + \frac{\gamma + 1}{2}u\right)\frac{\partial u}{\partial r} + \left(a_0 + \frac{\gamma - 1}{u}u\right)\frac{u}{r} = 0. \tag{9.55}$$

This is almost the same as (9.35) with $\beta = 1$ and can be handled similarly. The comparison of (9.55) with (6.83) for the plane case should also be noted. The exact solution is

$$\frac{u}{a_0}\left(1 + \frac{\gamma - 1}{2}\frac{u}{a_0}\right)^{2/(\gamma - 1)} = \frac{F(\tau)}{r}, \tag{9.56}$$

where τ is the characteristic variable to be determined from

$$\frac{\partial t}{\partial r} = \left(a_0 + \frac{\gamma + 1}{2}u\right)^{-1} \simeq \frac{1}{a_0} - \frac{\gamma + 1}{2}\frac{u}{a_0^2}. \tag{9.57}$$

A uniformly valid approximation is

$$\frac{u}{a_0} = \frac{F(\tau)}{r}, \qquad t = \frac{r}{a_0} - \frac{\gamma+1}{2a_0} F(\tau)\log r + T(\tau),$$

and, from (9.54),

$$\frac{2}{\gamma-1} \frac{a-a_0}{a_0} = \frac{u}{a_0} = \frac{F(\tau)}{r}.$$

This is the proposed solution (9.24). It should be noted, however, that in this approach as opposed to the last one we obtain only the geometrical acoustics form for u and a. This is perfectly satisfactory for the behavior in $a_0\tau/r \ll 1$, but it would require supplementing by other methods to determine $F(\tau)$.

When a shock is required at the head of the wave, the jumps in the entropy and the Riemann invariant (9.54) are of third order in the strength and do not affect the lowest order approximation.

N Wave Expansion.

After the shocks are included, the typical asymptotic behavior of the final waveform at large distances is an N wave centered around a limiting characteristic τ_0. For spherical waves, with details of the coefficients added to (9.34), we have

$$\frac{2}{\gamma-1} \frac{a-a_0}{a_0} \sim \frac{u}{a_0} \sim -\frac{2a_0}{\gamma+1}\left\{t - \frac{r}{a_0} - T(\tau_0)\right\}(r\log r)^{-1}. \quad (9.58)$$

This suggests that the final N wave form could be generated directly by looking for solutions in the form of expansions

$$u = v_1(r)(\zeta - \zeta_0) + v_2(r)(\zeta - \zeta_0)^2 + \cdots,$$

$$a = a_0 + b_1(r)(\zeta - \zeta_0) + b_2(r)(\zeta - \zeta_0)^2 + \cdots, \quad (9.59)$$

where $\zeta = t - r/a_0$ and ζ_0 refers to the asymptotically straight characteristic between the shocks. If these expansions are substituted in (9.45)–(9.46), the successive powers of $(\zeta - \zeta_0)$ lead to a hierarchy of equations for (v_1, b_1),

$(v_2, b_2), \ldots$. The first pair of equations gives

$$b_1 = \frac{\gamma - 1}{2} v_1,$$
(9.60)

$$\frac{dv_1}{dr} = \frac{\gamma + 1}{2} \frac{v_1^2}{a_0^2} - \frac{v_1}{r}.$$
(9.61)

The first equation confirms the relation between a and u. The second equation may be written

$$\frac{d}{dr} \left(\frac{1}{v_1 r} \right) + \frac{\gamma + 1}{2a_0^2} \frac{1}{r} = 0,$$

which integrates to

$$v_1 = - \frac{2a_0^2}{\gamma + 1} \frac{1}{r \log r}$$
(9.62)

and confirms the dependence on r noted in (9.58).

This gives a very simple approach to the asymptotic behavior, which is one of the outstanding results of the theory. It may be continued further to include the shock determination. If $\zeta - \zeta_0 = G(r)$ at the front shock, (9.59) provides power series in $G(r)$ for the flow quantities at the shock. The shock condition (9.28) in this case is

$$\frac{dG}{dr} = - \frac{1}{2} \frac{u + a - a_0}{a_0^2}$$

and, from (9.60)–(9.62), we find

$$\frac{dG}{dr} = \frac{1}{2} \frac{G}{r \log r} + O(G^2).$$

Therefore

$$G(r) \propto \log^{1/2} r.$$
(9.63)

This agrees with (9.31) for spherical waves.

Although this final method is perhaps the simplest of all, and one in which higher order corrections may be easily found, it cannot predict how the coefficients in the shock equation and shock strength depend on the initial source.

9.3 Sonic Booms

The central problem for sonic booms is to determine the shocks produced by an axisymmetrical body in steady supersonic flight. The effects of different body shapes, acceleration, curved flight paths, and nonuniform atmosphere are all developed in various ways from this basic problem.

For the basic problem it is convenient to take a frame of reference in which the flow is steady. The linearized theory has been discussed in detail in Section 7.5 and the nonlinearization can now proceed in close analogy with the techniques developed here for unsteady waves. The corresponding problem of plane flow treated in Section 6.17 also contributes to the background of ideas.

If U is the mainstream velocity parallel to the x axis and the perturbed velocity components in the x and r directions are now denoted by $U(1 + u)$ and Uv, we have

$$u = -\frac{1}{2\pi} \int_0^{x - Br} \frac{S''(\eta)\, d\eta}{\sqrt{(x-\eta)^2 - B^2 r^2}} \tag{9.64}$$

$$v = \frac{1}{2\pi r} \int_0^{x - Br} \frac{S''(\eta)\, d\eta}{\sqrt{(x-\eta)^2 - B^2 r^2}} \tag{9.65}$$

where $B = \sqrt{M^2 - 1}$ and $S(x)$ is the cross-sectional area at a distance x from the nose. The disturbance is confined behind the Mach cone $x - Br = 0$, which makes the Mach angle

$$\mu_0 = \sin^{-1} \frac{1}{M} \tag{9.66}$$

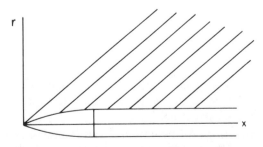

Fig. 9.2. Linear characteristic pattern in supersonic flow past a body.

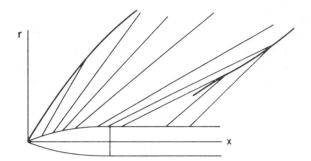

Fig. 9.3. Nonlinear characteristic pattern with shocks in supersonic flow past a body.

with the stream direction. The quantity $x - Br$ is the linear characteristic variable and corresponds to $t - r/c_0$ in the discussion of unsteady cylindrical waves. In the (x, r) diagram, the linear characteristics are a family of parallel straight lines making an angle μ_0 to the x axis as shown in Fig. 9.2. In the region $(x - Br)/Br \ll 1$, the approximations in (7.45)–(7.47) may be used. This region includes the front shock and the main part of the far field and it is here that the nonlinear corrections are crucial. The nonlinear effects modify the characteristics and introduce shocks as indicated in Fig. 9.3. Following the nonlinearization technique, we replace $x - Br$ by $\xi(x, r)$, where ξ is to be determined so that the curves $\xi = $ constant are an adequate approximation to the exact characteristics. The modified expressions for the flow quantities are

$$u = -\frac{F(\xi)}{\sqrt{2Br}}, \qquad v = \frac{BF(\xi)}{\sqrt{2Br}}, \tag{9.67}$$

$$\frac{p - p_0}{p_0} = \gamma M^2 \frac{F(\xi)}{\sqrt{2Br}}, \qquad \frac{a - a_0}{a_0} = \frac{\gamma - 1}{2} M^2 \frac{F(\xi)}{\sqrt{2Br}}, \tag{9.68}$$

where

$$F(\xi) = \frac{1}{2\pi} \int_0^\xi \frac{S''(\eta)}{\sqrt{\xi - \eta}} \, d\eta. \tag{9.69}$$

A typical F curve was given in Fig. 7.3. As noted in (7.48), $F \rightarrow 0$ as $\xi \rightarrow 0$ and the linear theory with $\xi = x - Br$ does not predict a shock. The nonlinearization is clearly crucial here.

The exact equations for irrotational axisymmetric flow are the same as for plane flow, (6.158)–(6.159), with y replaced by r and an additional term

$-a^2v/r$ in (6.158). Since the highest derivatives are unaffected, the characteristic directions are still given by $\theta \pm \mu$, where θ is the flow direction and μ is the exact Mach angle defined by $\mu = \sin^{-1}a/q$. Accordingly, on $\xi = $ constant,

$$\frac{dx}{dr} = \cot(\mu + \theta).$$

Just as in the unsteady wave problems, the first order perturbation approximation to this is sufficient and we use

$$\frac{\partial x}{\partial r} = \cot \mu_0 - (\mu - \mu_0 + \theta)\operatorname{cosec}^2 \mu_0.$$

To the same order

$$\theta \approx v, \qquad \mu - \mu_0 \approx \frac{a_0}{U}\left(\frac{a - a_0}{a_0} - u\right)\sec \mu_0;$$

hence

$$\frac{\partial x}{\partial r} = \frac{1}{B} - \frac{M^2}{B}\left(\frac{a - a_0}{a_0} - u\right) - M^2 v.$$

On substitution from (9.67) and (9.68), the equation becomes

$$\frac{\partial x}{\partial r} = \frac{1}{B} - \frac{(\gamma + 1)M^4}{(2B)^{3/2}}\frac{F(\xi)}{r^{1/2}},$$

and we have

$$x = Br - kF(\xi)r^{1/2} + \xi, \tag{9.70}$$

where

$$k = \frac{(\gamma + 1)M^4}{2^{1/2}B^{3/2}}. \tag{9.71}$$

The nonlinearized solution in the region $\xi/Br \ll 1$ is given by (9.67), (9.68), and (9.70).

The Shocks.

The characteristics overlap and a shock is required when $F'(\xi) > 0$. For a finite body,

$$\int_0^\infty F(\xi)d\xi = 0$$

as noted earlier, so there will be two such regions in general and two shocks. The counterpart to (9.28) is that the shock slope is the mean of the characteristic slopes on the two sides, and the shock determination is completely analogous. If a shock is described by

$$x = Br - G(r),$$

we have

$$G(r) = kF(\xi_1)r^{1/2} - \xi_1,$$

$$= kF(\xi_2)r^{1/2} - \xi_2,$$

where

$$\frac{1}{2}\{F(\xi_1) + F(\xi_2)\}(\xi_2 - \xi_1) = \int_{\xi_1}^{\xi_2} F(\xi)d\xi.$$

For the front shock which has undisturbed flow ahead of it, $F(\xi_1) = 0$ and ξ_1 may be eliminated from the determination. Then, dropping the subscript on ξ_2, we have

$$\frac{1}{2}kF^2(\xi)r^{1/2} = \int_0^\xi F(\xi')d\xi', \tag{9.72}$$

$$x = Br - kF(\xi)r^{1/2} + \xi, \tag{9.73}$$

for the determination of the shock. The flow quantities immediately behind the shock are given by (9.67) and (9.68) with $\xi(r)$ determined from (9.72).

Flow Past a Slender Cone.

For a cone with semiangle ϵ, $S(x) = \pi\epsilon^2 x^2$ and the F function (9.69) is

$$F(\xi) = 2\epsilon^2 \xi^{1/2}.$$

In that case the relation (9.72) between ξ and r for points on the shock is

$$\xi^{1/2} = \frac{3}{2}k\epsilon^2 r^{1/2}.$$

The shock equation (9.73) reduces to

$$x = Br - \frac{3}{4}k^2\epsilon^4 r.$$

This corresponds to a conical shock with a semiangle of

$$\mu_0 + \frac{3}{8} \frac{(\gamma+1)^2 M^6}{(M^2-1)^{3/2}} \epsilon^4. \tag{9.74}$$

The shock strength obtained from (9.68) is

$$\frac{p-p_0}{p_0} = \frac{3\gamma(\gamma+1)M^6}{2(M^2-1)} \epsilon^4. \tag{9.75}$$

For a cone, dimensional arguments show that the *exact* solution is a similarity solution with the flow quantities functions of r/x. The exact nonlinear equations may then be reduced to ordinary differential equations and integrated numerically. This is the famous Taylor-Maccoll solution (1933), which was a landmark in the development of the theory of supersonic flow. The results (9.74) and (9.75) were deduced for slender cones within the similarity theory by Lighthill (1948). It provides a valuable check on the more general approach for slender bodies. Numerical results show that (9.74)–(9.75) are very good approximations for cones up to 10° semiangle over a range of Mach numbers from about 1.1 to 3.0.

For general slender bodies, these formulas give the initial behavior of the shock. It should be noted that whereas the disturbances near the body are $O(\epsilon^2)$, the shock strength is $O(\epsilon^4)$. This explains in a sense why it is missed in the linear theory.

Behavior at Large Distances for Finite Bodies.

According to (9.72), for points on the shock $\xi \to \xi_0$ as $r \to \infty$, where $F(\xi_0) = 0$. Then (9.72) is asymptotically

$$F(\xi) \sim \left\{ \frac{2}{k} \int_0^{\xi_0} F(\xi')d\xi' \right\}^{1/2} r^{-1/4}. \tag{9.76}$$

The shock is asymptotic to

$$x \sim Br - \left\{ 2k \int_0^{\xi_0} F(\xi)d\xi \right\} r^{1/4} - \xi_0, \tag{9.77}$$

and the shock strength is

$$\frac{p-p_0}{p_0} \sim \frac{\gamma M^2}{(2B)^{1/2}} \left\{ \frac{2}{k} \int_0^{\xi_0} F(\xi) d\xi \right\}^{1/2} r^{-3/4}$$

$$= \frac{2^{1/4}\gamma}{(\gamma+1)^{1/2}} (M^2-1)^{1/8} \left\{ \int_0^{\xi_0} F(\xi) d\xi \right\}^{1/2} r^{-3/4}. \qquad (9.78)$$

This is the most important formula for sonic boom work. It shows that the boom at the ground depends very weakly on the Mach number, depends on distance like $r^{-3/4}$, and depends on the body shape through the factor

$$K = \left\{ \int_0^{\xi_0} F(\xi) d\xi \right\}^{1/2}. \qquad (9.79)$$

If the length of the body is l and the ratio of maximum diameter to length is the thickness ratio δ, the shape factor $K \propto \delta l^{3/4}$. For a body shape

$$R(x) = \begin{array}{ll} \delta l \left\{ 1 - \left(1 - \dfrac{x}{l} \right)^3 \right\}, & 0 \leqslant x \leqslant l, \\[2mm] \delta l, & l \leqslant x, \end{array}$$

we find $K = 1.04\delta l^{3/4}$.

The asymptotic wave profile is a balanced N wave. Between the shocks $\xi \sim \xi_0$, $F(\xi) \sim 0$, so that from (9.70)

$$F(\xi) \sim \frac{Br - x + \xi_0}{kr^{1/2}},$$

and from (9.68) and (9.71) the pressure ratio is

$$\frac{p-p_0}{p_0} \sim \frac{\gamma}{\gamma+1} \frac{(M^2-1)^{1/2}}{M^2} \frac{(Br-x+\xi_0)}{r}. \qquad (9.80)$$

The flow behind the rear shock is not completely undisturbed but is of smaller order than the disturbance in the N wave. For this and other details reference may be made to the original account (Whitham, 1952).

Extensions of the Theory.

Axisymmetric bodies might seem a far cry from real aircraft, but it is known that the far flow field in any direction away from a finite body can be represented as the flow due to an equivalent body of revolution. That is, in any direction the expressions (9.67)–(9.69) apply but the F function will be different for different directions. In the linear theory from which one starts, the contributions from fuselage, wings, lift distribution, and the like can be treated separately and superposed to give the final F function for each direction. The nonlinear expressions then apply with this F function. The volume contribution is related to a distribution of cross-sectional area $S(x)$, where the cuts are made by planes at an angle to the stream in accordance with the supersonic area rule. For details of the method and the nonlinear results, see Whitham (1956). When various protuberances such as wings are included $S'(x)$ becomes discontinuous and (9.69) must be modified appropriately (Whitham, 1952).

The effects of the lift distribution are of equal importance with the volume effects. In the linear theory the lift distribution $L(x)$ provides a contribution

$$\Phi_1 = -\frac{1}{2\pi\rho_0 U}\frac{\cos\tilde{\omega}}{r}\int_0^{x-Br}\frac{(x-\eta)L(\eta)}{\sqrt{(x-\eta)^2 - B^2 r^2}}\,d\eta \qquad (9.81)$$

to the velocity potential, where $\tilde{\omega}$ is the angle of a meridian plane through the flight path and is measured from the downward vertical. This may be approximated for $(x-Br)/Br \ll 1$ as before and the perturbations are again given by (9.67)–(9.68) with

$$F(\xi) = \frac{1}{2\pi}\frac{B\cos\tilde{\omega}}{\rho_0 U^2}\int_0^{\xi}\frac{L'(\eta)}{\sqrt{\xi-\eta}}\,d\eta. \qquad (9.82)$$

This is an interesting illustration of the "equivalent body" concept for asymmetric distributions. It should be noted that the approximations (9.67)–(9.68) are valid for $\xi/Br \ll 1$ and they are sufficient to determine the shocks. However, the pressure distribution behind the main N wave makes important contributions to the total lift transmitted to the ground. The full form (9.81) and, when necessary, its nonlinearization are required for a detailed accounting of the lift. This has sometimes caused confusion in the literature where it has been remarked that the pressure distribution given by (9.68), when integrated over the ground, does not give the total lift

$\int_0^\infty L(x)dx$. The expression (9.68) applies only in the region of the main N wave. The full formula derived from (9.81) integrates correctly to give the total lift.

The remaining extensions to accelerating bodies and nonuniform atmospheres, the latter being always important in the real situation, can be handled analytically to some extent and the theory leans heavily on geometrical acoustics (see Friedman, Kane, and Signalla, 1963, and references given there). Further developments and comparisons with wind tunnel and observational data are reviewed in a series of papers published in the *Journal of the Acoustical Society of America* (1965). Similar comparisons have been made by various government laboratories and aircraft companies. (A popular account for laymen which contains some interesting checks of the theory with reality is presented in Boeing Document D6A10598-1). The conclusion seems to be that the theory provides good results and valuable insight in an extremely complicated practical problem.

CHAPTER 10

Wave Hierarchies

The study of a single set of hyperbolic waves, including the various effects of geometry, diffusion, and damping, has now been covered in considerable detail. To complete this first part, we discuss the situation when waves of different orders appear in the same problem. Typical examples arose in Chapter 3 and some preliminary comments were made there. In traffic flow, for instance, the equations

$$\rho_t + (\rho v)_x = 0,$$

$$\tau(v_t + vv_x) + \frac{\nu}{\rho}\rho_x + v - V(\rho) = 0 \tag{10.1}$$

were proposed at a certain level of description. This system has two families of characteristics and the characteristic velocities are found to be

$$v + \sqrt{\frac{\nu}{\tau}}, \qquad v - \sqrt{\frac{\nu}{\tau}}. \tag{10.2}$$

Consequently, waves with these velocities will have their important roles to play. Yet the reduced equations

$$\rho_t + (\rho v)_x = 0, \qquad v = V(\rho), \tag{10.3}$$

which are expected to be a good approximation for sufficiently small values of ν and τ, have a single family of characteristics and the characteristic velocity is neither of the two in (10.2); it is in fact

$$V(\rho) + \rho V'(\rho). \tag{10.4}$$

If there is to be no inconsistency between the two levels of description, waves with velocity (10.4) must also play an important role in the solutions to (10.1), even though they no longer correspond to the characteristics. The aim here is to clarify further the roles of the "higher order waves" (10.2)

339

and the "lower order waves" (10.4), and to see how each set is modified by the presence of the other.

We consider first the linearized versions of systems like (10.1), since general solutions to typical problems can be obtained by transforms and used to exhibit the salient features explicitly. Similar analytic solutions are rarely available for the full nonlinear systems, but the linear results may be used to infer the corresponding nonlinear behavior of the various waves and to suggest simplifying approximations for their description.

When systems such as (10.1) are linearized it is more convenient to work with the equivalent single second order equation. It takes the general form

$$\eta\left(\frac{\partial}{\partial t}+c_1\frac{\partial}{\partial x}\right)\left(\frac{\partial}{\partial t}+c_2\frac{\partial}{\partial x}\right)\varphi+\left(\frac{\partial}{\partial t}+a\frac{\partial}{\partial x}\right)\varphi=0, \qquad (10.5)$$

where the coefficients are constants and, for definiteness, we choose $c_1>c_2$. In different notation, this was (3.4) for traffic flow; c_1 and c_2 are the linearized forms of (10.2), namely the values in the undisturbed flow, and a is the linearized form of (10.4). For flood waves, the nonlinear system (3.37) is similar to (10.1). The characteristic velocities are $v\pm\sqrt{g'h}$, but the reduced system (3.38) shows also the presence of lower order waves with velocity $3v/2$. The linearized equation (3.41) is in the same form as (10.5). For the chemical exchange processes formulated in (3.74), the linear equation is exact and is covered by (10.5) with one of the c's zero. Other examples will be mentioned later. If the systems concerned are higher than second order, there will be corresponding increases in the number of factors that make up the operators in (10.5).

The waves of different order are clearly displayed by the factored operators in (10.5). Indeed if the lower order terms were absent ($\eta=\infty$), the general solution would be ·

$$\varphi=\varphi_1(x-c_1t)+\varphi_2(x-c_2t). \qquad (10.6)$$

Conversely, if the higher order terms were absent ($\eta=0$), the solution would be

$$\varphi=\varphi_0(x-at). \qquad (10.7)$$

The latter corresponds, of course, to the reduced level of description whose linearized version is the equation

$$\frac{\partial\varphi}{\partial t}+a\frac{\partial\varphi}{\partial x}=0. \qquad (10.8)$$

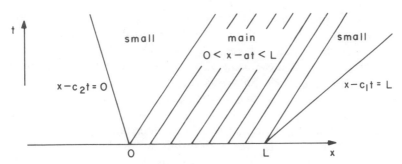

Fig. 10.1. The (x,t) diagram for the initial value problem.

Our questions concern the combined system, the different roles played by the waves in the two levels, and the modifications to (10.6) and (10.7).

We can sketch out in advance what must happen. Since the characteristics of (10.5) are determined by the higher order terms, the first signals and wavefronts *must* travel with the velocities c_1 and c_2. But to fit in with the reduced description, some of the disturbance must travel with velocity a. This is indicated in the (x,t) diagram in Fig. 10.1. As the parameter η is reduced, the first signals must become small; the main disturbance must be moving with velocity a and be reasonably well approximated by (10.7).

This picture makes sense only if a lies between c_1 and c_2. But as we saw in Chapter 3, exactly this ordering of velocities is necessary for stability, and the stability condition ties in nicely with the ideas on propagation. One is tempted to say that the instability that arises when a does not lie between c_1 and c_2 is a consequence of an unresolvable competition between the two sets of waves.

The question of appropriate boundary conditions also arises here, since the number of boundary conditions is determined by the number of characteristics pointing into the (x,t) region of interest. Since the number of characteristics can change in going from (10.1) to (10.3), or from (10.5) to (10.8), a rationalization of this apparent disagreement is also required. In view of the inequality

$$c_1 > a > c_2 \tag{10.9}$$

required for stability, (10.5) can only require *more* boundary conditions than (10.8). When this is the case, there will be no inconsistency in the two levels of description if the additional information for (10.5) affects the solution only in a layer next to the boundary, which is thin for small η, and outside this layer the solution of (10.5) is well approximated by a solution of (10.8). The appropriate solution of (10.8) will satisfy only some of the boundary data, and the remaining adjustment to the additional boundary data occurs in the boundary layer.

The details of all these arguments are substantiated from complete solutions of (10.5). The relevant ideas can then be taken over piecemeal to the nonlinear situation. In the nonlinear case, the possibility of shocks arises and different types of shock will appear as appropriate discontinuities in different levels of description. An understanding of the relations between them leads to a simple criterion to predict when the shock structure will still contain a discontinuity. Instances of this were noted in the inequalities (3.17) and (3.52). We shall now be able to give a more general view of these and add further examples.

10.1 Exact Solutions for the Linearized Problem

First we check the stability requirements on (10.5). An elementary solution is

$$\varphi = Ae^{ikx - i\omega t}$$

provided that

$$\eta(\omega - kc_1)(\omega - kc_2) + i(\omega - ka) = 0. \tag{10.10}$$

For relatively short waves, $kc_1\eta \gg 1$, we have

$$\omega \simeq \begin{cases} kc_1 - \dfrac{i}{\eta}\dfrac{c_1 - a}{c_1 - c_2}, & (10.11) \\[2ex] kc_2 - \dfrac{i}{\eta}\dfrac{a - c_2}{c_1 - c_2}. & (10.12) \end{cases}$$

One or the other of these has positive imaginary part showing instability unless

$$\eta > 0, \qquad c_1 > a > c_2. \tag{10.13}$$

It is then easily verified that $\Im\,\omega < 0$ for all k under these conditions, so they provide the complete stability requirements. We now suppose that the inequalities in (10.13) are satisfied and consider more general solutions.

The main points can be made equally well on the solution of the initial value problem by Fourier transforms or the solution of the signaling problem by Laplace transforms. The latter is chosen since it offers a richer number of cases depending on the signs of c_1, c_2, and a.

Case $c_1 > a > 0,\ c_2 < 0.$

This is the simplest case; with $c_2 < 0$, only the c_1 family of higher order waves is generated and with $a > 0$ there is no conflict in the number of boundary conditions posed on $x = 0$. For (10.5), a well-posed problem is then

$$\varphi = \varphi_t = 0, \quad x > 0, \quad t = 0,$$

$$\varphi = f(t), \quad x = 0, \quad t > 0. \tag{10.14}$$

For the reduced equation (10.8), the initial condition $\varphi_t = 0$ would be dropped, but in either case the solution remains identically zero in $x > 0$ for a finite time so the difference does not show up. The solution of (10.5) by Laplace transforms may be taken to be

$$\varphi(x,t) = \frac{1}{2\pi i} \int_{\mathscr{B}} \frac{\tilde{\varphi}(p,x)e^{pt}}{p} dp, \quad t > 0, \tag{10.15}$$

where $\mathscr{B}$ is a Bromwich path $\mathscr{R}p = \text{constant}$ to the right of all the singularities of the integrand in the complex p plane. On substitution in (10.5), we have

$$\eta c_1 c_2 \tilde{\varphi}_{xx} + \{\eta(c_1 + c_2)p + a\}\tilde{\varphi}_x + p(\eta p + 1)\tilde{\varphi} = 0,$$

and the possible solutions for $\tilde{\varphi}$ are

$$\tilde{\varphi} = F(p)e^{xP_1(p)} + G(p)e^{xP_2(p)}, \tag{10.16}$$

where P_1, P_2 are the roots of

$$\eta c_1 c_2 P^2 + \{\eta(c_1 + c_2)p + a\}P + p(\eta p + 1) = 0, \tag{10.17}$$

and F, G are arbitrary functions. For large p,

$$P_1 \sim -\frac{p}{c_1}, \quad P_2 \sim -\frac{p}{c_2}.$$

In this first case, with $c_1 > 0$, $c_2 < 0$, the second term in (10.16) is unbounded for large $\mathscr{R}p$, so we must take $G(p) = 0$; this term would correspond to incoming waves with velocity $c_2 < 0$ and is excluded. The remaining function $F(p)$ is determined from the *single* boundary condition $\varphi = f(t)$ at $x = 0$. In fact the requirement is simply that $F(p)$ be the Laplace transform of $f(t)$. The final solution therefore is

$$\varphi = \frac{1}{2\pi i} \int_{\mathscr{B}} \frac{F(p)}{p} e^{pt + P_1(p)x} dp, \tag{10.18}$$

where

$$F(p) = p \int_0^\infty f(t) e^{-pt} dt, \tag{10.19}$$

$$f(t) = \frac{1}{2\pi i} \int_{\mathscr{B}} \frac{F(p) e^{pt}}{p} dp,$$

and $P_1(p)$ is the root of (10.17), which is asymptotic to $-p/c_1$ as $p \to \infty$.

When $t - x/c_1 < 0$, the contour can be closed by a large semicircle to the right to show that $\varphi = 0$. Thus the wavefront is $x - c_1 t = 0$. The behavior of φ near the wavefront is obtained from the more detailed asymptotic behavior of the integrand of (10.18) as $p \to \infty$. If the contour $\mathscr{B}$ is taken far enough to the right, we may substitute the expansion

$$P_1 = -\frac{p}{c_1} - \frac{1}{\eta c_1} \frac{c_1 - a}{c_1 - c_2} + O\left(\frac{1}{p}\right)$$

in (10.18) and deduce the approximation

$$\varphi \approx f\left(t - \frac{x}{c_1}\right) \exp\left\{ -\frac{c_1 - a}{c_1 - c_2} \frac{x}{c_1 \eta} \right\}. \tag{10.20}$$

This is in fact the first term in the geometrical optics expansion (cf. Section 7.7); further terms in the series are obtained by continuing the expansion of $e^{P_1 x}$ for large p. The general form can be found by substitution of the geometrical optics expansion directly in (10.5), but (10.20) also relates the function of $t - x/c_1$ to the boundary condition. The expression (10.20) is valid near the wavefront. It shows that the first disturbance propagates with the c_1 waves, but this disturbance damps out exponentially and becomes negligible in a distance of order $c_1 \eta$. As $\eta \to 0$, this disturbance becomes negligible for all $x > 0$, in agreement with the reduced description.

We ask, then, where the main disturbance described by (10.18) is to be found. To obtain this information, we investigate the behavior on the family of lines $x/t = \text{constant}$ in the (x, t) plane, since each one of these represents the path of a wave moving with constant velocity. We shall need to be reasonably careful about the various limits involved and, accordingly, introduce nondimensional quantities

$$q = \eta p, \qquad Q(q) = \eta c_1 P_1(p), \qquad m = \frac{x}{c_1 t}.$$

In general, the boundary function $f(t)$ will introduce another time scale T, say, and $F(p)$ will take the form

$$F(p) = \mathscr{F}\left(q\frac{T}{\eta}\right).$$

Then (10.18) becomes

$$\varphi = \frac{1}{2\pi i} \int_{\mathscr{B}} \frac{\mathscr{F}(qT/\eta)}{q} e^{(q + mQ)t/\eta} dq, \tag{10.21}$$

where $Q(q)$ is the appropriate root of

$$\frac{c_2}{c_1} Q^2 + \left\{\left(1 + \frac{c_2}{c_1}\right)q + \frac{a}{c_1}\right\}Q + q(q+1) = 0.$$

We now consider the asymptotic behavior of (10.21) as $t/\eta \to \infty$, with m fixed. According to the saddle point method, the dominant contribution comes from the neighborhood of the point $q = q^*$ for which

$$\frac{d}{dq}(q + mQ) = 0,$$

that is,

$$1 + mQ'(q^*) = 0. \tag{10.22}$$

The first term of the asymptotic expansion is found by deforming the contour into the path of steepest descent $\mathcal{C}$ through $q = q^*$ and expanding $q + mQ$ as far as quadratic terms in $q - q^*$. Thus we have

$$\varphi \sim \exp\left(\frac{t}{\eta}\{q^* + mQ(q^*)\}\right)$$

$$\times \frac{1}{2\pi i} \int_{\mathcal{C}} \frac{\mathscr{F}(qT/\eta)}{q} \exp\left\{\frac{1}{2}\frac{t}{\eta}mQ''(q^*)(q - q^*)^2\right\}dq \tag{10.23}$$

as $t/\eta \to \infty$.

In the usual application of the saddle point method, the remaining part of the integrand would also be expanded in Taylor series about $q = q^*$

and $\mathcal{F}(qT/\eta)/q$ would be replaced by $\mathcal{F}(q^*T/\eta)/q^*$. This further step would be valid for the limit $t/\eta \to \infty$, T/η held fixed, and is relevant when $t \gg \eta$, $t \gg T$. But we are interested in the case $t \gg \eta$, $T \gg \eta$, independent of the magnitude of t/T. To include this case, the possibility $T/\eta \to \infty$ must be allowed in (10.23) and the more general form must be retained.

In discussing the behavior of (10.23), it is convenient to revert to the original variables. We have

$$\varphi \sim \exp\{tp^* + xP_1(p^*)\} \frac{1}{2\pi i} \int_{\mathcal{C}} \frac{F(p)}{p} \exp\left\{\frac{1}{2}xP_1''(p^*)(p-p^*)^2\right\} dp, \quad (10.24)$$

where p^* is the function of x and t determined by

$$t + xP_1'(p^*) = 0; \tag{10.25}$$

this provides the asymptotic behavior of φ as $t/\eta \to \infty$ keeping $x/c_1 t$ fixed. For simplicity we shall assume that $\int_0^\infty f(t)\, dt$ is convergent so that $F(p)/p$ is finite as $p \to 0$ and there is no pole. [The case in which $f(t)$ approaches a constant as $t \to \infty$ is of interest, but this is most easily handled by reformulating the problem in terms of φ_t.] The asymptotic expression (10.24) is itself dominated by the exponential factor outside the integral. The exponent is stationary when

$$\frac{\partial}{\partial x}\{ tp^* + xP_1(p^*)\} = 0,$$

and, in view of (10.25) which determines $p^*(x, t)$, this condition reduces to

$$P_1(p^*) = 0.$$

From (10.17), $P_1(p^*) = 0$ must correspond to either $p^* = 0$ or $p^* = -1/\eta$, and it is a simple matter to check that $p^* = 0$ is the correct choice for P_1. Therefore the exponential factor in (10.24) has a stationary value, in fact a local maximum, for those x and t which give $p^* = 0$ as the solution of (10.25). Thus the maximum is found on

$$t + P_1'(0)x = 0.$$

One checks from (10.17) that $P_1'(0) = -a^{-1}$. Hence the maximum of the exponential factor is to be found on

$$x = at,$$

and the maximum value is one. The disturbance is exponentially small (in this limit) except in the neighborhood of $x = at$; this result shows that *the main part of the disturbance eventually travels with velocity a.* Since the approximation is for $t \gg \eta$, the result applies increasingly earlier as $\eta \to 0$.

We can extract further information on the behavior of the main disturbance. In the neighborhood of $x = at$, the corresponding values of $p*$ are small. The details of the disturbance may then be obtained by making further expansions of (10.24) about $p* = 0$. But we would then have made an approximation to (10.18) in two stages: first an expansion of $pt + P_1(p)x$ about $p = p*$ and then an expansion of the resulting expression about $p* = 0$. Obviously, the final result must be included by simply expanding $pt + P_1(p)x$ about $p = 0$. We have

$$P_1(p) \sim -\frac{p}{a} + \frac{p^2 \eta (c_1 - a)(a - c_2)}{a^3} + \cdots .$$

Hence

$$\varphi \sim \frac{1}{2\pi i} \int_{\mathcal{C}} \frac{F(p)}{p} \exp\left\{ p\left(t - \frac{x}{a}\right) + \frac{p^2 \eta (c_1 - a)(a - c_2)x}{a^3} \right\} dp \quad (10.26)$$

applies in the neighborhood of $x - at = 0$ as $t/\eta \to \infty$. The first approximation is just

$$\varphi \sim \frac{1}{2\pi i} \int_{\mathcal{C}} \frac{F(p)}{p} e^{p(t - x/a)} dp = f\left(t - \frac{x}{a}\right),$$

which is exactly the prediction of the lower order equation 10.8. Thus the lower order formulation is shown to give a correct description of the main disturbance.

To see the effect of the quadratic term in the exponential in (10.26), it is more instructive to find the equation satisfied by (10.26) rather than to intepret the integral. The expression is, in fact, identical with the solution of the equation

$$\varphi_t + a\varphi_x = \frac{\eta (c_1 - a)(a - c_2)}{a^2} \varphi_{tt}, \quad (10.27)$$

for the same boundary condition $\varphi = f(t)$ on $x = 0$. The right side of (10.27) is already a small correction (of order η/t compared with the other terms), so it is consistent to use the first approximation $\partial/\partial t \simeq -a(\partial/\partial x)$ in it and to take an equivalent form

$$\varphi_t + a\varphi_x = \eta (c_1 - a)(a - c_2)\varphi_{xx}. \quad (10.28)$$

This is more familiar; it shows that the main part of the disturbance propagates with velocity a and is diffused by the effects of the higher order terms in the equation. But the latter effect is small when η is small. To summarize then in the case $c_1 > a > 0$, $c_2 < 0$: The first signals propagate out with velocity c_1 but damp out as shown in (10.20). The main disturbance lags behind and moves with the lower order wave speed a. In this case there is no disagreement in the number of boundary conditions to be prescribed at $x = 0$; $\varphi = f(t)$ is appropriate for both (10.5) and (10.8). After a time of order η, the first signals are exponentially small and the main part of the solution to (10.5) is well described by (10.8) using the same boundary condition at $x = 0$. The effect of the higher order terms is to produce a diffusion of the lower order waves as shown by (10.28) but this is small when η is appropriately small.

Case $c_1 > 0$, $c_2 < a < 0$.

In this case the maximum value of the exponential factor in (10.24) still occurs at $x = at$, but with $a < 0$ this is outside the region $x > 0$. The expression (10.24) is exponentially small throughout the region $x > 0$. The saddle point formula does not apply at $x = 0$, since from (10.25) there is clearly no saddle point, but it is easily shown from (10.18) that the solution falls exponentially from the value $\varphi = f(t)$ on $x = 0$, and the disturbance is confined to a boundary layer of thickness η / c_1. In this case the first signal decays exponentially and the main disturbance does not propagate.

The reduced equation (10.8) does not permit data to be specified at $x = 0$, and its solution is $\varphi \equiv 0$. This agrees with the preceding description outside the boundary layer, and the difference in boundary conditions is accommodated by the boundary layer.

Case $c_1 > a > c_2 > 0$.

In this case both characteristics of (10.5) point into the region $x > 0$, and it is appropriate to give the two conditions

$$\varphi = f(t), \qquad \varphi_x = g(t) \qquad \text{for } x = 0, \quad t > 0, \tag{10.29}$$

However, we note that only one of these or possibly a combination of them could be posed with (10.8). The two conditions (10.29) correspond precisely to the fact that with c_1 and c_2 positive, both terms in (10.16) must be retained and there are two arbitrary functions to be determined. If $\tilde{f}(p)$

and $\tilde{g}(p)$ are the transforms of $f(t)$ and $g(t)$, then the arbitrary functions in (10.16) are determined by

$$F + G = \tilde{f}, \qquad P_1 F + P_2 G = \tilde{g}. \tag{10.30}$$

The discussion of the term (10.18) in the solution is exactly the same as before with the same conclusion that the first signals travel with velocity c_1 but are damped out; the main disturbance travels with velocity a and is diffused by the higher order effects. The main disturbance is again well described by (10.8) and the only new question concerns which boundary condition is in fact adopted. The function $F(p)$ which appears in the corresponding solution (10.26) is obtained from (10.30) as

$$F = \frac{P_2 \tilde{f} - \tilde{g}}{P_2 - P_1}. \tag{10.31}$$

But when (10.26) applies, P_1 is approximated for small values of ηp, and P_2 must be approximated in the same way. It is easily seen from (10.17) that

$$\eta P_1 = -\frac{\eta p}{a} + O(\eta^2 p^2), \qquad \eta P_2 \sim -\frac{a}{c_1 c_2} + O(\eta p);$$

hence (10.31) reduces to

$$F = \tilde{f}$$

in this approximation. Therefore the boundary condition $\varphi = f(t)$ is in fact satisfied and must be used with the reduced equation.

The second term in the complete solution is

$$\frac{1}{2\pi i} \int_{\mathcal{B}} \frac{G(p)}{p} e^{pt + P_2(p)x} dp. \tag{10.32}$$

Since $P_2 \sim -p/c_2$ as $p \to \infty$, this expression is zero for $x > c_2 t$, corresponding to the second wavefront provided by the waves with velocity c_2. It is easily shown, as in the earlier case, that these waves are damped exponentially and become negligible when $x/c_2 \eta \gg 1$. The saddle point investigation then shows that the contribution of (10.32) is small except near $x = 0$. Near $x = 0$, we may use the asymptotic expansion for

$$\frac{t}{\eta} \to \infty, \qquad \frac{x}{c_2 \eta} \text{ fixed}$$

to see how the solution behaves. According to Watson's lemma, this is found from the behavior of the integrand in (10.32) for small ηp. We have

$$P_2 \sim -\frac{a}{c_1 c_2 \eta}, \qquad G(p) \sim -\left(\tilde{g} + \frac{p}{a}\tilde{f}\right)\frac{c_1 c_2 \eta}{a}.$$

Therefore (10.32) is asymptotic to

$$-\left\{g(t) + \frac{1}{a}f'(t)\right\}\frac{c_1 c_2}{a}\eta \exp\left(-\frac{a}{c_1 c_2}\frac{x}{\eta}\right). \tag{10.33}$$

Thus the first contribution from (10.32) travels with the c_2 waves but damps out and its main contribution is the boundary layer given by (10.33).

A composite solution for $t/\eta \gg 1$ is obtained by adding the two main contributions to give

$$\varphi = f\left(t - \frac{x}{a}\right) - \left\{g(t) + \frac{1}{a}f'(t)\right\}\frac{c_1 c_2}{a}\eta \exp\left(-\frac{a}{c_1 c_2}\frac{x}{\eta}\right). \tag{10.34}$$

This satisfies both boundary conditions to the first order. The second term is needed to satisfy the second boundary condition, but it decays rapidly away from the boundary in a layer whose thickness is of order $\eta c_1 c_2 / a$. The various results are conveniently represented in an (x, t) diagram as shown in Fig. 10.2.

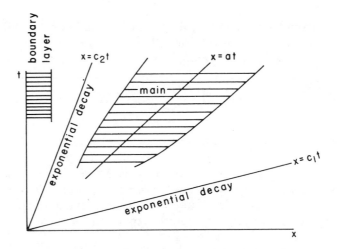

Fig. 10.2. The (x, t) diagram for the signaling problem.

10.2 Simplified Approaches

Other linear equations with constant coefficients can alwa s be treated in similar fashion by transforms and suitable asymptotic expansions. However, the details are tedious and the main ingredients in any solution can be seen very simply, and with greater insight, by more intuitive arguments. We indicate the techniques on the previous problem.

First, in any wave profile moving approximately with speed V, the t and x derivatives are related approximately by

$$\frac{\partial}{\partial t} \simeq -V\frac{\partial}{\partial x}. \tag{10.35}$$

We can use this in (10.5) to examine the waves moving with velocities c_1, c_2, a, in turn. For c_1 waves, we use $\partial/\partial t \simeq -c_1 \, \partial/\partial x$ in all derivatives except the sensitive one where the factor $\partial/\partial t + c_1 \, \partial/\partial x$ itself appears. We have

$$\eta(c_2 - c_1)\frac{\partial}{\partial x}\left(\frac{\partial}{\partial t} + c_1\frac{\partial}{\partial x}\right)\varphi + (a - c_1)\frac{\partial\varphi}{\partial x} = 0.$$

The $\partial/\partial x$ operator can be integrated out without loss since it corresponds to the remnants of the other waves. Accordingly we take

$$\frac{\partial\varphi}{\partial t} + c_1\frac{\partial\varphi}{\partial x} + \frac{c_1 - a}{\eta(c_1 - c_2)}\varphi = 0. \tag{10.36}$$

The solution is exactly the one found in (10.20). Similarly for the c_2 waves, we have

$$\frac{\partial\varphi}{\partial t} + c_2\frac{\partial\varphi}{\partial x} + \frac{a - c_2}{\eta(c_1 - c_2)}\varphi = 0;$$

the solution is similar in form to (10.20) and may be verified in detail from (10.32).

For the lower order waves propagating with velocity a, we use $\partial/\partial t \simeq -a \, \partial/\partial x$ in the second order terms in (10.5) and the approximate form is

$$\varphi_t + a\varphi_x = \eta(c_1 - a)(a - c_2)\varphi_{xx}. \tag{10.37}$$

This in exact agreement with (10.28). If the t derivatives are preferred in the second order term we obtain the alternative form (10.27).

The possibility of boundary layers near $x=0$ may be investigated in the same spirit by arguing that x derivatives will be much larger than t derivatives so that the approximation

$$\frac{\partial}{\partial x} \gg \frac{\partial}{\partial t} \qquad (10.38)$$

should be made. This may be interpreted as the special case of (10.35) with $V=0$; it corresponds to nonpropagating waves. Under this approximation, (10.5) is reduced to

$$\eta c_1 c_2 \varphi_{xx} + a\varphi_x = 0 \qquad (10.39)$$

and the general solution is

$$\varphi = A(t) + B(t) \exp\left(-\frac{a}{c_1 c_2} \frac{x}{\eta}\right), \qquad (10.40)$$

in agreement with (10.34). Of course the exponential solution is omitted unless $a/c_1 c_2 \eta > 0$ and it is only in the case of exponential decrease that the existence of a boundary layer is inferred.

For initial layers, where data for φ and φ_t are imposed at $t=0$ in the full equation, we consider the form of the equation under the approximation $\partial/\partial t \gg \partial/\partial x$. For (10.5) we have

$$\eta \varphi_{tt} + \varphi_t = 0, \qquad \varphi = C(x) + D(x)e^{-t/\eta}.$$

This shows how the solution adjusts to the approximate form where only the initial value of φ is relevant.

These techniques allow a quick assessment of the various regions of interest and of the relevant approximations. They are easily made the basis of more formal perturbation procedures. For example, the straightforward expansion

$$\varphi = \varphi_0(x,t) + \eta \varphi_1(x,t) + \eta^2 \varphi_2(x,t) + \cdots$$

gives (10.8) for φ_0; the expansion

$$\varphi = \varphi_0(\xi,t) + \eta^{1/2} \varphi_1(\xi,t) + \cdots,$$

$$\xi = \eta^{-1/2}(x - at)$$

leads to (10.28) for φ_0; the expansion

$$\varphi = \varphi_0(X,t) + \eta \varphi_1(X,t) + \cdots,$$

$$X = \eta^{-1}x$$

leads to the boundary layer equation (10.39) for φ_0.

10.3 Higher Order Systems, Nonlinear Effects, and Shocks

For a nonlinear system of equations which exhibits plane waves of different orders, we do not normally have the luxury of complete exact solutions and for analytic work have to rely on the counterparts of the approximations in the last section. Just how to proceed in detail will vary from problem to problem but we may note a few guidelines. For ease of reference suppose we call the complete set of equations the system I and the reduced set obtained by setting some parameter η equal to zero the system II. The theory of characteristics will provide the characteristic velocities $c_1,\ldots,c_n$ for I, and characteristic velocities $a_1,\ldots,a_m$ ($m < n$) for II. In general, for nonlinear problems these will be functions of the dependent variables. However, the linearized theory for small perturbations about some uniform state will be useful to set the scene and give information about stability. If just two orders are present the linearized theory for plane waves in a uniform medium may be reduced to a single equation

$$\eta\left(\frac{\partial}{\partial t} + c_1\frac{\partial}{\partial x}\right)\cdots\left(\frac{\partial}{\partial t} + c_n\frac{\partial}{\partial x}\right)\varphi + \left(\frac{\partial}{\partial t} + a_1\frac{\partial}{\partial x}\right)\cdots\left(\frac{\partial}{\partial t} + a_m\frac{\partial}{\partial x}\right)\varphi = 0,$$

$$(10.41)$$

for some perturbation potential φ, where the propagation speeds now take their constant values in the uniform state. The standard stability argument turns up the interesting result that the orders must satisfy either $m = n-1$ or $m = n-2$ for stability. In the first case the complete requirements are

$$m = n-1, \qquad \eta \geqslant 0, \qquad c_1 > a_1 > c_2 > a_2 \cdots > a_{n-1} > c_n. \quad (10.42)$$

These are just the conditions that allow a satisfactory interpretation of how the solutions to II approximate to the full set I. The second case, $m = n-2$, introduces effects more typical of dispersive waves and its discussion is relegated to the second part of the book. (The stability conditions for this case were given by Wu, 1961, correcting an earlier misstatement by the author that (10.42) was the only case to consider.)

Equation 10.41 may be solved by transforms, but the overall picture may be outlined by the techniques of the last section. Each of the c_i waves satisfies an approximate equation

$$\eta(\varphi_t + c_i\varphi_x) + \gamma_i\varphi = 0, \qquad (10.43)$$

where γ_i is determined in terms of the a's and c's similar to (10.36). The stability conditions ensure that $\gamma_i > 0$, so we have exponential decay.

Similarly, the a_i waves satisfy an approximate equation

$$\varphi_t + a_i \varphi_x = \eta \alpha_i \varphi_{xx}, \tag{10.44}$$

analogous to (10.28), and $\alpha_i > 0$.

When nonlinear effects are included, we should aim toward an equation of the form

$$\varphi_t + c_i(\varphi)\varphi_x + \beta_i \varphi = 0 \tag{10.45}$$

to replace (10.43). Within the waves of any order there is now nontrivial interaction between the different waves and extracting (10.45) would require something akin to the "simple wave argument." In nonlinear problems it would normally be more convenient to work with n dependent variables and a system of first order equations. The technique of using $\partial / \partial t \simeq - c_i \partial / \partial x$ for c_i waves is roughly equivalent to taking the $(n-1)$ Riemann variables of the other $(n-1)$ c waves to be constants.

Equation 10.45 can then be solved by the methods of Chapter 2. Waves carrying an increase of $c_i(\varphi)$ *may* break [see the discussion of (2.72)] and shocks will be required. These shocks are discontinuities of the system I and the shock conditions are deduced in the standard manner from the relevant system of conservation equations. For $\beta_i > 0$, even when shocks are produced the disturbance normally decays and the main disturbance is eventually carried by the lower order waves.

If $\eta = 0$, a nonlinear simple wave on the a_i characteristics will satisfy an equation of the form

$$\varphi_t + a_i(\varphi)\varphi_x = 0. \tag{10.46}$$

Shocks required in the solution of this equation will satisfy the discontinuity conditions derived from the reduced system II and will be different from the shock conditions for system I.

When higher order diffusive effects are introduced one should aim for an equation

$$\varphi_t + a_i(\varphi)\varphi_x = \nu_i \varphi_{xx} \tag{10.47}$$

corresponding to (10.37). This can be related to Burgers' equation for a_i, within the same type of approximation, and analyzed from the results of Chapter 4.

10.4 Shock Structure

A shock of system II, let us call it an S_{II}, will be smoothed out to some extent when viewed within system I. It is just the shock structure problem. However, we can now comment more decisively on the occurrence of discontinuities in the structure. An S_{II} will always be associated with one particular family of a waves, and each family of a waves is sandwiched between two families of c waves. But the S_{II} shock will go faster than the a waves ahead of it and slower than the a waves behind it. Even in a stable situation, it may overtake the next c wave ahead or lag behind the c wave following. This would violate the characteristic property of the c waves *if the solution remains continuous*. But S_I discontinuities can do this. Therefore, whenever this situation occurs, the S_{II} profile will require discontinuities in the shock structure. These discontinuities are S_I's satisfying the S_I jump conditions, and we may regard the complete profile as a combined S_I-S_{II} wave.

If the velocity of an S_{II} associated with the a_i family is U_i, and if superscripts 1 and 2 now denote the values of quantities ahead and behind it, the criterion for a continuous shock structure is

$$c_{i+1}^{(2)} < U_i < c_i^{(1)}. \qquad (10.48)$$

If this is violated, there will be a discontinuity in the profile, and this may be the case even though the states on the two sides are stable. If we consider a nonlinear version of Fig. 10.1, the main disturbance remains continuous provided its velocity is well away from the higher order wave velocities on the two sides.

In the nonlinear case all these velocities have a certain range of values, so it is possible for the main wave to combine with the c waves ahead or behind. When this happens the exponential decay of the c waves ceases and, since the c waves break, an S_I shock is picked up in the shock structure. The criterion (10.48) on the velocities gives a very simple method of prediction which avoids the much more involved analysis of the integral curves for the shock structure equations. Its use will be noted in the following examples.

10.5 Examples

Flood Waves.

This case was studied in great detail in Chapter 3. We note that, with the current notation,

$$c_1 = v + \sqrt{g'h} , \qquad c_2 = v - \sqrt{g'h} ,$$

$$a = \frac{3v}{2} .$$

According to (10.42) a uniform flow $h = h_0, v = v_0$ is stable (as noted in Section 3.2) provided

$$v_0 - \sqrt{g'h_0} < \frac{3v_0}{2} < v_0 + \sqrt{g'h_0} ,$$

the lower limit being no restriction. According to (10.48) the S_{II} shock structure will be continuous provided

$$v^{(2)} - \sqrt{g'h^{(2)}} < U < v^{(1)} + \sqrt{g'h^{(1)}} .$$

The S_{II} shock conditions show that $U > v^{(2)}$, so the lower inequality is always satisfied. The criterion for the appearance of a discontinuity in the S_{II} structure is

$$U > v^{(1)} + \sqrt{g'h^{(1)}} .$$

This is precisely the result found in (3.52) from the detailed analysis.

Magnetogasdynamics.

The equations for magnetogasdynamic waves are given in Examples 10 and 10′ of Section 5.2. The first set is the system I and we have

$$c_1 = (\epsilon_0 \mu)^{-1/2}, \qquad c_2 = u + a, \qquad c_3 = u,$$

$$c_4 = u - a, \qquad c_5 = -(\epsilon_0 \mu)^{-1/2}.$$

The second set is the system II and

$$a_1 = u + \left(a^2 + \frac{B^2}{\mu\rho} \right)^{1/2} , \qquad a_2 = a_3 = u,$$

$$a_4 = u - \left(a^2 + \frac{B^2}{\mu\rho} \right)^{1/2} .$$

The genuine waves moving relative to the fluid have alternating velocities as required by (10.42) and are stable. The confluence of a_2, a_3, with c_3 on the particle paths is easily checked to be a stable situation.

An S_{II} shock of the a_1 family moving with velocity U has a continuous structure provided

$$u^{(2)} + a^{(2)} < U < (\epsilon_0 \mu)^{-1/2}.$$

The speed of light is effectively infinite, so we deduce that a discontinuity appears at the back of the profile when

$$U < u^{(2)} + a^{(2)}.$$

This is a simple derivation of the result found by Marshall (1955) by a detailed analysis of the shock structure. Further discussion of this case may be found in Whitham (1959a).

Relaxation Effects in Gases.

The equations of inviscid gas dynamics (Chapter 6) may be written

$$\rho_t + u\rho_x + \rho u_x = 0,$$

$$u_t + uu_x + \frac{1}{\rho}p_x = 0,$$

$$e_t + ue_x + \frac{p}{\rho}u_x = 0.$$

During rapid changes in the flow the internal energy e may lag behind the equilibrium value corresponding to the ambient pressure and density. The translational energy adjusts quickly, but the rotational and vibrational energy may take an order of magnitude longer. If we suppose that α of the degrees of freedom adjust instantaneously but a further α_r degrees of freedom take longer to relax, we may take

$$e = \frac{\alpha}{2}\frac{p}{\rho} + E,$$

where E is the energy in the lagging degrees of freedom. *In equilibrium* [see (6.42)] E would have the value

$$E_{equil} = \frac{\alpha_r}{2}\frac{p}{\rho}.$$

A simple overall equation to represent the relaxation is

$$E_t + uE_x = -\tau\left(E - \frac{\alpha_r}{2}\frac{p}{\rho}\right),$$

where τ is the relaxation time. After some minor manipulation, the set of equations may be written

$$\rho_t + u\rho_x + \rho u_x = 0,$$

$$u_t + uu_x + \frac{1}{\rho}p_x = 0,$$

$$\frac{\alpha}{2}(p_t + up_x) - \left(1 + \frac{\alpha}{2}\right)\frac{p}{\rho}(\rho_t + u\rho_x) + \rho(E_t + uE_x) = 0,$$

$$E_t + uE_x + \tau\left(E - \frac{\alpha_r}{2}\frac{p}{\rho}\right) = 0.$$

The characteristic velocities are

$$c_1 = u + a_f, \qquad c_2 = c_3 = u, \qquad c_4 = u - a_f,$$

where a_f is the "frozen" sound speed defined by

$$a_f^2 = \left(1 + \frac{2}{\alpha}\right)\frac{p}{\rho} = \gamma_f\frac{p}{\rho}.$$

This is the system I for this case. However, if the relaxation time τ is taken to be so short that $E = (\alpha_r/2)(p/\rho)$ is an adequate approximation to the last equation of the set, we have the equilibrium theory:

$$\rho_t + u\rho_x + \rho u_x = 0,$$

$$u_t + uu_x + \frac{1}{\rho}p_x = 0,$$

$$\frac{\alpha + \alpha_r}{2}(p_t + up_x) - \left(1 + \frac{\alpha + \alpha_r}{2}\right)\frac{p}{\rho}(\rho_t + u\rho_x) = 0.$$

This is the reduced system II for our problem. The characteristics are

$$a_1 = u + a_e, \qquad a_2 = u, \qquad a_3 = u - a_e,$$

where a_e is the equilibrium sound speed defined by

$$a_e^2 = \left(1 + \frac{2}{\alpha + \alpha_r}\right)\frac{p}{\rho} = \gamma_e\frac{p}{\rho}.$$

Since $\gamma_e < \gamma_f$, the various velocities alternate and there is stability; again confluence of velocities with the particle velocity is stable.

Viewed from the full system the S_{II} shock structure, which takes the flow between two uniform states in equilibrium, will be continuous if

$$u^{(2)} - a_f^{(2)} < U < u^{(1)} + a_f^{(1)}. \qquad (10.49)$$

Since the S_{II} shock conditions show that $U > u^{(2)}$, only the upper restriction arises. A frozen S_I shock will appear at the front of the profile and will be followed by a continuous relaxation region when

$$U > u^{(1)} + a_f^{(1)}.$$

This criterion may be written as

$$M = \frac{U - u^{(1)}}{a_e^{(1)}} > \frac{a_f^{(1)}}{a_e^{(1)}} = \left(\frac{\gamma_f}{\gamma_e} \right)^{1/2}. \qquad (10.50)$$

For a diatomic molecule the two rotational modes may lag behind the three translation modes, and we may describe this by taking $\alpha = 3$, $\alpha_r = 2$. The frozen and equilibrium sound speeds are given by

$$a_f^2 = \frac{5}{3} \frac{p}{\rho}, \qquad a_e^2 = \frac{7}{5} \frac{p}{\rho}.$$

The criterion (10.50) predicts a fully relaxed smooth profile for

$$M < 1.091,$$

and a discontinuity followed by a relaxation region when M exceeds this critical value. The discontinuity would itself be resolved into a thin sublayer by the inclusion of viscosity and heat conduction.

Griffith, Brickl, and Blackman (1956) and Griffith and Kenny (1957) report on experimental observations for vibrational relaxations in CO_2. The appropriate choice for α is 5 to include translational and rotational modes. The vibrational modes take considerably longer to adjust. At 300°K, the appropriate choice* for α_r is 2. The critical value for M is 1.043, and the experimental observations in the references cited bear this out very accurately. (Further details may be found in the above papers and in the excellent article by Lighthill, 1956.)

We see that a careful assessment of the waves involved and the roles they must play leads to relatively simple predictions of important phenomena in quite complicated situations.

---

*At this temperature the four modes have only one half of their classical energies so that $\alpha_r = 2$ is appropriate.

Part II

DISPERSIVE WAVES

CHAPTER 11

Linear Dispersive Waves

The discussion in Part I was concerned primarily with hyperbolic systems. Most wave motions, including the familiar one of water waves, are not described by hyperbolic equations in the first instance. At a later stage there is some link-up with hyperbolic equations in describing the propagation of important average quantities associated with the disturbance. But a different set of basic ideas and mathematical techniques must be developed.

These nonhyperbolic wave motions can be grouped largely into a second main class which we call *dispersive*. In general the classification is less precise than that for hyperbolic waves, since it is made on the type of solution rather than on the equations themselves. But it may be made precise in a restricted class of problems, and one can make extensions in a natural way or proceed by analogy. It should be added that a few special equations exhibit both hyperbolic and dispersive behavior, the different behaviors appearing in different regions of the solution. But these are exceptional.

The first two chapters develop the general ideas for linear systems. Chapter 13 is devoted to water waves, a subject which is fascinating for its own sake as well as being the subject in which many of the ideas of dispersive waves were first developed. In Chapter 13 a start is made on nonlinear dispersive waves in this specific context to serve as background for the general developments of nonlinear theory in Chapters 14 and 15; Chapter 16 contains miscellaneous applications. The final chapter covers the recent work on solitary waves and special equations.

11.1 Dispersion Relations

For linear problems, dispersive waves usually are recognized by the existence of elementary solutions in the form of sinusoidal wavetrains

$$\varphi(\mathbf{x}, t) = A e^{i\mathbf{\kappa} \cdot \mathbf{x} - i\omega t}, \tag{11.1}$$

where κ is the wave number, ω is the frequency, and A is the amplitude. In the elementary solution, κ, ω, A are constants. Since the equations are linear, A factors out and is arbitrary. But to satisfy the equations, κ and ω have to be related by an equation

$$G(\omega, \kappa) = 0.$$

The function G is determined by the particular equations of the problem. For example, if φ satisfies the beam equation

$$\varphi_{tt} + \gamma^2 \varphi_{xxxx} = 0,$$

we require

$$\omega^2 - \gamma^2 \kappa^4 = 0.$$

The relation between ω and κ is called the *dispersion relation*, and as will become evident below, we can dispense with the equations once we know the dispersion relation; conversely, we can construct the equation from the dispersion relation.

We assume that the dispersion relation may be solved in the form of real roots

$$\omega = W(\kappa). \tag{11.2}$$

There will be a number of such solutions, in general, with different functions $W(\kappa)$. We refer to these as different *modes*. The beam equation, for example, allows two modes:

$$\omega = \gamma \kappa^2, \qquad \omega = -\gamma \kappa^2.$$

For the present we study one mode; in linear problems the modes can be superposed to make up the complete solution. The linearity also allows us to work with the complex form (11.1) with the understanding that the real part should be taken when necessary. The actual solution is

$$\Re \varphi = |A| \cos(\kappa \cdot x - \omega t + \eta), \qquad \eta = \arg A.$$

The quantity

$$\theta = \kappa \cdot x - \omega t \tag{11.3}$$

is the *phase*; it determines the position on the cycle between a crest, where $\Re \varphi$ is maximum, and a trough, where $\Re \varphi$ is a minimum. In this plane wave solution, phase surfaces $\theta = $ constant are parallel planes. The gradient of θ in space is the wave number κ, whose direction is normal to the planes and whose magnitude κ is the average number of crests per 2π units of distance in that direction. Similarly, $-\theta_t$ is the frequency ω, the average number of crests per 2π units of time. (The normalization to 2π units is convenient in working with the trigonometric functions.) The wavelength is $\lambda = 2\pi/\kappa$ and the period is $\tau = 2\pi/\omega$.

The wave motion is recognized from (11.3). Any particular phase surface is moving with normal velocity ω/κ in the direction of κ. We therefore introduce the *phase velocity*,

$$\mathbf{c} = \frac{\omega}{\kappa}\,\hat{\kappa},\qquad(11.4)$$

where $\hat{\kappa}$ is the unit vector in the κ direction. For any particular mode $\omega = W(\kappa)$, the phase velocity is a function of κ. For the wave equation $\varphi_{tt} = c_0^2 \nabla^2 \varphi$, the dispersion relation gives $\omega = \pm c_0\kappa$ and $c = \pm c_0$: the phase velocity agrees with the usual propagation speed. In general, c is not independent of κ. Different wave numbers will lead to different phase speeds. This accounts for the term "dispersion." In a Fourier synthesis of more general solutions by superposition, components with different wave numbers disperse as time goes on. This crucial process is discussed in detail in the next section.

As regards classification, we must exclude the case $\mathbf{c} = $ constant from the class of dispersive waves, since the dispersion would be absent in that case. It is also clear that the solutions (11.2) must be real for these effects. The heat equation $\varphi_t = \nabla^2 \varphi$ has solutions (11.1) with $\omega = -i\kappa^2$, but the solutions are not wavelike. To eliminate the unwanted cases, we restrict the term dispersive, in the first instance, to those cases for which

$$W(\kappa) \text{ is real, } \quad \text{and}$$

$$\text{determinant}\left|\frac{\partial^2 W}{\partial \kappa_i\, \partial \kappa_j}\right| \neq 0. \qquad (11.5)$$

For one dimensional problems, the second condition is just

$$W''(\kappa) \neq 0.$$

This is slightly stronger than $c'(\kappa) \neq 0$, since it eliminates $W = a\kappa + b$ as well. The reason for this is amplified below, but for the one dimensional case we note in advance that the group velocity $W'(\kappa)$ is the more important propagation velocity and the condition $W''(\kappa) \neq 0$ ensures that it is not a constant. We can see directly that the excluded case $W = a\kappa + b$ is not really dispersive. The elementary solution is

$$e^{-ibt}e^{i\kappa(x-at)}$$

and Fourier superposition gives the general solution

$$e^{-ibt}f(x-at).$$

An initial waveform $f(x)$ is modified as it propagates but there is no dispersion. One can easily show that the governing equation is hyperbolic.

The determinant in (11.5) may vanish near some special value of κ, for example, as $\kappa \to 0$ or $\kappa \to \infty$, and these limiting values will require special consideration as singularities in the general formulas.

Examples.

Some typical examples which will be used as illustration in the development of the general theory are:

$$\varphi_{tt} - \alpha^2 \nabla^2 \varphi + \beta^2 \varphi = 0, \qquad \omega = \pm \sqrt{\alpha^2 \kappa^2 + \beta^2} \; ; \qquad (11.6)$$

$$\varphi_{tt} - \alpha^2 \nabla^2 \varphi = \beta^2 \nabla^2 \varphi_{tt}, \qquad \omega = \pm \frac{\alpha \kappa}{\sqrt{1 + \beta^2 \kappa^2}} \; ; \qquad (11.7)$$

$$\varphi_{tt} + \gamma^2 \varphi_{xxxx} = 0, \qquad \omega = \pm \gamma \kappa^2; \qquad (11.8)$$

$$\varphi_t + \alpha \varphi_x + \beta \varphi_{xxx} = 0, \qquad \omega = \alpha \kappa - \beta \kappa^3. \qquad (11.9)$$

The first of these is hyperbolic but has dispersive solutions satisfying (11.5) nevertheless. It represents vibrations for a displacement φ with additional restoring force proportional to φ; it is also the Klein-Gordon equation of quantum theory. The other equations are not hyperbolic, and these are the more typical cases. Equation 11.7 appears in elasticity for longitudinal waves in bars, in water waves in the Boussinesq approximation for long waves, and in plasma waves. Equation 11.8 is the equation for flexural vibrations of a beam. Equation 11.9 also applies to long water waves; it is the linearized form of the Korteweg-deVries equation. The water wave approximations will be investigated in detail later; the others are taken to be reasonably familiar.

Correspondence Between Equation and Dispersion Relation.

It is obvious from these examples, and clearly true in general, that equations with real coefficients will lead to real dispersion relations only if they consist entirely of even derivatives or they consist entirely of odd derivatives. Each differentiation brings out a factor i. Even derivatives will lead to a real coefficient, odd derivatives to a pure imaginary coefficient; they cannot be mixed if the final form is to be real. Schrödinger's equation,

$$i\hbar \frac{\partial \varphi}{\partial t} = -\frac{\hbar^2}{2m} \nabla^2 \varphi, \tag{11.10}$$

with mixed odd and even derivatives, has a real dispersion relation

$$\hbar \omega = \frac{\hbar^2 \kappa^2}{2m}$$

by allowing a complex coefficient.

We can take the correspondence between the equation and the dispersion relation much further. A single linear equation with constant coefficients may be written

$$P\left(\frac{\partial}{\partial t}, \frac{\partial}{\partial x_1}, \frac{\partial}{\partial x_2}, \frac{\partial}{\partial x_3} \right) \varphi = 0,$$

where P is a polynomial. When the elementary solution (11.1) is substituted in the equation, each $\partial/\partial t$ will produce a factor $-i\omega$, and each $\partial/\partial x_j$ produces a factor $i\kappa_j$. The dispersion relation must be

$$P(-i\omega, i\kappa_1, i\kappa_2, i\kappa_3) = 0, \tag{11.11}$$

and we have a direct correspondence between the equation and the dispersion relation through the correspondence

$$\frac{\partial}{\partial t} \leftrightarrow -i\omega, \qquad \frac{\partial}{\partial x_j} \leftrightarrow i\kappa_j.$$

From (11.11) we can recover the equation. This is the basis for the earlier remark that the equation can be dropped when the dispersion relation is known.

It is seen, however, that an equation of this type can yield only polynomial dispersion relations. A natural question to ask is what kind of operators yield more general dispersion relations. One possibility is that the oscillatory wave motion represented by (11.1) takes place in only some of the space coordinates while there is a more complicated behavior in the remaining coordinates. A typical example is the theory of deep water waves in which waves propagate horizontally, and the dependence on the vertical coordinate is not oscillatory; this will be seen later. A second possibility, which has wavelike behavior in all the variables, can be illustrated in one dimension by the integrodifferential equation

$$\frac{\partial \varphi}{\partial t}(x,t) + \int_{-\infty}^{\infty} K(x-\xi) \frac{\partial \varphi}{\partial \xi}(\xi,t)\, d\xi = 0, \qquad (11.12)$$

where the kernel $K(x)$ is a given function. This equation has elementary solutions $\varphi = A e^{i\kappa x - i\omega t}$ provided

$$-i\omega e^{i\kappa x} + \int_{-\infty}^{\infty} K(x-\xi) i\kappa e^{i\kappa \xi}\, d\xi = 0.$$

The condition can be rearranged as

$$c = \frac{\omega}{\kappa} = \int_{-\infty}^{\infty} K(\zeta) e^{-i\kappa \zeta}\, d\zeta. \qquad (11.13)$$

The right hand side is the Fourier transform of the given kernel $K(x)$ and, by the inversion theorem, we have

$$K(x) = \frac{1}{2\pi} \int_{-\infty}^{\infty} c(\kappa) e^{i\kappa x}\, d\kappa. \qquad (11.14)$$

Thus (11.12) can be constructed to give any desired $c(\kappa)$ and consequently any desired dispersion function: simply choose $K(x)$ as the Fourier transform (11.14) of the desired phase velocity $c(\kappa)$. In particular, if

$$c(\kappa) = c_0 + c_2 \kappa^2 + \cdots + c_{2m} \kappa^{2m},$$

then

$$K(x) = c_0 \delta(x) - c_2 \delta''(x) + \cdots + (-1)^m c_{2m} \delta^{(2m)}(x),$$

and (11.12) reduces to the differential equation

$$\frac{\partial \varphi}{\partial t} + c_0 \frac{\partial \varphi}{\partial x} - c_2 \frac{\partial^3 \varphi}{\partial x^3} + \cdots + (-1)^m c_{2m} \frac{\partial^{2m+1} \varphi}{\partial x^{2m+1}} = 0.$$

When $c(\kappa)$ is a more general function with an infinite Taylor series in powers of κ, we can take the corresponding differential equation with an infinite series of derivatives. This is effectively summed by (11.12).

Definition of Dispersive Waves.

We can now introduce a restricted definition of dispersive linear systems as those for which there are solutions (11.1), (11.2), with (11.5) satisfied. There is some overlap with hyperbolic systems as the example (11.6) illustrates, but the system is usually not hyperbolic. We need not confine the considerations to differential equations, as the last paragraph shows.

It is immediately clear that the definition is too restrictive. Even for linear differential equations, it is limited to constant coefficients. For example, if γ is a function of x in the beam equation, that is,

$$\varphi_{tt} + \gamma^2(x)\varphi_{xxxx} = 0,$$

(11.1) is not a solution. Yet unless $\gamma(x)$ is a particularly violent function of x, we would expect the solution to have many features similar to the case of constant γ; in some sense the structure is the same. We would think of this as a problem of dispersive waves in a nonuniform medium. Again, an equation may have separable solutions, say,

$$X(\kappa x)e^{-i\omega t}, \qquad \omega = W(\kappa),$$

where X is an oscillatory function such as a Bessel function. This would be dispersive in a similar way, but it would be hard to include in an overall definition. We seem to be left at present with the looser idea that whenever oscillations in space are coupled with oscillations in time through a dispersion relation, we expect the typical effects of dispersive waves.

The situation is similar for nonlinear systems: a restricted class can be identified and then the ideas are extrapolated in a natural way.

A more comprehensive answer lies perhaps in the variational formulations to be developed in the later chapters. These allow the theory of the solutions to be developed in a general way, and presumably they provide the appropriate general framework for many questions, including the classification. But this is still an open question.

11.2 General Solution by Fourier Integrals

If (11.1)–(11.2) is an elementary solution for a linear equation, then, formally at least,

$$\varphi(\mathbf{x}, t) = \int_{-\infty}^{\infty} F(\boldsymbol{\kappa})e^{i\boldsymbol{\kappa}\cdot\mathbf{x} - iW(\boldsymbol{\kappa})t}d\boldsymbol{\kappa} \tag{11.15}$$

is also a solution. The arbitrary function $F(\kappa)$ may be chosen to fit arbitrary initial or boundary data, provided the data are reasonable enough to admit Fourier transforms. If there are n modes with n different choices of $W(\kappa)$, there will be n terms like (11.15) with n arbitrary functions $F(\kappa)$. It will then be appropriate to give n initial conditions to determine the solution. The examples (11.6)–(11.8) all have two modes and it is appropriate to prescribe φ and φ_t at $t=0$. As happens in these cases, the two modes will often be $\omega = \pm W(\kappa)$ and, in a typical one dimensional problem, we have then

$$\varphi = \int_{-\infty}^{\infty} F_1(\kappa) e^{i\kappa x - iW(\kappa)t} d\kappa + \int_{-\infty}^{\infty} F_2(\kappa) e^{i\kappa x + iW(\kappa)t} d\kappa \qquad (11.16)$$

with initial conditions

$$\varphi = \varphi_0(x), \qquad \varphi_t = \varphi_1(x)$$

at $t=0$. If $W(\kappa)$ is odd in κ, as in (11.7), the first term in (11.16) represents waves moving to the right and the second term represents waves moving to the left. If $W(\kappa)$ is even, as in (11.6) and (11.8), waves moving to the right and left appear in both contributions. Applying the initial conditions, we require

$$\varphi_0(x) = \int_{-\infty}^{\infty} \{ F_1(\kappa) + F_2(\kappa) \} e^{i\kappa x} d\kappa,$$

$$\varphi_1(x) = -i \int_{-\infty}^{\infty} W(\kappa) \{ F_1(\kappa) - F_2(\kappa) \} e^{i\kappa x} d\kappa.$$

The inverse formulas give

$$F_1(\kappa) + F_2(\kappa) = \Phi_0(\kappa) = \frac{1}{2\pi} \int_{-\infty}^{\infty} \varphi_0(x) e^{-i\kappa x} dx,$$

$$-iW(\kappa) \{ F_1(\kappa) - F_2(\kappa) \} = \Phi_1(\kappa) = \frac{1}{2\pi} \int_{-\infty}^{\infty} \varphi_1(x) e^{-i\kappa x} dx,$$

and $F_1(\kappa), F_2(\kappa)$, are determined to be

$$F_1(\kappa) = \frac{1}{2} \left\{ \Phi_0(\kappa) + \frac{i\Phi_1(\kappa)}{W(\kappa)} \right\},$$

$$F_2(\kappa) = \frac{1}{2} \left\{ \Phi_0(\kappa) - \frac{i\Phi_1(\kappa)}{W(\kappa)} \right\}.$$

Since $\varphi_0(x), \varphi_1(x)$ are real, $\Phi_0(-\kappa) = \Phi_0^*(\kappa)$ and $\Phi_1(-\kappa) = \Phi_1^*(\kappa)$ where asterisks denote complex conjugates. It follows that for $W(\kappa)$ odd

$$F_1(-\kappa) = F_1^*(\kappa),$$

$$F_2(-\kappa) = F_2^*(\kappa); \tag{11.17}$$

and for $W(\kappa)$ even

$$F_1(-\kappa) = F_2^*(\kappa),$$

$$F_2(-\kappa) = F_1^*(\kappa). \tag{11.18}$$

In either case, the solution (11.16) is real: real initial conditions must lead to real solutions for real equations.

A standard solution from which others can be reconstructed is obtained by taking

$$\varphi_0(x) = \delta(x), \qquad \varphi_1(x) = 0.$$

Then $F_1(\kappa) = F_2(\kappa) = 1/4\pi$ and (11.16) reduces to

$$\varphi = \frac{1}{\pi} \int_0^\infty \cos \kappa x \cos W(\kappa) t \, d\kappa; \tag{11.19}$$

of course this is a formal integral to be interpreted as a generalized function.

11.3 Asymptotic Behavior

Although the Fourier integrals give exact solutions, the content is hard to see. It becomes clearer and one starts to understand the main features of dispersive waves by considering the asymptotic behavior for large x and t. We consider first the typical integral

$$\varphi(x, t) = \int_{-\infty}^\infty F(\kappa) e^{i\kappa x - iW(\kappa)t} \, d\kappa$$

in the one dimensional case. For wave motions we are interested in the behavior for both large x and t; the interesting limit is $t \to \infty$ with x/t held fixed. (A particular choice of x/t allows us to examine waves moving with that velocity.) Accordingly, the integral is written

$$\varphi(x, t) = \int_{-\infty}^\infty F(\kappa) e^{-i\chi t} \, d\kappa, \tag{11.20}$$

where

$$\chi(\kappa) = \bar{W}(\bar{\kappa}) - \kappa \frac{x}{t} \, .$$

For the present, x/t is a fixed parameter and only the dependence on κ is displayed in χ. The integral in (11.20) may then be studied by the method of stationary phase; this is, in fact, the problem for which Kelvin developed the method. Kelvin argued that for large t, the main contribution to the integral is from the neighborhood of stationary points $\kappa = k$ such that

$$\chi'(k) = W'(k) - \frac{x}{t} = 0. \tag{11.21}$$

Otherwise, the contributions oscillate rapidly and make little net contribution. The later development of the method of steepest descents (or saddle point method) is easier to justify and to assess errors. A full discussion of the methods is given, for example, in Jeffreys and Jeffreys (1956, Sections 17.04–17.05). It will be sufficient here to derive the first term in the asymptotic expansion following Kelvin's argument.

The functions $F(\kappa), \chi(\kappa)$ in (11.20) are expanded in Taylor series in the neighborhood of $\kappa = k$. The dominant contribution comes from the terms

$$F(\kappa) \simeq F(k),$$

$$\chi(\kappa) \simeq \chi(k) + (\kappa - k)^2 \chi''(k),$$

provided $\chi''(k) \neq 0$. With these approximations, the contribution is

$$F(k) \exp\left\{ -i\chi(k)t \right\} \int_{-\infty}^{\infty} \exp\left\{ -\frac{i}{2}(\kappa - k)^2 \chi''(k) t \right\} d\kappa.$$

The remaining integral is reduced to the real error integral

$$\int_{-\infty}^{\infty} e^{-\alpha z^2} dz = \left(\frac{\pi}{\alpha} \right)^{1/2}$$

by rotating the path of integration* through $\pm \pi/4$; the sign should be chosen to be the same as the sign of $\chi''(k)$. Then we have

$$F(k) \sqrt{ \frac{2\pi}{t |\chi''(k)|} } \ \exp\left\{ -i\chi(k)t - \frac{i\pi}{4} \operatorname{sgn} \chi'' \right\}$$

---

*This corresponds to changing to the path of steepest descents.

If there is more than one stationary point $\kappa = k$ satisfying (11.21), each one contributes a similar term, and we have,

$$\varphi \sim \sum_{\substack{\text{stationary} \\ \text{points } k}} F(k) \sqrt{\frac{2\pi}{t|W''(k)|}} \ \exp\left\{ikx - iW(k)t - \frac{i\pi}{4}\operatorname{sgn} W''(k)\right\}. \quad (11.22)$$

The next term in the asymptotic behavior requires the Taylor series continued as far as the $(\kappa - k)^2$ term in $F(\kappa)$ and the $(\kappa - k)^4$ term in $\chi(\kappa)$. *Two* further terms are required because the odd powers integrate out eventually. When this is done, preferably using the method of steepest descents, the additional term may be written as a factor

$$1 - \frac{i}{t|W''|}\left(\frac{F''}{2F} - \frac{1}{2}\frac{W'''}{W''}\frac{F'}{F} + \frac{5}{24}\frac{W'''^2}{W''^2} - \frac{1}{8}\frac{W^{iv}}{W''}\right) \quad (11.23)$$

multiplying the term in (11.22). The complicated form comes from the necessity of working to two further terms in the Taylor series for F and χ. In general, this series continues in inverse powers of t with coefficients which are functions of k.

Earlier, the precise meaning of "large t" was left vague. The requirement may now be taken to be that the correction term in (11.23) be small; t must be large compared with time scales derived from the dispersion relation and from the length scale in the initial conditions. For sharply peaked initial conditions with small length scale, F' and F'' are small and the requirement is that t be large relative to the typical period in $W(k)$, which in turn is given by parameters in the equation. For the extreme case of a δ function initial condition, F is constant and $F' = F'' = 0$.

For the special case of two modes with $\omega = \pm W(\kappa)$, the full solution is given by (11.16). We make the further assumption that $W'(\kappa)$ is monotonic and positive for $\kappa > 0$ (this is usually the case), and we consider the asymptotic behavior of (11.16) for $x > 0$. If $W(\kappa)$ is odd, $W'(\kappa)$ is even and (11.21) has two roots $\pm k$. These two contributions in (11.22) can be combined, since $F_1(-k) = F_1^*(k)$ from (11.17), and we have

$$\varphi \sim 2\Re\left(F_1(k)\sqrt{\frac{2\pi}{t|W''(k)|}} \ \exp\left\{ikx - iW(k)t - i\frac{\pi}{4}\operatorname{sgn} W''(k)\right\}\right),$$

$$t \to \infty, \frac{x}{t} > 0, \quad (11.24)$$

where $k(x,t)$ is the positive root of (11.21) defined by

$$k(x,t): \qquad W'(k) = \frac{x}{t}, \quad k > 0, \quad \frac{x}{t} > 0. \quad (11.25)$$

For odd $W(k)$, the second integral in (11.16) does not contribute to the solution in $x > 0$; it gives a corresponding expression for $x < 0$.

When $W(\kappa)$ is even, $W'(\kappa)$ is odd; (11.21) has one root k for $x > 0$ and it is positive. There is only one contribution from the first integral in (11.16). However, for the second integral in (11.16), the stationary points satisfy

$$W'(\kappa) = -\frac{x}{t},$$

and $-k$ is a solution of this for $x > 0$. That is, with k defined as in (11.25), there is a stationary point $\kappa = k$ in the first integral of (11.16) and a stationary point $\kappa = -k$ in the second one. In view of (11.18), the contributions can again be combined and the net result is the same formula as (11.24).

The significance of the condition $W''(\kappa) \neq 0$ in the definition of dispersive waves for linear systems is now clear. If $W'(\kappa)$ is a constant, there are no stationary points for general x/t and the whole asymptotic analysis is different. Of course it is not necessary, since the Fourier integrals can be simplified immediately. The significance of $W''(k) \neq 0$ appears also in the denominator of (11.24) and in the error term (11.23). If $W''(\kappa)$ is not identically zero, but vanishes for some particular stationary point k, the correct asymptotic behavior is found by going to further terms in the Taylor series for χ. If $\chi''(k) = 0$, but $\chi'''(k) \neq 0$, the contribution to (11.20) is

$$F(k)\exp\{-i\chi(k)t\} \int_{-\infty}^{\infty} \exp\left\{-\frac{i}{6}t\chi'''(k)(\kappa - k)^3\right\} d\kappa$$

$$= (\tfrac{1}{3})! 3^{5/6} 2^{1/3} \frac{F(k)}{(t|W'''(k)|)^{1/3}} \exp\{ikx - iW(k)t\}. \tag{11.26}$$

Since k is a function of x/t this would indicate a singular behavior on the corresponding line $x/t = W'(k)$ and in its neighborhood.

We now take up a detailed discussion of (11.24) and (11.25).

11.4 Group Velocity; Wave Number and Amplitude Propagation

At any point (x, t), (11.25) determines a certain wave number $k(x, t)$, and the dispersion relation $\omega = W(k)$ gives also a frequency $\omega(x, t)$ at that point. We may introduce a phase

$$\theta(x, t) = xk(x, t) - t\omega(x, t),$$

and (11.24) may then be written

$$\varphi = \mathscr{R}\{A(x,t)e^{i\theta(x,t)}\}, \tag{11.27}$$

where the complex amplitude is

$$A(x,t) = 2F_1(k)\sqrt{\frac{2\pi}{t|W''(k)|}}\; e^{-(\pi i/4)\mathrm{sgn}\,W''}. \tag{11.28}$$

The expression (11.27) is in the *form* of the elementary solution, but A, k, ω are no longer constants. However, the solution still represents an oscillatory wavetrain, with a phase θ describing the variations between local maxima and minima. The difference is that the wavetrain is not uniform; the distance and time between successive maxima are not constant, nor is the amplitude.

It is natural to generalize the concept of wave number and frequency in this nonuniform case by defining them as θ_x and $-\theta_t$, respectively. Counting the number of maxima in unit distance would obviously be a clumsy and ill-defined quantity, whereas θ_x is straightforward and does correspond to the intuitive idea of a local wave number. Moreover, in the present case, we have

$$\theta(x,t) = kx - W(k)t,$$

$$\frac{\partial\theta}{\partial x} = k + \{x - W'(k)t\}\frac{\partial k}{\partial x}, \tag{11.29}$$

$$\frac{\partial\theta}{\partial t} = -W(k) + \{x - W'(k)t\}\frac{\partial k}{\partial t};$$

the stationary condition (11.25) eliminates the terms in k_x, k_t, and we have just

$$\frac{\partial\theta}{\partial x} = k(x,t), \tag{11.30}$$

$$\frac{\partial\theta}{\partial t} = -W(k) = -\omega(x,t). \tag{11.31}$$

Thus the wave number k, which was first introduced as a particular value of the wave number in the Fourier integral, agrees with our extended definition of a local wave number θ_x in an oscillatory nonuniform wavetrain. The same is true of the corresponding frequency. *Furthermore, the local wave number and local frequency satisfy the dispersion relation even in the nonuniform wavetrain.*

These extensions work so neatly because the nonuniformity is not too great. If an oscillation is too irregular, we might find a phase function θ and then define θ_x as the wave number, but the intuitive interpretation would be lost if θ_x itself varied rapidly in the course of *one* oscillation. In our case, $k(x,t)$ is a slowly varying function. From (11.25),

$$\frac{k_x}{k} = \frac{W'}{kW''}\frac{1}{x}, \qquad \frac{k_t}{k} = -\frac{W'}{kW''}\frac{1}{t},$$

and x, t are both relatively large. Therefore the relative change in one wavelength or one period is small. Thus k is a slowly varying function in this sense; the same is true of ω. [Again we note the singular behavior near any points with $W''(k) = 0$.] From (11.28), it is easily seen that A is also slowly varying.

With these interpretations of the quantities appearing in (11.27), we return to the determination of k and ω as functions of (x,t) from (11.25), and the determination of A from (11.28). Equation 11.25 determines k as a function of x and t, but to appreciate its content we turn it around and ask where a particular value k_0 will be found. The answer is: at points

$$x = W'(k_0)t.$$

That is, an observer moving with the velocity $W'(k_0)$ will always see waves with wave number k_0 and frequency $W(k_0)$. The quantity

$$W'(k) = \frac{d\omega}{dk}$$

is the *group velocity*; it is the important velocity for a "group" of waves with a distribution of wave number. The interpretation of (11.25) shows that *different wave numbers propagate with the group velocity*; any particular wave number k_0 is found displaced a distance $W'(k_0)t$ in time t.

Any particular value θ_0 of the phase propagates according to

$$\theta(x,t) = \theta_0.$$

Hence it moves according to

$$\theta_x \frac{dx}{dt} + \theta_t = 0,$$

that is,

$$\frac{dx}{dt} = -\frac{\theta_t}{\theta_x} = \frac{\omega}{k}.$$

Thus the phase velocity c is still given by ω/k even though the meanings of ω and k have been extended. But it is different from the group velocity. An observer following any particular crest moves with the local phase velocity but sees the local wave number and frequency changing; that is, neighboring crests get farther away. An observer moving with the group velocity sees the same local wave number and frequency, but crests keep passing him.

As illustration of this important distinction, consider the beam equation. The dispersion relation is

$$W(k) = \gamma k^2.$$

Hence (11.25) becomes

$$W'(k) = 2\gamma k = \frac{x}{t},$$

and we have

$$k = \frac{x}{2\gamma t}, \qquad \omega = \frac{x^2}{4\gamma t^2}, \qquad \theta = \frac{x^2}{4\gamma t}.$$

The group lines of constant k and ω are

$$\frac{x}{2\gamma t} = \text{constant};$$

the phase lines of constant θ are

$$\frac{x^2}{4\gamma t} = \text{constant}.$$

These are shown in the $(x; t)$ diagram in Fig. 11.1. In this case the group velocity $2\gamma k$ is greater than the phase velocity γk.

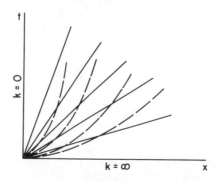

Fig. 11.1. Group lines (solid lines) and phase lines (dashed lines) for waves in a beam.

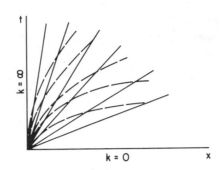

Fig. 11.2. Group lines (solid lines) and phase lines (dashed lines) for deep water waves.

For water waves in deep water (see Chapter 12), the dispersion relation is $W = \sqrt{gk}$. Hence $W'(k) = x/t$ leads to

$$k = \frac{gt^2}{4x^2}, \qquad \omega = \frac{gt}{2x}, \qquad \theta = -\frac{gt^2}{4x}.$$

The group velocity $\frac{1}{2}\sqrt{g/k}$ is less than the phase velocity $\sqrt{g/k}$ and we have the situation shown in Fig. 11.2.

In all cases (for the uniform media considered so far) the group lines are straight lines whereas the phase lines are not; each wave number propagates with a constant velocity; each phase accelerates or decelerates as it passes through different wave numbers.

If we take δ function initial conditions so that $F_1(k) = 1/4\pi$, the amplitudes $A(x,t)$ can be given explicitly as well and the complete asymptotic solutions are

$$\varphi \sim \frac{1}{\sqrt{4\pi\gamma t}} \cos\left(\frac{x^2}{4\gamma t} - \frac{\pi}{4}\right) \tag{11.32}$$

for the beam, and

$$\varphi \sim \frac{1}{2}\left(\frac{g}{\pi}\right)^{1/2} \frac{t}{x^{3/2}} \cos\left(\frac{gt^2}{4x} - \frac{\pi}{4}\right) \tag{11.33}$$

for water waves. [In fact, (11.32) is exact since $W(\kappa)$ is exactly quadratic for the beam.]

A second important role of the group velocity appears in studying the distribution of $A(x,t)$. The form of (11.28) suggests that $|A|^2$ is the interesting quantity to consider, and this is also natural on physical grounds since it is an energylike quantity. The relation of $|A|^2$ to the true energy density and to the so-called "wave action" will be pursued later. For the present, $A(x,t)$ is a well-defined quantity given by (11.28) and we may consider $|A|^2$.

The integral of $|A|^2$ between any points $x_2 > x_1 > 0$ is given from (11.28) by

$$Q(t) = \int_{x_1}^{x_2} AA^* \, dx$$

$$= 8\pi \int_{x_1}^{x_2} \frac{F_1(k) F_1^*(k)}{t|W''(k)|} \, dx.$$

In this integral k is given by (11.25). Since k appears in the arguments of the integrand and x does not, it is natural to introduce k as a new variable of integration through the transformation

$$x = W'(k)t.$$

For $W''(k) > 0$, we have

$$Q(t) = 8\pi \int_{k_1}^{k_2} F_1(k) F_1^*(k) \, dk, \tag{11.34}$$

where k_1 and k_2 are defined by

$$x_1 = W'(k_1)t, \qquad x_2 = W'(k_2)t. \tag{11.35}$$

If $W''(k) < 0$, the order of the limits is reversed.

Now, if k_1 and k_2 are held fixed as t varies, $Q(t)$ remains constant. According to (11.35), the points x_1 and x_2 are then moving with the corresponding group velocities. We have therefore proved that the total amount of $|A|^2$ between any pair of group lines remains the same. *In this sense, $|A|^2$ propagates with the group velocity*. The group lines diverge with a separation increasing like t; hence $|A|$ decreases like $t^{-1/2}$.

In the special case of a wave packet where the initial disturbance is localized in space and contains appreciable amplitude only in wave numbers close to some particular value k^*, say, the resulting disturbance is confined to the neighborhood of the particular group line k^* and the wave packet as a whole moves with the particular group velocity $W'(k^*)$. Accounts of group velocity in the literature are frequenctly limited to this case. The foregoing arguments are more general, however, and allow a general distribution over all wave numbers with the full dispersion shown in Figs. 11.1 and 11.2 as the entire range of values of $W'(k)$ come into play.

11.5 Kinematic Derivation of Group Velocity

The concept of group velocity is so fundamental in understanding the wave motion that one feels it should not be only the end product of Fourier analysis and stationary phase. In nonuniform media, or for nonlinear problems, where the Fourier analysis cannot be carried out in this way, surely the same concepts must appear and be equally important. How do we free the ideas from the Fourier analysis?

To see how to generalize the results, we look at their derivation by more intuitive arguments. The arguments can always be checked against the previous discussion, or justified eventually by direct asymptotic methods. The advantages are tremendous, since we can then make progress on approximate treatments of problems for which the exact solutions are not known. At the same time, we obtain quicker and fuller insight even in those problems where the exact solution can be found.

We first consider the role of group velocity in determining the propagation of wave number and frequency. On reexamination of the arguments we see that very little was required. First, if we assume there is a slowly varying wavetrain and that a phase function $\theta(x,t)$ exists, we can define local wave number and frequency by

$$k = \theta_x, \qquad \omega = -\theta_t. \tag{11.36}$$

If, further, we know or can propose a dispersion relation

$$\omega = W(k), \tag{11.37}$$

we have an equation for θ and we could proceed to solve it to determine the geometry of the wave pattern. It is usually more convenient, however, to eliminate θ in (11.36) to give

$$\frac{\partial k}{\partial t} + \frac{\partial \omega}{\partial x} = 0, \tag{11.38}$$

and use this in conjuction with (11.37) to solve first for $k(x,t)$ and $\omega(x,t)$. Notice that this formulation is the basic one for the nonlinear waves discussed in Chapter 2. Indeed, k is the density of waves, ω is the flux of waves, and (11.38) is a statement of the conservation of waves! Substituting the dispersion relation (11.37) in (11.38), we have

$$\frac{\partial k}{\partial t} + C(k)\frac{\partial k}{\partial x} = 0, \qquad C(k) = W'(k). \tag{11.39}$$

The group velocity C(k) is the propagation velocity for the wave number k.

According to the analysis in Chapter 2, the general solution of (11.39) for an initial distribution $k = f(x)$ at $t = 0$ is

$$k = f(\xi), \qquad x = \xi + \mathcal{C}(\xi)t, \tag{11.40}$$

where

$$\mathcal{C}(\xi) = C\{f(\xi)\}.$$

The special case of a centered simple wave arises when the range of values of k is initially concentrated at the origin. Then $k(x,t)$ is determined from

$$x = C(k)t.$$

This is just the determination of k given in (11.25) and represented in Figs. 11.1 and 11.2. The validity of the asymptotic expansion (11.24), noted from (11.23), requires x and t so large that the initial disturbance appears to be concentrated at the origin.

But, the concepts are already extended. The slowly varying wavetrain defined by $\theta(x,t)$ need not have originated from a relatively concentrated disturbance at the origin, and the distribution of $k(x,t)$ may be more general, as in (11.40). Moreover, there is no necessity for the solution φ to be sinusoidal in θ; any oscillatory wavetrain with a definable θ and a dispersion relation between k and ω is included.

It is interesting and significant that (11.39) for k is nonlinear, even though the original problem is linear, and that it is hyperbolic, even though the original equation for φ is not in general. This is the first instance of hyperbolic equations arising for the propagation of important overall quantities like k. In this sense one can preserve the association of wave propagation with hyperbolic equations, but there is a considerable non-hyperbolic substructure.

Extensions.

The simplified derivation of the group velocity is readily extended further to linear problems in more dimensions and in nonuniform media. The extension to nonlinear problems must await further developments, because the dispersion relation then involves also the amplitude. For multidimensional equations with constant coefficients, the exact solution can still be obtained from the use of multiple Fourier integrals and the asymptotic expansion can then be obtained from stationary phase. For n

space dimensions, it is easy to show that

$$\varphi = \int_{-\infty}^{\infty} F(\kappa) e^{i\kappa \cdot \mathbf{x} - iW(\kappa)t} d\kappa$$

$$\sim F(\mathbf{k}) \left(\frac{2\pi}{t} \right)^{n/2} \left(\det \left| \frac{\partial W}{\partial k_i \partial k_j} \right| \right)^{-1/2} \exp\{ i\mathbf{k} \cdot \mathbf{x} - iW(\mathbf{k})t + i\zeta \} \quad (11.41)$$

where

$$\frac{x_i}{t} = \frac{\partial W(\mathbf{k})}{\partial k_i},$$

and ζ depends on the number of factors $\pi i/4$ arising from path rotations. However, let us use the simpler kinematic derivation and at the same time include nonuniform media.

The description of a slowly varying wave in, say, three dimensions involves a phase $\theta(\mathbf{x}, t)$ where $\mathbf{x} = (x_1, x_2, x_3)$. We define the frequency ω and vector wave number $\mathbf{k}$ by

$$\omega = -\frac{\partial \theta}{\partial t}, \qquad k_i = \frac{\partial \theta}{\partial x_i}. \quad (11.42)$$

We assume that a dispersion relation is known, and that it may be written

$$\omega = W(\mathbf{k}, \mathbf{x}, t).$$

For a uniform medium, this would be obtained from the elementary solution (11.1). For a slightly nonuniform medium, it would appear reasonable to find the dispersion relation first for constant values of the parameters of the medium and then reinsert their dependence on $\mathbf{x}, t$. For example, if α, β, γ were slowly varying functions of $\mathbf{x}$ or t in the problems (11.6)–(11.9), we would use the same dispersion relations displayed there but with α, β, γ, taken to be the specified functions of $\mathbf{x}$ and t. Intuitively, this would appear to be a satisfactory procedure provided that α, β, γ, vary little in a typical wavelength and period. This will be substantiated in Sections 11.7 and 11.8.

If we eliminate θ from (11.42), we have

$$\frac{\partial k_i}{\partial t} + \frac{\partial \omega}{\partial x_i} = 0, \qquad \frac{\partial k_i}{\partial x_j} - \frac{\partial k_j}{\partial x_i} = 0. \quad (11.43)$$

Then, if $\omega = W(\mathbf{k}, \mathbf{x}, t)$ is introduced into the first of these,

$$\frac{\partial k_i}{\partial t} + \frac{\partial W}{\partial k_j}\frac{\partial k_j}{\partial x_i} = -\frac{\partial W}{\partial x_i}.$$

Since $\partial k_j/\partial x_i = \partial k_i/\partial x_j$, this may be modified to

$$\frac{\partial k_i}{\partial t} + C_j\frac{\partial k_i}{\partial x_j} = -\frac{\partial W}{\partial x_i}, \tag{11.44}$$

where

$$C_j(\mathbf{k}, \mathbf{x}, t) = \frac{\partial W(\mathbf{k}, \mathbf{x}, t)}{\partial k_j}. \tag{11.45}$$

The three dimensional group velocity $\mathbf{C}$ is defined by (11.45) and is the propagation velocity in (11.44) for the determination of k_i. Equation 11.44 may be written in characteristic form as

$$\frac{dk_i}{dt} = -\frac{\partial W}{\partial x_i} \quad \text{on} \quad \frac{dx_i}{dt} = \frac{\partial W}{\partial k_i}. \tag{11.46}$$

We note that $\mathbf{k}$ is constant on each characteristic when the medium is uniform in $\mathbf{x}$, and then the characteristics are straight lines in the $(\mathbf{x}, t)$ space. Each value of $\mathbf{k}$ propagates with the corresponding constant group velocity $\mathbf{C}(\mathbf{k})$. But this is not true in a nonhomogeneous medium, for then the values of $\mathbf{k}$ vary as they propagate along the characteristics and the characteristics are no longer straight. It might also be remarked that

$$\frac{d\omega}{dt} = \frac{\partial \omega}{\partial t} + C_j\frac{\partial \omega}{\partial x_j} = \frac{\partial W}{\partial t};$$

the frequency is constant on each characteristic when the medium is time independent but not otherwise.

It is interesting that the equations of (11.46) are identical with Hamilton's equations in mechanics if $\mathbf{x}$ and $\mathbf{k}$ are interpreted as coordinates and momenta and $W(\mathbf{k}, \mathbf{x}, t)$ is taken to be the Hamiltonian! If instead of eliminating θ, we substitute for ω and $\mathbf{k}$ in the dispersion relation, we have

$$\frac{\partial \theta}{\partial t} + W\left(\frac{\partial \theta}{\partial \mathbf{x}}, \mathbf{x}, t\right) = 0. \tag{11.47}$$

This is the Hamilton-Jacobi equation, with the phase θ as the action.

If W is independent of $\mathbf{x}$ and t, the solution of (11.46) for an initial distribution $k_i = f_i(\mathbf{x})$ is

$$k_i = f_i(\boldsymbol{\xi}), \qquad x_i = \xi_i + \mathcal{C}_i(\boldsymbol{\xi})t, \qquad (11.48)$$

where

$$\mathcal{C}_i(\boldsymbol{\xi}) = C_i\{\mathbf{f}(\boldsymbol{\xi})\}.$$

Again the centered solution corresponding to a whole range of $\mathbf{k}$ released at the origin at $t = 0$ is found by determining $\mathbf{k}(\mathbf{x}, t)$ from

$$x_i = C_i(\mathbf{k})t. \qquad (11.49)$$

This is the special case obtained in the asymptotic expansion (11.41) of the multiple Fourier integral.

Examples of the use of these various equations will be given in Chapter 12.

11.6 Energy Propagation

The preceding kinematic derivation shows one role of the group velocity and determines the geometry of the waves. The second role of the group velocity is in connection with the amplitude distribution $A(x, t)$ in (11.27)–(11.28). We would like to have direct access to the behavior of A and its involvement with the group velocity in much the same spirit. It looks feasible since energy is apparently involved and we expect to be able to make a direct statement of energy balance. This is the case. However, recent work using variational formulations has not only improved and generalized the derivations, it has also shown that "wave action" rather than energy is perhaps the more fundamental concept in this connection. The variational approach is subtle and it is useful to prepare the ground with a more traditional discussion of energy propagation.

We start as before with the one dimensional problem for a uniform medium and see how to obtain information on the amplitude distribution without using the Fourier integral solution. In this first approach we are forced to work with specific cases. The Klein-Gordon equation

$$\varphi_{tt} - \alpha^2 \varphi_{xx} + \beta^2 \varphi = 0$$

is one of the simplest to use, since it keeps the order of derivatives as low as possible. It is hyperbolic and is exceptional in that respect, but we are concerned only with oscillatory parts of the solution, not with wavefront behavior. The accompanying energy equation is easily obtained, and for constant coefficients α, β, it is

$$\frac{\partial}{\partial t}\left(\frac{1}{2}\varphi_t^2 + \frac{1}{2}\alpha^2\varphi_x^2 + \frac{1}{2}\beta^2\varphi^2\right) + \frac{\partial}{\partial x}\left(-\alpha^2\varphi_t\varphi_x\right) = 0. \qquad (11.50)$$

We now consider a slowly varying wavetrain in which

$$\varphi \sim \mathcal{R}(Ae^{i\theta}) = a\cos(\theta + \eta),$$

$$a = |A|, \qquad \eta = \arg A,$$

and we compute the energy density and energy flux. A term like $\frac{1}{2}\varphi_t^2$ will have

$$\frac{1}{2}\omega^2 a^2 \sin^2(\theta + \eta)$$

together with terms involving a_t and η_t. Because a and η are assumed to be slowly varying these latter terms are neglected in the first approximation. Treating the other terms similarly we see that the energy density is given approximately by

$$\frac{1}{2}(\omega^2 + \alpha^2 k^2)a^2 \sin^2(\theta + \eta) + \frac{1}{2}\beta^2 a^2 \cos^2(\theta + \eta), \qquad (11.51)$$

and the energy flux by

$$\alpha^2 \omega k a^2 \sin^2(\theta + \eta). \qquad (11.52)$$

In cases where higher derivatives occur, extra terms in the derivatives of ω and k would also arise but be neglected because ω and k are also slowly varying quantities.

Since we are concerned with variations of the overall quantities ω, k, a, and not with the details of the oscillations, we consider the average values of (11.51) and (11.52). The average values of $\cos^2(\theta + \eta)$ and $\sin^2(\theta + \eta)$ over one period are both equal to one half, so we have

$$\mathcal{E} = \frac{1}{4}(\omega^2 + \alpha^2 k^2 + \beta^2)a^2, \qquad (11.53)$$

$$\mathcal{F} = \frac{1}{2}\alpha^2 \omega k a^2, \qquad (11.54)$$

for the average values of the energy density and the energy flux. In this particular problem, the dispersion relation is

$$\omega = \sqrt{\alpha^2 k^2 + \beta^2} \; ; \qquad (11.55)$$

hence

$$\mathcal{E} = \frac{1}{2}(\alpha^2 k^2 + \beta^2)a^2, \qquad \mathcal{F} = \frac{1}{2}\alpha^2 k\sqrt{\alpha^2 k^2 + \beta^2}\, a^2. \qquad (11.56)$$

The group velocity is

$$C(k) = \frac{\alpha^2 k}{\sqrt{\alpha^2 k^2 + \beta^2}}, \qquad (11.57)$$

and we observe the important result that

$$\mathcal{F} = C(k)\mathcal{E}. \qquad (11.58)$$

This turns out to be general.

It is now tempting, on the intuitive grounds that energy must be balanced overall, to propose the "averaged" energy equation

$$\frac{\partial \mathcal{E}}{\partial t} + \frac{\partial}{\partial x}(C\,\mathcal{E}) = 0 \qquad (11.59)$$

as the equation to determine a. This is the differential form of the statement that the *total energy between any two group lines remains constant*. For if we consider the energy

$$E(t) = \int_{x_1(t)}^{x_2(t)} \mathcal{E}\, dx \qquad (11.60)$$

between points x_1, x_2, moving with the group velocities $C(k_1)$, $C(k_2)$, respectively, we have

$$\frac{dE}{dt} = \int_{x_1}^{x_2} \frac{\partial \mathcal{E}}{\partial t}\, dx + C_2\mathcal{E}_2 - C_1\mathcal{E}_1, \qquad (11.61)$$

and from (11.59) this is zero. Conversely, (11.59) is just the limit of (11.61) as $x_2 - x_1 \to 0$.

This behavior was found in Section 11.4 for a^2 itself rather than for $\mathcal{E}$. However, $\mathcal{E} = f(k)a^2$, and when this is substituted in (11.59) the resulting equation can be expanded to

$$f(k)\left\{\frac{\partial a^2}{\partial t} + \frac{\partial}{\partial x}(Ca^2)\right\} + f'(k)a^2\left\{\frac{\partial k}{\partial t} + C\frac{\partial k}{\partial x}\right\} = 0. \qquad (11.62)$$

Since

$$\frac{\partial k}{\partial t} + C\frac{\partial k}{\partial x} = 0 \qquad (11.63)$$

from (11.39), we have

$$\frac{\partial a^2}{\partial t} + \frac{\partial}{\partial x}(Ca^2) = 0. \tag{11.64}$$

We see that any function of k can be slipped in or out of equations (11.59) and (11.64), provided (11.63) holds. Now, by the same argument given for $\mathscr{E}$, (11.64) is the differential form of the result found in Section 11.4 that

$$Q(t) = \int_{x_1(t)}^{x_2(t)} a^2 \, dx \tag{11.65}$$

remains constant between group lines. Hence (11.64) and (11.59) are confirmed. Direct justification will appear later.

We might also note that the characteristic forms of (11.63) and (11.64) are

$$\frac{dk}{dt} = 0, \qquad \frac{da^2}{dt} = -C'(k)k_x a^2, \qquad \frac{dx}{dt} = C(k). \tag{11.66}$$

[In the second equation k_x can be treated as a known quantity because $k(x,t)$ would be determined first; this is the exceptional case of Example 7 in Section 5.2.] The group velocity $C(k)$ appears as a *double characteristic velocity*, corresponding to the dual role noted in Section 11.4.

The asymptotic solution obtained in Section 11.3 is the special case of a centered wave, where $k(x,t)$ is the function of x/t determined from

$$\frac{x}{t} = C(k).$$

In this case the amplitude equation is

$$\frac{da^2}{dt} = -\frac{a^2}{t}.$$

Since k itself may be used as a characteristic variable, the solution may be written

$$a = t^{-1/2}\mathscr{Q}(k)$$

where $\mathscr{Q}(k)$ is an arbitrary function. This agrees with (11.28) and again confirms the validity of the approach. Of course the function $\mathscr{Q}(k)$ cannot be determined without some tie to the initial conditions, and this cannot be found from asymptotic discussion alone.

In this initial value problem, we know in fact that $\mathscr{Q}(k)$ is given by (11.28), and it is interesting to note that the energy $E(t)$ between the group lines $k = k_1$ and $k = k_2$ as given by (11.60) is therefore

$$E(t) = 8\pi \int_{x_1}^{x_2} f(k) \frac{F_1(k)F_1^*(k)}{t|W''(k)|} dx$$

$$= 8\pi \int_{k_1}^{k_2} f(k)F_1(k)F_1^*(k) dk, \qquad (11.67)$$

where $f(k)$ is the factor $\frac{1}{2}(\alpha^2 k^2 + \beta^2)$ appearing in (11.56). From (11.50), the *exact* total energy is

$$E_{\text{tot}} = \int_{-\infty}^{\infty} \left(\frac{1}{2}\varphi_t^2 + \frac{1}{2}\alpha^2\varphi_x^2 + \frac{1}{2}\beta^2\varphi^2 \right) dx$$

and from the *exact* solution (11.16), together with relation (11.18), this may be put in the form

$$E_{\text{tot}} = 8\pi \int_{-\infty}^{\infty} f(\kappa)F_1(\kappa)F_1^*(\kappa) d\kappa. \qquad (11.68)$$

This applies at all times both before and after the dispersion into a wavetrain; it shows the distribution of energy over the wave number range. But *after dispersion* the wave number range is spread out explicitly as a distribution over x. The form of $E(t)$ in (11.67) shows that the same amount of energy is still associated with the range $k_1 < \kappa < k_2$. That is, the energy put into any wave number range remains there.

The energy arguments leading to (11.58) and (11.59) are easily extended to more dimensions. For the Klein-Gordon example, the energy equation becomes

$$\frac{\partial}{\partial t}\left(\frac{1}{2}\varphi_t^2 + \frac{1}{2}\alpha^2\varphi_{x_j}^2 + \frac{1}{2}\beta^2\varphi^2 \right) + \frac{\partial}{\partial x_j}\left(-\alpha^2\varphi_t\varphi_{x_j} \right) = 0,$$

and, for a slowly varying wavetrain $\varphi \sim a\cos(\theta + \eta)$, its averaged form is

$$\frac{\partial \mathcal{E}}{\partial t} + \frac{\partial \mathcal{F}_j}{\partial x_j} = 0,$$

where

$$\mathcal{E} = \frac{1}{4}(\omega^2 + \alpha^2 k_j^2 + \beta^2)a^2, \qquad \mathcal{F}_j = \frac{1}{2}\alpha^2\omega k_j a^2.$$

It may be verified from the dispersion relation that

$$\mathcal{F}_j = C_j \mathcal{E}, \tag{11.69}$$

and then the averaged energy equation becomes

$$\frac{\partial \mathcal{E}}{\partial t} + \frac{\partial}{\partial x_j}(C_j \mathcal{E}) = 0. \tag{11.70}$$

The total energy in any volume moving with the group lines remains constant. For

$$\frac{d}{dt} \int_{V(t)} \mathcal{E} \, dV = \int_{V(t)} \frac{\partial \mathcal{E}}{\partial t} \, dV + \int_{S(t)} \mathcal{E} \, C_j n_j \, dS,$$

where $S(t)$ is the surface of $V(t)$, n_j is the outward normal to $S(t)$, and $C_j n_j$ is its normal velocity. From the divergence theorem, (11.70) shows this to be zero. The characteristic form of (11.70) is

$$\frac{d \mathcal{E}}{dt} = - \frac{\partial C_j}{\partial x_j} \mathcal{E} \qquad \text{on} \qquad \frac{dx_i}{dt} = C_i(\mathbf{k});$$

the energy density decays due to the divergence $\partial C_j / \partial x_j$ of the group lines. For the uniform medium, $\mathbf{k}$ remains constant on the group lines [see (11.46)]. Therefore, since $\mathcal{E} = f(\mathbf{k}) a^2$, a^2 satisfies the same equations. This may also be verified directly from (11.70) with appropriate extension of (11.62). For the centered wave corresponding to (11.41), $\mathbf{k}$ is determined from

$$\frac{x_i}{t} = C_i(\mathbf{k});$$

hence

$$\frac{da^2}{dt} = - \frac{n a^2}{t},$$

where n is the number of dimensions. This agrees with the amplitude variation in (11.41).

We see that the averaged energy equation does indeed give a correct description of the amplitude distribution in accordance with the behavior found earlier. It is satisfactory in that it provides an approach that avoids the Fourier transforms and so offers the hope of generalization, but in the present form it is not completely satisfactory in that the seemingly general results (11.69) and (11.70) appear only at the end of manipulations using the specific equation. If we repeat the same type of argument on the other linear examples in (11.7)–(11.9), exactly the same final results (11.69) and (11.70) are found.

For example, the energy equation for (11.7) has an energy density

$$\frac{1}{2}\varphi_t^2 + \frac{1}{2}\alpha^2\varphi_{x_j}^2 + \frac{1}{2}\beta^2\varphi_{x_jt}^2$$

and energy flux vector

$$-\alpha^2\varphi_t\varphi_{x_j} - \beta^2\varphi_t\varphi_{ttx_j}.$$

The average values obtained by (1) substituting $\varphi \sim a\cos(\theta + \eta)$, (2) neglecting derivatives of a, η, k_i, and ω, and (3) replacing $\cos^2(\theta + \eta)$ and $\sin^2(\theta + \eta)$ by their average values of one-half, are

$$\mathcal{E} = \frac{1}{4}(\omega^2 + \alpha^2 k_j^2 + \beta^2\omega^2 k_j^2)a^2,$$

$$\mathcal{F}_j = \frac{1}{2}(\alpha^2\omega k_j - \beta^2\omega^3 k_j)a^2.$$

From the dispersion relation

$$\omega = \frac{\alpha k}{\sqrt{1 + \beta^2 k^2}}, \qquad k = |\mathbf{k}|,$$

it is verified that

$$\mathcal{F}_j = C_j\mathcal{E},$$

and the average energy equation can again be written as

$$\frac{\partial \mathcal{E}}{\partial t} + \frac{\partial}{\partial x_j}(C_j\mathcal{E}) = 0.$$

The same results are found for the remaining examples in (11.8) and (11.9).

It seems clear that these important general results should be established once and for all by general arguments without pursuing the detailed derivation each time. Such arguments (and much more) are provided by the variational approach.

11.7 The Variational Approach

This approach was originally developed for the much more difficult case of nonlinear wavetrains and it has many ramifications. A full account must await the further development of these topics, but we can take it far enough to complete the above discussion.

We consider first the variational principle

$$\delta J = \delta \int \int_R L(\varphi_t, \varphi_x, \varphi) \, dt \, dx = 0 \qquad (11.71)$$

for a function $\varphi(\mathbf{x}, t)$. The variational principle means that the integral $J[\varphi]$ over any finite region R should be stationary to small changes of φ in the following sense. Consider two neighboring functions $\varphi(\mathbf{x}, t)$ and $\varphi(\mathbf{x}, t) + h(\mathbf{x}, t)$ where h is "small"; since first derivatives appear in (11.71), both functions are taken to be continuously differentiable. The smallness of h is measured in this context by the "norm":

$$\|h\| = \max|h| + \max|h_t| + \max|h_{x_i}|.$$

The function L is usually some rather simple function and certainly we can suppose that it has bounded continuous second derivatives. Then by Taylor's expansion

$$J[\varphi + h] - J[\varphi] = \int \int_R \{ L_{\varphi_t} h_t + L_{\varphi, j} h_{x_j} + L_\varphi h \} \, dt \, dx + O(\|h\|^2), \quad (11.72)$$

where $\varphi_{,j}$ denotes $\partial\varphi/\partial x_j$. The expression linear in h is the first variation $\delta J[\varphi, h]$. The variational principle (11.71) requires that $\delta J[\varphi, h] = 0$ for all admissible h. By integration by parts (the divergence theorem) we have

$$\delta J[\varphi, h] = \int \int_R \left\{ -\frac{\partial}{\partial t} L_{\varphi_t} - \frac{\partial}{\partial x_j} L_{\varphi, j} + L_\varphi \right\} h \, dt \, dx, \qquad (11.73)$$

if we choose, in particular, functions h that vanish on the boundary of R. We now require (11.73) to vanish for all such h. This implies

$$\frac{\partial}{\partial t} L_{\varphi_t} + \frac{\partial}{\partial x_j} L_{\varphi, j} - L_\varphi = 0, \qquad (11.74)$$

by the usual continuity argument. [If (11.74) were nonzero, say positive, at any point, then there would be a small neighborhood in which it remained positive; a choice of h positive in this neighborhood and zero elsewhere would violate the requirement that (11.73) vanish.]

The argument extends in natural fashion if L includes second or higher derivatives in φ. The corresponding variational equation is

$$L_\varphi - \frac{\partial}{\partial t} L_{\varphi_t} - \frac{\partial}{\partial x_j} L_{\varphi,j} + \frac{\partial^2}{\partial t^2} L_{\varphi_{tt}} + \frac{\partial^2}{\partial t \partial x_j} L_{\varphi t,j} + \frac{\partial^2}{\partial x_j \partial x_k} L_{\varphi,jk} - \cdots = 0,$$

(11.75)

which is easily recognized as the end result of the repeated integration by parts. Equations 11.74 and 11.75 are partial differential equations for $\varphi(\mathbf{x}, t)$, and equations in this form can be given the equivalent variational formulation. A variational principle involving a number of functions $\varphi^{(\alpha)}(\mathbf{x}, t)$ would lead to (11.75) for each $\varphi^{(\alpha)}(\mathbf{x}, t)$ (since they could be varied independently), and hence to a system of equations. The question of finding a variational principle for a given system of equations can be a difficult one, but it is usually straightforward when only a single equation is involved. We note that Lagrangians for the examples (11.6)–(11.8) are

$$L = \frac{1}{2}\varphi_t^2 - \frac{1}{2}\alpha^2\varphi_{x_i}^2 - \frac{1}{2}\beta^2\varphi^2,$$

$$L = \frac{1}{2}\varphi_t^2 - \frac{1}{2}\alpha^2\varphi_{x_i}^2 + \frac{1}{2}\beta^2\varphi_{tx_i}^2,$$

(11.76)

$$L = \frac{1}{2}\varphi_t^2 - \frac{1}{2}\gamma^2\varphi_{xx}^2,$$

respectively, and (11.9) is included by substituting $\varphi = \psi_x$ and taking

$$L = \frac{1}{2}\psi_t\psi_x + \frac{1}{2}\alpha\psi_x^2 - \frac{1}{2}\beta\psi_{xx}^2.$$

To study slowly varying wavetrains in which

$$\varphi \sim a\cos(\theta + \eta),$$

(11.77)

we now calculate the Lagrangian L in exactly the same way as the energy density and flux were calculated in the last section. That is, (11.77) is substituted, derivatives of a, η, ω, $\mathbf{k}$ are all neglected as being small, and the result is averaged over one period. The result in each case is a function $\mathcal{L}(\omega, \mathbf{k}, a)$; in particular, for the examples in (11.76) we have

$$\mathcal{L} = \frac{1}{4}(\omega^2 - \alpha^2 k^2 - \beta^2)a^2,$$

$$\mathcal{L} = \frac{1}{4}(\omega^2 - \alpha^2 k^2 + \beta^2\omega^2 k^2)a^2,$$

(11.78)

$$\mathcal{L} = \frac{1}{4}(\omega^2 - \gamma^2 k^4)a^2.$$

We now propose the "average variational principle",

$$\delta \iint \mathcal{L}(-\theta_t, \theta_x, a)\, dt\, dx = 0, \tag{11.79}$$

for the functions $a(\mathbf{x}, t)$, $\theta(\mathbf{x}, t)$. This is similar to the proposal in (11.59), but it certainly is a much more subtle affair which will have to be examined in detail later. However, accepting it for the present, we shall see immediately that it provides a general and extremely powerful approach.

Since derivatives of a do not occur, the variational equation (11.75) for variations in a is merely

$$\delta a: \qquad \mathcal{L}_a = 0.$$

The variational equation for θ is

$$\delta\theta: \qquad \frac{\partial}{\partial t}\, \mathcal{L}_{\theta_t} + \frac{\partial}{\partial x_j}\, \mathcal{L}_{\theta, j} = 0.$$

In these expressions the dependence on θ involves only its derivatives. Accordingly, *once the variational equations have been obtained*, it is usually convenient to work again with $\omega, \mathbf{k}, a$, and take the set of equations

$$\mathcal{L}_a = 0, \tag{11.80}$$

$$\frac{\partial}{\partial t}\, \mathcal{L}_\omega - \frac{\partial}{\partial x_j}\, \mathcal{L}_{k_j} = 0, \tag{11.81}$$

$$\frac{\partial k_i}{\partial t} + \frac{\partial \omega}{\partial x_i} = 0, \qquad \frac{\partial k_i}{\partial x_j} - \frac{\partial k_j}{\partial x_i} = 0, \tag{11.82}$$

the latter being just the consistency equations for the existence of θ.

Equation 11.80 is a functional relation between $\theta, \mathbf{k}, a$, so it can be no other but the dispersion relation. We check from (11.78) that this is the case. Indeed in any linear problem, it is clear that L must be quadratic in φ and its derivatives, and that as a consequence $\mathcal{L}$ must always take the form

$$\mathcal{L} = G(\omega, \mathbf{k})a^2. \tag{11.83}$$

Then from (11.80) the dispersion relation must be

$$G(\omega, \mathbf{k}) = 0, \tag{11.84}$$

and the function $G(\omega, \mathbf{k})$ in $\mathcal{L}$ is nothing but the dispersion function. We did not even need to calculate $\mathcal{L}$ for each case!

This is all an unexpected bonus. The aim was to find a general argument for the amplitude equation, but we have actually included the kinematic theory proposed in Section 11.5 for the geometry of the wave pattern. Equations 11.80 and 11.82 provide just that theory.

We note that the stationary value of $\mathcal{L}$ is in fact zero. In those simple cases in which L is the difference of kinetic and potential energies, this proves [subject to the eventual justification of (11.79)] that their average values are equal. This is the well-known equipartition of energy for linear problems.

Turning next to the amplitude equation (11.81), we note that it may now be written

$$\frac{\partial}{\partial t}(G_\omega a^2) - \frac{\partial}{\partial x_j}(G_{k_j} a^2) = 0. \tag{11.85}$$

In principle (11.84) can be solved in the form $\omega = W(\mathbf{k})$ so that

$$G\{W(\mathbf{k}), \mathbf{k}\} = 0$$

is an identity. Therefore

$$G_\omega \frac{\partial W}{\partial k_j} + G_{k_j} = 0$$

and the group velocity

$$C_j = \frac{\partial W}{\partial k_j} = -\frac{G_{k_j}}{G_\omega}. \tag{11.86}$$

If we denote $G_\omega(W, \mathbf{k})$ by $g(\mathbf{k})$, (11.85) may be written

$$\frac{\partial}{\partial t}\{g(\mathbf{k}) a^2\} + \frac{\partial}{\partial x_j}\{g(\mathbf{k}) C_j(\mathbf{k}) a^2\} = 0. \tag{11.87}$$

From (11.82) we have

$$\frac{\partial k_i}{\partial t} + C_j \frac{\partial k_i}{\partial x_j} = 0, \qquad \frac{\partial k_i}{\partial x_j} - \frac{\partial k_j}{\partial x_i} = 0;$$

by using these, the factor $g(\mathbf{k})$ can be slipped out of (11.87) and we have the amplitude equation

$$\frac{\partial a^2}{\partial t} + \frac{\partial}{\partial x_j}(C_j a^2) = 0.$$

Thus the variational set (11.80)–(11.82) does give precisely the set of equations discussed in the last two sections.

At first sight one might expect that (11.87) is the averaged energy equation (11.70). But a check on examples shows that the factor $f(\mathbf{k})$ in $\mathcal{E}$ and the factor $g(\mathbf{k})$ are not the same. However, there is a standard procedure for associating an energy equation with a variational principle. Noether's theorem shows that there is a conservation equation corresponding to any group of transformations for which the variational principle is invariant (see Gelfand and Fomin, 1963, p. 177). If the principle is invariant to a translation in t, the corresponding equation is always the energy equation or a multiple of it. Since (11.79) is invariant to a translation in t, this applies and the corresponding energy equation is found to be

$$\frac{\partial}{\partial t}(\omega \mathcal{L}_\omega - \mathcal{L}) + \frac{\partial}{\partial x_j}(-\omega \mathcal{L}_{k_j}) = 0. \tag{11.88}$$

Here, rather than pursue the detailed application of Noether's theorem, it is sufficient to note that (11.88) follows from the system (11.80)–(11.82). This is the energy equation. One can easily verify that the previous examples agree.

In the linear problems considered here, we found that the stationary value of $\mathcal{L}$ is zero. Hence the energy density $\mathcal{E}$ and flux $\mathcal{F}$ are given by

$$\mathcal{E} = \omega \mathcal{L}_\omega, \qquad \mathcal{F}_j = -\omega \mathcal{L}_{k_j}. \tag{11.89}$$

We see therefore that the quantity $\mathcal{L}_\omega$ is in fact

$$\mathcal{L}_\omega = \frac{\mathcal{E}}{\omega}, \tag{11.90}$$

and (11.81) or (11.87) may be written

$$\frac{\partial}{\partial t}\left(\frac{\mathcal{E}}{\omega}\right) + \frac{\partial}{\partial x_j}\left(\frac{C_j \mathcal{E}}{\omega}\right) = 0. \tag{11.91}$$

From (11.83) and (11.89), we have

$$\mathcal{E} = \omega G_\omega a^2$$

$$\mathcal{F}_j = -\omega G_{k_j} a^2 = C_j \mathcal{E},$$

which gives the general proof of the relation between $\mathcal{F}$ and $\mathcal{E}$.

But we have another bonus from the general approach. It draws prime attention to the quantity (11.90) and to (11.81) and (11.91). The quantity $\mathcal{E}/\omega$ is well known in ordinary mechanics as the adiabatic invariant for slow modulations of a linear vibrating system. We shall show later that $\mathcal{L}_\omega$

is the appropriate quantity in the nonlinear case. Thus these concepts have been extended to the case of waves. Instead of an invariant we have the conservation equation (11.81) governed by a timelike adiabatic quantity $\mathcal{L}_\omega$ and spacelike quantities $-\mathcal{L}_{k_j}$. This conservation equation has become known as the conservation of "wave action".

There is also a "wave momentum" equation which is the counterpart to (11.88) with the roles of x_i and t interchanged:

$$\frac{\partial}{\partial t}(k_i \mathcal{L}_\omega) + \frac{\partial}{\partial x_j}(-k_i \mathcal{L}_{k_j} + \mathcal{L}\delta_{ij}) = 0. \tag{11.92}$$

This is easily verified from the set (11.80)–(11.82). We note that the momentum density is

$$k_i \mathcal{L}_\omega = \frac{k_i}{\omega}\mathcal{E}; \tag{11.93}$$

it is a vector in the direction of **k** with magnitude $\mathcal{E}/c$, where c is the phase speed. We again have the general proof of a familar result which is hard to establish by other means.

Nonuniform Media.

Another advantage of the variational approach is that there is no change in the basic equations (11.80)–(11.82) if the medium varies slowly with **x** and t. This would be the case, for example, if the parameters α, β, γ in (11.76) were functions of $(\mathbf{x}, t)$. If the change in one period is small, the average Lagrangian can be formed as before, neglecting the changes α, β, γ in one period along with the contributions of the derivatives of $\omega, \mathbf{k}, a, \eta$. Then (11.79) is proposed as before, the only difference being that $\mathcal{L}$ now depends explicitly on **x** and t as well as through the functions $a(\mathbf{x}, t)$ and $\theta(\mathbf{x}, t)$. However, the variational equations (11.80)–(11.82) are unchanged; one has only to be careful to include further derivatives in manipulating and expanding the equations. In particular, the energy equation now becomes

$$\frac{\partial}{\partial t}(\omega \mathcal{L}_\omega - \mathcal{L}) + \frac{\partial}{\partial x_j}(-\omega \mathcal{L}_{k_j}) = -\mathcal{L}_t, \tag{11.94}$$

as is easily verified. Similarly, the momentum equation (11.92) picks up a term $\mathcal{L}_{x_i}$ on the right hand side. If the medium depends on t, energy is no longer conserved. If it depends on **x**, momentum is no longer conserved. But notice that wave action is conserved in all cases. This again shows the preferred position of (11.81) over the energy equation in modulation theory.

Nonlinear Wavetrains.

Finally, the variational approach requires very little modification to study modulations on nonlinear wavetrains. The main questions will be the functional form to replace (11.77), the details of the averaging to find the function $\mathcal{L}$, and, in general, the appearance of further overall functions similar to $\omega, \mathbf{k}, a$ in the complete description. In the simplest cases, however, the latter do not arise and once $\mathcal{L}(\omega, \mathbf{k}, a)$ has been found, the set (11.80)–(11.82) still apply. The major difference is the crucially important one that $\mathcal{L}$ is no longer simply proportional to a^2, and (11.80) and (11.82) do not uncouple from (11.81). These questions and the careful justification of the theory so far will be taken up in Chapter 14.

11.8 The Direct Use of Asymptotic Expansions

A more obvious way to avoid the Fourier integrals and open the way for extension to problems of nonuniform media and nonlinear systems is to substitute asymptotic series' of the appropriate form directly into the equations of the problem. For the linear problems discussed so far, the required form of expansion is

$$\varphi \sim e^{i\theta(\mathbf{x}, t)} \sum_{n=0}^{\infty} A_n(\mathbf{x}, t), \tag{11.95}$$

where the A_n are terms of successively smaller order in the relevant small parameter. In the present context, this form is suggested by the first term obtained in (11.27). It is also an extension of the geometrical optics series discussed in (7.62). For the earlier hyperbolic problems the relation between θ_t and $\theta_{\mathbf{x}}$ would be homogeneous, so that for fixed frequency ω one may choose $\theta(\mathbf{x}, t) = \omega S(\mathbf{x}, t)$ as was done in that discussion. Here the dispersion relation between θ_t and $\theta_{\mathbf{x}}$ is more general and we allow a continuous distribution of frequency.

The approach using (11.95) is satisfactory as far as it goes and it can be applied to problems of nonuniform media. But to a greater extent than was found in the discussion via the average energy equation in Section 11.6, one works with the specific expressions for each particular problem only to find at the very end that the results are general. In the case of extensions to nonlinear problems, the correct form of expansion is not immediately clear, the manipulation of expressions may become horrendous, and again general results are hidden in the specific details. These weaknesses are remedied by applying expansions such as (11.95) directly in the variational formulation of the problem. This is, essentially, how the variational approach is justified. But some ingenuity is involved and it is useful to include here as background some discussion of the direct application of (11.95) to the equations. It is sufficient to treat the one dimensional case.

The expansion discussed in Section 11.3 is valid for $t \to \infty$, with x/t fixed. In that case $\theta(x,t)$ and $A_n(x,t)$ take the form

$$\theta(x,t) = t\tilde{\theta}\left(\frac{x}{t}\right), \qquad A_n(x,t) = t^{-n-1/2}B_n\left(\frac{x}{t}\right). \qquad (11.96)$$

The expansion (11.95) is in increasing powers of t^{-1} (or strictly speaking, in powers of τ/t where τ is a typical time scale introduced by the parameters in the equations and initial conditions). To keep the technique flexible and see the common features in the use of (11.95) in different circumstances, we do not introduce (11.96) explicitly, but we note rather that

$$\frac{\partial A_n}{\partial t}, \frac{\partial A_n}{\partial x} = O(A_{n+1}), \qquad \frac{\partial^2 A_n}{\partial t^2} = O(A_{n+2}), \cdots. \qquad (11.97)$$

That is, each differentiation increases the order by one. Similarly, θ_x and θ_t are $O(1)$ quantities and any further differentiation increases their order by one each time. The increase of order on taking derivatives indicates that θ_t, θ_x, and the A_n are *slowly varying functions*. This is a general feature in using (11.95) whether the expansion is in τ/t or some other quantity.

As an illustrative example we take the one dimensional Klein-Gordon equation

$$\varphi_{tt} - \alpha^2\varphi_{xx} + \beta^2\varphi = 0.$$

The expansion (11.95) is substituted and terms of successive orders are equated to zero. We have

$$(\theta_t^2 - \alpha^2\theta_x^2 - \beta^2)A_0 = 0,$$

$$(\theta_t^2 - \alpha^2\theta_x^2 - \beta^2)A_1 - \{2i\theta_t A_{0t} - 2i\alpha^2\theta_x A_{0x} + i(\theta_{tt} - \alpha^2\theta_{xx})A_0\} = 0,$$

$$(\theta_t^2 - \alpha^2\theta_x^2 - \beta^2)A_2 - \{2i\theta_t A_{1t} - 2i\alpha^2\theta_x A_{1x} + i(\theta_{tt} - \alpha^2\theta_{xx})A_1\} = A_{0tt} - \alpha^2 A_{0xx},$$

and so on. The first equation eliminates the corresponding term in the later equations. If, further, we introduce

$$k = \theta_x, \qquad \omega = -\theta_t,$$

the hierarchy becomes

$$\omega^2 - \alpha^2 k^2 - \beta^2 = 0, \tag{11.98}$$

$$2\omega A_{0t} + 2\alpha^2 k A_{0x} + (\omega_t + \alpha^2 k_x) A_0 = 0, \tag{11.99}$$

$$2\omega A_{1t} + 2\alpha^2 k A_{1x} + (\omega_t + \alpha^2 k_x) A_1 = -i(A_{0tt} - \alpha^2 A_{0xx}), \tag{11.100}$$

and so on.

The first equation is the dispersion relation between ω and k, and if we prefer to work with these quantities rather than θ itself, the consistency relation

$$k_t + \omega_x = 0 \tag{11.101}$$

is added. This is exactly the determination of θ, ω, k described in Section 11.5.

The equation for A_0 may be written

$$\frac{\partial}{\partial t}\left(\frac{1}{2}\omega A_0 A_0^*\right) + \frac{\partial}{\partial x}\left(\frac{1}{2}\alpha^2 k A_0 A_0^*\right) = 0. \tag{11.102}$$

Since $|A_0|^2 = a^2$, and in this case

$$\mathcal{L} = \frac{1}{4}\left(\omega^2 - \alpha^2 k^2 - \beta^2\right)a^2,$$

this is just the wave action equation (11.81). It is interesting that the wave action equation rather than the energy equation arises most obviously, although of course the energy equation can also be obtained from (11.99). Notice that this point might pass unrecognized without the Lagrangian theory.

One is usually interested only in the first term of the expansion and therefore in the first two equations (11.98) and (11.99). However, once θ and A_0 are determined, A_1 is obtained by solving (11.100), A_2 from the next equation in the hierarchy, and so on.

As a special case, it is easy to check that the equations have solutions of the form (11.96), and the expansion then agrees with the one obtained from the Fourier integrals in Section 11.3. The relevant solution of (11.98) and (11.101) is the function $k(x/t)$ determined from

$$\frac{x}{t} = C(k).$$

Then the various forms of (11.102) all give

$$A_0 = t^{-1/2} B_0 \left(\frac{x}{t} \right).$$

Since k is a function of x/t, this may also be written

$$A_0 = t^{-1/2} \mathcal{B}_0(k)$$

in agreement with (11.28). Of course the function $\mathcal{B}_0(k)$ is only determined by use of the initial conditions. In this particular case the expansion does not apply in the early stages and Fourier transforms or some equivalent bridge is unavoidable. When A_0 has been determined, (11.100) may be solved for A_1 and the result is (11.23). In fact it is much simpler to determine subsequent terms in the expansion by this direct method instead of carrying out stationary phase to high orders.

The expansions are not limited to the centered wave solution; they apply to any wavetrain that is slowly varying in the sense of (11.97). For example, we might consider a wavetrain produced by a modulated source that provides slow changes in the frequency and amplitude. If x and t are normalized variables obtained by dividing the original x and t by a typical wavelength and typical period, respectively, the modulations provided at the source are functions of ϵt and the appropriate forms of θ and A_n are

$$\theta = \epsilon^{-1} \tilde{\theta}(\epsilon x, \epsilon t), \qquad A_n = \epsilon^n \tilde{A}_n(\epsilon x, \epsilon t), \qquad (11.103)$$

where ϵ is the ratio of the typical period to the time scale of the modulations. The variables are slowly varying in the sense of (11.97) with ϵ as the relevant small parameter. For the Klein-Gordon example, the resulting equations are (11.98)–(11.100). These correspond to the successive terms of order $1, \epsilon, \epsilon^2$, respectively, but there is no need to display the dependence on ϵ explicitly if we follow the ordering in (11.97).

Nonuniform Media.

A more interesting case, similar to the preceding one, arises when the modulations are produced by slow variations in the medium. For example, we might consider an initially uniform wavetrain entering a nonuniform medium in which the parameters of the medium change slowly over a length scale L. If λ is a typical wavelength (say the value in the initial wavetrain) the small parameter is $\epsilon = \lambda / L$. In normalized variables, the medium will be described by functions of ϵx, and the form in (11.103) is appropriate to describe the modulated wavetrain. A similar formulation applies if the medium changes slowly in time.

As an illustrative example, we again take the Klein-Gordon equation. In a nonuniform medium the equation would usually arise in the self-adjoint form

$$\frac{\partial^2 \varphi}{\partial t^2} - \frac{\partial}{\partial x} \left\{ \alpha^2(x,t) \frac{\partial \varphi}{\partial x} \right\} + \beta^2(x,t)\varphi = 0. \qquad (11.104)$$

We suppose that x,t have already been normalized with respect to a typical wavelength and period (typical values of $\alpha\beta^{-1}$ and β^{-1} could be used), and to include space and time variations in the same analysis we suppose that

$$\alpha = \tilde{\alpha}(\epsilon x, \epsilon t), \qquad \beta = \tilde{\beta}(\epsilon x, \epsilon t). \qquad (11.105)$$

As before, we do not introduce the dependence on ϵ explicitly, but work directly with (11.95) and (11.104) with the understanding that

$$k = \theta_x, \qquad \omega = -\theta_t, \qquad A_0, \qquad \alpha, \qquad \beta$$

are all $O(1)$ quantities and any increase of derivative or increase in the subscript of A increases the order by one. The resulting hierarchy begins with

$$\omega^2 - \alpha^2 k^2 - \beta^2 = 0, \qquad (11.106)$$

$$2\omega A_{0t} + 2\alpha^2 k A_{0x} + (\omega_t + \alpha^2 k_x + 2k\alpha\alpha_x)A_0 = 0, \qquad (11.107)$$

and we add the consistency relation

$$\frac{\partial k}{\partial t} + \frac{\partial \omega}{\partial x} = 0. \qquad (11.108)$$

The determination ω, k, and θ from (11.106) and (11.108) is exactly the procedure proposed on more intuitive grounds in Section 11.5, and subsequently obtained from the variational approach in Section 11.7. The consequences were already examined in Section 11.5. In particular we noted that values of k propagate with the group velocity $\partial\omega/\partial k$ obtained from (11.106), but neither the group velocity nor the value of k need remain constant on a group line.

In this case the group velocity may be written $\alpha^2 k/\omega$, so that the characteristics for (11.107) are the same as for (11.108) and the values of A_0 may be obtained in principle by integration along these characteristics.

The main point to be noted, however, is that (11.107) may still be written in conservation form as

$$\left(\frac{1}{2}\omega A_0 A_0^*\right)_t + \left(\frac{1}{2}\alpha^2 k A_0 A_0^*\right)_x = 0. \qquad (11.109)$$

That is, the wave action equation (11.102) still holds in the nonuniform case in which α and the relation between ω and k depend on x and t. This substantiates the claim made in the variational approach.

Indeed if we form

$$\frac{\partial \mathcal{E}}{\partial t} + \frac{\partial \mathcal{F}}{\partial x}$$

with the energy density and flux from (11.53)–(11.54) and calculate it from (11.106)–(11.108), we have

$$\frac{\partial}{\partial t}\left\{\frac{1}{4}(\omega^2 + \alpha^2 k^2 + \beta^2)A_0 A_0^*\right\} + \frac{\partial}{\partial x}\left\{\frac{1}{2}\alpha^2 \omega k A_0 A_0^*\right\} =$$

$$\frac{1}{4}\left\{k^2\frac{\partial \alpha^2}{\partial t} + \frac{\partial \beta^2}{\partial t}\right\}A_0 A_0^*. \qquad (11.110)$$

This checks with (11.94), since

$$\mathcal{L} = \frac{1}{4}(\omega^2 - \alpha^2 k^2 - \beta^2)a^2.$$

The direct use of (11.95) in the equations leads to the required results but without the generality and insight of the variational approach. The two will be combined in the discussion of Chapter 14. We first consider applications of the theory so far, and amplify the ideas on specific problems.

CHAPTER 12

Wave Patterns

Some of the most interesting wave patterns are found in water waves. Some of them, such as the V-shaped ship wave pattern or the pattern of rings spreading out from a stone thrown in a pond, are familiar to everyone, and others are relatively easily observed. We start with these. Here the dispersion relation, the only input required, will be merely quoted. We shall need to look further into the subject of water waves later, since it was the first and most fruitful source of ideas on nonlinear dispersive waves. The derivation of the dispersion relation will be included then.

12.1 The Dispersion Relation for Water Waves

In still water elementary solutions for the perturbation η in the height of the surface take the basic form (11.1):

$$\eta = A e^{i\mathbf{k}\cdot\mathbf{x} - i\omega t},$$

provided

$$\omega^2 = (gk \tanh kh)\left(1 + \frac{T}{\rho g}k^2\right), \qquad k = |\mathbf{k}|. \qquad (12.1)$$

Here h is the undisturbed depth, g is the gravitational acceleration, ρ is the density, T is the surface tension. In still water the waves are isotropic and the dispersion involves only the magnitude k of the wave number vector. There are a number of interesting limits which are conveniently used as approximations in appropriate circumstances.

Gravity Waves.

In c.g.s. units, $g = 981$, $\rho = 1$, and $T = 74$, so that $\lambda_m = 2\pi(T/\rho g)^{1/2} = 1.73$ cm. Thus the surface tension effects become negligible for wavelengths

403

several times greater than this value. Then we have the usual formula for gravity waves:

$$\omega^2 = gk \tanh kh, \qquad \lambda \gg \lambda_m. \tag{12.2}$$

For these, the phase and group velocities are

$$c(k) = \left(\frac{g}{k} \tanh kh \right)^{1/2}, \tag{12.3}$$

$$C(k) = \frac{\partial \omega}{\partial k} = \frac{1}{2} c(k) \left(1 + \frac{2kh}{\sinh 2kh} \right). \tag{12.4}$$

Within this approximation, we have the limiting cases

$$\omega \sim (gk)^{1/2}, \quad c \sim \left(\frac{g}{k} \right)^{1/2}, \quad C \sim \frac{1}{2} \left(\frac{g}{k} \right)^{1/2}, \quad kh \to \infty, \tag{12.5}$$

$$\omega \sim (gh)^{1/2} k, \quad c \sim (gh)^{1/2}, \quad C \sim (gh)^{1/2}, \quad kh \to 0. \tag{12.6}$$

For fixed h, both c and C are increasing functions of $\lambda = 2\pi/k$, with $C < c$; in the long wave limit (12.6), $C \to c$ and the dispersive effects become small. Of course, the approximation (12.5) is appropriate for short waves when $\lambda_m \ll \lambda \ll h$.

Capillary Waves.

For $\lambda \ll \lambda_m$, the surface tension effect may be dominant and (12.1) is then approximated by

$$\omega^2 = \frac{T}{\rho} k^3 \tanh kh. \tag{12.7}$$

In this case

$$c(k) = \left(\frac{T}{\rho} k \tanh kh \right)^{1/2}, \tag{12.8}$$

$$C(k) = \frac{3}{2} c \left(1 + \frac{2kh}{3 \sinh 2kh} \right). \tag{12.9}$$

The further limits of these are

$$\omega \sim \left(\frac{T}{\rho}\right)^{1/2} k^{3/2}, \qquad c \sim \left(\frac{T}{\rho}\right)^{1/2} k^{1/2}, \qquad C \sim \frac{3}{2}\left(\frac{T}{\rho}\right)^{1/2} k^{1/2}, \qquad kh \to \infty,$$

$$(12.10)$$

and

$$\omega \sim \left(\frac{Th}{\rho}\right)^{1/2} k^{2}, \qquad c \sim \left(\frac{Th}{\rho}\right)^{1/2} k, \qquad C \sim 2\left(\frac{Th}{\rho}\right)^{1/2} k, \qquad kh \to 0.$$

$$(12.11)$$

For capillary waves, c and C are decreasing functions of λ, with $C > c$.

Combined Gravity and Surface Tension Effects.

When both effects are important, it is usually sufficient to consider relatively short waves, $kh \gg 1$, and take

$$\omega^2 = gk + \frac{T}{\rho} k^3. \qquad (12.12)$$

The phase and group velocities are

$$c = \left(\frac{g}{k} + \frac{T}{\rho} k\right)^{1/2}, \qquad (12.13)$$

$$C = \frac{1}{2} c \frac{1 + (3T/\rho g) k^2}{1 + (T/\rho g) k^2}. \qquad (12.14)$$

The phase velocity has a minimum at $k = k_m$, where

$$k_m = \left(\frac{\rho g}{T}\right)^{1/2}, \qquad \lambda_m = \frac{2\pi}{k_m} = 1.73 \text{ cm}; \qquad (12.15)$$

the corresponding values of c and C are equal with

$$c_m = 23.2 \text{ cm/sec.}$$

For $\lambda > \lambda_m$, often known as the gravity branch, $C < c$; whereas for $\lambda < \lambda_m$, known as the capillary branch, $C > c$. For any given value of $c > c_m$, there are two possible wavelengths. The minimum group velocity is attained at $\lambda = 2.54\lambda_m = 4.39$ cm, $C = 0.77c_m = 17.9$ cm/sec.

Shallow Water with Dispersion.

In the limit $kh \to 0$, (12.1) may be expanded as

$$\omega^2 \sim ghk^2 \left\{ 1 + \left(\frac{T}{\rho gh^2} - \frac{1}{3} \right) k^2h^2 + \cdots \right\}, \tag{12.16}$$

and we have

$$c \sim (gh)^{1/2} \left\{ 1 + \frac{1}{2} \left(\frac{T}{\rho gh^2} - \frac{1}{3} \right) k^2h^2 + \cdots \right\}. \tag{12.17}$$

When dispersion is neglected altogether, the equations for *nonlinear* shallow water theory are hyperbolic and similar to those of gas dynamics; this so-called hydraulic analogy has been exploited for experiments. The dispersion must be kept to a minimum so h is chosen such that

$$\frac{T}{\rho gh^2} - \frac{1}{3} = 0,$$

that is,

$$h = \left(\frac{3T}{\rho g} \right)^{1/2} = 0.48 \text{ cm.}$$

Magnetohydrodynamic Effects.

In a conducting liquid a third vertical restoring force may be introduced when a horizontal magnetic field is applied and horizontal currents flow through the liquid. This has been investigated by Shercliff (1969), who finds that the dispersion relation is

$$\rho \omega^2 = k \tanh kh (\rho g + k^2 T + J_s B_n),$$

where B_n is the magnetic field normal to the wave crests and J_s the current along them. The term $J_s B_n$ is the vertical component of the Lorentz force. It is interesting that the propagation depends on the orientation of the waves to the field and becomes anisotropic. Details of the phase and group velocities, and of the various limiting cases, are to be found in the paper quoted. We shall not pursue this case further here, although the various wave patterns can be studied by the methods developed below.

12.2 Dispersion from an Instantaneous Point Source

The waves from a point source spread out isotropically and the different values of k introduced initially propagate out with the corresponding group velocities $C(\hat{k})$. At time t any particular value k will be found at $r = C(k)t$. Hence $k(r,t)$ is the solution of

$$C(k) = \frac{r}{t}. \tag{12.18}$$

For deep water gravity waves (12.5), we have therefore

$$k = \frac{gt^2}{4r^2}, \qquad \omega = \frac{gt}{2r}. \tag{12.19}$$

This is the axisymmetric counterpart of the one dimensional problem noted in Section 11.4. This very simple formula for ω has been checked by Snodgrass et al, (1966) against observational data of the swells produced by storms in the South Pacific. At distances of the order of 2000 miles, the frequency was found to vary linearly with t, and the constant of proportionality gave a very accurate determination of the distance of the storm.

On a smaller scale, the typical rings spreading from a stone or other splash in a pond satisfy (12.18) with $C(k)$ given by (12.14). Since $C(k)$ has a minimum value of about 18 cm/sec, there is a quiescent circle of radius $18t$ cm. Beyond that. there are two values of k for each r/t, one on the gravity branch and one on the capillary branch, so there are two superimposed wavetrains. Of course, the energy in the different wave numbers is determined by the initial disturbance. Waves with wavelength of the same order as the size of the object will have the largest amplitudes and will be most accentuated. For an object which is large compared with $\lambda_m = 1.73$ cms, there would be little energy in the capillary branch.

12.3 Waves on a Steady Stream

The waves produced by an obstacle on a steady stream U in the x_1 direction may be viewed as the waves produced by an obstacle moving with speed U in the negative x_1 direction. For a two dimensional obstacle, with flow independent of x_2, the only waves that can keep up with the obstacle and appear steady when viewed from the obstacle must satisfy

$$c(k) = U. \tag{12.20}$$

We again take the situation when (12.12)–(12.14) apply. There will be no solutions to (12.20) and hence no steady wavetrain if $U < c_m$. In this case there will be local disturbances dying out away from the obstacle but no contribution to the asymptotic wave pattern. If $U > c_m$ there will be two solutions of (12.20): one of them, k_g say, on the gravity branch, and one, k_T say, on the capillary branch. Now $k_g < k_m$ and $k_T > k_m$; hence from the properties of (12.13)–(12.14), we have

$$C(k_g) < c(k_g) = U, \tag{12.21}$$

$$C(k_T) > c(k_T) = U. \tag{12.22}$$

Therefore, the gravity waves k_g have group velocity less than U and will appear behind the obstacle; the capillary waves k_T have group velocity greater than U and will be ahead of the obstacle. The resulting pattern is shown in Fig. 12.1.

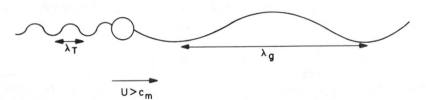

Fig. 12.1.   Sketch of capillary waves (upstream) and gravity waves (downstream) produced by an obstacle on the surface of a stream.

This is an interesting use of the group velocity concepts to determine the correct radiation condition in a steady flow problem. For this reason it is also interesting to derive the result in detail from the exact Fourier transform solution and see how the group velocity condition comes out of the usual type of radiation condition used in the techniques of solving boundary value problems. At the same time, the full solution gives the amplitudes of the waves. The asymptotic analysis tells us only that the amplitude remains constant in each wavetrain; the detailed initial conditions have to be analyzed to determine their values. It would interrupt the present discussion of kinematics to give the details here. They will be given in Section 13.9.

12.4 Ship Waves

For an obstacle that is finite in the x_2 direction we have a two dimensional wave pattern on the surface of the water and the analysis is more complicated. We shall study only the gravity wave problem for deep water and use the dispersion relation (12.5). This covers the pattern produced by any object of dimension $l \gg \lambda_m$ moving on water with depth $h \gg l$; this is the usual situation for ship waves.

The most striking result, originally due to Kelvin, is that in deep water the waves are confined to a wedge shaped region of wedge semiangle $\sin^{-1}\frac{1}{3} = 19.5°$. This result is independent of the velocity provided the velocity is constant; it is independent of the shape of the object, and it depends only on the fact that $C/c = \frac{1}{2}$ for deep water.

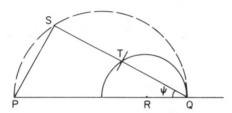

Fig. 12.2. Construction of wave elements in ship wave problem.

A concise form of the argument is given by Lighthill (1957). Consider the "ship" to move from Q to P in Fig. 12.2, in time t, and let its speed be U. For a wave crest to keep a stationary position relative to the ship,

$$U\cos\psi = c(k), \tag{12.23}$$

where ψ is the inclination of the normal (direction of **k**) to the line of motion QP. This condition is most easily seen by taking the frame of reference in which the stream of velocity U flows past a stationary ship; the stream has component $U\cos\psi$ normal to the wave element and this must be balanced by the phase velocity of the element. The condition tells us the value of k to be found in the direction ψ. It may be represented geometrically in Fig. 12.2 by constructing the semicircle with diameter PQ and noting that $PQ = Ut$, $SQ = Ut\cos\psi = ct$. Therefore wave elements parallel to PS will have $ct = QS$. Now c is the phase speed and is appropriate for condition (12.23), but the group velocity $C = \frac{1}{2}c$ determines the location of these waves. The waves produced at Q will have traveled a

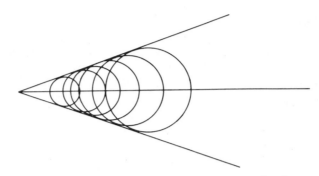

Fig. 12.3. Envelope of the disturbance emitted at successive times.

distance $Ct = \frac{1}{2}ct$. Therefore in the direction ψ they will be found at T, the midpoint of QS. Including all values of ψ, we deduce that those waves produced at Q which can contribute to a stationary pattern lie on a circle of radius $\frac{1}{4}Ut$ centered at R, where $PR = \frac{3}{4}Ut$. Finally, varying t for a fixed point P, we have the pattern of circles of Fig. 12.3. From the construction in Fig. 12.2, each circle has a radius one third the distance of its center from P. Hence they fill a wedge-shaped region with semiangle $\sin^{-1}\frac{1}{3} = 19.5°$. It is amusing to note that the construction in Fig. 12.3 is the same as for supersonic flow with Mach number 3; all swimming objects have effective Mach number 3.

Further Details of the Pattern.

In discussing the pattern in more detail, it is convenient to take the reference frame in which the source is fixed at P and there is a uniform stream U in the x_1 direction (see Fig. 12.4). This raises some general points about handling steady patterns which are also useful in other contexts. The dispersion relations in Section 12.1 apply to waves propagating into still water, but we may transfer to any other reference frame moving with relative velocity $-U$ by noting that the frequency ω relative to the moving frame is given in terms of the frequency ω_0 in the stationary frame by

$$\omega = U \cdot k + \omega_0(k). \qquad (12.24)$$

This is the dispersion relation between ω and k for waves superimposed on a stream U. Of course the propagation is no longer isotropic since the direction of U enters. For a steady wave pattern in this frame, $\omega = 0$ and (12.24) becomes a relation between the components k_1 and k_2 of the wave number vector $\mathbf{k}$. With $\omega_0(\mathbf{k}) = \sqrt{gk}$, we have

$$G(k_1, k_2) = Uk_1 + \sqrt{gk} = 0. \tag{12.25}$$

Since $\cos\psi = -k_1/k$ and $c(k) = \sqrt{g/k}$, this is the same as (12.23). We may also write it using the polar coordinates (k, ψ) for $\mathbf{k}$ as

$$\mathcal{G}(k, \psi) = Uk\cos\psi - \sqrt{gk} = 0. \tag{12.26}$$

Since the frequency ω is zero and $\mathbf{k}$ is independent of t, the kinematic description (11.43) reduces to the consistency relation

$$\frac{\partial k_2}{\partial x_1} - \frac{\partial k_1}{\partial x_2} = 0. \tag{12.27}$$

From (12.25), $k_1 = f(k_2)$, say, and (12.27) gives

$$\frac{\partial k_2}{\partial x_1} - f'(k_2)\frac{\partial k_2}{\partial x_2} = 0.$$

Hence k_2 and k_1 are constant on characteristics

$$\frac{dx_2}{dx_1} = -f'(k_2).$$

For a point source P, the characteristics carrying disturbances pass through P and we have a centered wave

$$\frac{x_2}{x_1} = -f'(k_2); \tag{12.28}$$

this gives k_2 as a function of x_2/x_1, and $k_1 = f(k_2)$ completes the solution for $\mathbf{k}$.

The basic relation (12.28) can be written symmetrically in k_1 and k_2. If $k_1 = f(k_2)$ satisfies (12.25) identically,

$$f'(k_2)G_{k_1} + G_{k_2} = 0,$$

and (12.28) may be written

$$\frac{x_2}{x_1} = \frac{G_{k_2}(k_1, k_2)}{G_{k_1}(k_1, k_2)}, \qquad G(k_1, k_2) = 0. \tag{12.29}$$

These are to be solved to give k_1 and k_2 as functions of $\mathbf{x}$. The distribution of $\mathbf{k}$ is sufficient to sketch out the pattern, but the phase $\theta(x_1, x_2)$ can also be deduced to give the equations of the crests.

It might be noted that (12.29) is the limit of the nonstationary centered wave solution as $\omega \to 0$. For the centered wave solution in (11.49) we have

$$\frac{x_i}{t} = C_i = -\frac{G_{k_i}}{G_\omega}, \qquad G(\mathbf{k}, \omega) = 0.$$

If we take the ratios of the first set to eliminate t and G_ω, (12.29) follows as the limit $\omega \to 0$. We can think of the disturbance propagating out with the group velocity C_i even though its form is unchanging and there is no change in the appearance of the pattern. We may refer to the group velocity in this sense even though it is only its direction $\partial G / \partial k_i$ that appears in the formulas.

A further remark is that polar coordinates are sometimes useful, as in (12.26). In polar coordinates the gradient $\partial G / \partial \mathbf{k}$ has a component $\partial \mathcal{G} / \partial k$ in the direction of $\mathbf{k}$ and a component $\partial \mathcal{G} / k \partial \psi$ perpendicular to $\mathbf{k}$. Hence the angle μ in Fig. 12.4 is given by

$$\tan \mu = \frac{1}{k} \frac{\partial \mathcal{G} / \partial \psi}{\partial \mathcal{G} / \partial k}. \tag{12.30}$$

The content of (12.29) is then equivalent to

$$\xi = \pi - \mu - \psi, \qquad \mathcal{G}(k, \psi) = 0. \tag{12.31}$$

Equations 12.30 and 12.31 determine k and ψ (and hence $\mathbf{k}$) for the direction ξ.

These formulations apply to any steady two dimensional pattern, and

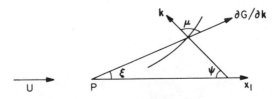

Fig. 12.4. Geometry of wave crests in ship wave problem.

Fig. 12.5. A complete wave crest in ship wave problem.

we now apply them to ship waves. Using (12.25) in (12.29), we have

$$\tan\xi = \frac{x_2}{x_1} = \frac{\dfrac{k_2}{2k}\sqrt{\dfrac{g}{k}}}{U + \dfrac{k_1}{2k}\sqrt{\dfrac{g}{k}}}, \qquad Uk_1 + \sqrt{gk} = 0.$$

It is clearly more convenient to switch to the (k,ψ) description of **k** and reduce these to

$$\tan\xi = \frac{\tan\psi}{1 + 2\tan^2\psi}, \qquad k = \frac{g}{U^2\cos^2\psi}. \tag{12.32}$$

If the approach via (12.30)–(12.31) is preferred, we have

$$\tan\mu = -2\tan\psi, \tag{12.33}$$

and (12.32) follows.

We may now sketch out a typical wave crest as ψ varies. From (12.25) or (12.26), $k_1 < 0$ and $\cos\psi > 0$, so that only the range $-\pi/2 < \psi < \pi/2$ is permissible. The pattern is clearly symmetrical and it is sufficient to take the range $0 < \psi < \pi/2$. From (12.32) we see that the values $\psi \to 0$ and

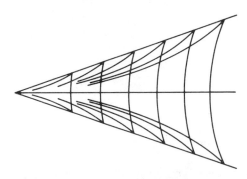

Fig. 12.6. Final ship wave pattern.

$\psi \to \pi/2$ are both to be found on $\xi = 0$, and there must be a maximum value of ξ in the range. It is easily verified that the maximum value is

$$\xi_m = \tan^{-1}\frac{1}{2\sqrt{2}} = 19.5° \qquad at \quad \psi_m = \tan^{-1}\frac{1}{\sqrt{2}} = 35.3°.$$

This agrees with the wedge angle found earlier and shows that the wave pattern is confined to this wedge. At the maximum $\psi \neq \pi/2$ therefore the wave crest can not turn back smoothly; there must be a cusp at $\psi = \psi_m$ on the boundary of the wedge. We may then complete the shape as given in Fig. 12.5 and the whole pattern must appear as in Fig. 12.6.

The formula for the phase function $\theta(\mathbf{x})$ may be found from

$$\theta = \int_0^{\mathbf{x}} \mathbf{k} \cdot d\mathbf{x}, \tag{12.34}$$

using any convenient path since $\mathbf{k}$ is irrotational. Obviously the rays $\xi = $ constant are convenient since $\mathbf{k}$ remains constant on them. We have

$$\theta = (k\cos\mu)r, \tag{12.35}$$

where $r = |\mathbf{x}|$ is distance from the origin. Here k and μ are functions of ξ given by (12.32) and (12.33). A phase curve $\theta = $ constant is given parametrically in terms of ψ by

$$r = \frac{\theta}{k\cos\mu} = -\frac{U^2\theta}{g}\cos^2\psi\{1 + 4\tan^2\psi\}^{1/2},$$

$$\tan\xi = \frac{\tan\psi}{1 + 2\tan^2\psi},$$

and θ is negative. These may also be written

$$x_1 = -\frac{U^2\theta}{g}\cos\psi(1 + \sin^2\psi),$$

$$x_2 = -\frac{U^2\theta}{g}\cos^2\psi\sin\psi. \tag{12.36}$$

12.5 Capillary Waves on Thin Sheets

One can study steady patterns of capillary waves in similar fashion, and a particularly interesting setting is Taylor's study (1959) of waves on thin sheets of water. Surface tension is the dominant effect and the sheet is thin enough to make the approximation $h/\lambda \ll 1$ appropriate. In one mode the sheet deforms as a whole, keeping roughly constant thickness, and in this mode (antisymmetric disturbance of the two surfaces) the waves are not dispersive. For the symmetric mode, however, in which the two surfaces oscillate symmetrically away from the central plane, there is dispersion. We may apply (12.7) taking $2h$ to be the thickness of the undisturbed sheet, since the plane of symmetry is equivalent to a solid surface for each half of the sheet. Since the sheet is very thin the approximations (12.11) may be used. The wave pattern from a point source in a stream U may be analyzed from the general formulation in the last section using now the dispersion relation

$$G = Uk_1 + \left(\frac{Th}{\rho}\right)^{1/2} k^2 = 0. \tag{12.37}$$

For a uniform sheet with uniform flow, U and h are constants. In that case, we immediately find from (12.29) that

$$\frac{x_2}{x_1} = \tan\xi = \frac{2\alpha k_2}{U + 2\alpha k_1}, \qquad \alpha = \left(\frac{Th}{\rho}\right)^{1/2}. \tag{12.38}$$

From (12.37), $k = U\cos\psi/\alpha$ in the (k,ψ) description of $\mathbf{k}$ and the characteristic relation (12.38) reduces to

$$\tan\xi = \frac{2\tan\psi}{\tan^2\psi - 1}, \qquad \text{that is,} \quad \xi = \pi - 2\psi; \tag{12.39}$$

it follows from (12.31) that the angle μ is equal to ψ. This time, ψ ranges from 0 to $\pi/2$, ξ ranges from $-\pi$ to 0, and we infer the roughly parabolic crests shown in Fig. 12.7. The phase function is

$$\theta = (k\cos\mu)r = \left(\frac{\rho U^2}{Th}\right)^{1/2} r \sin^2\frac{\xi}{2}; \tag{12.40}$$

successive crests are curves $r\sin^2\xi/2 = \text{constant}$.

Taylor also conducted experiments and developed the relevant theory for waves on a radially expanding sheet. In the undisturbed sheet the radial velocity, V say, may be taken constant since the pressure gradients are at most $O(h)$ and may be neglected. As a consequence, the semithick-

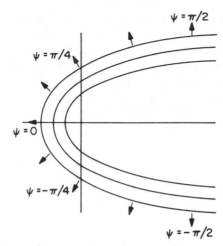

Fig. 12.7.   Pattern of wave crests on a thin sheet of water.

ness h is a function of the distance R from the center of symmetry (the source of the flow) given by

$$h = \frac{Q}{4\pi VR}, \tag{12.41}$$

where Q is the total volume flow. Because h depends on R, we have an example of waves on a nonuniform medium. Well away from $R=0$, the medium is slowly varying relative to a typical wavelength and we may apply and illustrate the ideas of Sections 11.5 and 11.7 on nonuniform media. In polar coordinates $(R, \tilde{\omega})$ based on the source of flow, *not* on the source of the waves (see Fig. 12.8),

$$\mathbf{k} = \left(\theta_R, \frac{1}{R}\theta_{\tilde\omega} \right), \qquad \mathbf{U} = (V, 0),$$

and to adjust for the radial flow the dispersion relation (12.37) is modified to

$$V\theta_R + \left\{ \frac{Th(R)}{\rho} \right\}^{1/2} \left(\theta_R^2 + \frac{1}{R^2}\theta_{\tilde\omega}^2 \right) = 0.$$

From (12.41), this dispersion relation may be written

$$G = \frac{1}{2}\left(\theta_R^2 + \frac{1}{R^2}\theta_{\tilde\omega}^2 \right) + \beta R^{1/2}\theta_R = 0, \qquad \beta = \left(\frac{\pi\rho V^3}{TQ} \right)^{1/2}. \tag{12.42}$$

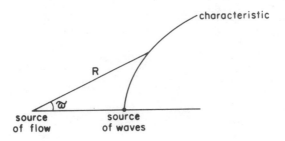

Fig. 12.8. Construction detail for waves on a radially expanding sheet.

It may be solved by the method of characteristics, but the details are more complicated than in the corresponding steps (12.27)–(12.29) since k is not constant on the characteristics nor are the characteristics straight lines. However, we may find the characteristic form from the general formulas of Section 2.13. If we let $p = \theta_R$, $q = \theta_{\tilde{\omega}}$, we have

$$\frac{dR}{d\tau} = G_p = p + \beta R^{1/2}, \qquad \frac{d\tilde{\omega}}{d\tau} = G_q = \frac{q}{R^2},$$

$$\frac{dp}{d\tau} = -G_R - pG_\theta = \frac{q^2}{R^3} - \frac{\beta p}{2R^{1/2}},$$

$$\frac{dq}{d\tau} = -G_{\tilde{\omega}} - qG_\theta = 0,$$

where τ is a parameter on the characteristic. Since q is constant on each characteristic, it becomes a convenient characteristic variable and, from (12.42),

$$p = \beta R^{1/2}\left(1 - \frac{q^2}{\beta^2 R^3}\right)^{1/2} - \beta R^{1/2}. \qquad (12.43)$$

The characteristic curves are given by

$$\frac{d\tilde{\omega}}{dR} = \frac{q/R^2}{p + \beta R^{1/2}} = \frac{q}{\beta R^{5/2}(1 - q^2/\beta^2 R^3)^{1/2}}.$$

These pass through the source, at $R = R_0$, $\tilde{\omega} = 0$, say, and the appropriate integral is

$$\left(\frac{R_0}{R}\right)^{3/2} = \frac{\sin(\sigma - (3/2)\tilde{\omega})}{\sin\sigma}, \qquad \sin\sigma = \frac{q}{\beta R_0^{3/2}}. \qquad (12.44)$$

This equation for the characteristics may be solved to express q as a function of R and $\tilde{\omega}$, and then (12.43) used to obtain p. We have

$$q = \theta_{\tilde{\omega}} = \frac{\beta R^{3/2} \sin(3/2)\tilde{\omega}}{\left\{ (R/R_0)^3 - 2(R/R_0)^{3/2} \cos(3/2)\tilde{\omega} + 1 \right\}^{1/2}},$$

$$p = \theta_R = \frac{\beta R^{1/2}\left\{ (R/R_0)^{3/2} - \cos(3/2)\tilde{\omega} \right\}}{\left\{ (R/R_0)^3 - 2(R/R_0)^{3/2} \cos(3/2)\tilde{\omega} + 1 \right\}^{1/2}} - \beta R^{1/2},$$

and, finally,

$$\theta = \frac{2}{3}\beta R_0^{3/2}\left\{ \left(\frac{R}{R_0}\right)^3 - 2\left(\frac{R}{R_0}\right)^{3/2} \cos\frac{3}{2}\tilde{\omega} + 1 \right\}^{1/2} + \frac{2}{3}\beta R_0^{3/2}\left\{ 1 - \left(\frac{R}{R_0}\right)^{3/2} \right\}.$$

$$(12.45)$$

This result was first found by Ursell (1960b) in an amplification of Taylor's arguments. For small $\tilde{\omega}$, we have

$$\theta \simeq \frac{3}{4}\beta R_0^{3/2}\tilde{\omega}^2\left\{ 1 - \left(\frac{R_0}{R}\right)^{3/2} \right\}^{-1}; \qquad (12.46)$$

this agrees with Taylor's equation for the crests and it compares well with the experiments.

This particular case shows the power of the kinematic arguments, since any direct attack on the boundary value problem involved here would be a formidable undertaking.

12.6 Waves in a Rotating Fluid

For small perturbations in an incompressible fluid which has a basic flow velocity U along the x_3 axis and a solid body rotation with angular velocity Ω about that axis, the linearized equations are

$$\frac{Du_1}{Dt} - 2\Omega u_2 = -\frac{\partial P}{\partial x_1}, \qquad \frac{Du_2}{Dt} + 2\Omega u_1 = -\frac{\partial P}{\partial x_2},$$

$$\frac{Du_3}{Dt} = -\frac{\partial P}{\partial x_3}, \qquad \frac{\partial u_1}{\partial x_1} + \frac{\partial u_2}{\partial x_2} + \frac{\partial u_3}{\partial x_3} = 0,$$

where

$$P = \frac{p - p_0}{\rho} - \frac{1}{2}\Omega^2(x_1^2 + x_2^2), \qquad \frac{D}{Dt} = \frac{\partial}{\partial t} + U\frac{\partial}{\partial x_3}.$$

The velocity perturbations may be eliminated in favor of P and the single equation

$$\left(\frac{\partial}{\partial t} + U\frac{\partial}{\partial x_3}\right)^2 \nabla^2 P + 4\Omega^2 \frac{\partial^2 P}{\partial x_3^2} = 0 \qquad (12.47)$$

obtained.

When $U = 0$, the reduced equation for periodic disturbances $P = \mathcal{P}e^{-i\omega t}$ is

$$\frac{\partial^2 \mathcal{P}}{\partial x_1^2} + \frac{\partial^2 \mathcal{P}}{\partial x_2^2} + \left(1 - \frac{4\Omega^2}{\omega^2}\right)\frac{\partial^3 \mathcal{P}}{\partial x_3^2} = 0. \qquad (12.48)$$

The change of type from elliptic for $\omega > 2\Omega$ to hyperbolic for $\omega < 2\Omega$ leads to both interesting phenomena and interesting mathematical problems. For $\omega > 2\Omega$ the disturbance from a source will have the typical $1/r^2$ decay of a doublet solution of Laplace's equation, whereas for $\omega < 2\Omega$ it will be confined inside the characteristic cone of semiangle $\tan^{-1}(4\Omega^2/\omega^2 - 1)^{-1/2}$ around the x_3 axis. For flow inside a container the boundary conditions are of elliptic type, which leads to unusual eigenvalue problems in the hyperbolic case $\omega < 2\Omega$. Solutions have been found for special shapes (Greenspan, 1968; Barcilon, 1968; Franklin, 1972).

When the stream U is included, the dispersion relation for (12.47) is

$$(\omega - Uk_3)^2 k^2 - 4\Omega^2 k_3^2 = 0. \qquad (12.49)$$

Waves are possible only when $(\omega - Uk_3)^2 < 4\Omega^2$; for $U = 0$ this checks with the condition for (12.48) to be hyperbolic. We have two modes satisfying

$$\omega = Uk_3 \pm \frac{2\Omega k_3}{k}, \qquad (12.50)$$

and the group velocity has components

$$C_1 = \mp 2\Omega\frac{k_1 k_3}{k^3}, \qquad C_2 = \mp 2\Omega\frac{k_2 k_3}{k^3}, \qquad C_3 = U \pm 2\Omega\frac{(k_1^2 + k_2^2)}{k^3}. \qquad (12.51)$$

For a point source of constant frequency ω on the x_3 axis, the distribution of k is determined from

$$\frac{x_3}{\left(x_1^2 + x_2^2\right)^{1/2}} = \frac{C_3}{\left(C_1^2 + C_2^2\right)^{1/2}}. \tag{12.52}$$

When $U=0$ this reduces to

$$\frac{x_3}{\left(x_1^2 + x_2^2\right)^{1/2}} = \pm \frac{\left(k^2 - k_3^2\right)^{1/2}}{k_3}$$

$$= \pm \left(\frac{4\Omega^2}{\omega^2} - 1\right)^{1/2}. \tag{12.53}$$

The disturbance is found on the characteristic cone in agreement with (12.48).

When $U \neq 0$, there is dispersion even for fixed ω and different values of k satisfying (12.49) are dispersed over different cones. Complete wave patterns can be worked out by the techniques developed here; the results can be found in the paper by Nigam and Nigam (1962). But perhaps the most interesting questions concern the wave propagation view of the Taylor column.

In a famous experiment Taylor (1922) found that when a sphere is pulled slowly along the axis of rotation a whole cylindrical column of fluid circumscribing the sphere is pushed along with it. The complete analysis of the phenomenon is difficult (see Greenspan, 1968, p. 192), but some information can be obtained from the wave kinematics. We take the steady frame of reference with main stream U. For waves to appear upstream, they must have $C_3 < 0$; hence

$$\frac{2\Omega(k_1^2 + k_2^2)}{k^3} > U.$$

The most favorable case is for $k_3 = 0$, which would correspond to the surface of the Taylor column. To be upstream the waves must have $2\Omega/k > U$, or equivalently, $\lambda > U\pi/\Omega$. We should expect the dominant wavelengths produced to have $\lambda = O(a)$, where a is the radius of the sphere. Indeed Taylor found a column when $\Omega a / \pi U > 1$ and this would fit exactly with the choice $\lambda = a$. Subsequent experiments and theory indicate that the transition is not sharp, and this result should be taken as an estimate of when the Taylor column will be reasonably strong.

12.7 Waves in Stratified Fluids

Gravity waves in a density-stratified fluid are of great interest in meteorology and oceanography. The basic density gradient may be established by heating or salt content or other effects, but it is often desirable to eliminate compressibility and sound waves in the subsequent motion. To achieve this the continuity equation is split into two parts:

$$\frac{D\rho}{Dt} = 0, \qquad \nabla \cdot \mathbf{u} = 0,$$

and both are retained! The density is not constant but is assumed to be unchanged following a particle path in the wave motion. To these equations are added the momentum equations

$$\rho \frac{D\mathbf{u}}{Dt} = -\nabla p - \rho \mathbf{g}.$$

The double use of the continuity equation is in lieu of an energy equation and we have a complete system. The results and approximations can be checked against more complete descriptions, the main requirement being that the sound speed should be much greater than the wave speeds found in this theory.

For two dimensional flow in the (x,y) plane with stratification in the y direction, we take the undisturbed distributions to be $u = v = 0$, $\rho = \rho_0(y)$, $p = p_0(y)$ with

$$\frac{dp_0}{dy} + \rho_0 g = 0, \tag{12.54}$$

and we linearize for small perturbations about these values. If the perturbations of ρ and p are denoted by ρ_1 and p_1, respectively, the linearized equations are

$$\rho_{1t} + v\rho_0' = 0, \qquad u_x + v_y = 0,$$

$$\rho_0 u_t + p_{1x} = 0, \qquad \rho_0 v_t + p_{1y} + g\rho_1 = 0.$$

A single equation can be deduced; in terms of a stream function Ψ defined by $u = \Psi_y$, $v = -\Psi_x$, it is

$$\rho_0 \Psi_{xxtt} + (\rho_0 \Psi_{yt})_{yt} - g\rho_0' \Psi_{xx} = 0. \tag{12.55}$$

It is convenient for the wave motion to have an equation with even derivatives only, which may be achieved by the substitution $\Psi = \rho_0^{-1/2}\chi$ to give

$$\chi_{xxtt} + \chi_{yytt} + \left(\frac{\rho_0'^2}{4\rho_0^2} - \frac{\rho_0''}{2\rho_0} \right)\chi_{tt} - \frac{g\rho_0'}{\rho_0}\chi_{xx} = 0. \qquad (12.56)$$

The special case of an exponential distribution, $\rho_0 \propto e^{-\alpha y}$, has the attraction of constant coefficients and can be matched reasonably well to other distributions. In that case the dispersion relation is

$$\omega^2 \left(k_1^2 + k_2^2 + \frac{1}{4}\alpha^2 \right) - \alpha g k_1^2 = 0. \qquad (12.57)$$

In many situations the interesting wavelengths are in the range $k \gg \alpha$, and (12.57) is approximated by

$$\omega^2 = \frac{\omega_0^2 k_1^2}{k_1^2 + k_2^2}, \qquad \omega_0^2 = \alpha g = -\frac{g\rho_0'}{\rho_0}. \qquad (12.58)$$

The frequency ω_0 is the Vaisala-Brunt frequency; it is constant for the exponential distribution and would be a function of y in the more general case.

We note that waves are possible only in the case $\omega < \omega_0$, and the situation is somewhat similar to the example of rotating fluids. For a source with fixed frequency $\omega < \omega_0$, one solution of (12.58) may be taken to be

$$k_1 = k\cos\psi, \qquad k_2 = -k\sin\psi, \qquad (12.59)$$

where

$$\psi = \cos^{-1}\frac{\omega}{\omega_0}; \qquad (12.60)$$

whatever the magnitude of k, the waves all have the fixed inclination ψ with the x axis. The corresponding group velocity has components

$$C_1 = \frac{\omega_0 k_2^2}{k^3}, \qquad C_2 = -\frac{\omega_0 k_1 k_2}{k^3}. \qquad (12.61)$$

Since $\mathbf{k} \cdot \mathbf{C} = 0$, the phase and group velocities are perpendicular. Hence the group velocity is at an angle $\xi = \pi/2 - \psi$ to the x axis. The direction of the group velocity determines where the waves will be found. In view of

(12.60), this direction is the same for all the waves and is at the angle

$$\xi = \frac{\pi}{2} - \psi = \sin^{-1} \frac{\omega}{\omega_0} \qquad (12.62)$$

with the x axis. When all the possible signs of k_1 and k_2 are included, the disturbance forms an X-shaped pattern. The crests also lie on the X but move locally normal to the arms, continually dying out as they leave the pattern but being replenished by new ones appearing behind. (Of course each arm has a finite thickness in reality because the source is finite.)

Excellent photographs were obtained by Mowbray and Rarity (1967a) and are reproduced in Fig. 12.9. The source is an oscillating cylinder normal to the plane of the photograph and oscillating horizontally; the vertical rod is a probe. The source introduced also a faint but discernible second harmonic with frequency 2ω. This gives the pattern with angle $\sin^{-1} 2\omega/\omega_0$. In this paper and later ones (Rarity, 1967; Mowbray and Rarity, 1967b), these authors investigate the theoretical patterns in detail and also study the pattern produced by a moving sphere. In reality the distribution of density is not quite exponential and the bending of the group lines due to the dependence of ω_0 on y can be seen in some of the photographs. This illustrates the effects of a nonhomogeneous medium, and the variations could be analyzed by the kinematic methods developed in the last chapter and applied in Section 12.5.

12.8 Crystal Optics

In crystals the anisotropic properties of the medium produce striking effects in the wave patterns. The structure of the crystal produces directional effects in the dielectric properties and the relation between the displacement vector $\mathbf{D}$ and the electric field $\mathbf{E}$ must be described by a tensor relation. The usual relation $\mathbf{B} = \mu_0 \mathbf{H}$ suffices for the magnetic vectors. The effects are described by taking the constituitive relations

$$D_i = \epsilon_{ij} E_j, \qquad B_i = \mu_0 H_i \qquad (12.63)$$

in Maxwell's equations. In general the dielectric tensor ϵ_{ij} will depend on the frequency ω. This may be accommodated by using the relations (12.63) only after the time dependence $e^{-i\omega t}$ has been taken out. For a plane wave with all components of the field vectors proportional to $e^{i\mathbf{k}\cdot\mathbf{x} - i\omega t}$, Maxwell's equations reduce to

$$-i\omega\mathbf{B} + i\mathbf{k} \times \mathbf{E} = 0,$$

$$i\omega\mathbf{D} + i\mathbf{k} \times \mathbf{H} = 0. \qquad (12.64)$$

Fig. 12.9. (1) The image of the undisturbed fluid. (2) $\omega/\omega_0 = 0.318$. (3) $\omega/\omega_0 = 0.366$.
(4) $\omega/\omega_0 = 0.419$. (5) $\omega/\omega_0 = 0.615$. (6) $\omega/\omega_0 = 0.699$. (Mowbray and Rarity, 1967a.)

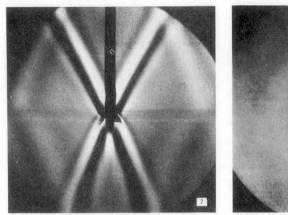 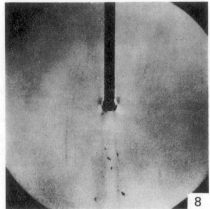

Fig. 12.9. (*Continued*) (7) $\omega/\omega_0 = 0.900$. (8) $\omega/\omega_0 = 1.11$. (Mowbray and Rarity, 1967a.)

Since $\mathbf{B} \propto \mathbf{H}$, it follows that $\mathbf{k}$, $\mathbf{D}$, and $\mathbf{H}$ are mutually orthogonal, so that $\mathbf{D}$ and $\mathbf{H}$ are transverse to the direction of propagation. Since $\mathbf{E}$ is orthogonal to $\mathbf{B}$, it lies in the same plane as $\mathbf{D}$ and $\mathbf{k}$ but is not transverse to the direction of propagation in general. When $\mathbf{B}$ and $\mathbf{H}$ are eliminated from (12.64), we have

$$\omega^2 \mu_0 \mathbf{D} + \mathbf{k} \times (\mathbf{k} \times \mathbf{E}) = 0.$$

On substitution for $\mathbf{D}$ in terms of $\mathbf{E}$, this gives

$$\omega^2 \mu_0 \epsilon_{ij} E_j + k_i k_j E_j - k^2 E_i = 0.$$

The dispersion relation then follows from the condition that the determinant

$$G(\omega, \mathbf{k}) \equiv |\omega^2 \mu_0 \epsilon_{ij} + k_i k_j - k^2 \delta_{ij}| = 0. \tag{12.65}$$

In pursuing details of the wave patterns it is convenient to choose coordinates along the principal axes of ϵ_{ij}. If the principal values are $\epsilon_1, \epsilon_2, \epsilon_3$, (12.65) may be expanded to

$$G(\omega, \mathbf{k}) \equiv \omega^6 \mu_0^3 \epsilon_1 \epsilon_2 \epsilon_3 - \omega^4 \mu_0^2 \left\{ \epsilon_2 \epsilon_3 (k_2^2 + k_3^2) + \epsilon_3 \epsilon_1 (k_3^2 + k_1^2) + \epsilon_1 \epsilon_2 (k_1^2 + k_2^2) \right\}$$

$$+ \omega^2 \mu_0 k^2 \left\{ \epsilon_1 k_1^2 + \epsilon_2 k_2^2 + \epsilon_3 k_3^2 \right\} = 0. \tag{12.66}$$

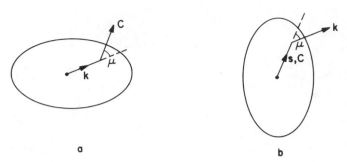

Fig. 12.10. (a) Dispersion surface in **k** space. (b) Phase surface in **x** space.

For a source of fixed frequency ω, (12.66) describes the surface in **k** space which determines the possible wave numbers **k** of the wave elements produced. For an admissible value of **k**, the corresponding group velocity is

$$C_i(\mathbf{k}) = -\frac{G_{k_i}}{G_\omega};$$ \hfill (12.67)

it is in the direction of the normal to the surface (12.66), as shown in Fig. 12.10a. This geometrical correspondence between **C** and **k** is useful in determining wave patterns.

A dual surface is also useful. It may be constructed in terms of the phase surfaces produced by a periodic point source at the origin. This particular problem is not the one of most interest in crystal optics, since one would not normally envisage a source imbedded in the crystal, but it is a convenient route to the construction and the analysis applies to anisotropic waves generally.

Wave elements with wave number **k** are found in the direction **C(k)** from the source. Therefore in each direction **C** from the source we can determine the corresponding value of **k** for that direction (see Fig. 12.10b). But the phase surfaces generally do not move out on the group lines with speed C. The phase velocity has magnitude ω/k in the direction **k**. Hence the point of intersection between a phase surface and a group line moves out with speed $\omega/(k\cos\mu)$, where μ is the angle between **C** and **k**. The phase surface leaving the origin at time $t=0$ will be at

$$\mathbf{x} = \frac{\omega}{k\cos\mu}\frac{\mathbf{C}}{C}t = \frac{\omega}{\mathbf{k}\cdot\mathbf{C}}\mathbf{C}t$$ \hfill (12.68)

at time t. Varying **k** over all values satisfying (12.66) gives the phase surface.

An alternative derivation is to note that the phase function $\theta(\mathbf{x}, t)$ is given by

$$\theta(\mathbf{x}, t) = -\omega t + \int_0^{\mathbf{x}} \mathbf{k} \cdot d\mathbf{x}, \tag{12.69}$$

and the integral can be taken along the group line to x. Therefore

$$\theta = -\omega t + k \cos \mu |\mathbf{x}|. \tag{12.70}$$

The phase surface $\theta = 0$, which left the origin at $t = 0$, will be a distance

$$|\mathbf{x}| = \frac{\omega}{k \cos \mu} t$$

out along the group line with direction $\mathbf{C}$. Hence (12.68) follows.

In optics it is usual to introduce the *ray vector* s defined by

$$\mathbf{s} = \frac{\omega}{c_0 \mathbf{k} \cdot \mathbf{C}} \mathbf{C}, \tag{12.71}$$

where c_0 is the speed of light *in vacuo*. Then the phase surfaces are given by

$$\mathbf{x} = \mathbf{s} c_0 t. \tag{12.72}$$

Thus s is proportional to the group velocity but is reduced in magnitude to give the phase propagation along the ray (group line) as a fraction of c_0. Since s is a function of $\mathbf{k}$ and conversely, the dispersion relation (12.66) may be used to find the corresponding surface in s space. Since this s surface is the canonical shape for the phase surface, the normal at any point is in the direction of $\mathbf{k}$. Thus, using s in place of $\mathbf{C}$, we have dual properties between the $\mathbf{k}$ and s surfaces. In optics it is also usual to work with the refractive index $\mathbf{n} = c_0 \mathbf{k}/\omega$ in place of $\mathbf{k}$. Then we have in addition $\mathbf{s} \cdot \mathbf{n} = 1$. For any point on the n surface, the corresponding s has the direction of the normal and its magnitude is the inverse of the perpendicular distance from the origin to the tangent plane at the point. Conversely, on the dual s surface, the corresponding n is in the direction of the normal and its magnitude is the inverse of the perpendicular distance to the tangent plane.

In the special case when the dispersion relation is homogeneous in ω, k_1, k_2, k_3, the dispersion function has the property

$$G(\rho \omega, \rho k_1, \rho k_2, \rho k_3) = 0$$

for arbitrary ρ. Hence differentiating with respect to ρ and then setting $\rho = 1$, we have

$$\omega G_\omega + k_i G_{k_i} = 0.$$

Therefore $\mathbf{k} \cdot \mathbf{C} = \omega$. In this case, *but only in this case*, (12.68) reduces to

$$\mathbf{x} = \mathbf{C}t,$$

and the phase surfaces move out along the group lines with speed C. The difference between the group and phase velocities is just compensated for by the inclination factor $\cos \mu$. This special case applies to (12.66), for example, if the ϵ_{ij} are taken to be independent of ω.

Uniaxial Crystals.

In the case of a uniaxial crystal with symmetry about the x_1 direction, we have $\epsilon_2 = \epsilon_3$ and the common value will be denoted by $\epsilon_\perp$. The dispersion relation (12.66) then factors to give the two possibilities

$$\omega^2 = \frac{k^2}{\epsilon_\perp \mu_0}, \tag{12.73}$$

$$\omega^2 = \frac{k_1^2}{\epsilon_\perp \mu_0} + \frac{k_2^2 + k_3^2}{\epsilon_1 \mu_0}. \tag{12.74}$$

One would expect anisotropy to distort the waves, but it is perhaps unexpected that splitting would occur and one family remain isotropic. The interesting phenomena in crystal optics stem primarily from this splitting.

We now have two surfaces in $\mathbf{k}$ space as shown in Fig. 12.11a. The waves described by (12.73) are isotropic with speed $(\epsilon_\perp \mu_0)^{-1/2}$ and they are called *ordinary waves*. The surface in $\mathbf{k}$ space is a sphere, the group velocity is parallel to $\mathbf{k}$, and dispersion arises only if $\epsilon_\perp$ depends on ω. The other family (12.74) is suitably called *extraordinary* and the surface in $\mathbf{k}$ space is an ellipsoid; these waves are dispersive. The surfaces are shown in Fig. 12.11a for the case $\epsilon_1 > \epsilon_\perp$. There are two group velocities $\mathbf{C}_0$ and $\mathbf{C}_e$ for each $\mathbf{k}$. As a consequence there will be *two* phase surfaces as shown in Fig. 12.11b. For the ordinary waves we have $\mathbf{C}_0 \propto \mathbf{k}$ and the phase equation (12.68) is

$$\mathbf{x} = \frac{\mathbf{k}}{\epsilon_\perp \mu_0 \omega} t. \tag{12.75}$$

Using (12.73) to eliminate $\mathbf{k}$, we have the phase surfaces

$$\epsilon_\perp \mu_0 (x_1^2 + x_2^2 + x_3^2) = t^2; \tag{12.76}$$

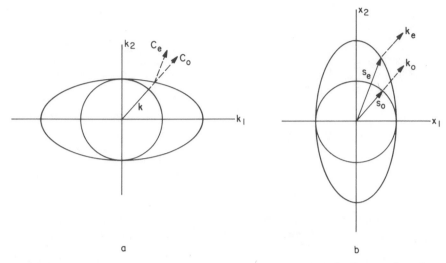

Fig. 12.11. (a) Dispersion surfaces in **k** space for a uniaxial crystal. (b) Phase surfaces in **x** space for a uniaxial crystal.

these are just ordinary spherical waves with speed $(\epsilon_\perp \mu_0)^{-1/2}$. For the extraordinary waves,

$$\mathbf{C}_e \propto \left(\frac{k_1}{\epsilon_\perp}, \frac{k_2}{\epsilon_1}, \frac{k_3}{\epsilon_1} \right) \tag{12.77}$$

and the phase equation (12.68) is

$$\mathbf{x} = \left(\frac{k_1}{\epsilon_\perp \mu_0 \omega}, \frac{k_2}{\epsilon_1 \mu_0 \omega}, \frac{k_3}{\epsilon_1 \mu_0 \omega} \right) t. \tag{12.78}$$

From (12.74), the phase surfaces are ellipses

$$\epsilon_\perp \mu_0 x_1^2 + \epsilon_1 \mu_0 (x_2^2 + x_3^2) = t^2. \tag{12.79}$$

The canonical **s** surfaces are obtained from (12.76) and (12.79) with $\mathbf{x} = s c_0 t$.

For the extraordinary waves the direction of propagation for waves of wave number **k** is given by (12.77). Waves with **k** at an angle ψ to the axis of symmetry propagate at an angle ξ given by

$$\tan \xi = \frac{\epsilon_\perp}{\epsilon_1} \tan \psi. \tag{12.80}$$

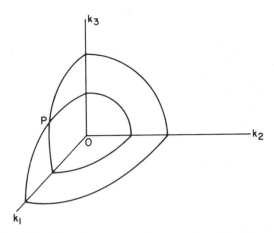

Fig. 12.12. Dispersion surfaces in **k** space for a biaxial crystal.

As a consequence of the splitting, a beam incident on a uniaxial crystal will usually be refracted into two separate beams. The refracted beams are determined by the continuity of the tangential component $\mathbf{k}_t$ of $\mathbf{k}$. But for the given incident $\mathbf{k}_t$ there will be two possible wave vectors $\mathbf{k}$ that satisfy (12.73) and (12.74), respectively. The refracted beams travel in the directions of the corresponding group velocities.

Biaxial Crystals.

For biaxial crystals with $\epsilon_1, \epsilon_2, \epsilon_3$ unequal, the surface (12.66) consists of two sheets with four isolated points of intersection instead of the *circle* of contact between the sphere and ellipsoid of Fig. 12.11a. One octant is indicated in Fig. 12.12 for the case $\epsilon_1 < \epsilon_2 < \epsilon_3$. The point P is one point of intersection and there are three others symmetrically placed in the other quadrants of the k_1, k_3 plane. At a singular point the normal can take any value lying on a cone of directions at the point. If a beam of light enters the crystal normally in this direction, a cone of refracted rays is produced.

Further details of this become complicated and require a lengthy account. These and other questions may be pursued in the excellent accounts by Sommerfeld (1954, Chapter 4) and Landau-Lifshitz (1960a, Chapter 11.)

CHAPTER 13

Water Waves

Many of the general ideas about dispersive waves originated in the problems of water waves. This is a fascinating subject because the phenomena are familiar and the mathematical problems are various. We now turn explicitly to this topic. First we substantiate results referred to earlier, amplify specific details, and include a few problems special to the subject. Then we take up the nonlinear theory which first provided some insight into the questions of how nonlinearity affects dispersive waves; this eventually led to a general point of view on such questions. It will serve the same purpose here, providing motivation for the general discussion and for the study of similar phenomena in different contexts.

13.1 The Equations for Water Waves

We consider an inviscid incompressible fluid (water) in a constant gravitational field. The space coordinates are denoted by (x_1, x_2, y) and the corresponding components of the velocity vector $\mathbf{u}$ by (u_1, u_2, v). The gravitational acceleration g is in the negative y direction. The inviscid equations are given in (6.49); we now assume in addition that the density ρ remains constant and that there is an external force $\mathbf{F} = -\rho g \mathbf{j}$, where $\mathbf{j}$ is the unit vector in the y direction. The equations are

$$\nabla \cdot \mathbf{u} = 0, \tag{13.1}$$

$$\frac{D\mathbf{u}}{Dt} = \frac{\partial \mathbf{u}}{\partial t} + (\mathbf{u} \cdot \nabla)\mathbf{u} = -\frac{1}{\rho}\nabla p - g\mathbf{j}. \tag{13.2}$$

In the main problems of water waves, the flow may be taken to be irrotational, $\operatorname{curl}\mathbf{u} = 0$, and a velocity potential φ introduced with $\mathbf{u} = \nabla\varphi$. This may be argued, as usual, from the equation for the vorticity $\omega = \operatorname{curl}\mathbf{u}$. Equation 13.2 is first rewritten in the form

$$\frac{\partial \mathbf{u}}{\partial t} + \nabla\left(\frac{1}{2}\mathbf{u}^2\right) + \omega \times \mathbf{u} = -\frac{1}{\rho}\nabla p - g\mathbf{j}. \tag{13.3}$$

431

Then if the curl of this equation is taken to eliminate the pressure, we have Helmholtz' equation

$$\frac{\partial \omega}{\partial t} + \nabla \times (\omega \times \mathbf{u}) = 0. \tag{13.4}$$

Since $\nabla \cdot \mathbf{u} = 0$, it may also be written

$$\frac{D\omega}{Dt} = \frac{\partial \omega}{\partial t} + (\mathbf{u} \cdot \nabla)\omega = (\omega \cdot \nabla)\mathbf{u}. \tag{13.5}$$

Now $\omega = 0$ is a possible solution, and the solution is unique provided, say, that all components of $\nabla \mathbf{u}$ are bounded. Therefore if $\omega = 0$ initially, it remains so for all time. In water waves, typical problems concern propagation into water at rest or through a uniform stream. In both cases $\omega = 0$ initially and the argument applies. We restrict the discussion to irrotational flows.

When $\mathbf{u} = \nabla \varphi$, (13.3) may be integrated to

$$\frac{p - p_0}{\rho} = B(t) - \varphi_t - \frac{1}{2}(\nabla \varphi)^2 - gy, \tag{13.6}$$

where $B(t)$ is an arbitrary function, and p_0 is an arbitrary constant separated from $B(t)$ for convenience in applying the free surface condition. Clearly, $B(t)$ can be absorbed into φ by choosing a new potential $\varphi' = \varphi - \int B(t)\,dt$. Usually we assume this is done and take

$$\mathbf{u} = \nabla \varphi,$$

$$\frac{p - p_0}{\rho} = -\varphi_t - \frac{1}{2}(\nabla \varphi)^2 - gy. \tag{13.7}$$

From (13.1) the equation for φ is Laplace's equation

$$\nabla^2 \varphi = 0. \tag{13.8}$$

When the solution of (13.8) has been found for the relevant boundary conditions, the interesting physical quantities $\mathbf{u}, p$ are given by (13.7). This sounds simple enough, and it appears to have little to do with waves since Laplace's equation is involved. Both reactions are wrong because of the curious effects of the free surface conditions.

    We consider the case of a body of water with air above it (although clearly the interface could be between any two fluids) and let the interface be described by

$$f(x_1, x_2, y, t) = 0. \tag{13.9}$$

The interface is *defined* by the property that fluid does not cross it. Hence the velocity of the fluid normal to the interface must be equal to the velocity of the interface normal to itself. The normal velocity of a surface defined by (13.9) is

$$\frac{-f_t}{\sqrt{f_{x_1}^2 + f_{x_2}^2 + f_y^2}}.$$

The normal velocity of the fluid is

$$\frac{u_1 f_{x_1} + u_2 f_{x_2} + v f_y}{\sqrt{f_{x_1}^2 + f_{x_2}^2 + f_y^2}}.$$

The condition that these be equal therefore is

$$\frac{Df}{Dt} = f_t + u_1 f_{x_1} + u_2 f_{x_2} + v f_y = 0. \tag{13.10}$$

This shows that particles in the surface remain there, and the condition is often introduced directly on these grounds. It is easy to have misgivings about the direct statement, however, and it seems preferable to derive it as above from the basic property of an interface.

In working with the equations it is convenient to describe the surface by $y = \eta(x_1, x_2, t)$, and choose

$$f(x_1, x_2, y, t) \equiv \eta(x_1, x_2, t) - y$$

in (13.10). This gives the boundary condition in the form

$$\frac{D\eta}{Dt} = \eta_t + u_1 \eta_{x_1} + u_2 \eta_{x_2} = v. \tag{13.11}$$

Equation 13.10 or 13.11 is a kinematic condition on the boundary. There is also a dynamic condition. Since the interface has no mass, the forces in the fluids on the two sides must be equal. Hence if surface tension is neglected for the present, the pressure in the water and the pressure in the air must be equal at the surface. Any disturbance of the surface clearly implies some motion of the air. But the argument is made that the change in the pressure in the air due to the motion is negligible, and the air pressure may be approximated by its undisturbed value. This is because the density of air is very small compared with that of water, and changes of pressure are of order ρu^2. The assumption can be confirmed in detail by including the motion of the air in typical examples (see Section 13.7). If the motion of the air is neglected on this basis, the second boundary condition becomes $p = p_0$ where p is the pressure in the water, given by (13.7), and p_0 is the constant value in the undisturbed air. The two boundary conditions at the free surface are then

$$\left.\begin{array}{c} \eta_t + \varphi_{x_1}\eta_{x_1} + \varphi_{x_2}\eta_{x_2} = \varphi_y, \\[2mm] \varphi_t + \dfrac{1}{2}\left(\varphi_{x_1}^2 + \varphi_{x_2}^2 + \varphi_y^2\right) + g\eta = 0. \end{array}\right\} \text{ on } y = \eta(x_1, x_2, t). \quad (13.12)$$

Usually one boundary condition is given for Laplace's equation, but that is when the boundary is known. Two conditions are needed at a *free* surface because the surface position η has to be determined as well as φ.

On a solid fixed boundary, the normal velocity of the fluid must vanish, that is, $\mathbf{n} \cdot \nabla \varphi = 0$. In particular, if the bottom is $y = -h_0(x_1, x_2)$, we have

$$\varphi_y + \varphi_{x_1}h_{0x_1} + \varphi_{x_2}h_{0x_2} = 0 \qquad \text{on} \quad y = -h_0(x_1, x_2). \quad (13.13)$$

This is the special case of the interface condition (13.10) since we take $f(x_1, x_2, y, t) = y + h_0(x_1, x_2)$. For a horizontal flat bottom h_0 is a constant and

$$\varphi_y = 0 \qquad \text{on} \quad y = -h_0. \quad (13.14)$$

13.2 Variational Formulation

In view of the general use of variational principles introduced in Chapter 11, it is important to have a variational principle for water waves. This does not seem to have been noted explicitly until the relatively recent paper by Luke (1967). It is of course well known that Laplace's equation follows from

$$\delta \int \int \int \frac{1}{2}(\nabla \varphi)^2 \, dx \, dy \, dt = 0, \tag{13.15}$$

but Luke points out that the variational principle

$$\delta \int\!\!\int_R L \, dx \, dt = 0, \tag{13.16}$$

$$L = -\rho \int_{-h_0}^{\eta} \left\{ \varphi_t + \frac{1}{2}(\nabla \varphi)^2 + gy \right\} dy, \tag{13.17}$$

also gives the all-important boundary conditions. Here R is an arbitrary region in the (x, t) space. When (13.17) is substituted in (13.16) the integration is over the region R_1 of the (x, y, t) space consisting of points with (x, t) in R and $-h_0 \leqslant y \leqslant \eta$. The extra terms φ_t and gy in (13.17), compared with Dirichlet's principle (13.15), affect only the boundary conditions, since they may be integrated out and contribute only to terms from the boundary of R_1.

For a small change $\delta\varphi$ in φ,

$$-\delta \int\!\!\int \frac{L}{\rho} \, dx \, dt = \int\!\!\int_R \left\{ \int_{-h_0}^{\eta} (\delta\varphi_t + \nabla \varphi \cdot \nabla \delta\varphi) \, dy \right\} dx \, dt$$

$$= \int\!\!\int_R \left\{ \frac{\partial}{\partial t} \int_{-h_0}^{\eta} \delta\varphi \, dy + \frac{\partial}{\partial x_i} \int_{-h_0}^{\eta} \varphi_{x_i} \delta\varphi \, dy \right\} dx \, dt$$

$$- \int\!\!\int_R \left\{ \int_{-h_0}^{\eta} (\varphi_{x_i x_i} + \varphi_{yy}) \delta\varphi \, dy \right\} dx \, dt$$

$$- \int\!\!\int_R \left[(\eta_t + \varphi_{x_i} \eta_{x_i} - \varphi_y) \delta\varphi \right]_{y=\eta} dx \, dt$$

$$+ \int\!\!\int_R \left[(\varphi_{x_i} h_{0 x_i} + \varphi_y) \delta\varphi \right]_{y=-h_0} dx \, dt. \tag{13.18}$$

(Repeated subscripts i are summed over $i = 1, 2$.) The first term integrates out to the boundaries of R and vanishes if $\delta\varphi$ is chosen to vanish on the boundaries of R. If (13.18) is to vanish for *all* such $\delta\varphi$, it follows that

$$\varphi_{x_i x_i} + \varphi_{yy} = 0, \qquad -h_0 < y < \eta,$$

$$\eta_t + \varphi_{x_i}\eta_{x_i} - \varphi_y = 0, \qquad y = \eta, \tag{13.19}$$

$$\varphi_{x_i}h_{0 x_i} + \varphi_y = 0, \qquad y = -h_0.$$

The first is obtained by choosing $\delta\varphi = 0$ on $y = \eta$ and $y = -h_0$, and applying the usual variational argument. Then, with the first two terms of (13.18) eliminated, an appropriate choice of $\delta\varphi > 0$ on $y = \eta$, $\delta\varphi = 0$ on $y = -h_0$, gives the boundary condition on $y = \eta$; similarly the choice $\delta\varphi = 0$ on $y = \eta$, $\delta\varphi > 0$ on $y = -h_0$, gives the boundary condition on $y = -h_0$.

For a variation $\delta\eta$ in (13.16)–(13.17), it is immediate that

$$\delta \iint_R L \, d\mathbf{x}\, dt = -\rho \iint_R \left[\varphi_t + \frac{1}{2}(\nabla\varphi)^2 + gy \right]_{y=\eta} \delta\eta \, d\mathbf{x}\, dt = 0,$$

and by the usual argument

$$\left[\varphi_t + \frac{1}{2}(\nabla\varphi)^2 + gy \right]_{y=\eta} = 0 \tag{13.20}$$

Equations 13.19–13.20 are the equations established in the previous section, and we see that this formulation is contained in (13.16)–(13.17).

The significance of (13.17) is that the quantity in braces is $p - p_0$: the principle is one of stationary pressure! The relation of this to Hamilton's principle is discussed in detail by Seliger and Whitham (1968).

LINEAR THEORY

13.3 The Linearized Formulation

For small perturbations on water initially at rest, η and φ are small and the equations may be linearized for a first investigation. The linearized free surface conditions (13.12) are

$$\eta_t = \varphi_y, \qquad \varphi_t + g\eta = 0, \tag{13.21}$$

and we may linearize further by applying these conditions on $y = 0$ rather than on $y = \eta$. After this further linearization, η can be eliminated to give

$$\varphi_{tt} + g\varphi_y = 0 \quad \text{on} \quad y = 0.$$

Laplace's equation and the boundary condition (13.13) on the bottom are already linear and independent of η. Thus we have the linear problem for φ alone:

$$\varphi_{x_1 x_1} + \varphi_{x_2 x_2} + \varphi_{yy} = 0, \qquad -h_0 < y < 0,$$

$$\varphi_{tt} + g\varphi_y = 0, \qquad y = 0, \tag{13.22}$$

$$\varphi_y + h_{0x_1}\varphi_{x_1} + h_{0x_2}\varphi_{x_2} = 0, \qquad y = -h_0.$$

After the solution for φ has been found, the surface is given from (13.21) by

$$\eta(x_1, x_2, t) = -\frac{1}{g}\varphi_t(x_1, x_2, 0, t). \tag{13.23}$$

The problem in (13.22) has to be supplemented by appropriate initial conditions.

13.4 Linear Waves in Water of Constant Depth

In the case of water waves, the waves propagate horizontally in that the elementary sinusoidal solutions take the form

$$\eta = A e^{i\boldsymbol{\kappa} \cdot \mathbf{x} - i\omega t}, \qquad \varphi = Y(y) e^{i\boldsymbol{\kappa} \cdot \mathbf{x} - i\omega t};$$

they are oscillatory in $\mathbf{x}, t$ but not in y. From Laplace's equation, this form of φ is a solution provided

$$Y'' - \kappa^2 Y = 0, \qquad \kappa = |\boldsymbol{\kappa}| = (\kappa_1^2 + \kappa_2^2)^{1/2}.$$

For water of constant depth h_0, the boundary condition on $y = -h_0$ requires $Y'(y) = 0$. Hence

$$Y \propto \cosh k(h_0 + y).$$

From (13.23),

$$A = \frac{i\omega}{g} Y(0)$$

is the amplitude of η, so we take

$$Y(y) = -\frac{ig}{\omega} A \frac{\cosh \kappa (h_0 + y)}{\cosh \kappa h_0}.$$

Then

$$\eta = A e^{i\kappa \cdot x - \omega t},$$

$$\varphi = -\frac{ig}{\omega} A \frac{\cosh \kappa (h_0 + y)}{\cosh \kappa h_0} e^{i\kappa \cdot x - \omega t}. \tag{13.24}$$

The remaining condition $\varphi_{tt} + g\varphi_y = 0$ on $y = 0$ gives the dispersion relation

$$\omega^2 = g\kappa \tanh \kappa h_0. \tag{13.25}$$

It was noted in Section 11.1 that differential equations must lead to polynomial dispersion relations provided that the dependence is sinusoidal in all the independent variables. The transcendental equation (13.25) is obtained here because the variation in y is not sinusoidal. One might say that the waves are in (x, t) space and the y dependence provides a coupling between the wave motions at different depths.

13.5 Initial Value Problem

The dispersion relation (13.25) has two modes $\omega = \pm W(\kappa)$, where

$$W(\kappa) = \sqrt{g\kappa \tanh \kappa h_0}. \tag{13.26}$$

The possible branch point at $\kappa = 0$ is spurious, since $g\kappa \tanh \kappa h_0 \sim g h_0 \kappa^2$ as $\kappa h_0 \to 0$. The function W chosen to have $W \sim \kappa \sqrt{g h_0}$ near the origin is single valued and analytic on the real κ axis. It has branch points at the other zeros and infinities of $\tanh \kappa h_0$, at $\kappa h_0 = \pm n\pi i$, $\pm (n - \frac{1}{2})\pi i$, $n = 1, 2, 3, \ldots$. The functions $W(\kappa)$ and $-W(\kappa)$ are both single valued analytic functions of κ in the complex κ plane cut from $-\infty i$ to $-\pi i/2 h_0$ and from $\pi i/2 h_0$ to ∞i.

The general solution is obtained from the Fourier transforms corre-sponding to (13.24) with two modes corresponding to $\omega = \pm W(\kappa)$. Two initial conditions are needed to determine the arbitrary functions $F(\kappa)$ in the transforms. Of course any prescribed function φ must satisfy Laplace's equation, otherwise compressibility effects will come into play and change the initial distribution rapidly to some new *effective* initial distribution. For simplicity we take the case of fluid initially at rest with $\varphi = 0$. Then, from (13.21), $\eta_t = 0$ initially. To this we add a prescribed initial surface

$$\eta(\mathbf{x}, 0) = \eta_0(\mathbf{x}), \qquad t = 0. \tag{13.27}$$

For this problem the solution is

$$\eta(\mathbf{x}, t) = \int_{-\infty}^{\infty} F(\boldsymbol{\kappa}) e^{i\boldsymbol{\kappa} \cdot \mathbf{x} - iWt} \, d\boldsymbol{\kappa} + \int_{-\infty}^{\infty} F(\boldsymbol{\kappa}) e^{i\boldsymbol{\kappa} \cdot \mathbf{x} + iWt} \, d\boldsymbol{\kappa}, \tag{13.28}$$

where $F(\boldsymbol{\kappa})$ is the Fourier transform of $\frac{1}{2}\eta_0(\mathbf{x})$.

For the one dimensional problem, $\boldsymbol{\kappa}$ and $\mathbf{x}$ are scalars in (13.28), and

$$F(\kappa) = \frac{1}{4\pi} \int_{-\infty}^{\infty} \eta_0(x) e^{-i\kappa x} \, dx. \tag{13.29}$$

The general solution can be reconstructed from the special case $\eta_0(x) = \delta(x)$, $F(\kappa) = 1/4\pi$, which is famous in water waves as the Cauchy-Poisson problem. Its solution can be put in the form

$$\eta(x, t) = \frac{1}{\pi} \int_0^{\infty} \cos \kappa x \cos W(\kappa) t \, d\kappa \tag{13.30}$$

noted in (11.19).

For axial symmetry about a vertical line the two dimensional form of (13.28) can be reduced to

$$\eta(r, t) = 2 \int_0^{\infty} \int_0^{2\pi} \kappa F(\kappa) e^{i\kappa r \cos \xi} \cos W(\kappa) t \, d\kappa \, d\xi,$$

where $r = |\mathbf{x}|$, $\kappa = |\boldsymbol{\kappa}|$, and ξ is the angle between $\mathbf{x}$ and $\boldsymbol{\kappa}$. An integral representation for the Bessel function J_0 is

$$J_0(\kappa r) = \frac{1}{2\pi} \int_0^{2\pi} e^{i\kappa r \cos \xi} \, d\xi,$$

and the solution may be written

$$\eta(r, t) = 4\pi \int_0^{\infty} \kappa F(\kappa) J_0(\kappa r) \cos W(\kappa) t \, d\kappa. \tag{13.31}$$

The inverse formula may be written similarly as

$$F(\kappa) = \frac{1}{4\pi} \int_0^{\infty} r \eta_0(r) J_0(\kappa r) \, dr. \tag{13.32}$$

Of course these may be obtained also by separation of variables in polar coordinates and Fourier-Bessel transforms. For a δ function initial condition $\eta_0(r) = \delta(r)/2\pi r$ and the solution is

$$\eta(r,t) = \frac{1}{2\pi} \int_0^\infty \kappa J_0(\kappa r) \cos W(\kappa) t \, d\kappa. \tag{13.33}$$

The asymptotic results of Chapter 11 may be applied to these solutions. In particular, from (11.24)–(11.25), the asymptotic behavior for the one dimensional solution is

$$\eta \sim 2\Re\left(F(k) \sqrt{\frac{2\pi}{t|W''(k)|}} \, \exp\left\{ ikx - iW(k)t + \frac{\pi i}{4} \right\} \right), \qquad t \to \infty, \frac{x}{t} > 0, \tag{13.34}$$

where $k(x,t)$ is the positive root of

$$W'(k) = \frac{x}{t}, \tag{13.35}$$

and $F(k)$ is given by (13.29). The interpretation was discussed in detail in Chapter 11, and the properties of the group velocity $C(k)$ were discussed in (12.4)–(12.6). Since $C(k)$ is a decreasing function of k, the longest waves appear at the head of the disturbance and are followed by successively shorter waves. The group lines of constant k and the phase lines of constant θ are indicated in Fig. 13.1; a typical wavetrain is shown in Fig. 13.2.

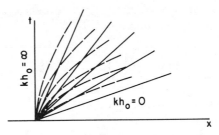

Fig. 13.1. Group lines (solid lines) and phase lines (dashed lines) for water waves.

For finite depth there is a finite maximum group velocity $\sqrt{gh_0}$ for $kh_0 \to 0$, so that the head of the disturbance moves with velocity $\sqrt{gh_0}$. This is a sharp wavefront only in the approximation (13.34), not in the full

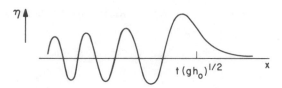

Fig. 13.2. Wavetrain near the front of a disturbance in water waves.

solution. In the full solution the disturbance falls off exponentially, without oscillation, ahead of this front and the disturbance is relatively small. Since $C(k) = W'(k) \to \sqrt{gh_0}$ and $W''(k) \to 0$ as $kh_0 \to 0$, (13.34) is not valid in the neighborhood of the transition region. We now investigate the true behavior.

13.6 Behavior Near the Front of the Wavetrain

Exactly on the line $x = \sqrt{gh_0}\, t$, the correct asymptotic behavior can be found from the extended form of the stationary phase argument in (11.26), since we have $W'''(0) \neq 0$. If $F(0)$ is finite and nonzero, this gives an amplitude decay

$$\eta \propto t^{-1/3}, \tag{13.36}$$

to replace the decay $\eta \propto t^{-1/2}$ away from the front. Since

$$F(0) = \frac{1}{4\pi} \int_{-\infty}^{\infty} \eta_0(x)\, dx, \tag{13.37}$$

this applies when the total initial elevation is finite and nonzero.

However, we would like to have a uniformly valid solution through the whole transition region. It may be obtained by noting that the entire transition region corresponds to small values of k. Both (13.34) for small k and (13.36) for $k=0$ may be included by expanding the exponent in the Fourier transform about $\kappa = 0$, rather than the stationary point, and retaining up to third powers of κ, From (13.26),

$$W(\kappa) \sim c_0\kappa - \gamma\kappa^3 + \cdots, \tag{13.38}$$

where

$$c_0 = \sqrt{gh_0}\,, \qquad \gamma = \frac{1}{6}h_0^2\sqrt{gh_0}\,, \tag{13.39}$$

Therefore, near the head of the wave moving to the right, we take

$$\eta \sim \int_{-\infty}^{\infty} F(\kappa) \exp\left\{ i\kappa(x - c_0 t) + i\gamma\kappa^3 t \right\} d\kappa. \tag{13.40}$$

It is also consistent to expand $F(\kappa)$ in its Taylor series and retain only the first term. If $\int_{-\infty}^{\infty} \eta_0(x)\, dx$ is finite and chosen to be equal to one, the first term is $F(0) = 1/4\pi$ and the solution is

$$\eta \sim \eta_f = \frac{1}{4\pi} \int_{-\infty}^{\infty} \exp\left\{ i\kappa(x - c_0 t) + i\gamma\kappa^3 t \right\} d\kappa. \tag{13.41}$$

This may be expressed in terms of the standard Airy integral

$$Ai(z) = \frac{1}{2\pi} \int_{-\infty}^{\infty} \exp\left\{ i\left(sz + \frac{1}{3}s^3 \right) \right\} ds = \frac{1}{\pi} \int_{0}^{\infty} \cos\left(sz + \frac{1}{3}s^3 \right) ds,$$

by a change of variable $s = (3\gamma t)^{1/3}\kappa$. Then we have

$$\eta_f = \frac{1}{2(3\gamma t)^{1/3}} Ai\left\{ \frac{x - c_0 t}{(3\gamma t)^{1/3}} \right\}. \tag{13.42}$$

The Airy function $Ai(z)$ has the general form shown in Fig. 13.2. Its asymptotic behavior is

$$Ai(z) \sim \begin{cases} \dfrac{1}{2\sqrt{\pi}} z^{-1/4} \exp\left(-\dfrac{2}{3} z^{3/2} \right), & z \to +\infty \\[2ex] \dfrac{1}{\sqrt{\pi}} |z|^{-1/4} \sin\left(\dfrac{2}{3} |z|^{3/2} + \dfrac{\pi}{4} \right), & z \to -\infty. \end{cases}$$

From these, we see that η_f decays exponentially ahead of $x = c_0 t$ and becomes oscillatory behind. Exactly on $x = c_0 t$, $\eta_f \propto t^{-1/3}$ in agreement with (13.36). The transition region is of width proportional to $(\gamma t)^{1/3}$ about $x = c_0 t$. Away from the transition region, as $(x - c_0 t)/(3\gamma t)^{1/3} \to -\infty$,

$$\eta_f \sim (4\pi)^{-1/2} \left\{ 3\gamma t(c_0 t - x) \right\}^{-1/4} \sin\left\{ \frac{2}{3} \frac{(c_0 t - x)^{3/2}}{(3\gamma t)^{1/2}} + \frac{\pi}{4} \right\}. \tag{13.43}$$

It may be verified that this merges correctly with (13.34)–(13.35).

If $F(\kappa) \sim F_n \kappa^n$ for integer n as $\kappa \to 0$, the solution (13.40) can be written in terms of η_f by taking the appropriate number of x derivatives if $n > 0$, or the appropriate number of integrals with respect to x if $n < 0$. For example, the solution for the step function

$$\eta_0(x) = \begin{cases} 0, & x > 0, \\ 1, & x < 0, \end{cases}$$

is just

$$\eta \sim \int_x^\infty \eta_f(x)\, dx$$

$$= \frac{1}{2} \int_z^\infty \mathrm{Ai}(s)\, ds, \qquad z = \frac{x - c_0 t}{(3\gamma t)^{1/3}}. \tag{13.44}$$

The Airy function has the property

$$\int_{-\infty}^\infty \mathrm{Ai}(s)\, ds = 1.$$

The factor $\frac{1}{2}$ appears in (13.42) and (13.44) because these represent only the waves moving to the right; those moving to the left complete the full initial condition.

A simpler view of these solutions is to note that the dispersion relation (13.38) corresponds to the equation

$$\eta_t + c_0 \eta_x + \gamma \eta_{xxx} = 0. \tag{13.45}$$

We are solving this equation for $\eta_0(x) = \frac{1}{2}\delta(x)$ in (13.42) and for $\eta_0 = \frac{1}{2}H(-x)$ in (13.44). The solutions are members of the family of similarity solutions

$$\eta = (3\gamma t)^{-m} f_m(z), \qquad z = \frac{(x - c_0 t)}{(3\gamma t)^{1/3}}; \tag{13.46}$$

after substitution it is easy to relate $f_m(z)$ to Airy's equation

$$\mathrm{Ai}''(z) = z\,\mathrm{Ai}(z) \tag{13.47}$$

and construct the solutions.

Equation 13.45 is the linearized Korteweg-deVries equation, which will play an important role later. We might note that for any dispersion relation with an expansion in the form (13.38), the long waves are described by (13.45) in the linear theory, and the solutions (13.42) and (13.44) apply.

A restriction on the linear theory might also be noted. In (13.42) the amplitude of the first few crests decays proportional to $t^{-1/3}$, whereas the dispersive effects (of relative order k^2) decrease like $t^{-2/3}$. Thus in the final decay nonlinear effects eventually become as important as the dispersion. Under appropriate conditions there is an intermediate range in which the asymptotic linear theory applies. The nonlinear effects require an extra term proportional to $\eta\eta_x$ in (13.45). The equation is then the full Korteweg-deVries equation and we shall see later that the decay is eventually halted and a series of solitary waves is formed. This nonuniform validity of the linear theory near the front of a wavetrain is similar in a general way to that discussed in Chapter 2 for hyperbolic equations.

13.7 Waves on an Interface between Two Fluids

The theory discussed so far ignores the changes in pressure above the water surface due to the motion of the air. We next confirm this in detail in a typical case. The argument can be usefully combined with a discussion of other effects on the interface between two fluids including the case of comparable densities. We consider a fluid with density ρ' above one with density ρ, and, for simplicity, the fluids are infinitely deep. The flows are irrotational with velocity potentials φ', φ, respectively, and the interface is $y = \eta$. The pressures in the two fluids are

$$p' - p_0 = -\rho'\left\{ \varphi_t' + \frac{1}{2}(\nabla\varphi')^2 + gy \right\},$$

$$p - p_0 = -\rho\left\{ \varphi_t + \frac{1}{2}(\nabla\varphi)^2 + gy \right\},$$

where p_0 is the common undisturbed value, and the conditions on the free surface are

$$\left. \begin{array}{l} p' = p, \\[4pt] \eta_t + \varphi_{x_1}'\eta_{x_1} + \varphi_{x_2}'\eta_{x_2} - \varphi_y' = 0, \\[4pt] \eta_t + \varphi_{x_1}\eta_{x_1} + \varphi_{x_2}\eta_{x_2} - \varphi_y = 0, \end{array} \right\} \quad \text{on } y = \eta.$$

It is interesting to consider perturbations to main streams U', U in the two fluids. If we consider only one dimensional waves and linearize the boundary conditions, we set

$$\varphi' = U'x - \frac{1}{2}U'^2 t + \Phi', \qquad \varphi = Ux - \frac{1}{2}U^2 t + \Phi,$$

and retain only the first order terms in Φ', Φ, η. The boundary conditions become

$$\left. \begin{array}{l} \rho'(\Phi'_t + U'\Phi'_x + g\eta) = \rho(\Phi_t + U\Phi_x + g\eta), \\[2mm] \eta_t + U'\eta_x - \Phi'_y = 0, \\[2mm] \eta_t + U\eta_x - \Phi_y = 0, \end{array} \right\} \qquad \text{on } y = 0. \qquad (13.48)$$

Since Φ', Φ satisfy Laplace's equation and tend to zero as $y \to +\infty$, $y \to -\infty$, respectively, the elementary solution takes the form

$$\Phi' = B'e^{i(\kappa x - \omega t) - \kappa y}, \qquad \Phi = Be^{i(\kappa x - \omega t) + \kappa y}, \qquad \eta = Ae^{i(\kappa x - \omega t)}.$$

The boundary conditions (13.48) then give the dispersion relation

$$\frac{\omega}{\kappa} = \frac{\rho U + \rho' U'}{\rho + \rho'} \pm \left\{ \frac{g}{\kappa} \frac{\rho - \rho'}{\rho + \rho'} - \frac{\rho \rho'}{(\rho + \rho')^2}(U - U')^2 \right\}^{1/2}. \qquad (13.49)$$

For the case $U = U' = 0$, it is noted that

$$\omega = \left\{ g\kappa \left(\frac{\rho - \rho'}{\rho + \rho'} \right) \right\}^{1/2}.$$

In the limit $\rho'/\rho \to 0$, this confirms the elementary solution neglecting motion of the air and provides the small correction.

But it is also interesting that various instabilities are indicated when ω has an imaginary part. The following cases may be singled out:

1. $U = U' = 0$. This is unstable if $\rho' > \rho$, as would be expected.
2. $g = 0, U \neq U'$. This case is always unstable and is known as Helmholtz instability.
3. $\rho = \rho', U \neq U'$. This is the same as 2 above, since the gravitational effects can arise only for different densities.
4. $\rho \neq \rho', U \neq U'$. The solution is always unstable for sufficiently short waves, but this result would be modified by the stabilizing influence

of surface tension for extremely short waves. The effects of surface tension are easily incorporated, using the boundary condition given in (13.50).

13.8 Surface Tension

Surface tension acts like a stretched membrane on the surface of the water. For small deviations $y = \eta(x_1, x_2, t)$ from a plane surface the net effect is a normal force $T\eta_{x_i x_i}$ per unit area. When this is included, the pressure condition at the surface becomes

$$p + T\eta_{x_i x_i} = p_0, \tag{13.50}$$

and the linearized conditions (13.21) are modified to

$$\eta_t = \varphi_y, \qquad \varphi_t + g\eta - \frac{T}{\rho}\eta_{x_i x_i} = 0 \qquad \text{on} \quad y = 0. \tag{13.51}$$

The functional forms of η and φ are the same as before, but the revised boundary conditions at the surface lead to the dispersion relation

$$\omega^2 = g\kappa \tanh \kappa h_0 \left(1 + \frac{T}{\rho g}\kappa^2\right). \tag{13.52}$$

The properties and consequences of this were discussed and applied to some extent in Chapter 12.

It was noted that the group velocity now has a minimum value at $\kappa = k_m$. At the minimum $W''(k_m) = 0$ and there is again a transition region in which (13.34) does not apply. The behavior of the solution in the transition region can be found by an approach similar to that in Section 13.6. The expansion

$$W = W(k_m) + (\kappa - k_m)W'(k_m) + \frac{1}{3!}(\kappa - k_m)^3 W'''(k_m) + \cdots$$

is used in the Fourier transform solution, and the resulting form related to the Airy function. The details are given, for example, in Jeffreys and Jeffreys (1956, Section 17.09).

13.9 Waves on a Steady Stream

The geometry of various steady wave patterns was determined in Chapter 12. The variation of the amplitude along each group line can be determined from the general group velocity concepts. However, as pointed out earlier, the initial distribution of amplitude among the different group lines can be obtained only from a more complete solution. We now study the Fourier transform solution for the uniform stream case, show how the simple kinematic description ties in with the full transform solution, and determine the amplitudes. We shall take the source to be an external steady pressure applied to the surface of the stream, rather than a pre-scribed displacement, since this more nearly represents the effect of a floating body.

When solving steady wave problems by transforms, care is needed in applying a suitable radiation condition to ensure uniqueness. It is interest-ing to see how the radiation condition in this context parallels the use of group velocity arguments in determining where the waves will appear. The nonuniqueness arises because the steady state is artificial in that no flow situation can have existed for all time. In principle, an ideal way to correct this is to solve a more realistic unsteady problem with suitable initial conditions applied at some finite time in the past, $t = -t_0$, say, then let $t_0 \to \infty$ in the solution. This program may be difficult to carry through, however, with a lot of unnecessary detail that does not figure in the final answer. A simpler version of the radiation condition follows this idea in principle and yet requires only a minimum extension of the problem. In the present case we take the applied pressure on $y = 0$ to be

$$\frac{p - p_0}{\rho} = f(x_1, x_2) e^{\epsilon t}, \qquad \epsilon > 0. \tag{13.53}$$

It corresponds to a source that was effectively zero a long time ago and then grew gradually until it is close to $f(x_1, x_2)$ at current times. This device has the desirable features of the initial value problem, having reasonable starting conditions, and yet keeps the time dependence simple. After the problem is solved we take the limit $\epsilon \to 0$ to get the steady state solution.

The boundary conditions (13.12) are modified by the inclusion of the applied pressure distribution, and the contribution of surface tension is included since it was seen in Section 12.3 to have particularly interesting consequences for the appearance of upstream waves. For small perturba-tions about a main stream U in the x_1 direction, the velocity potential is taken to be

$$\varphi = U x_1 - \frac{1}{2} U^2 t + \Phi \tag{13.54}$$

where $\nabla \Phi$ is small compared with U. The linearized boundary conditions at the free surface are then

$$\left.\begin{array}{c} \eta_t + U\eta_{x_1} - \Phi_y = 0, \\[2mm] \Phi_t + U\Phi_{x_1} + g\eta - \dfrac{T}{\rho}\eta_{x_i x_i} = -f(\mathbf{x})e^{\epsilon t} \end{array}\right\} \quad \text{on } y = 0. \quad (13.55)$$

The perturbation potential Φ still satisfies Laplace's equation and, for simplicity, we take the case of infinite depth so that $\Phi \rightarrow 0$ as $y \rightarrow -\infty$. The Fourier transform solution which satisfies the latter two, and which varies like $e^{\epsilon t}$, is

$$\Phi = e^{\epsilon t}\int_{-\infty}^{\infty} B(\boldsymbol{\kappa})e^{i\boldsymbol{\kappa}\cdot\mathbf{x}+\kappa y}d\boldsymbol{\kappa}, \qquad \kappa = |\boldsymbol{\kappa}|,$$

$$\eta = e^{\epsilon t}\int_{-\infty}^{\infty} A(\boldsymbol{\kappa})e^{i\boldsymbol{\kappa}\cdot\mathbf{x}}d\boldsymbol{\kappa}.$$

The boundary conditions (13.55) then relate A and B to each other and to the Fourier transform $F(\boldsymbol{\kappa})$ of $f(\mathbf{x})$. The result is

$$\eta(\mathbf{x}, t) = e^{\epsilon t}\int_{-\infty}^{\infty}\frac{\kappa F(\boldsymbol{\kappa})e^{i\boldsymbol{\kappa}\cdot\mathbf{x}}d\boldsymbol{\kappa}}{(\kappa_1 U - i\epsilon)^2 - \omega_0^2(\kappa)}, \qquad (13.56)$$

where

$$\omega_0^2 = g\kappa + \frac{T}{\rho}\kappa^3 \qquad (13.57)$$

is the dispersion relation for waves moving into water at rest.

The role of ϵ can now be seen. For $\epsilon = 0$ there would be poles satisfying

$$\kappa_1^2 U^2 - \omega_0^2(\kappa) = 0 \qquad (13.58)$$

on the path of integration. The presumption would be that the path should go one way or the other around each pole, but the open choice leads to nonuniqueness. With $\epsilon > 0$, the roots of the denominator in (13.56) are complex; the poles have been pushed off the path and the ambiguity does not arise. Whether a pole contributes will then be determined by further deformations of the path in evaluating the integrals and will correspond to a particular choice of path in going around the pole. The values of κ satisfying (13.58) determine which values of κ can make a strong contribution; the position of the pole above or below the path will determine where the contribution appears. The condition (13.58) is essentially the condition

(12.23) [or (12.24) with $\omega = 0$] that determines which waves can stand steady in the stream and so make a large contribution. The analysis showing whether the pole is to be included corresponds to the group velocity determination of where these waves appear.

One Dimensional Gravity Waves

For simplicity we consider first the one dimensional case and neglect surface tension. Then (13.56) becomes

$$\eta(x,t) = e^{\epsilon t} \int_{-\infty}^{\infty} \frac{\kappa F(\kappa_1) e^{i\kappa_1 x_1}}{(\kappa_1 U - i\epsilon)^2 - g\kappa} d\kappa_1, \qquad \kappa = |\kappa_1|. \qquad (13.59)$$

The appearance of $\kappa = |\kappa_1|$ should be carefully noted; it stems from the fact that both positive and negative values of κ_1 appear in the transform, and in the one dimensional case the exponential in the expression for Φ must be

$$\exp(i\kappa_1 x_1 + |\kappa_1| y)$$

to ensure $\Phi \to 0$ as $y \to -\infty$.

In evaluating (13.59), we shall choose the special case of $f(x_1) = P\delta(x_1)$, $F(\kappa_1) = P/2\pi$, since the more general case can always be reconstructed. Then we have

$$\eta(x,t) = \frac{P e^{\epsilon t}}{2\pi} \int_{-\infty}^{\infty} \frac{\kappa e^{i\kappa_1 x_1} d\kappa_1}{(\kappa_1 U - i\epsilon)^2 - g\kappa}.$$

Since $\kappa = |\kappa_1|$ appears in the integral, it is convenient to break the integral into the ranges $\kappa_1 > 0$ and $\kappa_1 < 0$, and use κ as the variable of integration in each. Then

$$\frac{2\pi\eta}{P} = e^{\epsilon t} \int_0^{\infty} \frac{\kappa e^{i\kappa x} d\kappa}{(\kappa U - i\epsilon)^2 - g\kappa} + e^{\epsilon t} \int_0^{\infty} \frac{\kappa e^{-i\kappa x} d\kappa}{(\kappa U + i\epsilon)^2 - g\kappa}.$$

Since $\epsilon \to 0$ ultimately, the ϵ^2 term in the denominators may be neglected and the $e^{\epsilon t}$ factor, which has now served its purpose, may be dropped. We have

$$\frac{2\pi\eta}{P} = \lim_{\epsilon \to 0} \left(\int_0^{\infty} \frac{e^{i\kappa x_1} d\kappa}{\kappa U^2 - 2i\epsilon U - g} + \int_0^{\infty} \frac{e^{-i\kappa x_1} d\kappa}{\kappa U^2 + 2i\epsilon U - g} \right). \qquad (13.60)$$

The poles of the two integrands are at

$$\kappa = \frac{g}{U^2} + \frac{2i\epsilon}{U}, \qquad \kappa = \frac{g}{U^2} - \frac{2i\epsilon}{U},$$

respectively, The paths can be rotated into either the positive or negative imaginary axis. For $x_1 > 0$, the path of the first integral can be rotated into the positive imaginary axis and the path of the second into the negative imaginary axis; both poles contribute. We have

$$\frac{U^2\eta}{P} = \frac{i}{2\pi} \int_0^\infty \frac{e^{-mx_1}dm}{im - k} + ie^{ikx_1} + \frac{i}{2\pi} \int_0^\infty \frac{e^{-mx_1}dm}{im + k} - ie^{-ikx_1}$$

$$= -2\sin kx_1 + \frac{1}{\pi} \int_0^\infty \frac{m}{m^2 + k^2} e^{-mx_1}dm, \qquad x_1 > 0, \qquad (13.61)$$

where

$$k = \frac{g}{U^2}. \qquad (13.62)$$

For $x_1 < 0$, the paths may be rotated the other way and the poles do not contribute:

$$\frac{U^2\eta}{P} = \frac{1}{\pi} \int_0^\infty \frac{m}{m^2 + k^2} e^{mx_1}dm, \qquad x_1 < 0. \qquad (13.63)$$

Relation 13.62 may be written

$$U = \sqrt{\frac{g}{k}} = c(k);$$

it determines the wave number k of waves that can keep a steady position against the oncoming stream. Since $\sin kx_1$ is a solution of the steady free surface problem for all x_1, the radiation condition is crucial in showing that such standing waves appear only downstream ($x_1 > 0$) of the applied pressure disturbance. The integrals in (13.61) and (13.63) are needed in the complete solution, but they become small for $|x_1| \gg 1$. The asymptotic expansions of these integrals are found by formally expanding $(m^2 + k^2)^{-1}$ in ascending powers of m^2 and integrating the resulting series term by term. This procedure gives

$$\int_0^\infty \frac{me^{-m|x_1|}}{m^2+k^2}\, dm \sim \frac{1}{k^2 x_1^2} - \frac{3!}{k^4 x_1^4} + \frac{5!}{k^6 x_1^6} - \cdots,$$

and it may be justified by Watson's lemma.

The conclusion is that away from the source there is a standing wave pattern *downstream only* given by

$$\eta = -\frac{2P}{U^2}\sin kx_1, \qquad k = \frac{g}{U^2}. \tag{13.64}$$

One Dimensional Waves with Surface Tension.

When surface tension is included, the expression corresponding to (13.60) is

$$\frac{2\pi\eta}{P} = \lim_{\epsilon\to 0}\left\{ \int_0^\infty \frac{e^{i\kappa x_1}\, d\kappa}{\kappa U^2 - 2i\epsilon U - g - (T/\rho)\kappa^2} + \int_0^\infty \frac{e^{-i\kappa x_1}\, d\kappa}{\kappa U^2 + 2i\epsilon U - g - (T/\rho)\kappa^2}\right\}. \tag{13.65}$$

The poles are close to the zeros of

$$kU^2 = g + \frac{T}{\rho}k^2; \tag{13.66}$$

this is again the condition

$$U = c(k)$$

used in (12.20) to determine which waves can stand in the stream.

When $U < c_m$, where c_m is the minimum wave speed, the zeros of (13.66) are not real. Hence there are no poles close to the real axis in (13.65), all the contributions to the integrals fall off rapidly with x_1, and there is no standing wave pattern. This agrees with the conclusion drawn in Section 12.3.

When $U > c_m$, there are two real roots of (13.66), and they are the values k_g and k_T referred to in (12.21)–(12.22), with $k_T > k_g$. In this case, the integrands in (13.65) have poles close to k_g and k_T. They are located at

$$\kappa = k_g + \frac{2\rho U}{(k_T - k_g)T}i\epsilon, \qquad \kappa = k_T - \frac{2\rho U}{(k_T - k_g)T}i\epsilon$$

for the first integral, and at the conjugate points for the second integral. For $x_1 > 0$, the first integral may be rotated onto the positive imaginary axis picking up the gravity pole near k_g but not the other. The second integral is rotated on to the negative imaginary axis and also picks up its gravity pole. Thus downstream of the source, the gravity waves appear. Similarly, for $x_1 < 0$ the opposite rotations must be used and the capillary poles make the contribution. The conclusions agree with the group velocity arguments used in (12.21)–(12.22): gravity waves downstream and capillary waves upstream. The asymptotic wavetrains are found to be

$$
\eta \sim \begin{cases} \dfrac{-2P\rho}{(k_T - k_g)T} \sin k_g x_1, & x_1 > 0, \\[3mm] \dfrac{-2P\rho}{(k_T - k_g)T} \sin k_T x_1, & x_1 < 0. \end{cases}
$$

Ship Waves.

For the two dimensional problem of gravity waves produced by a point source $f(x_1, x_2) = P\delta(x_1, x_2)$ we take $F(\kappa) = P/4\pi^2$ in (13.56) and $\omega_0^2 = g\kappa$. Then (13.56) reduces to

$$
\frac{4\pi^2 U^2 \eta}{P} = e^{\epsilon t} \int_{-\infty}^{\infty} \int_{-\infty}^{\infty} \frac{\kappa \exp\{i(\kappa_1 x_1 + \kappa_2 x_2)\} \, d\kappa_1 \, d\kappa_2}{(\kappa_1 - i\epsilon/U)^2 - g\kappa/U^2}, \quad (13.67)
$$

where $\kappa = (\kappa_1^2 + \kappa_2^2)^{1/2}$. It is convenient to introduce polar coordinates

$$
x_1 = r\cos\xi, \qquad x_2 = r\sin\xi,
$$

$$
\kappa_1 = -\kappa\cos\chi, \qquad \kappa_2 = \kappa\sin\chi.
$$

The contribution from the range $\pi/2 < \chi < 3\pi/2$ is the conjugate of the range $-\pi/2 < \chi < \pi/2$, and in the limit $\epsilon \to 0$, (13.67) may be taken as

$$
\frac{2\pi^2 U^2 \eta}{P} = \mathcal{R}\left(\lim_{\epsilon \to 0} \int_{-\pi/2}^{\pi/2} \frac{d\chi}{\cos^2\chi} \int_0^{\infty} \frac{\kappa \exp\{-i\kappa r \cos(\xi + \chi)\}}{\kappa - \kappa_0} \, d\kappa \right), \quad (13.68)
$$

where

$$
\kappa_0 = \frac{g}{U^2 \cos^2\chi} - \frac{2i\epsilon}{U\cos\chi}. \quad (13.69)
$$

Since $\cos\chi>0$, the pole $\kappa=\kappa_0$ is in the lower half of the complex κ plane.

The pattern is symmetrical about the x_1 axis, so it is sufficient to consider the range $0<\xi<\pi$. When $\cos(\xi+\chi)>0$, that is, $-\pi/2<\chi<\pi/2-\xi$, the path of the integral in the κ plane can be rotated into the negative imaginary axis and the pole contributes; when $\cos(\xi+\chi)<0$, that is, $\pi/2-\xi<\chi<\pi/2$, it can be rotated into the positive imaginary axis and the pole does not contribute.

The further details are quite extensive if one keeps careful track of all the terms. Here we note the contribution of the pole and gloss over the remainder. The contribution of the pole is

$$\eta\sim\frac{gP}{\pi U^4}\mathcal{g}\left(\int_{-\pi/2}^{\pi/2-\xi}\frac{\exp\{-i\kappa_0 r\cos(\xi+\chi)\}}{\cos^4\chi}\,d\chi\right),$$

where the ϵ term in κ_0 can now be dropped. The exponent

$$s(\chi)=\kappa_0\cos(\xi+\chi)=\frac{g\cos(\xi+\chi)}{U^2\cos^2\chi}$$

has a stationary point at $\chi=\psi$ where

$$\tan(\xi+\psi)=2\tan\psi. \tag{13.70}$$

Hence by the standard stationary phase formula we have

$$\eta\sim\frac{gP}{\pi\,U^4\cos^4\psi}\mathcal{g}\left[\left(\frac{2\pi}{r|s''(\psi)|}\right)^{1/2}\exp\left\{-irs(\psi)-\frac{\pi i}{4}\operatorname{sgn}s''(\psi)\right\}\right].$$

After some simplification, this reduces to

$$\eta\sim-\left(\frac{2g}{\pi r}\right)^{1/2}\frac{P}{U^3\cos^3\psi}\frac{(1+4\tan^2\psi)^{1/4}}{|1-2\tan^2\psi|^{1/2}}\sin\left\{kr\cos(\xi+\psi)+\frac{\pi}{4}\operatorname{sgn}s''(\psi)\right\},$$

$$\tag{13.71}$$

where

$$k=\kappa_0(\psi)=\frac{g}{U^2\cos^2\psi}, \tag{13.72}$$

and $\psi(\xi)$ is determined from (13.70).

The wave number determination of k and ψ from (13.70) and (13.72) agrees with (12.32); the phase $kr\cos(\xi+\psi)$ agrees with (12.35). The amplitude is singular at the maximum wedge angle $\xi=\xi_m=19.5°$ where $\tan\psi$ $=2^{-1/2}$. This is where the lateral and transverse crests meet at a cusp and corresponds in the analysis to the confluence of the two stationary points of (13.70). Since $s''(\psi)\to0$, we have a transition region of the same general type as that in Section 13.6. At the wedge the phase jumps by $\pi/2$. The singular behavior on the x_1 axis, where $\psi\to0$, is related to the assumption of a point disturbance. The singular regions are studied in detail by Ursell (1960a).

NONLINEAR THEORY

13.10 Shallow Water Theory; Long Waves

For gravity waves with $\kappa h_0\to0$, the dispersion relation is approximately

$$\omega^2\sim gh_0\kappa^2 \tag{13.73}$$

and the phase speed $c_0=\sqrt{gh_0}$ becomes independent of κ. The dispersive effects drop out in this limit and, in one space dimension, the Fourier superposition of solutions for η leads to

$$\eta=\int_{-\infty}^{\infty}F(\kappa)e^{i\kappa(x-c_0t)}d\kappa+\int_{-\infty}^{\infty}G(\kappa)e^{i\kappa(x+c_0t)}d\kappa$$

$$=f(x-c_0t)+g(x+c_0t).$$

This is the general solution of the linear wave equation

$$\eta_{tt}-c_0^2\eta_{xx}=0. \tag{13.74}$$

Clearly there must be some direct way of extracting this equation from the full equations and, in fact, it was already obtained in a more general setting in Section 3.2. There, in the study of river flow, nonlinearity and friction were included. Here we neglect frictional effects but are concerned with nonlinearity.

First we recall the previous derivation. The key step is to approximate the vertical component of the momentum equation (13.2) by

$$-\frac{1}{\rho}\frac{\partial p}{\partial y}-g=0.$$

Then

$$p - p_0 = \rho g(\eta - y).$$
(13.75)

The horizontal components of (13.2) become

$$\frac{\partial u_i}{\partial t} + u_j \frac{\partial u_i}{\partial x_j} + v \frac{\partial u_i}{\partial y} = -g \frac{\partial \eta}{\partial x_i}$$
(13.76)

[using now the mixed notation $\mathbf{u} = (u_1, u_2, v)$, $\mathbf{x} = (x_1, x_2, y)$ so that $i = 1, 2$ and the summation is for $j = 1, 2$ in (13.76)]. Since the right hand side is independent of y, the rate of change of u_i following a particle is independent of y. Hence if u_i is independent of y initially, it remains so. We consider this to be the case. Then (13.76) becomes

$$\frac{\partial u_i}{\partial t} + u_j \frac{\partial u_i}{\partial x_j} + g \frac{\partial \eta}{\partial x_i} = 0.$$
(13.77)

Although the vertical acceleration was neglected in (13.75) compared with the terms retained, there is no reason to neglect $\partial v / \partial y$ in (13.1). However, we may use the integrated form of (13.1), which *must* give the conservation equation

$$\frac{\partial h}{\partial t} + \frac{\partial}{\partial x_i}(h u_i) = 0,$$
(13.78)

where

$$h = h_0 + \eta$$

is the total depth from $y = -h_0$ at the bottom to $y = \eta$ at the top. In detail

$$0 = \int_{-h_0}^{\eta} \left\{ \frac{\partial u_i}{\partial x_i} + \frac{\partial v}{\partial y} \right\} dy$$

$$= \frac{\partial}{\partial x_i} \int_{-h_0}^{\eta} u_i \, dy + [v]_{y=-h_0}^{y=\eta} - [u_i]_{y=\eta} \frac{\partial \eta}{\partial x_i} - [u_i]_{y=-h_0} \frac{\partial h_0}{\partial x_i},$$

and from the boundary conditions (13.11) and (13.13), this reduces to

$$\frac{\partial}{\partial x_i} \int_{-h_0}^{\eta} u_i \, dy + \frac{\partial \eta}{\partial t} = 0;$$

since u_i is independent of y in this approximation and $\eta_t = h_t$, (13.78) follows. Equations 13.77 and 13.78 are the shallow water equations for $\eta(x,t)$, $\mathbf{u}(x,t)$. [Equations 3.37 follow by writing $g \, \partial\eta / \partial x = g \, \partial h / \partial x - gS$ in (13.77) where $S = \partial h_0 / \partial x$ is the slope of the bottom, and adding a friction term].

It is easy to make an order of magnitude estimate of the approximation. The error for p in (13.75) is of order $\rho h_0 v_t$ and, from (13.1), $v \approx - h_0 u_x$; hence the relative error in (13.77) is of order

$$\frac{-p_x}{\rho u_t} \approx \frac{h_0^2 u_{xxt}}{u_t} \approx \frac{h_0^2}{l^2},$$

where l is a length scale for the waves in the x direction. This fits with the approximation $(\kappa h_0)^2 \ll 1$ in obtaining (13.73). Thus (13.77)–(13.78) provide a nonlinear set for relatively shallow water or, equivalently, for relatively long waves. The effects of dispersion do not appear in this approximation. In the next section the shallow water equations will be derived as the first terms in expansions in $(h_0/l)^2$; small dispersive effects will be included by going to the next order.

The linearized equations lead to (13.74), but we may obtain nonlinear solutions using the theory of Part I, since the system is hyperbolic. In particular, for one dimensional waves on a horizontal bottom we may take

$$h_t + (uh)_x = 0,$$

$$u_t + uu_x + gh_x = 0. \tag{13.79}$$

The characteristic velocities are $u \pm \sqrt{gh}$, the Riemann invariants are $u \pm 2\sqrt{gh}$, and a simple wave moving to the right into water with $h = h_0$ is given by

$$h = H(\xi), \qquad u = 2\sqrt{gH} - 2\sqrt{gh_0},$$

$$x = \xi + \left\{ 3\sqrt{gH(\xi)} - 2\sqrt{gh_0} \right\} t. \tag{13.80}$$

All such waves carrying an increase of elevation break. Then a discontinuity may be fitted in, and the shock conditions, derived from the conservation forms of (13.79), are

$$-U[uh] + \left[u^2 h + \frac{1}{2} gh^2 \right] = 0,$$

$$-U[h] + [uh] = 0. \tag{13.81}$$

as noted in (3.53) and (3.54). This is the turbulent *bore*.

This breaking phenomenon is one of the most intriguing long-standing problems of water wave theory. First, when gradients are no longer small the approximation h_0^2/l^2 is no longer valid, so the solution (13.80) should cease to apply long before breaking occurs. Yet breaking certainly does occur and in some circumstances does not seem to be too far away from the description given by (13.80); moreover, bores, breakers, and hydraulic jumps are sometimes reasonably well described by (13.81). But the shallow water theory goes too far: it predicts that *all* waves carrying an increase of elevation break. Observations have long since established that some waves do not break. So an invalid theory seems to be right sometimes and wrong sometimes! It is not hard to see how the neglected dispersion effects inhibit breaking, but in the simple theories including some of these effects (see the next section), they in turn go too far and show that no waves break! We postpone further discussion until the dispersion has been included. Before doing this, however, a few more details of shallow water theory will be noted.

Dam Break Problem.

First, a classic solution for the breaking of a dam does not involve bores (strangely enough) and is easily solved by simple wave theory. The problem is formulated with initial conditions

$$
\left.
\begin{array}{ll}
u = 0, & -\infty < x < \infty, \\
h = 0, & 0 < x < \infty, \\
h = H_0 > 0, & -\infty < x < 0,
\end{array}
\right\} \quad \text{at } t = 0.
$$

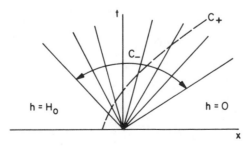

Fig. 13.3. Characteristic pattern in the dam-break problem.

Then, on any positive characteristic C_+ from the region $h = H_0$ (see Fig. 13.3), the Riemann invariant

$$u + 2\sqrt{gh} = 2\sqrt{gH_0} \,. \tag{13.82}$$

In the region covered by these characteristics, the solution is a simple wave on straight C_- characteristic through the origin. They are given by

$$\frac{x}{t} = u - \sqrt{gh} \,. \tag{13.83}$$

But (13.82) and (13.83) provide the whole solution:

$$\left.\begin{aligned}
\sqrt{gh} &= \frac{1}{3}\left(2\sqrt{gH_0} - \frac{x}{t}\right), \\[2mm]
u &= \frac{2}{3}\left(\sqrt{gH_0} + \frac{x}{t}\right),
\end{aligned}\right\} \quad -\sqrt{gH_0} \leqslant \frac{x}{t} \leqslant 2\sqrt{gH_0} \,. \tag{13.84}$$

The free surface is parabolic between the front $h = 0$, $x = 2t\sqrt{gH_0}$, traveling with speed $2\sqrt{gH_0}$, and the undisturbed dam level $h = H_0$ at $x = -t\sqrt{gH_0}$.

Again the shallow water theory is not strictly valid in the initial instants since the horizontal length scale l is small, but as the flow develops H_0^2/l^2 becomes small and the real flow is reasonably well described. It should be noted that $h = 4H_0/9$, $u = 2\sqrt{gH_0}/3$, remain constant at the dam position $x = 0$ for all $t > 0$. The speed of the front is modified considerably by friction; for attempts to estimate this see Dressler (1952) and Whitham (1955).

Bore Conditions.

The bore conditions (13.81) conserve mass and momentum across the bore and one naturally asks what happens to the energy. The conservation of energy for equations (13.79) is

$$\left(\frac{1}{2}u^2h + \frac{1}{2}gh^2\right)_t + \left(\frac{1}{2}u^3h + ugh^2\right)_x = 0. \tag{13.85}$$

This would provide a third potential shock condition, but only two conditions can be used with the system (13.79). Rayleigh proposed that in fact energy is not conserved across a bore, attributing the loss to the observed turbulence, so that the shock condition corresponding to (13.85) should not be used. It is easy to show that whereas (13.79) imply (13.85), the jump conditions (13.81) imply

$$\left[\frac{1}{2}u^3h + ugh^2\right]_1^2 - U\left[\frac{1}{2}u^2h + \frac{1}{2}gh^2\right]_1^2 < 0 \qquad \text{for} \quad h_2 > h_1. \quad (13.86)$$

Since bores are required only when $h_2 > h_1$, the *loss* of energy agrees in sign. The energy plays a role similar to the entropy in gas dynamics; in gas dynamics any internal energy is included in the detailed description, so energy *is* conserved and the extra variable in the description allows an extra shock condition. Turbulent energy is not included in (13.85).

A convenient form of the bore conditions (13.81) is

$$U = u_1 + \left\{\frac{gh_2(h_1 + h_2)}{2h_1}\right\}^{1/2}, \qquad (13.87)$$

$$u_2 = u_1 + \frac{h_2 - h_1}{h_2}\left\{\frac{gh_2(h_1 + h_2)}{2h_1}\right\}^{1/2}. \qquad (13.88)$$

Further Conservation Equations.

It is interesting that the shallow water equations (13.79) admit an infinite number of conservation equations of the general form

$$\frac{\partial}{\partial t}P(u,h) + \frac{\partial}{\partial x}Q(u,h) = 0.$$

It is only necessary that

$$Q_u = uP_u + hP_h, \qquad Q_h = gP_u + uP_h.$$

Thus any solution of

$$gP_{uu} = hP_{hh}$$

will lead to a conservation equation. The most interesting are polynomial in u and h. They may be obtained consistently by taking

$$P = \sum_{m=0}^{n} P_m(u) h^m,$$

from which it follows that

$$g p''_{m-1} = m(m-1) p_m, \qquad m = 2, \cdots, n,$$

$$p''_0 = 0, \qquad p''_n = 0.$$

The first few are:

P	Q
u	$\frac{1}{2} u^2 + gh$
h	uh
uh	$u^2 h + \frac{1}{2} gh^2$
$\frac{1}{2} u^2 h + \frac{1}{2} gh^2$	$\frac{1}{2} u^3 h + ugh^2$
$\frac{1}{3} u^3 h + ugh^2$	$\frac{1}{3} u^4 h + \frac{3}{2} u^2 gh^2 + \frac{1}{3} g^2 h^3$

The second, third, and fourth correspond to mass, momentum, and energy, respectively. The others have no obvious interpretation. However, since each can be used to give a constant integral

$$\int_{-\infty}^{\infty} \{ P(u,h) - P(0,h_0) \} \, dx = \text{constant}$$

in any problem in which $u \to 0$, $h \to h_0$ at $\pm \infty$, an infinite number of integrals of the solution are known. One might therefore expect to be able to find the solution analytically. Indeed, by means of the hodograph transformation (13.79) can be converted into a linear equation and solved in principle; the analysis is identical to that of Section 6.12 with $\gamma = 2$.

13.11 The Korteweg-deVries and Boussinesq Equations

We next consider how dispersive effects may be incorporated into the shallow water theory. This may be done by setting up a more formal expansion in the small parameter $(h_0/l)^2$ and including the next order terms beyond shallow water theory. Before doing this, however, it is advantageous for flexibility and insight to use a quicker and more intuitive procedure. We consider the case of one dimensional waves with h_0 constant. The linearized version of the equations we seek must give the dispersion relation (13.25) in the next approximation beyond (13.73):

$$\omega^2 = c_0^2 \kappa^2 - \frac{1}{3} c_0^2 h_0^2 \kappa^4. \tag{13.89}$$

An equation for η with this dispersion relation is

$$\eta_{tt} - c_0^2 \eta_{xx} - \frac{1}{3} c_0^2 h_0^2 \eta_{xxxx} = 0. \tag{13.90}$$

The shallow water equations (13.79) linearize to (13.74). If we can add an extra linear term to (13.79) so that they linearize to (13.90), we should have a system which includes both the nonlinear effects of relative order a/h_0 (where a is a typical amplitude) and the dispersive effects of relative order h_0^2/l^2. This is easily done. There are various forms equivalent in the desired approximation. If we choose to add a term νh_{xxx} in the second of (13.79), the linearized equations are

$$\eta_t + h_0 u_x = 0,$$

$$u_t + g\eta_x + \nu \eta_{xxx} = 0,$$

and u may be eliminated to give

$$\eta_{tt} - c_0^2 \eta_{xx} - \nu h_0 \eta_{xxxx} = 0.$$

Therefore we choose $\nu = \frac{1}{3} c_0^2 h_0$ to agree with (13.90). The argument then is that the system

$$h_t + (uh)_x = 0,$$

$$u_t + uu_x + gh_x + \frac{1}{3} c_0^2 h_0 h_{xxx} = 0, \tag{13.91}$$

reduces correctly to (13.90) in the limit $a/h_0 \to 0$, correctly to (13.79) in the limit $h_0^2/l^2 \to 0$, and so combines the first order corrections to (13.74) in both a/h_0 and in h_0^2/l^2.

One can always substitute the lowest approximation (13.74) into the correction term. Hence an equivalent system to the order considered is

$$h_t + (uh)_x = 0,$$

$$u_t + uu_x + gh_x + \frac{1}{3}h_0 h_{xtt} = 0. \tag{13.92}$$

This is the system favored by Boussinesq, who first formulated these equations. The linearized version of (13.92) leads to

$$\eta_{tt} - c_0^2 \eta_{xx} - \frac{1}{3} h_0^2 \eta_{xxtt} = 0,$$

and the dispersion relation is

$$\omega^2 = \frac{c_0^2 \kappa^2}{1 + (1/3)\kappa^2 h_0^2}. \tag{13.93}$$

The expansion of this dispersion relation for small $(\kappa h_0)^2$ agrees with (13.89) to the first two terms; hence the two systems are equivalent in this approximation. However, if the equations are in fact used when h_0^2/l^2 is not small, (13.92) is superior to (13.91). According to (13.89) small perturbations with $(\kappa h_0)^2 > 3$ will in fact amplify because ω becomes imaginary, whereas (13.93) retains a real ω even though it is inaccurate in the range concerned. In numerical work, the various effects of finite differencing and truncation introduce small oscillations of small wavelength even if the formulated analytic problem satisfies $h_0^2/l^2 \ll 1$, so (13.92) is preferred.

Boussinesq's equations include waves moving to both left and right. By following the same steps restricted to waves moving to the right only, we obtained the Korteweg-deVries equation. For waves moving to the right the first two terms in the dispersion relation are

$$\omega = c_0 \kappa - \gamma \kappa^3, \qquad \gamma = \frac{1}{6} c_0 h_0^2, \tag{13.94}$$

corresponding to the equation

$$\eta_t + c_0 \eta_x + \gamma \eta_{xxx} = 0. \tag{13.95}$$

In the nonlinear shallow water equations (13.79), waves moving to the right into undisturbed water of depth h_0 satisfy the Riemann invariant

$$u = 2\sqrt{g(h_0 + \eta)} - 2\sqrt{gh_0}, \tag{13.96}$$

and after substitution in either of the equations in (13.79) we have

$$\eta_t + \left\{ 3\sqrt{g(h_0 + \eta)} \ - 2\sqrt{gh_0} \ \right\} \eta_x = 0. \tag{13.97}$$

Combining (13.95) and (13.97), we have

$$\eta_t + \left\{ 3\sqrt{g(h_0 + \eta)} \ - 2\sqrt{gh_0} \ \right\} \eta_x + \gamma\eta_{xxx} = 0. \tag{13.98}$$

If the nonlinear terms are approximated to the first order in a/h_0, we take

$$\eta_t + c_0\left(1 + \frac{3}{2}\frac{\eta}{h_0}\right)\eta_x + \gamma\eta_{xxx} = 0. \tag{13.99}$$

This is the Korteweg-deVries equation. There is no reason to believe that retaining (13.98) is preferable, since other terms, proportional to the product of a/h_0 and h_0^2/l^2, for example, may be as important as the nonlinear terms in a^2/h_0^2. Again one may use $\eta_t \simeq - c_0\eta_x$ in the dispersive correction term and adopt

$$\eta_t + c_0\left(1 + \frac{3}{2}\frac{\eta}{h_0}\right)\eta_x - \frac{\gamma}{c_0}\eta_{xxt} = 0.$$

The linearized equation then has the dispersion relation

$$\omega = \frac{c_0\kappa}{1 + \gamma\kappa^2/c_0}.$$

This agrees with (13.94) for small κ, and, in contrast to (13.94), has bounded phase and group velocities if κ becomes large. Since ω remains real in each case, the modification is less compelling than in the Boussinesq case but nevertheless is probably desirable for some purposes (Benjamin et al., 1972). However, many fascinating exact analytic solutions have been found for (13.99), and in general it may be transformed into a linear integral equation; at present these features dominate the other issues.

The preceding derivations allow great flexibility and the approach naturally allows the various alternatives that were discussed. It also becomes clear that the equations apply to many dispersive wave problems quite apart from water waves. Any dispersion relation with an odd function for $\omega(\kappa)$ may be expanded to the two terms in (13.89) or in (13.94), and then (13.90) and (13.95) cover the linearized theory. It is then only necessary to have some access to the form of the nonlinear terms and those in (13.91) or (13.99) are rather typical. In this way the equations have arisen in plasma physics, for example.

Of course a confirming formal expansion is also desirable and inform-ative. If, temporarily, Y is measured from the horizontal bottom, we have to solve Laplace's equation

$$\varphi_{xx} + \varphi_{YY} = 0$$

with $\varphi_Y = 0$ on $Y = 0$. The shallow water theory, with φ_x approximately independent of Y, and the small total depth both suggest an expansion

$$\varphi = \sum_0^\infty Y^n f_n(x, t).$$

When this is substituted in Laplace's equation and the boundary condition on $Y = 0$ is used, we have

$$\varphi = \sum_0^\infty (-1)^m \frac{Y^{2m}}{(2m)!} \frac{\partial^{2m} f}{\partial x^{2m}}, \tag{13.100}$$

where $f = f_0$. The final step is to substitute this expansion into the boundary conditions on the free surface. Because they are nonlinear and applied on $Y = h_0 + \eta$, it is here that the analysis becomes a little involved and terms have to be ordered in expansions with respect to the two parameters $\alpha = a/h_0$, $\beta = h_0^2/l^2$. In carrying through the details, it is best to normalize the variables from the start by taking the original variables (primed) to be

$$x' = lx, \qquad Y' = h_0 Y, \qquad t' = \frac{lt}{c_0},$$

$$\eta' = a\eta, \qquad \varphi' = \frac{gla\varphi}{c_0}.$$

The different stretchings in Y and x introduce the crucial step. In the normalized variables, the problem is formulated as

$$\beta \varphi_{xx} + \varphi_{YY} = 0, \qquad 0 < Y < 1 + \alpha\eta,$$

$$\varphi_Y = 0, \qquad Y = 0,$$

$$\left. \begin{array}{l} \eta_t + \alpha\varphi_x\eta_x - \dfrac{1}{\beta}\varphi_Y = 0, \\[4mm] \eta + \varphi_t + \dfrac{1}{2}\alpha\varphi_x^2 + \dfrac{1}{2}\dfrac{\alpha}{\beta}\varphi_Y^2 = 0, \end{array} \right\} \quad Y = 1 + \alpha\eta \ \cdot$$

The expansion for φ now appears as an expansion in powers of β, but from Laplace's equation and $\varphi_Y = 0$ on $Y = 0$, we again have

$$\varphi = \sum_{0}^{\infty} (-1)^m \frac{Y^{2m}}{(2m)!} \frac{\partial^{2m} f}{\partial x^{2m}} \beta^m.$$

On substitution in the surface conditions one finds

$$\eta_t + \{(1 + \alpha\eta)f_x\}_x - \left\{\frac{1}{6}(1 + \alpha\eta)^3 f_{xxxx} + \frac{1}{2}\alpha(1 + \alpha\eta)^2 \eta_x f_{xxx}\right\}\beta + O(\beta^2) = 0,$$

$$\eta + f_t + \frac{1}{2}\alpha f_x^2 - \frac{1}{2}(1 + \alpha\eta)^2 \{f_{xxt} + \alpha f_x f_{xxx} - \alpha f_{xx}^2\}\beta + O(\beta^2) = 0.$$

If all terms in β are dropped and the second equation differentiated with respect to x, we have the nonlinear shallow water equations:

$$\eta_t + \{(1 + \alpha\eta)w\}_x = 0,$$

$$w_t + \alpha w w_x + \eta_x = 0, \qquad w = f_x.$$

If the terms in the first power of β are retained, but to simplify them terms of $O(\alpha\beta)$ are dropped, we have

$$\eta_t + \{(1 + \alpha\eta)w\}_x - \frac{1}{6}\beta w_{xxx} + O(\alpha\beta, \beta^2) = 0,$$

$$w_t + \alpha w w_x + \eta_x - \frac{1}{2}\beta w_{xxt} + O(\alpha\beta, \beta^2) = 0, \qquad w = f_x.$$

$$(13.101)$$

These are just a variant of Boussinesq's equations. The quantity w is only the first term in the expansion of the velocity φ_x, which is

$$\varphi_x = w - \beta \frac{Y^2}{2} w_{xx} + O(\beta^2).$$

The value averaged over the depth is

$$\tilde{u} = w - \frac{1}{6}\beta w_{xx} + O(\alpha\beta, \beta^2);$$

the inverse is

$$w = \tilde{u} + \frac{1}{6}\beta \tilde{u}_{xx} + O(\alpha\beta, \beta^2).$$

If these are used in the equations, we have

$$\eta_t + \left\{ (1 + \alpha \eta) \tilde{u} \right\}_x + O(\alpha\beta, \beta^2) = 0,$$

$$\tilde{u}_t + \alpha \tilde{u} \tilde{u}_x + \eta_x - \frac{1}{3} \beta \tilde{u}_{xxt} + O(\alpha\beta, \beta^2) = 0.$$

Finally, substituting the lowest order $\tilde{u}_x = -\eta_t + O(\alpha, \beta)$ from the first equation into the $\tilde{u}_{xxt}$ term, we have Boussinesq's equations (13.92) in normalized form.

The Korteweg-deVries equation is derived from any of these systems by specializing to a wave moving to the right. To lowest order, neglecting the terms of order α and β, such a solution of (13.101) has

$$w = \eta, \qquad \eta_t + \eta_x = 0.$$

We look for a solution, corrected to first order in α and β, in the form

$$w = \eta + \alpha A + \beta B + O(\alpha^2 + \beta^2),$$

where A and B are functions of η and its x derivatives. Then equations (13.101) become

$$\eta_t + \eta_x + \alpha(A_x + 2\eta\eta_x) + \beta\left(B_x - \frac{1}{6}\eta_{xxx} \right) + O(\alpha^2 + \beta^2) = 0,$$

$$\eta_t + \eta_x + \alpha(A_t + \eta\eta_x) + \beta\left(B_t - \frac{1}{2}\eta_{xxt} \right) + O(\alpha^2 + \beta^2) = 0.$$

Since $\eta_t = -\eta_x + O(\alpha, \beta)$, all t derivatives in the first order terms may be replaced by minus the x derivatives. Then the two equations are consistent if

$$A = -\frac{1}{4}\eta^2, \qquad B = \frac{1}{3}\eta_{xx}.$$

Hence we have

$$w = \eta - \frac{1}{4}\alpha\eta^2 + \frac{1}{3}\beta\eta_{xx} + O(\alpha^2 + \beta^2),$$

$$\eta_t + \eta_x + \frac{3}{2}\alpha\eta\eta_x + \frac{1}{6}\beta\eta_{xxx} + O(\alpha^2 + \beta^2) = 0.$$

$$(13.102)$$

The second equation is the normalized form of the Korteweg-deVries equation (13.99). The first is similar to a Riemann invariant.

13.12 Solitary and Cnoidal Waves

The Korteweg-deVries equation was derived in 1895. Earlier, Stokes (1847) had found approximate expressions for nonlinear periodic waves in the case of infinitely deep water or water of moderate depth, while Boussinesq (1871) and Rayleigh (1876) had found approximate expressions for the solitary wave, a wave consisting of a single hump of constant shape and constant speed, which was first observed experimentally by Scott Russell (1844). The solitary wave is most easily obtained as a special solution of the equation obtained by Korteweg and deVries, and these authors went on to show that periodic solutions were also possible. Although Stokes' approximation does not apply when β becomes as small as α, his solutions overlap with those of Korteweg-deVries when $\alpha \ll \beta$. We consider solutions of the Korteweg-deVries equations first, because of their simplicity, even though they came much later.

Both the solitary waves and the periodic waves described by (13.99) are found as solutions of constant shape moving with constant velocity. They therefore may be described in the form

$$\eta = h_0 \zeta(X), \qquad X = x - Ut.$$

From (13.99) we then have

$$\frac{1}{6} h_0^2 \zeta''' + \frac{3}{2} \zeta\zeta' - \left(\frac{U}{c_0} - 1 \right) \zeta' = 0.$$

This integrates immediately to

$$\frac{1}{6} h_0^2 \zeta'' + \frac{3}{4} \zeta^2 - \left(\frac{U}{c_0} - 1 \right) \zeta + G = 0,$$

with a further integration to

$$\frac{1}{3} h_0^2 \zeta'^2 + \zeta^3 - 2\left(\frac{U}{c_0} - 1 \right) \zeta^2 + 4G\zeta + H = 0, \qquad (13.103)$$

after multiplication by ζ'; G and H are constants of integration.

In the special case when ζ and its derivatives tend to zero at ∞, $G = H = 0$. Then the equation may be written

$$\frac{1}{3} h_0^2 \left(\frac{d\zeta}{dX} \right)^2 = \zeta^2 (\alpha - \zeta),$$

$$\frac{U}{c_0} = 1 + \frac{\alpha}{2}. \qquad (13.104)$$

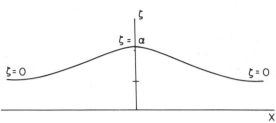

Fig. 13.4. Solitary wave.

It is clear qualitatively that ζ increases from $\zeta = 0$ at $X = \infty$, rises to a maximum $\zeta = \alpha$, and then returns symmetrically to $\zeta = 0$ at $X = -\infty$ (see Fig. 13.4). This is the solitary wave. The range in η is $\eta_0 = h_0 \alpha$, so α plays the same role as in the last section. The velocity of the solitary wave depends on the amplitude according to

$$U = c_0 \left(1 + \frac{1}{2} \frac{\eta_0}{h_0} \right). \tag{13.105}$$

The actual solution of (13.104) is

$$\zeta = \alpha \operatorname{sech}^2 \left(\frac{3\alpha}{4h_0^2} \right)^{1/2} X; \tag{13.106}$$

hence

$$\eta = \eta_0 \operatorname{sech}^2 \left\{ \left(\frac{3\eta_0}{4h_0^3} \right)^{1/2} (x - Ut) \right\}. \tag{13.107}$$

This is a solution of the Korteweg-deVries equation for all η_0/h_0; however, the equation is derived under the assumption $\eta_0/h_0 \ll 1$, and in fact solitary waves are found to peak at a maximum height; experimentally it is $\eta_0/h_0 \approx 0.7$, and theoretically $\eta_0/h_0 \approx 0.78$.

In the general case, $G, H \neq 0$,

$$\zeta'^2 = \mathcal{C}(\zeta)$$

where $\mathcal{C}(\zeta)$ is a cubic with simple zeros. For bounded solutions, the zeros must be real, and bounded solutions must oscillate periodically between two of the zeros of $\mathcal{C}$. We may, without loss of generality, choose the two zeros concerned to be $\zeta = 0$ (which fixes h_0) and $\zeta = \alpha$ (which is double the amplitude). The third zero must then be negative; if we take it to be $\alpha - \beta$, it will turn out that $\beta = h_0^2 / l^2$, where l is the horizontal length scale, and the parameters α and β have the roles appropriate to the discussion of the last section. With these choices, the equation for $\zeta(X)$ is

$$\frac{1}{3} h_0^2 \left(\frac{d\zeta}{dX} \right)^2 = \zeta(\alpha - \zeta)(\zeta - \alpha + \beta), \qquad 0 < \alpha < \beta, \quad (13.108)$$

$$\frac{U}{c_0} = 1 + \frac{2\alpha - \beta}{2}. \qquad (13.109)$$

The wave length is

$$\lambda = \frac{2h_0}{\sqrt{3}} \int_0^\alpha \frac{d\zeta}{\sqrt{\zeta(\alpha - \zeta)(\zeta - \alpha + \beta)}}. \qquad (13.110)$$

In this nonlinear problem U is the phase speed, since any point on the profile moves with this speed. The solution could have been written in the form

$$\zeta(X) = f(\theta) = f(\kappa x - \omega t)$$

and chosen so that f has period 2π in θ. Then

$$\omega = U\kappa = \left(1 + \frac{2\alpha - \beta}{2} \right) c_0 \kappa, \qquad (13.111)$$

and

$$\kappa = \frac{2\pi}{\lambda}.$$

From (13.110), β is a function of λ and α. Hence the dispersion relation (13.111) takes the form

$$\omega = \omega(\kappa, \alpha). \qquad (13.112)$$

This is the first instance of a most important result for nonlinear dispersive waves: *The dispersion relation between frequency ω and wave number κ involves the amplitude.*

For waves of infinitesimal amplitude, $\alpha \to 0$, (13.108) and (13.109) reduce to

$$\frac{1}{3} h_0^2 \left(\frac{d\zeta}{dX} \right)^2 \simeq \zeta(\alpha - \zeta)\beta,$$

$$\frac{U}{c_0} \simeq 1 - \frac{\beta}{2}.$$

The solution is

$$\zeta = \frac{\alpha}{2} + \frac{\alpha}{2} \cos \sqrt{3\beta} \, \frac{X}{h_0},$$

$$\frac{U}{c_0} = 1 - \frac{\beta}{2}.$$

If we introduce $\kappa = \sqrt{3\beta} \, / h_0$, we have

$$\zeta = \frac{\alpha}{2} + \frac{\alpha}{2} \cos(\kappa x - \omega t),$$

$$\omega = \kappa U = c_0 \kappa \left(1 - \frac{1}{6} \kappa^2 h_0^2 \right), \tag{13.113}$$

in agreement with the linear theory [see (13.94) and (13.95)]. *In this limit, the amplitude drops out of the dispersion relation.*

In the other limit of $\alpha \to \beta$, the wavelength λ, given by (13.110), tends to infinity and we have the solitary wave.

The solution (13.108) may be expressed in standard form in Jacobian elliptic functions. It is

$$\zeta = \alpha \, \mathrm{cn}^2 \left(\frac{3\beta}{4h_0^2} \right)^{1/2} X, \tag{13.114}$$

where the modulus s of the elliptic function is

$$s = \left(\frac{\alpha}{\beta} \right)^{1/2}, \tag{13.115}$$

and

$$\lambda = \frac{4h_0}{\sqrt{3\beta}} K(s), \tag{13.116}$$

where $K(s)$ is the complete integral of the first kind. In view of (13.114), Korteweg and deVries named these solutions *cnoidal waves*. The form of (13.114) confirms β as a measure of h_0^2 / l^2. The modulus s measures the relative importance of nonlinearity to dispersion. In the linear limit $s \to 0$, $\mathrm{cn}\,\xi \to \cos\xi$; in the solitary wave limit $s \to 1$, $\mathrm{cn}\,\xi \to \mathrm{sech}\,\xi$.

Again it should be noted that cnoidal waves are a solution of the Korteweg-deVries equation for all α and β restricted only by $0 \leqslant \alpha \leqslant \beta$, but the equations are only valid when α and β are small. Like the solitary waves, they really attain a maximum height at which crests peak. The theoretical analysis is not complete. As a guide to supplement the result for solitary waves, it may be noted that numerical calculations of periodic waves on deep water (Michell, 1893) peak when $a/\lambda = 0.142$. Strictly speaking this is outside the range of the Korteweg-deVries theory, but it perhaps gives some idea of the values involved. We should interpret it according to $a \approx \frac{1}{2}\alpha h_0$, $\lambda \approx (2\pi h_0)/(3\beta)^{1/2}$ (the linear values), as $\alpha\beta^{1/2} \approx 1$. The critical value for solitary waves $\alpha \approx \beta \approx 0.7$–$0.8$ fits in with this rough estimate. It was pointed out at the end of Section 13.6 that solutions of the linearized theory $\alpha/\beta \ll 1$ cease to be valid near the front of a wavetrain because the effective α/β increases with t as $t \to \infty$. Since the above periodic solutions tend to solitary waves as α/β increases to 1, one might expect that the end result is a series of solitary waves. This and other problems can now be studied analytically as a consequence of the remarkable investigations of Kruskal, Greene, Gardner, Miura, and co-workers. They developed an ingenious way of finding fairly general solutions of the Korteweg-deVries equation, and these include the main features of the resolution of an arbitrary finite initial distribution into a series of solitary waves. The explicit solution for the collision of two or more solitary waves can also be given. This work ties in with related results for other equations and other physical situations, so it is described separately (Chapter 17).

We now continue the discussion of periodic waves.

13.13 Stokes Waves

Stokes' investigations in water waves, with the first publication in 1847, are the starting point for the nonlinear theory of dispersive waves. It was in this work, and far ahead of other developments, that he found the crucial results that, first, periodic wavetrains are possible in nonlinear systems and, second, that the dispersion relation involves the amplitude. The dependence on amplitude produces important qualitative changes in the behavior and introduces new phenomena, not merely corrections to some of the numbers.

It is easiest to present Stokes' approach on the Korteweg-deVries equation and then quote the results without actually doing the more extensive details required for the full equations for arbitrary depth. Stokes' aim was to find the next approximation to the linear periodic wavetrain. For the Korteweg-deVries equation this corresponds to an expansion in powers of α with $\alpha \ll \beta$. It could be obtained from the exact solution (13.114), but it is both simpler and more instructive to proceed directly on the equation. We look for a solution of (13.99) in the expansion

$$\frac{\eta}{h_0} = \zeta = \epsilon\zeta_1(\theta) + \epsilon^2\zeta_2(\theta) + \epsilon^3\zeta_3(\theta) + \cdots, \qquad \theta = \kappa x - \omega t, \quad (13.117)$$

where ϵ is the small parameter a/h_0 (proportional to α). We then obtain the hierarchy

$$(\omega - c_0\kappa)\zeta_1' - \gamma\kappa^3\zeta_1''' = 0,$$

$$(\omega - c_0\kappa)\zeta_2' - \gamma\kappa^3\zeta_2''' = \frac{3}{2}c_0\kappa\zeta_1\zeta_1',$$

$$(\omega - c_0\kappa)\zeta_3' - \gamma\kappa^3\zeta_3''' = \frac{3}{2}c_0\kappa(\zeta_1\zeta_2)',$$

and so on. A solution of the first is

$$\zeta_1 = \cos\theta, \qquad \omega = \omega_0(\kappa),$$

where $\omega_0(\kappa)$ is the linear dispersion function

$$\omega_0(\kappa) = c_0\kappa - \gamma\kappa^3.$$

If these values are adopted, the right hand side of the second equation for ζ_2 is proportional to $\sin 2\theta$ and a solution $\zeta_2 \propto \cos 2\theta$ can be found. Then in the third equation, the right hand side can be expressed as a linear combination of $\sin\theta$ and $\sin 3\theta$. The term in $\sin 3\theta$ can be accommodated by a solution $\zeta_3 \propto \cos 3\theta$, but the $\sin\theta$ term resonates with the operator on the left, since on substituting $\zeta_3 \propto \cos\theta$ the left hand side is zero. There is a solution with $\zeta_3 \propto \theta\sin\theta$ but this "secular term" is unbounded in θ. Stokes argued that the periodic solution can be found if ω is also expanded in a power series

$$\omega = \omega_0(\kappa) + \epsilon\omega_1(\kappa) + \epsilon^2\omega_2(\kappa) + \cdots.$$

Since trouble arises only at the third order, we may take $\omega_1 = 0$ in advance. The hierarchy now reads

$$(\omega_0 - c_0\kappa)\zeta_1' - \gamma\kappa^3\zeta_1''' = 0,$$

$$(\omega_0 - c_0\kappa)\zeta_2' - \gamma\kappa^3\zeta_2''' = \frac{3}{2}c_0\kappa\zeta_1\zeta_1',$$

$$(\omega_0 - c_0\kappa)\zeta_3' - \gamma\kappa^3\zeta_3''' = \frac{3}{2}c_0\kappa(\zeta_1\zeta_2)' - \omega_2\zeta_1'.$$

If we again take

$$\zeta_1 = \cos\theta, \qquad \omega_0 = c_0\kappa - \gamma\kappa^3,$$

we find

$$\zeta_2 = \frac{c_0}{8\gamma\kappa^2}\cos 2\theta.$$

The right hand side of the equation for ζ_3 may be written

$$-\frac{3c_0^2}{32\gamma\kappa}(\sin\theta + 3\sin 3\theta) + \omega_2\sin\theta.$$

We now choose

$$\omega_2 = \frac{3c_0^2}{32\gamma\kappa},$$

and the resonant term is eliminated. The final result is

$$\frac{\eta}{h_0} = \epsilon\cos\theta + \frac{3\epsilon^2}{4\kappa^2 h_0^2}\cos 2\theta + \frac{27\epsilon^3}{64\kappa^4 h_0^4}\cos 3\theta + \cdots, \qquad (13.118)$$

$$\frac{\omega}{c_0\kappa} = 1 - \frac{1}{6}\kappa^2 h_0^2 + \frac{9\epsilon^2}{16\kappa^2 h_0^2} + \cdots. \qquad (13.119)$$

The second equation shows the crucial dependence of the dispersion relation on the amplitude.

It should be noted that the expansion parameter is really $\epsilon/\kappa^2 h_0^2$, and this corresponds to α/β. In comparing the results with the expansion of (13.114) for small $s^2 = \alpha/\beta$, the different choices of the origin for η, and hence different choices of h_0, must be included. The use of different origins is unavoidable; for the solitary wave limit it is convenient to have the origin at the troughs, whereas for the linear limit it is convenient to have it at the mean level.

It should also be noted that (13.118) could be motivated as the Fourier series of the periodic wavetrain and this general form assumed from the start.

Arbitrary Depth.

In the general case the aim is to find small amplitude expansions to periodic solutions of

$$\varphi_{xx} + \varphi_{yy} = 0, \qquad -h_0 < y < \eta,$$
$$\varphi_y = 0, \qquad y = -h_0,$$

$$\left.\begin{array}{l} \eta_t + \varphi_x \eta_x - \varphi_y = 0, \\[2mm] \varphi_t + \dfrac{1}{2}\varphi_x^2 + \dfrac{1}{2}\varphi_y^2 + g\eta = 0, \end{array}\right\} \qquad y = \eta.$$

We have $\eta = \eta(\theta)$, $\varphi = \varphi(\theta, y)$, $\theta = \kappa x - \omega t$, where η and φ are periodic in θ. We may choose the origin $y = 0$ so that η has zero mean value and the expansion for η is

$$\eta = a\cos\theta + \mu_2 a^2 \cos 2\theta + \cdots, \tag{13.120}$$

where μ_2 is a coefficient to be found. With the choice of mean value $\bar{\eta} = 0$, it is clear from the second condition on $y = \eta$ that the mean value $\bar{\varphi}_t$ cannot be zero and φ must at least have a term t in its expansion. One may also interpret this as a consequence of absorbing a constant of integration when the expression for $(p - p_0)/\rho$ was first derived. Alternatively, one could keep a nonzero mean value in η. Since only derivatives of φ occur in physical quantities, terms proportional to t or to x are acceptable in φ. A term proportional to x represents a nonzero mean in the horizontal velocity. Here it can be normalized to zero. Later in discussing modulated wavetrains we shall need both $\bar{\eta} \neq 0$ and $\bar{\varphi}_x \neq 0$, since they both vary in the modulations and cannot be normalized out. The extensions of the formula are straightforward, so we take $\bar{\eta} = 0$, $\bar{\varphi}_x = 0$ here. The expansion for φ may then be written

$$\varphi = \nu_0 a^2 t + \nu_1 a \cosh\kappa(y + h_0)\sin\theta + \nu_2 a^2 \cosh 2\kappa(y + h_0)\sin 2\theta + \cdots .$$
$$\tag{13.121}$$

Again to avoid secular terms at the third order, ω must also be expanded as

$$\omega = \omega_0(\kappa) + a^2 \omega_2(\kappa) + \cdots. \tag{13.122}$$

When all this is carried through it is found that

$$\frac{\omega^2}{g\kappa \tanh\kappa h_0} = 1 + \left(\frac{9\tanh^4\kappa h_0 - 10\tanh^2\kappa h_0 + 9}{8\tanh^4\kappa h_0}\right)\kappa^2 a^2 + \cdots, \tag{13.123}$$

and

$$\mu_2 = \frac{1}{2} \kappa \coth \kappa h_0 \left(1 + \frac{3}{2 \sinh^2 \kappa h_0} \right),$$

$$\nu_0 = -\frac{g\kappa}{2 \sinh 2\kappa h_0}, \qquad \nu_1 = \frac{\omega_0}{\kappa \sinh \kappa h_0}, \qquad \nu_2 = \frac{3}{8} \frac{\omega_0}{\sinh^4 \kappa h_0}.$$

The dispersion relation (13.123) is the main result. For $\kappa^2 h_0^2 \ll 1$ we recover (13.119), and for $\kappa^2 h_0^2 \gg 1$ we have Stokes' original result for deep water:

$$\omega^2 = g\kappa (1 + \kappa^2 a^2 + \cdots). \tag{13.124}$$

The details of the algebra in these derivations become extensive in the arbitrary depth case. It can be offset to some extent by substituting the Fourier series expansions in the variational principle (13.16)–(13.17) and obtaining the coefficients from the variational equations (see Whitham, 1967).

The Stokes expressions are limited to small amplitude and cannot exhibit the wavetrain of greatest height at which the crests are observed to become sharp. However, Stokes contributed a separate argument to show that if sharp crests are attained in a *steady profile* wave, the angle there must be 120°. The argument depends crucially on the restriction to a wave of constant shape propagating at a constant speed. In these circumstances the flow is steady in a frame moving with the crest. The solutions of Laplace's equation are then analytic functions of $z = x + iy$ and a reasonably general singular function (taking the origin at the crest) is

$$\varphi + i\psi \propto z^m.$$

In local polar coordinates $(r, \tilde{\omega})$, with $\tilde{\omega}$ measured from the downward vertical

$$x = r \sin \tilde{\omega}, \qquad y = -r \cos \tilde{\omega}, \qquad z = -ire^{i\tilde{\omega}},$$

and we have

$$\varphi = Cr^m \sin m\tilde{\omega}.$$

If $\eta = -r \cos \tilde{\omega}_0$ locally on the free surface, the pressure condition in steady flow requires

$$\varphi_r^2 + \frac{1}{r^2} \varphi_{\tilde{\omega}}^2 + g\eta = \text{constant}.$$

With the above expressions for φ and η, we require

$$C^2 m^2 r^{2m-2} - gr \cos \tilde{\omega}_0 = \text{constant.}$$

The powers of r show that

$$m = \frac{3}{2}.$$

The second boundary condition, which defines the free surface, is $\partial \varphi / \partial \tilde{\omega}$ $= 0$ on $\tilde{\omega} = \tilde{\omega}_0$. Hence

$$\cos m \tilde{\omega}_0 = 0, \qquad \tilde{\omega}_0 = \frac{\pi}{2m} = \frac{\pi}{3}.$$

The total angle $2\tilde{\omega}_0$ is $120°$.

Michell (1893) calculated the periodic wave of greatest height numerically for water of infinite depth and found the limiting height to be attained at $a/\lambda = 0.142$, as previously noted.

The result on the angle at the crest is not necessarily true in unsteady problems. For standing waves a theoretical argument proposed by Penney and Price (1952), but less conclusive than in Stokes' case, suggests $90°$; this value was confirmed experimentally by Taylor (1953).

13.14 Breaking and Peaking

It was remarked earlier that the nonlinear shallow water equations which neglect dispersion altogether lead to breaking of the typical hyperbolic kind, with the development of a vertical slope and a multivalued profile. It seems clear that the third derivative term in the Korteweg-deVries equation will prevent this ever happening in its solutions. But in both cases, the long wave assumption under which the equations were derived is no longer valid. Since some waves appear to break in this way, if the depth is small enough, one concludes that some dispersion is necessary but the Korteweg-deVries term is too strong for short wavelengths. This is not surpising in view of the fact that the κ^3 term makes (13.94) a bad approximation to the full dispersion relation when $(\kappa h_0)^2$ becomes large.

On the other hand the dispersion included in the Korteweg-deVries equation does allow the solitary and periodic waves which are not found in the shallow water theory. For these solutions, however, the Korteweg-deVries equation cannot describe the observed symmetrical "peaking" of the crests with a finite angle there. Again it might be argued that this is a small scale phenomenon where the small wavelength components become important and the assumption $(\kappa h)^2 \ll 1$ is no longer adequate. Doubtless this is combined with additional nonlinear effects.

Stokes waves include the full dispersive effects of $\omega_0^2 = g\kappa \tanh \kappa h_0$, but being limited to small amplitude, they do not show the solitary waves nor do they show the peaking.

Although both breaking and peaking, as well as criteria for the occurrence of each, are without doubt contained in the equations of the exact potential theory, it is intriguing to know what kind of simpler mathematical equation could include all these phenomena. In light of the foregoing comments it seems necessary to include at least the "breaking operator" of shallow water theory with the *full* dispersion relation of linear theory. Now the breaking of shallow water theory is given by

$$\eta_t + c_0 \eta_x + \frac{3c_0}{2h_0} \eta \eta_x = 0, \tag{13.125}$$

as noted in deriving (13.99). On the other hand, the linear equation corresponding to an arbitrary linear dispersion relation

$$\frac{\omega}{\kappa} = c(\kappa)$$

was given in (11.12) as

$$\eta_t + \int_{-\infty}^{\infty} K(x-\xi)\eta_\xi(\xi,t)\,d\xi = 0, \tag{13.126}$$

where

$$K(x) = \frac{1}{2\pi} \int_{-\infty}^{\infty} c(\kappa) e^{i\kappa x}\,d\kappa. \tag{13.127}$$

The two can be combined into

$$\eta_t + \frac{3c_0}{2h_0} \eta \eta_x + \int_{-\infty}^{\infty} K(x-\xi)\eta_\xi(\xi,t)\,d\xi = 0. \tag{13.128}$$

The Korteweg-deVries equation takes

$$c(\kappa) = c_0 - \gamma\kappa^2, \qquad K(x) = c_0\delta(x) + \gamma\delta''(x).$$

We now propose the improved description

$$c(\kappa) = \left(\frac{g}{\kappa}\tanh \kappa h_0\right)^{1/2}, \tag{13.129}$$

$$K_g(x) = \frac{1}{2\pi} \int_{-\infty}^{\infty} \left(\frac{g}{\kappa} \tanh \kappa h_0 \right)^{1/2} e^{i\kappa x} \, d\kappa. \qquad (13.130)$$

This combines full linear dispersion with long wave nonlinearity. In terms of the parameters α and β used in Section 13.11, we are retaining terms of all orders in β and the nonlinear term proportional to α, while neglecting all higher powers of α and all product terms. We could in fact keep terms of all order in α by taking the nonlinear operator from (13.97) and using the combined equation

$$\eta_t + \left\{ 3\sqrt{g(h_0 + \eta)} - 2\sqrt{gh_0} \right\} \eta_x + \int_{-\infty}^{\infty} K(x - \xi) \eta_\xi(\xi, t) \, d\xi = 0; \qquad (13.131)$$

however, the gain is probably not worth the extra complication. It is easy to show by standard methods that the normalized function K_g with $g = 1, h_0 = 1$ has the properties

$$K_g(x) = K_g(-x),$$

$$K_g(x) \sim (2\pi x)^{-1/2} \qquad \text{as} \quad x \to 0,$$

$$K_g(x) \sim \left(\frac{1}{2} \pi^2 x \right)^{-1/2} e^{-\pi x/2} \qquad \text{as} \quad x \to \infty,$$

$$\int_{-\infty}^{\infty} K_g(x) \, dx = 1.$$

The dispersive term is milder now, since $K_g(x)$ does not involve δ functions. Indeed the behavior of $c(\kappa)$ for large κ controls the behavior of $K(x)$ for small x; the change in the high wave number behavior replaces the $\delta''(x)$ by $x^{-1/2}$ as $x \to 0$. The analysis of such nonlinear integral equations is quite difficult, and the one with $K = K_g$ has not yet been analyzed completely. But the question was raised in the general terms of what kind of mathematical equation can describe waves with both peaking and breaking. We can show that (13.128) can do this for some simpler choices of $K(x)$ and infer that the same is true for $K = K_g$.

Steady profile solutions of (13.128) are obtained as usual by investigating solutions with $\eta = \eta(X)$, $X = x - Ut$. In normalized variables equivalent to taking $g = h_0 = 1$, we then have

$$\left(U - \frac{3}{2} \eta \right) \eta'(X) = \int_{-\infty}^{\infty} K(X - y) \eta'(y) \, dy, \qquad (13.132)$$

where y was set equal to $\xi - Ut$ in the integral. The possibility of peaking arises because the slope η' may be discontinuous at $\eta = 2U/3$, and this would in fact give the relation between velocity and amplitude for the wave of greatest height. The equation can be integrated to

$$A + U\eta - \frac{3}{4}\eta^2 = \int_{-\infty}^{\infty} K(X-y)\eta(y)\,dy. \tag{13.133}$$

Stokes waves for small amplitude can be found starting from the linear solution, and it seems reasonable to assume that in fact a critical height is reached when $\eta = 2U/3$. If $K(X)$ behaves like $|X|^p$ as $X \to 0$ and $\eta(X)$ behaves like $2U/3 - |X|^q$, a local argument in (13.132) suggests that

$$2q - 1 = p + q; \tag{13.134}$$

hence $q = p + 1$. According to this, the crest would be cusped with $\eta \sim \frac{2}{3} U - |X|^{1/2}$ for $K = K_g$; a delicate result like Stokes' $120°$ angle is too much to hope for in such a simplified model.

Although this is as far as the case $K = K_g$ has been taken, one can go further with an approximate kernel. A trick often used in such integral equations is to approximate the kernel by

$$K_0(x) = \mu e^{-\nu|x|},$$

or by a series of these exponentials. The kernel $K_0(x)$ is the Green's function for the operator $d^2/dx^2 - \nu^2$; hence the integral is eliminated in (13.133) when this operator is applied to both sides. We cannot match the singularity of $K_g(x)$ as $x \to 0$, but we can take $\nu = \pi/2$ to be reasonably close to the behavior as $x \to \infty$ and then choose μ so that

$$\int_{-\infty}^{\infty} K_0(x)\,dx = \frac{2\mu}{\nu} = 1;$$

that is, $\mu = \pi/4$, $\nu = \pi/2$. Since

$$\frac{d^2 K_0(x)}{dx^2} - \nu^2 K_0(x) = -\nu^2 \delta(x),$$

(13.133) becomes

$$\left(\frac{d^2}{dX^2} - \nu^2\right)\left(A + U\eta - \frac{3}{4}\eta^2\right) = -\nu^2\eta.$$

This may be integrated once to the form

$$\left(U - \frac{3}{2}\eta\right)^2 \eta'^2 = \text{quartic in } \eta.$$

Periodic solutions correspond to oscillations of η between two simple zeros of the quartic. Solitary waves are found in the limiting case when two zeros coalesce at $\eta = 0$; then η rises from this double zero (corresponding to $X = +\infty$) to a simple zero $\eta = \eta_0$, say, and back to $\eta = 0$ at $X = -\infty$. All this is true provided $\eta = 2U/3$ is outside the range. It can be shown that the crests become peaked with a finite angle when $\eta = 2U/3$ just coincides with the upper zero. All the details will not be given. We merely note that the solitary wave of greatest height has

$$\eta = \frac{8}{9} e^{-\pi|X|/4}, \qquad U = \frac{4}{3}.$$

(These are in units of length h_0 and velocity c_0). The values obtained by McCowan (1894) in an approximate treatment of the solitary wave problem are $\eta_0 = 0.78$, $U = 1.249$. One might say the comparison is reasonable. The crest has a finite angle of $110°$. The finite angle agrees with (13.134), since $p = 0$, $q = 1$ for the kernel K_0; the closeness to Stokes' $120°$ is surely fortuitous. However seriously one takes these numbers, it does show that an equation like (13.128) can describe periodic and solitary waves with the desired peaking.

Turning now to the other kind of breaking, we note that Seliger (1968) was able to show by a rather ingenious argument that for kernels like $K_0(x)$, a sufficiently asymmetric hump would break in the typical hyperbolic manner. He required $K(0)$ to be finite, however (as well as monotonic decreasing to zero as $x \to \infty$), and could not extend the argument to K_g. Briefly the argument is to consider

$$m_1(t) = \text{minimum } \eta_x \text{ at } x = X_1(t),$$

$$m_2(t) = \text{maximum } \eta_x \text{ at } x = X_2(t),$$

(see Fig. 13.5), where $m_1 < 0$ and $m_2 > 0$. Then, on differentiating the normalized form of (13.128) and setting $x = X_i(t)$, we have

$$\frac{dm_i}{dt} + \frac{3}{2}m_i^2 + \int_{-\infty}^{\infty} K(\xi)\eta_{xx}(X_i - \xi, t)\,d\xi = 0, \qquad i = 1, 2.$$

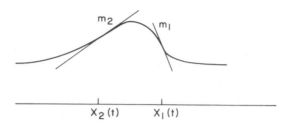

Fig. 13.5. Notation in discussion of breaking wave.

The integrals can be estimated in terms of m_1 and m_2 using the appropriate mean value theorem and we have

$$\frac{dm_1}{dt} \leqslant -\frac{3}{2}m_1^2 + (m_2 - m_1)K(0),$$

$$\frac{dm_2}{dt} \leqslant -\frac{3}{2}m_2^2 + (m_2 - m_1)K(0).$$

From the sum of these,

$$\frac{d}{dt}(m_1 + m_2) \leqslant (m_2 - m_1)\left\{2K(0) + \frac{3}{2}(m_1 + m_2)\right\} - 3m_2^2;$$

hence if $(m_1 + m_2) \leqslant -4K(0)/3$ initially, it remains so and

$$m_1 + m_2 \leqslant -\frac{4}{3}K(0) \tag{13.135}$$

for all t. Then

$$\frac{dm_1}{dt} \leqslant -\frac{3}{2}m_1^2 - 2m_1 K(0) - \frac{4}{3}K^2(0)$$

$$= -\frac{3}{2}\left\{m_1 + \frac{2}{3}K(0)\right\}^2 - \frac{2}{3}K^2(0). \tag{13.136}$$

The right hand side of this is negative, and in view of the m_1^2 term it is easy to show that $m_1 \to -\infty$ in a finite time; the details can be seen as follows. Let $M = -\frac{3}{2}m_1 - K(0)$, then $M = M_0 > 0$ initially [from (13.135) and $m_2 > 0$]; moreover,

$$\frac{dM}{dt} \geqslant M^2,$$

from (13.136). Therefore

$$\frac{d}{dt}\left(\frac{1}{M}\right) \leqslant -1, \qquad \frac{1}{M} \leqslant \frac{1}{M_0} - t, \qquad M \geqslant \frac{M_0}{1 - M_0 t};$$

$M \to \infty$ before t reaches $1/M_0$. The conclusion is that if (13.135) holds initially, that is, if the hump is asymmetric enough, it continually becomes more asymmetric and breaks with $m_1 \to -\infty$ in a time less than

$$\frac{1}{M_0} = \frac{1}{-\{(3/2)m_1(0) + K(0)\}}.$$

Again, whatever the validity of the model may be, we have shown that equations of the form (13.128) can indeed describe symmetric waves that propagate unchanged in shape and peak at a critical height, as well as asymmetric waves that break.

13.15 A Model for the Structure of Bores

In those cases where breaking rather than peaking occurs, the resulting bore takes two distinct forms. These are observed in tidal bores and can be produced experimentally by an analog of the shock tube used in gas dynamics. In the experiments, first documented by Favre (1935), a gate separating water at different levels is pulled up suddenly and bores of various strengths are produced by varying the levels. The weaker bores have a smooth but oscillatory structure as shown in Fig. 13.6, whereas the stronger ones have a rapid turbulent change with no coherent oscillation. In both cases the end states satisfy the jump conditions (13.81). The change in type seems to occur sharply at a depth ratio $h_2/h_1 \approx 1.28$, corresponding to a Froude number $U/\sqrt{gh_1} \approx 1.21$. A general discussion of the overall balance of mass, momentum, and energy, allowing radiation of energy under the wavetrain, is given by Benjamin and Lighthill (1954). The actual structure is obviously complex, but again one might ask what kind of description would embody both forms. The Korteweg-deVries equation is the natural starting place, but it has no solutions like Fig. 13.6 propagating unchanged in shape. Since dissipation is involved, it is natural to add a second derivative term and consider

$$\eta_t + c_0\left(1 + \frac{3}{2}\frac{\eta}{h_0}\right)\eta_x + \frac{1}{6}c_0 h_0^2 \eta_{xxx} - \mu\eta_{xx} = 0. \qquad (13.137)$$

This may not be a very close model to the frictional effects in water waves, but it is of interest in any case, since the Korteweg-deVries equation is a canonical one for the general study of dispersive waves.

Steady progressing solutions are found from

$$\eta = h_0 \zeta(X), \qquad X = x - Ut,$$

and the ordinary differential equation for ζ may be integrated to

$$\frac{1}{6} h_0^2 \zeta_{XX} - \frac{\mu}{c_0} \zeta_X + \frac{3}{4} \zeta^2 - \left(\frac{U}{c_0} - 1 \right) \zeta = 0, \qquad (13.138)$$

taking $\zeta \to 0$ as $X \to \infty$. A normalized form is

$$z_{\xi\xi} - m z_\xi + z^2 - z = 0, \qquad (13.139)$$

where

$$\xi = \{6(F-1)\}^{1/2} \frac{X}{h_0}, \qquad z = \frac{3}{4(F-1)} \zeta,$$

$$(13.140)$$

$$F = \frac{U}{c_0}, \qquad m = \left(\frac{6}{F-1} \right)^{1/2} \frac{\mu}{c_0 h_0}.$$

In the phase plane with $w = z_\xi$, we have

$$\frac{dw}{d\xi} = mw - z^2 + z, \qquad \frac{dz}{d\xi} = w. \qquad (13.141)$$

Solutions with end states $z = 0$, $w = 0$, and $z = 1$, $w = 0$ are possible. These are singular points in the (z, w) plane. The solution curve

$$\frac{dw}{dz} = \frac{mw - z^2 + z}{w}$$

must go from one to the other. In the neighborhood of $(0, 0)$ we have

$$w \sim \sigma_0 z, \quad z \propto e^{\sigma_0 \xi}, \qquad \sigma_0 = \frac{1}{2}(m - \sqrt{m^2 + 4});$$

this gives the exponential decay to zero as $\xi \to \infty$. In the neighborhood of $(1, 0)$, we have

$$w \sim \sigma_1(z - 1), \qquad (z - 1) \propto e^{\sigma_1 \xi}, \qquad \sigma_1 = \frac{1}{2}(m \pm \sqrt{m^2 - 4}).$$

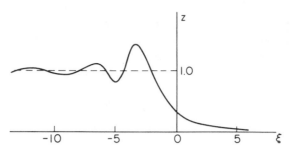

Fig. 13.6. Model bore structure. The solution of (13.139) for $m = \frac{1}{2}$.

The approach to $z = 1$ at $\xi \to -\infty$ is oscillatory or not accordingly as $m < 2$ or $m > 2$. Details may be pursued further, but we already see the two types of structure. For a fixed change of level, that is, for specified $F - 1$, small enough damping allows an oscillatory solution, but large damping suppresses the oscillation. For fixed μ, the criterion on F obtained from (13.140) would appear to be the wrong way around for water waves. However, in that case μ would be interpreted as an eddy viscosity dependent on the mean flow velocity. A simple estimate of the form of dependence is $\mu = b u_2 h_2$, where u_2, h_2 refer to conditions behind the bore and b is a numerical factor. Then from the bore conditions

$$\frac{\mu}{c_0 h_0} \approx \frac{4}{3} b (F - 1).$$

The criterion $m < 2$ for an oscillatory solution becomes $(F - 1) \lesssim 3/(8b^2)$. Favre's critical value of $F = 1.2$ would require $b \approx 1.4$, which is perhaps ten times greater than would be expected for an eddy viscosity. But we are making here only a crude model of the actual physical situation. The overall qualitative effect of dissipation seems to be mirrored correctly.

CHAPTER 14

Nonlinear Dispersion and the Variational Method

The nonlinear effects found in the study of water waves are typical of dispersive systems in general. Periodic wavetrains, similar to those of Stokes and Korteweg-deVries, are found in most systems and these are the basic solutions corresponding to the elementary solutions $ae^{ikx-i\omega t}$ of linear theory. In nonlinear theory, the solutions are no longer sinusoidal, but the existence of periodic solutions in $\theta = kx - \omega t$ can be shown explicitly in the simpler cases and inferred from the Stokes expansion in others. The main nonlinear effect is not the difference in functional form, it is the appearance of amplitude dependence in the dispersion relation. This leads to new qualitative behavior, not merely to the correction of linear formulas. Superposition of solutions is not available to generate more general wavetrains, but modulation theory can be studied directly. The theory can be developed in general using the variational approach of Section 11.7. The formulation will be studied in detail in this chapter and the justification as a formal perturbation method will be given to complete the earlier discussion. Detailed applications of the theory are then given in Chapters 15 and 16.

Another specific consequence of nonlinearity is the existence of solitary waves. Waves with these profiles would disperse in the linear theory, but the nonlinearity counterbalances the dispersion to produce waves of permanent shape. Solitary waves are found, in the first instance, as limiting cases of the periodic wavetrains, but recent work on their interactions and their production from arbitrary initial data has shown that their special structure is of separate importance. We shall return to these topics in Chapter 17.

For waves of moderately small amplitude in what might be called "near-linear theory," further results may be obtained by perturbation methods based on small amplitude expansions. In particular we may return to the Fourier analysis description and study the small nonlinear interactions of the Fourier components. The interactions transfer energy between different components and, through product terms in the equations,

485

generate new components from existing ones. These interactions can be followed effectively when only a few components are involved. We shall include typical results as appropriate, but the main emphasis is on methods that extend to the fully nonlinear case. From a Fourier analysis viewpoint, the nonlinear wavetrains and solitary waves are already quite complicated distributions of Fourier components with involved interactions maintaining a balance. The developments emphasized here build directly on these special structures without attempting to disentangle them into their components. However, in the near-linear case there are interesting and informative relations between the two points of view.

14.1 A Nonlinear Klein-Gordon Equation

It is useful to have a simple example to motivate and illustrate the steps in the development of the general theory. For this purpose a nonlinear version of the Klein-Gordon equation is particularly useful and is even simpler than the Korteweg-deVries equation, which would be the other obvious choice. We take the equation

$$\varphi_{tt} - \varphi_{xx} + V'(\varphi) = 0, \qquad (14.1)$$

where $V'(\varphi)$ is some reasonable nonlinear function of φ which is chosen as the derivative of a potential energy $V(\varphi)$ for later convenience. Equation 14.1 is not only a useful model; it arises in a variety of physical situations. This is especially true of the case $V'(\varphi) = \sin \varphi$, which almost inevitably has become known as the Sine-Gordon equation! An account of the physical problems in which this form arises is given by Barone et al. (1971), following a briefer version by Scott (1970, p. 250). Its first appearance is not in wave problems at all, but in the study of the geometry of surfaces with Gaussian curvature $K = -1$. In fact some of the transformation methods developed there have been remarkably valuable in finding solutions for interacting solitary waves, as will be discussed in Chapter 17. More recent problems listed by the same authors include:

1. Josephson junction transmission lines, where $\sin \varphi$ is the Josephson current across an insulator between two superconductors, the voltage being proportional to φ_t.
2. Dislocations in crystals, where the occurrence of $\sin \varphi$ is due to the periodic structure of rows of atoms.
3. The propagation in ferromagnetic materials of waves carrying rotations of the direction of magnetization.

4.   Laser pulses in two state media, where the variables can also be described in terms of a rotating vector.

Scott further describes his construction of a mechanical model with rigid pendula attached at close intervals along a stretched wire. Torsional waves propagating down the wire obey the wave equation and the pendula supply a restoring force proportional to sin φ, where φ is the angular displacement. Scott was able to generate the waves corresponding to many of the solutions of the Sine-Gordon equation.

Equation 14.1 has also been discussed by Schiff (1951), with a cubic nonlinearity, and by Perring and Skyrme (1962), with the sin φ term, in tentative investigations of elementary particles.

In this chapter the analysis applies for general $V(\varphi)$ with appropriate properties. The choice

$$V(\varphi) = \frac{1}{2}\varphi^2 + \sigma\varphi^4$$

is both the simplest to bear in mind and the correct expansion in the near-linear theory for even functions $V(\varphi)$. The small amplitude expansion of the Sine-Gordon equation has $\sigma = -1/24$.

We first check the existence of periodic wavetrains. They are obtained as usual by taking

$$\varphi = \Psi(\theta), \qquad \theta = kx - \omega t. \tag{14.2}$$

On substitution we have

$$(\omega^2 - k^2)\Psi_{\theta\theta} + V'(\Psi) = 0 \tag{14.3}$$

and the immediate integral

$$\frac{1}{2}(\omega^2 - k^2)\Psi_\theta^2 + V(\Psi) = A. \tag{14.4}$$

We use A for the constant of integration, although earlier it was used to denote the complex amplitude in linear problems. Only the real amplitude a will appear in the same context, so there should be no confusion. Here A is still an amplitude parameter; in the linear case, $V(\Psi) = \frac{1}{2}\Psi^2$, it is related to the actual amplitude a by $A = \frac{1}{2}a^2$.

The solution of (14.4) may be written

$$\theta = \left\{ \frac{1}{2}(\omega^2 - k^2) \right\}^{1/2} \int \frac{d\Psi}{\{A - V(\Psi)\}^{1/2}}, \tag{14.5}$$

and in the special cases where $V(\Psi)$ is either a cubic, a quartic, or trignometric, $\Psi(\theta)$ can be expressed in terms of standard elliptic functions. Periodic solutions are obtained when Ψ oscillates between two simple zeros of $A - V(\Psi)$. At the zeros $\Psi_\theta = 0$, and the solution curve has a maximum or a minimum; these points occur at finite values of θ, since (14.5) is convergent when the zeros are simple. If the zeros are denoted by Ψ_1 and Ψ_2, we shall take the case

$$\Psi_1 \leqslant \Psi \leqslant \Psi_2, \qquad A - V(\Psi) \geqslant 0, \qquad \omega^2 - k^2 > 0 \qquad (14.6)$$

for the present. The period in θ can be normalized to 2π (which is convenient in the linear limit) and we then have

$$2\pi = \left\{ \frac{1}{2}(\omega^2 - k^2) \right\}^{1/2} \oint \frac{d\Psi}{\{A - V(\Psi)\}^{1/2}}, \qquad (14.7)$$

where $\oint$ denotes the integral over a complete oscillation of Ψ from Ψ_1 up to Ψ_2 and back. The sign of the square root has to be changed appropriately in the two parts of the cycle. The integral may also be interpreted as a loop integral around a cut from Ψ_1 to Ψ_2 in the complex Ψ plane.

In the linear case $V(\Psi) = \frac{1}{2}\Psi^2$, the periodic solution is

$$\Psi = a \cos\theta, \qquad A = \frac{a^2}{2}; \qquad (14.8)$$

the amplitude a cancels out in (14.7), which becomes simply the linear dispersion relation

$$\omega^2 - k^2 = 1. \qquad (14.9)$$

In the nonlinear case, the amplitude parameter A does not drop out of (14.7) and we have the typical dependence of the dispersion relation on amplitude.

If the amplitude is small and V has the expansion

$$V = \frac{1}{2}\varphi^2 + \sigma\varphi^4 + \cdots, \qquad (14.10)$$

we have

$$\Psi = a \cos\theta + \frac{1}{8}\sigma a^3 \cos 3\theta + \cdots, \qquad (14.11)$$

$$\omega^2 - k^2 = 1 + 3\sigma a^2 + \cdots, \qquad (14.12)$$

$$A = \frac{1}{2}a^2 + \frac{9}{8}\sigma a^4 + \cdots . \tag{14.13}$$

These are the Stokes expansions, which may be obtained either by direct substitution in (14.3)–(14.4) or by expansion of the exact expressions (14.5) and (14.7) obtained above. It should be noted that a is the amplitude of the first term in (14.11); it differs slightly from the exact amplitude

$$a + \frac{1}{8}\sigma a^3 + \cdots .$$

14.2 A First Look at Modulations

In the basic case of one dimensional waves in a uniform medium, we saw in Chapter 11 that modulations on a *linear* wavetrain can be described by the equations

$$\frac{\partial k}{\partial t} + \frac{\partial \omega}{\partial x} = 0, \tag{14.14}$$

$$\frac{\partial a^2}{\partial t} + \frac{\partial}{\partial x}(C_0 a^2) = 0, \tag{14.15}$$

where $\omega = \omega_0(k)$ is given by the linear dispersion relation and $C_0 = \omega_0'(k)$ is the linear group velocity. (A subscript zero is added now to indicate the linear values.) The crucial qualitative change of nonlinearity is the dependence of ω on a, which couples (14.14) to (14.15). For moderately small amplitudes, ω may be expressed in Stokes fashion as

$$\omega = \omega_0(k) + \omega_2(k)a^2 + \cdots , \tag{14.16}$$

and (14.14) becomes

$$\frac{\partial k}{\partial t} + \{\omega_0'(k) + \omega_2'(k)a^2\}\frac{\partial k}{\partial x} + \omega_2(k)\frac{\partial a^2}{\partial x} = 0. \tag{14.17}$$

The important coupling term is $\omega_2(k)\partial a^2/\partial x$ because it introduces a term in the derivative of a; it leads to a correction $O(a)$ to the characteristic velocities. The other new term merely corrects the coefficient of the existing term in $\partial k/\partial x$ and consequently contributes only at the $O(a^2)$ level. Similarly, for small amplitudes, the nonlinear corrections to (14.15) would be various terms of order a^4 which would provide corrections of relative order a^2 to the coefficients of the existing terms in $\partial a^2/\partial x$ and $\partial k/\partial x$. Therefore in the first assessment of nonlinear effects we can proceed very simply, using only the new dispersion relation, and take

$$\frac{\partial k}{\partial t} + \omega_0'(k)\frac{\partial k}{\partial x} + \omega_2(k)\frac{\partial a^2}{\partial x} = 0, \tag{14.18}$$

$$\frac{\partial a^2}{\partial t} + \omega_0'(k)\frac{\partial a^2}{\partial x} + \omega_0''(k)a^2\frac{\partial k}{\partial x} = 0. \tag{14.19}$$

By the standard procedure of Chapter 5, the characteristic form of these coupled equations is found to be

$$\frac{1}{2}\left\{ \frac{\omega_0''(k)}{\omega_2(k)} \right\}^{1/2} (\text{sgn } \omega_0'')\, dk \pm da = 0 \tag{14.20}$$

on characteristics

$$\frac{dx}{dt} = \omega_0'(k) \pm \left\{ \omega_2(k)\omega_0''(k) \right\}^{1/2} a. \tag{14.21}$$

It may be verified that additional terms of relative order a^2 added to (14.18)–(14.19) contribute terms only of order a^2 to (14.20)–(14.21).

This simple formulation already shows some remarkable results. In the case $\omega_2\omega_0'' > 0$, the characteristics are real and the system is hyperbolic. The double characteristic velocity splits under the nonlinear correction and we have the *two* velocities given by (14.21). In general, an initial disturbance or modulating source will introduce disturbances on both families of characteristics. If the disturbance is initially finite in extent, for example a bulge on an otherwise uniform wavetrain, it will eventually split into two. This is completely different from the linear behavior where such a bulge may distort due to the dependence of $C_0(k)$ on k but would not split.

A second consequence of nonlinearity in the hyperbolic case is that "compressive" modulations will distort and steepen in the typical hyperbolic fashion discussed in Part I. This raises the question of multivalued solutions and breaking.

When $\omega_2\omega_0'' < 0$, the characteristics are imaginary and the system is elliptic. This leads to ill-posed problems in the wave propagation context. Among other things, it means that small perturbations will grow in time and in this sense the periodic wavetrain is *unstable*. The elliptic case turns out to be not uncommon and the modulation theory provides an interesting approach to some aspects of stability theory.

We might note that for Stokes waves in deep water, the dispersion relation (13.124) gives

$$\omega_0(k) = g^{1/2}k^{1/2}, \qquad \omega_2(k) = \frac{1}{2}g^{1/2}k^{3/2}, \tag{14.22}$$

so this is an unstable case with $\omega_0'' \omega_2 < 0$. This is surprising in view of the long history of the problem and the sometimes controversial arguments about the existence of periodic solutions; throughout these discussions the instability went unrecognized. For the Klein-Gordon example (14.12), we have

$$\omega_0(k) = (k^2 + 1)^{1/2}, \qquad \omega_2(k) = \frac{3}{2} \sigma (k^2 + 1)^{-1/2}. \tag{14.23}$$

The sign of $\omega_0'' \omega_2$ is the same as the sign of σ; the modulation equations are hyperbolic for $\sigma > 0$ and elliptic for $\sigma < 0$. For near-linear waves, the Sine-Gordon equation has $\sigma < 0$, so that in all the problems governed by this equation the near-linear wavetrains are unstable.

We shall return to all these questions after the formulation of the modulation equations has been studied in detail and extended to the fully nonlinear case.

14.3 The Variational Approach to Modulation Theory

The complete modulation equations are obtained in a particularly compact and significant form from the variational approach started in Chapter 11. We first see how to implement it for nonlinear problems using the Klein-Gordon equation as a typical example. General procedures then become apparent and we include these in the justification of the method.

In the Klein-Gordon case the periodic wavetrain is described by (14.4)–(14.5) and involves the parameters ω, k, and A. We need to find the equations satisfied by these parameters for a slowly varying wavetrain. Equation 14.1 is the Euler equation for the variational principle

$$\delta \int \int \left\{ \frac{1}{2} \varphi_t^2 - \frac{1}{2} \varphi_x^2 - V(\varphi) \right\} dx\, dt = 0, \tag{14.24}$$

as is easily verified from (11.74). The elementary solution corresponding to the solution $\varphi = a \cos(\theta + \eta)$ used in linear problems is $\varphi = \Psi(\theta)$. [A phase shift η can be added to (14.5), but it drops out of the modulation equations.] We therefore calculate the Lagrangian and its average value for $\varphi = \Psi(\theta)$; this is done keeping ω, k, and A constant. We have

$$L = \frac{1}{2} (\omega^2 - k^2) \Psi_\theta^2 - V(\Psi),$$

and the average value over one period in θ is

$$\bar{L} = \frac{1}{2\pi} \int_0^{2\pi} \left\{ \frac{1}{2}(\omega^2 - k^2)\Psi_\theta^2 - V(\Psi) \right\} d\theta. \tag{14.25}$$

In principle, the function Ψ is known completely from (14.5). However, we can avoid the integrated form and use (14.4) instead to express $\bar{L}$ as a function of ω, k, A. We note the successive steps

$$\bar{L} = \frac{1}{2\pi} \int_0^{2\pi} (\omega^2 - k^2)\Psi_\theta^2 \, d\theta - A$$

$$= \frac{1}{2\pi}(\omega^2 - k^2) \int_0^{2\pi} \Psi_\theta \, d\Psi - A$$

$$= \frac{1}{2\pi} \{2(\omega^2 - k^2)\}^{1/2} \oint \{A - V(\Psi)\}^{1/2} d\Psi - A. \tag{14.26}$$

The final loop integral is a well-defined function of A, in which Ψ is now merely a dummy variable of integration. The notation $\mathcal{L}(\omega, k, A)$ is reserved for the final form of $\bar{L}$.

When ω, k, A are allowed to be slowly varying functions of x and t, we propose the average variational principle,

$$\delta \iint \mathcal{L}(\omega, k, A) \, dx \, dt = 0, \tag{14.27}$$

as before. This is viewed as a variational principle for $\theta(x, t)$ and $A(x, t)$, with $\omega = -\theta_t$, $k = \theta_x$; the variational equations are

$$\delta A: \qquad\qquad \mathcal{L}_A = 0, \tag{14.28}$$

$$\delta\theta: \qquad \frac{\partial}{\partial t}\mathcal{L}_\omega - \frac{\partial}{\partial x}\mathcal{L}_k = 0. \tag{14.29}$$

After the variations have been taken, we again work with ω, k, A, and add the consistency relation

$$\frac{\partial k}{\partial t} + \frac{\partial \omega}{\partial x} = 0 \tag{14.30}$$

obtained by eliminating θ. The equations and their derivation from (14.27) are, of course, the same as in the linear case with the minor change of amplitude variable from a to A. The only new ingredient in the nonlinear theory is the calculation of $\mathcal{L}(\omega, k, A)$.

Equation 14.28 is a functional relation between ω, k, A, which can only be the dispersion relation. For the Klein-Gordon example with $\mathcal{L}$ given by (14.26), we confirm that it does indeed give the correct result (14.7). The system (14.28)–(14.30) is the exact nonlinear form for the modulation equations tentatively proposed in the approximation (14.18)–(14.19). Before discussing the properties of these equations and their various extensions we turn now to the question of how the variational approach may be justified.

14.4 Justification of the Variational Approach

It will be sufficient to consider in detail the case of one dimensional waves described by a variational principle

$$\delta \int \int L(\varphi_t, \varphi_x, \varphi) \, dx \, dt = 0. \tag{14.31}$$

The cases of more dimensions, more dependent variables and nonuniform media can all be treated similarly. The Euler equation for (14.31) is

$$\frac{\partial}{\partial t} L_1 + \frac{\partial}{\partial x} L_2 - L_3 = 0, \tag{14.32}$$

where the L_j denote the derivatives

$$L_1 = \frac{\partial L}{\partial \varphi_t}, \qquad L_2 = \frac{\partial L}{\partial \varphi_x}, \qquad L_3 = \frac{\partial L}{\partial \varphi}. \tag{14.33}$$

Equation 14.32 is a second order partial differential equation for $\varphi(x, t)$ and we assume that this has periodic wavetrain solutions of the appropriate type.

For problems of slow modulations a parameter ϵ will be introduced by the initial or boundary conditions (as discussed in various cases in Section 11.8); ϵ measures the ratio of a typical wavelength or period relative to a typical length or time scale of the modulation. We shall eventually suppose ϵ to be small, but we make no restriction on the magnitude of the amplitude, only that its variations are slow.

The first step is to describe a modulated wavetrain precisely. If x and t are measured on the scale of the typical wavelength and period, the slowly varying quantities are functions of $\epsilon x, \epsilon t$; modulation parameters such as k and ω should be functions of this type. Yet φ itself varies due to the relatively fast oscillations as well. To incorporate these requirements, φ is written explicitly as a function of a phase function θ *and* of $\epsilon x, \epsilon t$. Then θ is chosen as $\epsilon^{-1}\Theta(\epsilon x, \epsilon t)$ to provide the relatively fast oscillation and to give the correct dependence of $k = \theta_x$ and $\omega = -\theta_t$ on $\epsilon x, \epsilon t$. We therefore take

$$\varphi = \Phi(\theta, X, T; \epsilon), \tag{14.34}$$

$$\theta = \epsilon^{-1}\Theta(X, T), \qquad X = \epsilon x, T = \epsilon t. \tag{14.35}$$

We *define*

$$\nu(X, T) = -\omega(X, T) = \Theta_T, \qquad k(X, T) = \Theta_X \tag{14.36}$$

as the negative frequency and wave number. (In this general discussion we work with $\nu = -\omega$ to preserve the symmetry between x and t.) The scaling has been arranged so that

$$\frac{\partial \varphi}{\partial t} = \nu \frac{\partial \Phi}{\partial \theta} + \epsilon \frac{\partial \Phi}{\partial T}, \qquad \frac{\partial \varphi}{\partial x} = k \frac{\partial \Phi}{\partial \theta} + \epsilon \frac{\partial \Phi}{\partial X};$$

the variations due to the oscillations and to the slow modulations appear separately.

In vibration problems for ordinary mechanical systems, x is absent and the method amounts to distinguishing two time scales explicitly. It has become known as "two-timing," which is a colorful and convenient name even when "double-crossing" x variations are also involved. The art of two-timing lies in the fact that although one starts and ends with the correct number of independent variables, the expanded form can be used to advantage at intermediate steps. In the present case φ is ultimately a function of x and t through (14.35), but in appropriate parts of the analysis Φ is treated as a function of the *three* variables θ, X, T independently. In the usual two-timing procedures, the extra flexibility allows the suppression of secular and other terms. Its use in conjunction with variational principles will be different but equivalent.

The geometrical optics ($WKBJ$) type of expansion discussed in Section 11.8 is equivalent to choosing

$$\varphi(x, t) \sim e^{i\epsilon^{-1}\Theta(\epsilon x, \epsilon t)} \sum \epsilon^n A_n(\epsilon x, \epsilon t) \tag{14.37}$$

from the outset. The two-timing version would work with

$$\Phi(\theta, X, T; \epsilon) \sim e^{i\theta} \sum \epsilon^n A_n(X, T) \tag{14.38}$$

to the same ultimate ends. In either case, the exponential dependence on θ is limited to linear problems. For nonlinear problems, the counterpart would be to take an expansion

$$\Phi(\theta, X, T; \epsilon) \sim \sum \epsilon^n \Phi^{(n)}(\theta, X, T), \tag{14.39}$$

and determine the functions $\Phi^{(n)}$ successively. However, in the equivalent variational approach we make no such initial expansion; we work with (14.34)–(14.36) directly and avoid much of the tedious manipulation of the more standard perturbation procedures.

When (14.34) and (14.35) are substituted into the basic Euler equation (14.32) we have

$$\nu \frac{\partial L_1}{\partial \theta} + \epsilon \frac{\partial L_1}{\partial T} + k \frac{\partial L_2}{\partial \theta} + \epsilon \frac{\partial L_2}{\partial X} - L_3 = 0, \qquad (14.40)$$

where the arguments of the L_j are given by

$$L_j = L_j(\nu \Phi_\theta + \epsilon \Phi_T, k \Phi_\theta + \epsilon \Phi_X, \Phi). \qquad (14.41)$$

The relation $\theta = \epsilon^{-1}\Theta(X,T)$ was used to obtain (14.40), but it is now dropped. This is the crucial step in two-timing. Equation 14.40 is now considered as an equation for the function $\Phi(\theta, X, T)$ of three *independent* variables θ, X, T. The equation also involves the function $\Theta(X, T)$ through its derivatives $\nu = \Theta_T$, $k = \Theta_X$; the original relations of Θ, ν and k to the argument θ in Φ are also dropped. It is clear that if satisfactory solutions for $\Phi(\theta, X, T)$ and $\Theta(X, T)$ can be found, then $\Phi(\epsilon^{-1}\Theta, X, T)$ solves the original problem. The extra flexibility in the choice of $\Theta(X, T)$ is used to assure satisfactory behavior of $\Phi(\theta, X, T)$.

The choice of $\Theta(X, T)$ will appear in different ways depending on the particular variant of the method, but they are equivalent. Here we shall impose from the outset the requirement that Φ and its derivatives be periodic in θ. [Other variants leave $\Theta(X, T)$ open at first, find unwanted secular terms proportional to θ in the general expression for Φ, and eliminate them by the choice of Θ.] The period may be normalized to 2π, so we impose the condition that Φ and its derivatives be 2π-periodic in θ. To implement this condition, we note that (14.40) may be written in conservation form as

$$\frac{\partial}{\partial \theta}\{(\nu L_1 + k L_2)\Phi_\theta - L\} + \epsilon \frac{\partial}{\partial T}(\Phi_\theta L_1) + \epsilon \frac{\partial}{\partial X}(\Phi_\theta L_2) = 0. \quad (14.42)$$

Then, on integration from $\theta = 0$ to 2π, the contributions of the first term cancel, from the periodicity requirement, and we have

$$\frac{\partial}{\partial T} \frac{1}{2\pi} \int_0^{2\pi} \Phi_\theta L_1 \, d\theta + \frac{\partial}{\partial X} \frac{1}{2\pi} \int_0^{2\pi} \Phi_\theta L_2 \, d\theta = 0. \qquad (14.43)$$

Equations 14.40 and 14.43 are the two equations for $\Phi(\theta, X, T)$ and $\Theta(X, T)$.

It is a remarkable and surprising fact that these equations for Φ and Θ are just the variational equations for the variational principle

$$\delta \int \int \frac{1}{2\pi} \int_0^{2\pi} L(\nu\Phi_\theta + \epsilon\Phi_T, k\Phi_\theta + \epsilon\Phi_X, \Phi) d\theta \, dX \, dT = 0. \quad (14.44)$$

Variations $\delta\Phi$ lead to

$$\frac{\partial}{\partial\theta} L_{\Phi_\theta} + \frac{\partial}{\partial T} L_{\Phi_T} + \frac{\partial}{\partial X} L_{\Phi_X} - L_\Phi = 0$$

in the usual way, and with the particular form of L in (14.44) this is seen to be (14.40). Variations $\delta\Theta$ give

$$\frac{\partial}{\partial T} \bar{L}_\nu + \frac{\partial}{\partial X} \bar{L}_k = 0, \quad (14.45)$$

where

$$\bar{L} = \frac{1}{2\pi} \int_0^{2\pi} L(\nu\Phi_\theta + \epsilon\Phi_T, k\Phi_\theta + \epsilon\Phi_X, \Phi) \, d\theta; \quad (14.46)$$

this is (14.43). But, most striking of all, (14.44) is just an *exact* form of the average variational principle! Not only do we justify the variational approach, we obtain a powerful and compact basis for the entire perturbation analysis. Strangely enough, we have made no explicit assumption so far that ϵ is small. It is implicit, however, in the choice of the functional form of Φ and the requirement that Φ be periodic in θ.

In the lowest order approximation to (14.44), we have

$$\delta \int \int \bar{L}^{(0)} dX \, dT = 0, \quad (14.47)$$

$$\bar{L}^{(0)} = \frac{1}{2\pi} \int_0^{2\pi} L\{\nu\Phi_\theta^{(0)}, k\Phi_\theta^{(0)}, \Phi^{(0)}\} \, d\theta. \quad (14.48)$$

The variational equations are

$$\delta\Phi^{(0)}: \qquad \frac{\partial}{\partial\theta}\{\nu L_1^{(0)} + k L_2^{(0)}\} - L_3^{(0)} = 0, \quad (14.49)$$

$$\delta\Theta: \qquad \frac{\partial}{\partial T} \bar{L}_\nu^{(0)} + \frac{\partial}{\partial X} \bar{L}_k^{(0)} = 0; \quad (14.50)$$

these are the lowest order approximations to (14.40) and (14.45), of course. Since X, T derivatives of $\Phi^{(0)}$ do not occur in (14.49), it is effectively an ordinary differential equation for $\Phi^{(0)}$ as a function of θ. An immediate first integral [the corresponding approximation to (14.42)] is

$$\left\{ \nu L_1^{(0)} + L_2^{(0)} \right\} \Phi_\theta^{(0)} - L^{(0)} = A(X,T). \tag{14.51}$$

Equations 14.49 and 14.51 are just the ordinary differential equations describing the uniform periodic wavetrain, but with the difference that the parameters ν, k, A are now functions of X, T. The dependence on θ is exactly the same as in the periodic wavetrain; the dependence of ν, k, A on X, T provides the modulation. The explicit separation of θ from X, T automatically allows integrations with respect to θ in which ν, k, A are held fixed; integrations such as those in (14.25) and (14.26) are now seen to be in this sense.

When the solution of (14.51) is combined with (14.47)–(14.48), we have exactly the variational approach proposed earlier. It is now justified as the first approximation in a formal perturbation scheme.

In the actual use of the method there is an important question of technique to be explained in general terms. As it stands (14.51) can be used to determine both the function $\Phi^{(0)}$ *and* the dispersion relation between ν, k, A. [See (14.5) and (14.7) for the Klein-Gordon example.] The manipulations in (14.26) show that by limited use of (14.51) in (14.48) the explicit determination of $\Phi^{(0)}$ (which is just Ψ changed to the expanded notation) can be avoided and the dispersion relation can be incorporated as an additional variational equation derivable from (14.47). This is much to be preferred. For then the form of the average Lagrangian is simplified and, more importantly, *all* the equations relating the slowly varying parameters ν, k, A are collected in the variational principle. The question is how to describe this procedure in general terms. The problem is how to extract from (14.51) enough information on the functional form of $\Phi^{(0)}$ and not use complete information on the dispersion relation. We now show how this may be done.

14.5 Optimal Use of the Variational Principle

In the linear case there is no difficulty in separating the functional form of $\Phi^{(0)}$ from the dispersion relation. We know in advance that the solution of (14.49) or (14.51) will take the form

$$\Phi^{(0)} = a \cos(\theta + \eta),$$

where $a(X,T)$ is the amplitude related to $A(X,T)$ and used instead of it.

The phase parameter $\eta(X,T)$ will drop out in forming the average Lagrangian (14.48) and plays no role in this lowest order approximation. This rather trivial information on $\Phi^{(0)}$ is the only information extracted from (14.51) and does not include the dispersion relation. When this $\Phi^{(0)}$ is substituted in (14.48) we have the function

$$\mathcal{L}(\nu, k, a) = \frac{1}{2\pi} \int_0^{2\pi} L(-\nu a \sin\theta, -ka\sin\theta, a\cos\theta)\, d\theta$$

for the average Lagrangian.

In the near-linear case there is also no difficulty. We may use the Stokes expansion

$$\Phi^{(0)} = a\cos(\theta + \eta) + a_2\cos(2\theta + \eta_2) + a_3\cos(3\theta + \eta_3) + \cdots$$

as the required form of $\Phi^{(0)}$ without including the dispersion relation. The relations of $a_2, a_3, \ldots, \eta_2, \eta_3, \ldots$, to a and η may be taken from (14.49) or (14.51), or they may be left arbitrary and also determined from the variational principle. For example, in the Klein-Gordon problem with $V(\varphi)$ given by (14.10), we take

$$\Phi^{(0)} = a\cos\theta + a_3\cos 3\theta + a_5\cos 5\theta + \cdots.$$

(It is easy to see in advance that the odd cosine terms are sufficient.) Then*

$$\bar{L}^{(0)} = \frac{1}{2\pi} \int_0^{2\pi} \left\{ \frac{1}{2}(\nu^2 - k^2)\Phi_\theta^{(0)} - \frac{1}{2}\Phi^{(0)2} - \sigma\Phi^{(0)4} \right\} d\theta$$

$$= \frac{1}{4}(\nu^2 - k^2 - 1)a^2 - \frac{3\sigma a^4}{8} + \left(2a_3^2 - \frac{1}{2}\sigma a^3 a_3\right) + \cdots.$$

Variation with respect to a_3 shows that $a_3 = \frac{1}{8}\sigma a^3$ in agreement with (14.11). On resubstitution of this expression for a_3 in $\bar{L}^{(0)}$ we have

$$\mathcal{L}(\nu, k, a) = \frac{1}{4}(\nu^2 - k^2 - 1)a^2 - \frac{3}{8}\sigma a^4 - \frac{1}{32}\sigma^2 a^6 + \cdots. \tag{14.52}$$

Variation of a now gives the dispersion relation (14.12).

In the fully nonlinear case it is harder to disentangle the functional form of $\Phi^{(0)}$ from the dispersion relation. However, it can be done by use of a Hamiltonian version of the equations.

---

*A term proportional to $(\nu^2 - k^2 - 1)a_3^2$ is omitted because the subsequent equations show that $(\nu^2 - k^2 - 1) = 0(a^2)$.

Hamiltonian Transformation.

The transformation will be applied here only to the lowest order approximation in (14.47)–(14.51), so to ease the notation we drop the superscript zero on all quantities. The idea is to eliminate the quantity Φ_θ in favor of $\partial L / \partial \Phi_\theta$ just as $\dot{q}$ is eliminated in favor of a generalized momentum $p = \partial L / \partial \dot{q}$ in ordinary mechanics. A new variable is defined by

$$\Pi = \frac{\partial L}{\partial \Phi_\theta} = \nu L_1 + k L_2, \tag{14.53}$$

and the Hamiltonian $H(\Pi, \Phi; \nu, k)$ is defined by

$$H = \Phi_\theta \frac{\partial L}{\partial \Phi_\theta} - L = \Phi_\theta (\nu L_1 + k L_2) - L. \tag{14.54}$$

From the transformation alone we have

$$\frac{\partial \Phi}{\partial \theta} = \frac{\partial H}{\partial \Pi}, \tag{14.55}$$

and (14.49) provides

$$\frac{\partial \Pi}{\partial \theta} = - \frac{\partial H}{\partial \Phi}. \tag{14.56}$$

These replace the second order equation (14.49) for Φ by two first order equations for Φ and Π. The variational principle (14.47) may now be written with

$$\bar{L} = \frac{1}{2\pi} \int_0^{2\pi} (\Pi \Phi_\theta - H) \, d\theta. \tag{14.57}$$

Moreover, there is an important extension. In the original form the variation $\delta \Phi_\theta$ is tied to $\delta \Phi$; hence $\delta \Pi$ is tied to the variation of Φ through (14.53), and (14.55) is a consequence of the transformation not a variational equation. However, we simply observe that both (14.55) and (14.56) follow from (14.57) if Φ and Π are allowed to vary *independently*. We are therefore free to make this extension. The next thing to note is that (14.51) is just the energy integral

$$H(\Pi, \Phi; \nu, k) = A(X, T) \tag{14.58}$$

for (14.55) and (14.56). Moreover, in this form it provides only the function

$$\Pi(\Phi;\nu,k,A).$$

Without using the relation of Π to Φ_θ, which has now been turned into one of the variational equations, there is no way to deduce the dispersion relation as well. This achieves the required separation of (14.51) into information about the form of solutions (now provided by the dependence of Π on Φ) and the dispersion relation. Finally, since the stationary values of (14.57) are known to satisfy (14.58), we may restrict the variations to functions which already satisfy (14.58). Then (14.57) may be evaluated as

$$\mathcal{L}(\nu,k,A)=\frac{1}{2\pi}\oint\Pi(\Phi;\nu,k,A)\,d\Phi-A, \qquad (14.59)$$

and $\Pi(\Phi;\nu,k,A)$ is the function determined from (14.58). The variational principle becomes

$$\delta\int\int\mathcal{L}(\nu,k,A)\,dX\,dT=0.$$

The variation with respect to A is the only remnant of the variations of Φ and Π. The variational equations are now

$$\delta A: \qquad\qquad \mathcal{L}_A=0,$$

$$\delta\Theta: \qquad \frac{\partial}{\partial T}\mathcal{L}_\nu+\frac{\partial}{\partial X}\mathcal{L}_k=0,$$

and the consistency relation

$$\frac{\partial k}{\partial T}-\frac{\partial \nu}{\partial X}=0,$$

is added. These are the equations (14.28)–(14.30) with $\nu=-\omega$.

In the Klein-Gordon example,

$$L=\frac{1}{2}(\nu^2-k^2)\Phi_\theta^2-V(\Phi),$$

the Hamiltonian transformation is

$$\Pi=\frac{\partial L}{\partial \Phi_\theta}=(\nu^2-k^2)\Phi_\theta,$$

$$H=\Phi_\theta\frac{\partial L}{\partial \Phi_\theta}-L=\frac{1}{2}(\nu^2-k^2)^{-1}\Pi^2+V(\Phi).$$

The integral $H = A$ is solved as

$$\Pi = \{2(\nu^2 - k^2)\}^{1/2}\{A - V(\Phi)\}^{1/2},$$

and

$$\mathcal{L} = \frac{1}{2\pi} \oint \Pi\, d\Phi - A$$

$$= \frac{1}{2\pi} \{2(\nu^2 - k^2)\}^{1/2} \oint \{A - V(\Phi)\}^{1/2} d\Phi - A,$$

in agreement with (14.26).

Naturally the Hamiltonian transformation can be used also in the linear or near-linear cases. The expressions for $\mathcal{L}$ may then differ in form from those obtained previously, but of course the resulting variational equations are equivalent.

14.6 Comments on the Perturbation Scheme

The usual procedure in applying perturbation methods is to substitute suitable expansions in powers of ϵ directly into the differential equations of the problem, obtain a hierarchy of equations for the successive orders, and then take steps to ensure uniform validity. It was in this manner that the *results* of the variational approach were first verified by Luke (1966). The expansion (14.39) is substituted in the equation for φ to give equations that we may write schematically as

$$E_0\{\Phi^{(0)}\} = 0, \qquad E_1\{\Phi^{(1)}, \Phi^{(0)}\} = F_1\{\Phi^{(0)}\}, \qquad \text{and so on.}$$

The zeroth equation for $\Phi^{(0)}$ is equivalent to (14.49). It is solved for $\Phi^{(0)}$; the dispersion relation is obtained between ν, k, A, but their dependence on X, T is undetermined at this level. The equation for $\Phi^{(1)}$ involves only θ derivatives of $\Phi^{(1)}$ and so is effectively an ordinary differential equation. Its solution has unbounded terms proportional to θ, unless conditions are imposed on $F_1\{\Phi^{(0)}\}$. These "secular" terms must be suppressed to ensure uniform validity of the expansion for large θ. The required condition on $F_1\{\Phi^{(0)}\}$ leads to the further equation for ν, k, A, which completes the lowest order solution. In the subsequent equations for the $\Phi^{(n)}$, there are further parameters and further secular conditions.

The prior requirement that Φ be periodic is equivalent to the suppression of secular terms. Therefore the successive approximations to the periodicity condition (14.43) would appear as secular conditions in the more traditional procedure. We see the advantage of starting from (14.42) and (14.43) even if that procedure were to be followed. But, better still, since (14.42) and (14.43) correspond to the variational principle (14.44), the expansion can be substituted directly in (14.44) and the variational principle used to generate both the equations for $\Phi^{(n)}$ and the secular conditions. Thus the variational approach should not be considered as a separate method. It includes the usual expansion approach, for which it streamlines the details and allows general results to be formulated.

There are other advantages. The variational principle (14.44) has been established independently of any assumed form for the dependence on ϵ. Furthermore, Θ may also be allowed to depend on ϵ; it was taken independent of ϵ only for simplicity in the initial presentation. We may use expansions in powers of ϵ for Φ or Θ or both, but we are also free to take other forms. For example, in the near-linear case, we may use expansions in powers of the amplitude, or, what amounts to the same thing, Fourier series for Φ. This will be the choice in the discussion of higher order approximations in Section 15.5.

14.7 Extensions to More Variables

The extension to more space dimensions is immediate. The plane periodic wave solutions have $\varphi = \Psi(\theta)$ where $\theta = \theta(\mathbf{x}, t)$ depends on a vector $\mathbf{x}$ and the propagation is in the direction of the vector wave number $\mathbf{k} = \theta_{\mathbf{x}}$. The average Lagrangian becomes $\mathcal{L}(\omega, \mathbf{k}, A)$ and modulations in space (i.e., slowly curving phase surfaces) also become possible. The modulation equations are (11.80)–(11.82). The justification of the last section requires only the obvious changes of replacing x, X, k by x_i, X_i, k_i and performing the corresponding summations when necessary.

The case of a single higher order equation goes through simlarly with only minor extensions. There will be higher order derivatives in (14.31) and in all the later steps, but the extensions are straightforward.

The case of more dependent variables requires detailed comment. First, for a linear system in a set of functions $\varphi^{(\alpha)}(\mathbf{x}, t)$, periodic wavetrains may be described by

$$\varphi^{(\alpha)} = a_\alpha \cos \theta + b_\alpha \sin \theta.$$

The average Lagrangian calculated from this is a function of the two sets a_α, b_α, as well as ω and $\mathbf{k}$. The corresponding variational principle

$$\delta \int \int \mathcal{L}(\theta_t, \theta_{x_i}, a_\alpha, b_\alpha) \, d\mathbf{x} \, dt = 0 \tag{14.60}$$

leads to the variational equations

$$\mathcal{L}_{a_\alpha} = 0, \qquad \mathcal{L}_{b_\alpha} = 0,$$

$$\frac{\partial}{\partial t} \mathcal{L}_\omega - \frac{\partial}{\partial x_j} \mathcal{L}_{k_j} = 0 \qquad (14.61)$$

$$\frac{\partial k_i}{\partial t} + \frac{\partial \omega}{\partial x_i} = 0, \qquad \frac{\partial k_i}{\partial x_j} - \frac{\partial k_j}{\partial x_i} = 0.$$

The set of equations $\mathcal{L}_{a_\alpha} = \mathcal{L}_{b_\alpha} = 0$ are linear and homogeneous (since $\mathcal{L}$ is quadratic in a_α, b_α) and they may in general be solved to express the a_α and b_α in terms of single amplitude a. These expressions may be reinserted into the Lagrangian to give $\mathcal{L}$ as a function $\mathcal{L}_1(\omega, \mathbf{k}, a)$, and the modulation equations are the same as in the single variable case. The substitution is permissible, since the restricted choice for a_α and b_α does satisfy the stationary conditions. The equivalence may also be verified directly, since

$$\mathcal{L}_{1a} = \frac{\partial a_\alpha}{\partial a} \mathcal{L}_{a_\alpha} + \frac{\partial b_\alpha}{\partial a} \mathcal{L}_{b_\alpha} = 0,$$

$$\mathcal{L}_{1\omega} = \mathcal{L}_\omega + \frac{\partial a_\alpha}{\partial \omega} \mathcal{L}_{a_\alpha} + \frac{\partial b_\alpha}{\partial a} \mathcal{L}_{b_\alpha} = \mathcal{L}_\omega,$$

and similarly $\mathcal{L}_{1k_j} = \mathcal{L}_{k_j}$. In the course of the substitutions different expressions for $\mathcal{L}_1$ may result, depending on which relations are chosen, but the final equations are the same. The justification via two-timing proceeds as before.

For nonlinear problems, the usual situation concerns a system of equations with a corresponding Lagrangian $L\{\varphi_t^{(\alpha)}, \varphi_x^{(\alpha)}, \varphi^{(\alpha)}\}$ involving only the $\varphi^{(\alpha)}$ and their first derivatives. However, it is typical that for some of the φ, only the derivatives appear in L; they are "potentials" in that only the derivatives φ_t, φ_x represent physical quantities. This requires a highly nontrivial extension with important mathematical and physical consequences. In the uniform wavetrain solution any potential variable $\tilde{\varphi}$ must be expressed as

$$\tilde{\varphi} = \boldsymbol{\beta} \cdot \mathbf{x} - \gamma t + \tilde{\Phi}(\theta), \qquad \theta = \mathbf{k} \cdot \mathbf{x} - \omega t, \qquad (14.62)$$

in order to ensure complete generality. The physical quantities involve only

$$\tilde{\varphi}_t = -\gamma - \omega \tilde{\Phi}_\theta, \qquad \tilde{\varphi}_x = \boldsymbol{\beta} + \mathbf{k} \tilde{\Phi}_\theta, \qquad (14.63)$$

and $-\gamma, \beta$ represent the mean values. These are important physical quantities; in water waves they give the mean fluid velocity and mean height, for example. Moreover, a most important nonlinear effect is the coupling of modulations in the wavetrain to similar slow variations in these mean quantities. Thus in the modulation theory the term $\beta \cdot \mathbf{x} - \gamma t$ must be generalized to a function $\tilde{\theta}(\mathbf{x}, t)$ and γ, β defined by

$$\gamma = -\tilde{\theta}_t, \qquad \beta = \tilde{\theta}_{\mathbf{x}}. \tag{14.64}$$

The function $\tilde{\theta}$ is similar to θ and is a pseudo-phase appearing in the problem. The quantities γ and β are a pseudo-frequency and a pseudo-wave number. Furthermore, each potential $\tilde{\varphi}$ has the term $L_{\tilde{\varphi}}$ missing in its Euler equation

$$\frac{\partial}{\partial t} L_{\tilde{\varphi}_t} + \frac{\partial}{\partial \mathbf{x}} \cdot L_{\tilde{\varphi}_{\mathbf{x}}} = 0; \tag{14.65}$$

in the course of the analysis, this always allows an extra integral and an extra parameter B to be introduced similar to A. The triads (γ, β, B) are similar, although subsidiary, to the main triad $(\omega, \mathbf{k}, A)$.

The two-timed form corresponding to (14.62) is

$$\tilde{\varphi}(\mathbf{x}, t) = \epsilon^{-1} \tilde{\Theta}(\mathbf{X}, T) + \tilde{\Phi}(\theta, \mathbf{X}, T; \epsilon),$$

where

$$\gamma(\mathbf{X}, T) = -\tilde{\Theta}_T, \qquad \beta(\mathbf{X}, T) = \tilde{\Theta}_{\mathbf{X}}, \qquad \mathbf{X} = \epsilon \mathbf{x}, \qquad T = \epsilon t,$$

and $\tilde{\Phi}$ is chosen to be periodic in θ. For a Lagrangian

$$L(\varphi_t, \varphi_{\mathbf{x}}, \varphi, \tilde{\varphi}_t, \tilde{\varphi}_{\mathbf{x}}),$$

it may be shown that the two-timed equations *and the conditions that Φ and $\tilde{\Phi}$ be 2π-periodic in θ*, are equivalent to an exact variational principle similar to (14.44). To lowest order it is

$$\delta \int \int \frac{1}{2\pi} \int_0^{2\pi} L\left(-\omega \Phi_\theta, \mathbf{k}\Phi_\theta, \Phi, -\gamma - \omega \tilde{\Phi}_\theta, \beta + \mathbf{k}\tilde{\Phi}_\theta\right) d\theta \, d\mathbf{X} \, dT = 0. \tag{14.66}$$

The variational equations corresponding to $\delta \Phi$ and $\delta \tilde{\Phi}$ determine the functions Φ and $\tilde{\Phi}$ and we have the two integrals

$$(-\omega L_1 + \mathbf{k} \cdot L_2 - \omega L_4 + \mathbf{k} \cdot L_5)\Phi_\theta - L = A(\mathbf{X}, T), \tag{14.67}$$

$$-\omega L_4 + \mathbf{k} \cdot L_5 = B(\mathbf{X}, T). \tag{14.68}$$

The variations $\delta\Theta$ and $\delta\tilde{\Theta}$ lead to the two secular conditions

$$\frac{\partial}{\partial t}\bar{L}_\omega - \frac{\partial}{\partial \mathbf{X}}\cdot\bar{L}_\mathbf{k} = 0, \qquad \frac{\partial}{\partial T}\bar{L}_\gamma - \frac{\partial}{\partial \mathbf{X}}\cdot\bar{L}_\beta = 0.$$

Finally, a Hamiltonian transformation can be introduced as before, based on generalized momenta $\partial L/\partial\Phi_\theta$, $\partial L/\partial\tilde{\Phi}_\theta$, and (14.67)–(14.68) can be used to eliminate explicit dependence on Φ and $\tilde{\Phi}$ in favor of including the parameters A and B in the variational principle. We then have

$$\delta \int \int \mathcal{L}(\omega,\mathbf{k},A,\gamma,\beta,B)\,d\mathbf{X}\,dT = 0, \qquad (14.69)$$

and the variational equations are

$$\mathcal{L}_A = 0, \qquad \mathcal{L}_B = 0, \qquad (14.70)$$

$$\frac{\partial}{\partial T}\mathcal{L}_\omega - \frac{\partial}{\partial X_j}\mathcal{L}_{k_j} = 0, \qquad \frac{\partial}{\partial T}\mathcal{L}_\gamma - \frac{\partial}{\partial X_j}\mathcal{L}_{\beta_j} = 0, \qquad (14.71)$$

together with the consistency conditions

$$\frac{\partial k_i}{\partial T} - \frac{\partial \omega}{\partial X_i} = 0, \qquad \frac{\partial \beta_i}{\partial T} - \frac{\partial \gamma}{\partial X_i} = 0, \qquad (14.72)$$

$$\frac{\partial k_i}{\partial X_j} - \frac{\partial k_j}{\partial X_i} = 0, \qquad \frac{\partial \beta_i}{\partial X_j} - \frac{\partial \beta_j}{\partial X_i} = 0. \qquad (14.73)$$

Further details of the procedure and examples are given in the original papers (Whitham, 1965, 1967, 1970). An application to water waves on finite depth, where the extra parameters are crucially important, will be given in Chapter 16.

In this more general case, the energy equation that corresponds via Noether's theorem to the invariance of (14.69) with respect to shifts in T is now

$$\frac{\partial}{\partial T}(\omega\mathcal{L}_\omega + \gamma\mathcal{L}_\gamma - \mathcal{L}) + \frac{\partial}{\partial X_j}(-\omega\mathcal{L}_{k_j} - \gamma\mathcal{L}_{\beta_j}) = 0. \qquad (14.74)$$

The momentum equation which corresponds to the invariance with respect to shifts in X_i is

$$\frac{\partial}{\partial T}(k_i\mathcal{L}_\omega + \beta_i\mathcal{L}_\gamma) + \frac{\partial}{\partial X_j}(-k_i\mathcal{L}_{k_j} - \beta_i\mathcal{L}_{\beta_j} + \mathcal{L}\delta_{ij}) = 0. \qquad (14.75)$$

The final extension is to note that if the medium is not constant but depends on $\mathbf{X}, T$, these will appear explicitly in the Lagrangian and therefore in $\mathcal{L}$. But the variational equations (14.70)–(14.73) are unchanged. The conservation equations (14.74) and (14.75), however, have terms $-\mathcal{L}_T, \mathcal{L}_{X_i}$, respectively, on the right hand sides [as may be verified directly from (14.70)–(14.73)].

14.8 Adiabatic Invariants

It was remarked previously that the quantities $\mathcal{L}_\omega, \mathcal{L}_{k_j}$ are analogous to the adiabatic invariants of classical mechanics. This correspondence can now be explored. In mechanics the setting is the theory of slow modulations for vibrating systems. The only independent variable is time, so in this case the modulations can be produced only by externally imposed changes in some parameter $\lambda(t)$. (This corresponds to a varying medium in the case of waves.) The classical theory is usually developed by Hamiltonian methods, which are not directly applicable to waves, but we may instead derive the simplest of the classical results by the methods developed here. For an oscillator with one degree of freedom $q(t)$ and one slowly varying parameter $\lambda(t)$, the variational principle is

$$\delta \int_{t_1}^{t_2} L(q, \dot{q}, \lambda)\, dt = 0,$$

and the variational equation is

$$\frac{d}{dt} L_{\dot{q}} - L_q = 0. \tag{14.76}$$

This case is covered by the arguments of Sections 14.3 and 14.4 simply by dropping the dependence on x. But it is useful to note the steps separately in the usual notation of mechanics. We follow the simple intuitive approach of Section 14.3; it is justified by Section 14.4.

We first calculate the average Lagrangian for the periodic motion with λ held fixed. If the period is $\tau = 2\pi/\nu$, then

$$\mathcal{L} = \frac{\nu}{2\pi} \int_0^\tau L\, dt. \tag{14.77}$$

In the periodic motion ($\lambda = $ constant), (14.76) has the energy integral

$$\dot{q} L_{\dot{q}} - L = E. \tag{14.78}$$

This may be solved, in principle, to express $\dot{q}$ as a function of (q, E, λ) and then the generalized momentum $p = L_{\dot{q}}$ can also be expressed as

$$p = p(q, E, \lambda).$$

If (14.78) is used to replace L in (14.77), we have

$$\mathcal{L} = \frac{\nu}{2\pi} \int_0^\tau p\dot{q}\, dt - E$$

$$= \frac{\nu}{2\pi} \oint p(q, E, \lambda)\, dq - E, \tag{14.79}$$

where $\oint p\, dq$ means the integral over one complete period of oscillation [a closed loop in the (p, q) plane]. We now allow slow variations of λ, with consequent slow changes of ν and E, and use the average variational principle

$$\delta \int_{t_0}^{t_1} \mathcal{L}(\nu, E, \lambda)\, dt = 0. \tag{14.80}$$

It is again crucial to define ν as the derivative $\dot{\theta}$ of a phase $\theta(t)$ which increases by a constant normalized amount in one oscillation. This step looks perhaps less natural than in the waves case, but it becomes clear in the two-timing. The variations of (14.80) with respect to E and θ give

$$\mathcal{L}_E = 0, \qquad \frac{d}{dt}\mathcal{L}_\nu = 0, \tag{14.81}$$

respectively. The first of these corresponds to the dispersion relation (14.28) and the second corresponds to the conservation equation (14.29). In view of (14.79) we have

$$\mathcal{L}_\nu = \frac{1}{2\pi} \oint p\, dq = \text{constant}, \tag{14.82}$$

which is just the classical result of the adiabatic invariant. As the system is modulated, ν and E vary individually but

$$I(\nu, E) = \frac{1}{2\pi} \oint p\, dq \tag{14.83}$$

remains constant. From (14.79) and (14.81) the period is given by

$$\tau = \frac{2\pi}{\nu} = I_E, \tag{14.84}$$

which is also classical. (An excellent account of the usual theory may be found in Landau and Lifshitz, 1960b, p. 154.)

In the two-timed form (14.59), the quantity Π is defined as $\partial L / \partial \Phi_\theta$, whereas the generalized momentum p is $\partial L / \partial \varphi_t$. Since $\varphi_t = \nu \Phi_\theta$, to lowest order, we have $\Pi = \nu p$ and the expressions (14.59) and (14.79) agree.

It is clear from this comparison that in the case of waves $\mathcal{L}_\omega$ is akin to the adiabatic invariant and that the $\mathcal{L}_{k_j}$ are similar quantities for spatial modulations. In waves there is no need for an external drain of energy, since modulations in time can be balanced by modulations in space. If the medium is not constant, however, we have the additional effect of parameters analogous to λ, but the equation

$$\frac{\partial}{\partial t} \mathcal{L}_\omega - \frac{\partial}{\partial x_j} \mathcal{L}_{k_j} = 0 \qquad (14.85)$$

still holds. The equation has become known as the conservation of wave action.

In the special case of a wavetrain uniform in space but responding to changes of the medium in time we have

$$\mathcal{L}_\omega = \text{constant.}$$

Alternatively, for a wavetrain of fixed frequency moving into a medium dependent on one space dimension x, we have

$$\mathcal{L}_k = \text{constant.}$$

These provide simple determinations of the amplitude. In general, modulations in space and time balance according to (14.85) and produce a *propagation* of the modulations.

The quantities $\mathcal{L}_\gamma$ and $\mathcal{L}_{\beta_j}$ in (14.71) are similar to $\mathcal{L}_\omega$ and $\mathcal{L}_{k_j}$. They arise because of the extra dependent variables, just as ordinary dynamical systems (involving only the time) may have further adiabatic invariants when there are more degrees of freedom. These wave systems have only one genuine frequency and so correspond to the degenerate cases of equal frequencies in dynamics.

14.9 Multiple-Phase Wavetrains

The general case of multiply periodic motions in dynamics would be mirrored in wave theory by wavetrains with more than one genuine phase function. It is straightforward to extend the formalism to this case but questions of existence suggest caution. For a two-phase wavetrain, for example, the starting point would be a quasi-periodic solution

$$\varphi = \Psi(\theta_1, \theta_2), \qquad \theta_1 = k_1 x - \omega_1 t, \qquad \theta_2 = k_2 x - \omega_2 t, \qquad (14.86)$$

in which Ψ is 2π periodic in both θ_1 and θ_2. One would then go on to handle modulation theory as before. However, even in ordinary dynamics questions of the existence of quasi-periodic solutions are difficult ones in the nonlinear case, involving the well-known problems of small divisors, so there may be considerable difficulties hidden under the formalism. If the existence of solutions (14.86) and of neighboring modulated solutions is simply assumed, modulation equations can be developed as before. Ablowitz and Benney (1970) and Ablowitz (1971) have pursued some of the consequences. Delaney (1971) notes that the variational formalism goes through. If modulated wavetrains can be described by

$$\varphi = \Phi(\theta_1, \theta_2, X, T; \epsilon),$$

$$\theta_1 = \epsilon^{-1} \Theta_1(X, T), \qquad \theta_2 = \epsilon^{-1} \Theta_2(X, T),$$

it is straightforward to show that the two-timed equation for Φ and the *two* periodicity conditions follows from the variational principle

$$\delta \int \int \overline{L} \, dX \, dT = 0,$$

$$\overline{L} = \frac{1}{4\pi^2} \int_0^{2\pi} \int_0^{2\pi} L\left(\nu_1 \Phi_{\theta_1} + \nu_2 \Phi_{\theta_2} + \epsilon \Phi_T, k_1 \Phi_{\theta_1} + k_2 \Phi_{\theta_2} + \epsilon \Phi_X, \Phi\right) d\theta_1 \, d\theta_2.$$

Modulation equations can then be developed as before.

14.10 Effects of Damping

As in Hamiltonian dynamics, the variational formalism applies naturally to conservative systems; dissipative effects have to be tacked on a little awkwardly as nonzero right hand sides to the previous equations. However, various canonical forms can be maintained and the left hand sides can still be written in terms of the Lagrangian. To illustrate this, we consider as a specific example the equation

$$\varphi_{tt} - \varphi_{xx} + V'(\varphi) = -\epsilon D(\varphi, \varphi_t),$$

where $\epsilon D(\varphi, \varphi_t)$ represents small dissipative effects. The two-timed equation corresponding to (14.42) is

$$\frac{\partial}{\partial\theta}\left\{\frac{1}{2}(\nu^2-k^2)\Phi_\theta^2+V(\Phi)-\frac{1}{2}\epsilon^2\Phi_T^2+\frac{1}{2}\epsilon^2\Phi_X^2\right\}$$

$$+\epsilon\frac{\partial}{\partial T}\left\{\nu\Phi_\theta^2+\epsilon\Phi_\theta\Phi_T\right\}-\epsilon\frac{\partial}{\partial X}\left\{k\Phi_\theta^2+\epsilon\Phi_\theta\Phi_X\right\}$$

$$=-\epsilon\Phi_\theta D(\Phi,\nu\Phi_\theta+\epsilon\Phi_T).$$

To lowest order, we have

$$\frac{1}{2}(\nu^2-k^2)\Phi_\theta^2+V(\Phi)=A(X,T) \tag{14.87}$$

and the periodicity condition

$$\frac{\partial}{\partial T}\int_0^{2\pi}\nu\Phi_\theta^2 d\theta-\frac{\partial}{\partial X}\int_0^{2\pi}k\Phi_\theta^2 d\theta=-\int_0^{2\pi}\Phi_\theta D(\Phi,\nu\Phi_\theta)d\theta. \tag{14.88}$$

From (14.87), Φ_θ can be expressed as a function of Φ, ν, k, A, and the integrals in (14.88) can all be written as loop integrals. We have

$$\frac{\partial}{\partial T}\mathcal{L}_\nu+\frac{\partial}{\partial X}\mathcal{L}_k=-\mathcal{D}, \tag{14.89}$$

where

$$\mathcal{L}(\nu,k,A)=\frac{1}{2\pi}\left\{2(\nu^2-k^2)\right\}^{1/2}\oint\left\{A-V(\Phi)\right\}^{1/2}d\Phi-A,$$

as before, and

$$\mathcal{D}(\nu,k,A)=\frac{1}{2\pi}\oint D(\Phi,\Phi_\theta)d\Phi.$$

To (14.89) are added

$$\mathcal{L}_A=0,\qquad \frac{\partial k}{\partial T}-\frac{\partial\nu}{\partial X}=0, \tag{14.90}$$

to complete the set of equations for ν, k, A. Equation 14.89 shows the loss in wave action due to dissipation.

Here we have returned to two-timing on the equations but retained the canonical forms suggested by the Lagrangian for the conservative part. This is obviously less desirable than two-timing directly some extended principle. Recently Jimenez (1972) has had some success in deriving results such as (14.89) by Prigogine's approach to irreversible systems (Donnelly et al., 1966).

CHAPTER 15

Group Velocities, Instability, and Higher Order Dispersion

Most of Chapter 14 was concerned with questions of formulation. We now study the modulation equations and their solutions in some detail and emphasize the important differences between linear and nonlinear theory. In this chapter we consider the basic case of one dimensional waves in a uniform medium and, for simplicity, suppose that pseudo-frequencies and wave numbers do not arise. For the present the nonlinear Klein-Gordon equation and the problems noted in Section 14.1 can be taken as typical applications for the theory. More specific applications in nonlinear optics and water waves will be given in the next chapter. Extensions to more dimensions, nonuniform media, and higher order systems will be included in those particular contexts.

Modulated wavetrains are described to all orders of approximation by the variational principle (14.44). In the lowest order approximation we have (14.47)–(14.48), and by means of the Hamiltonian transformation we obtain the average variational principle

$$\delta \int \int \mathcal{L}(\omega, k, A)\, dx\, dt = 0,$$

where $\omega = -\theta_t$ and $\theta_x = k$. (We drop the two-timing notation and revert to the earlier form, except when the precise ordering of terms again becomes an issue.) In this lowest approximation the variational equations for A, ω, k are

$$\mathcal{L}_A = 0, \tag{15.1}$$

$$\frac{\partial}{\partial t} \mathcal{L}_\omega - \frac{\partial}{\partial x} \mathcal{L}_k = 0, \tag{15.2}$$

$$\frac{\partial k}{\partial t} + \frac{\partial \omega}{\partial x} = 0. \tag{15.3}$$

We first study these equations and then return to (14.44) to incorporate higher order dispersive effects.

511

15.1 The Near-Linear Case

Before the main discussion, we note how the near-linear equations obtained directly in Section 14.2 fit into the general formalism. The near-linear theory is obtained by expanding $\mathfrak{L}$ in powers of the amplitude. This expansion may be taken as

$$\mathfrak{L} = P(\omega,k)A + P_2(\omega,k)A^2 + \cdots. \tag{15.4}$$

but it would usually be derived from Fourier series, as in (14.52), in the equivalent form

$$\mathfrak{L} = G(\omega,k)a^2 + G_2(\omega,k)a^4 + \cdots.$$

The dispersion relation $\mathfrak{L}_a = 0$ is solved to give

$$\omega = \omega_0(k) + \omega_2(k)a^2 + \cdots,$$

where

$$G(\omega_0,k) = 0, \qquad \omega_2 = -\frac{2G_2(\omega_0,k)}{G_\omega(\omega_0,k)}.$$

Equations 15.2 and 15.3 become

$$\frac{\partial}{\partial t}\left\{ g(k)a^2 + \cdots \right\} + \frac{\partial}{\partial x}\left\{ g(k)\omega_0'(k)a^2 + \cdots \right\} = 0,$$

$$\frac{\partial k}{\partial t} + \frac{\partial}{\partial x}\left\{ \omega_0(k) + \omega_2(k)a^2 + \cdots \right\} = 0,$$

where $g(k) = G_\omega(\omega_0,k)$ and the relation

$$\omega_0'(k) = -\frac{G_k(\omega_0,k)}{G_\omega(\omega_0,k)}$$

for the linear group velocity has been used. The coefficient $g(k)$ can be factored out, in view of the second equation for k, and a sufficient approximation, as explained in Section 14.2, is

$$\frac{\partial a^2}{\partial t} + \frac{\partial}{\partial x}\left\{ \omega_0'(k)a^2 \right\} = 0,$$

$$\frac{\partial k}{\partial t} + \omega_0'(k)\frac{\partial k}{\partial x} + \omega_2(k)\frac{\partial a^2}{\partial x} = 0.$$

The characteristic equations were found to be

$$\frac{1}{2}\left\{\frac{\omega_0''(k)}{\omega_2(k)}\right\}^{1/2}(\operatorname{sgn}\omega_0'')\,dk\pm da=0,$$

$$\frac{dx}{dt}=\omega_0'(k)\pm\left\{\omega_2(k)\omega_0''(k)\right\}^{1/2}a.$$

(15.5)

If (15.4) is used, we have the equivalent results with a replaced by $A^{1/2}$ and G,G_2 replaced by P,P_2 in the definition of ω_2.

We now consider the exact equations (15.1)–(15.3).

15.2 Characteristic Form of the Equations

There are two useful forms of the characteristic equations depending on whether the symmetry between t and x variables is maintained. If symmetry is maintained, it is convenient on this occasion to work with θ rather than ω and k. Then (15.2) becomes

$$\mathcal{L}_{\omega\omega}\,\theta_{tt}-2\mathcal{L}_{\omega k}\,\theta_{tx}+\mathcal{L}_{kk}\,\theta_{xx}-\mathcal{L}_{\omega A}\,A_t+\mathcal{L}_{kA}\,A_x=0.$$

The derivatives A_t,A_x may be eliminated in favor of θ from (15.1) for, on taking its t and x derivatives, we have

$$-\mathcal{L}_{\omega A}\,\theta_{tt}+\mathcal{L}_{kA}\,\theta_{tx}+\mathcal{L}_{AA}\,A_t=0,$$

$$-\mathcal{L}_{\omega A}\,\theta_{tx}+\mathcal{L}_{kA}\,\theta_{xx}+\mathcal{L}_{AA}\,A_x=0.$$

The second order equation for θ becomes

$$p\theta_{tt}-2r\theta_{tx}+q\theta_{xx}=0,\qquad\qquad(15.6)$$

where

$$p=\mathcal{L}_{\omega\omega}\mathcal{L}_{AA}-\mathcal{L}_{\omega A}^2,$$

$$q=\mathcal{L}_{kk}\,\mathcal{L}_{AA}-\mathcal{L}_{kA}^2,$$

$$r=\mathcal{L}_{\omega k}\,\mathcal{L}_{AA}-\mathcal{L}_{\omega A}\,\mathcal{L}_{kA}.$$

From (15.1), A may be expressed in terms of θ_t,θ_x, and then (15.6) is a second order quasi-linear equation for θ. The characteristics are

$$\frac{dx}{dt}=\frac{-r\pm\sqrt{r^2-pq}}{p}.$$

In the linear case

$$\mathcal{L} = P(\omega, k)A,$$

$$p = -P_\omega^2, \qquad q = -P_k^2, \qquad r = -P_\omega P_k,$$

and we have the double characteristic velocity

$$\frac{dx}{dt} = -\frac{P_k}{P_\omega};$$

this is just the linear group velocity. The near-linear results can be recovered similarly.

If x, t symmetry is abandoned, one useful possibility is to choose k and $I = \mathcal{L}_\omega$ as the dependent variables and suppose that

$$I = \mathcal{L}_\omega, \qquad J = -\mathcal{L}_k, \qquad \mathcal{L}_A = 0$$

are solved for the functions

$$\omega(k, I), \qquad J(k, I), \qquad A(k, I).$$

Then $\mathcal{L}$ may also be evaluated as

$$\mathfrak{M}(k, I) = \mathcal{L}\{\omega(k, I), k, A(k, I)\}.$$

We have

$$\mathfrak{M}_k = \omega_k I - J, \qquad \mathfrak{M}_I = \omega_I I; \qquad (15.7)$$

hence from $\mathfrak{M}_{kI} = \mathfrak{M}_{Ik}$,

$$\omega_k = J_I. \qquad (15.8)$$

The system of equations (15.1)–(15.3) reduces to

$$k_t + \omega_k k_x + \omega_I I_x = 0,$$

$$I_t + \omega_k I_x + J_k k_x = 0. \qquad (15.9)$$

The characteristic equations are

$$\sqrt{J_k} \, dk \pm \sqrt{\omega_I} \, dI = 0 \qquad (15.10)$$

on

$$\frac{dx}{dt} = \omega_k \pm \sqrt{\omega_I J_k}. \qquad (15.11)$$

This choice of variables keeps a much closer relation with the earlier discussions of the linear and near-linear cases.

In the linear case, $\mathcal{L} = P(\omega, k)A$, the dispersion relation $P(\omega, k) = 0$ is solved in the form

$$\omega = \omega_0(k),$$

and we have

$$J = -P_k A = -\frac{P_k I}{P_\omega} = \omega_0'(k)I.$$

Since $\omega_I = 0$, the characteristic velocities (15.11) both reduce to $\omega_0'(k)$. The system is not strictly hyperbolic, as noted earlier, because there is only one differential form, $dk = 0$, provided by (15.10). However, once $k(x, t)$ has been found, I is obtained from integration of

$$I_t + \omega_0'(k)I_k + \omega_0''(k)Ik_x = 0$$

along the same characteristics.

In the near-linear case with $\mathcal{L}$ given by (15.4), we have

$$\mathcal{L}_A = P(\omega, k) + 2P_2(\omega, k)A + \cdots = 0,$$

$$I = \mathcal{L}_\omega = P_\omega(\omega, k)A + \cdots, \quad J = -\mathcal{L}_k = -P_k(\omega, k)A + \cdots.$$

These may be solved to give

$$\omega = \omega_0(k) + \tilde{\omega}_2(k)I + \cdots,$$
$$J = \omega_0'(k)I + \cdots. \tag{15.12}$$

The characteristic velocities (15.11) are

$$\frac{dx}{dt} = \omega_0'(k) \pm \sqrt{\omega_0''(k)\tilde{\omega}_2(k)I} + \cdots. \tag{15.13}$$

These check with (15.5), for if $I = g(k)a^2$, then

$$\omega = \omega_0(k) + \omega_2(k)a^2 + \cdots,$$

with

$$\omega_2(k) = g(k)\tilde{\omega}_2(k),$$

and (15.13) becomes the same as (15.5).

Hayes (1973) notes that if a partial Hamiltonian transformation

$$\mathcal{H}(k,I) = \omega \mathcal{L}_\omega - \mathcal{L} = \omega I - \mathcal{L} \tag{15.14}$$

is introduced at the same time as k and I, we have

$$J = \mathcal{H}_k, \qquad \omega = \mathcal{H}_I. \tag{15.15}$$

These may be seen also from (15.7) with $\mathcal{M} = \omega I - \mathcal{H}$. The equations become

$$\frac{\partial k}{\partial t} + \frac{\partial}{\partial x} \mathcal{H}_I = 0,$$

$$\frac{\partial I}{\partial t} + \frac{\partial}{\partial x} \mathcal{H}_k = 0. \tag{15.16}$$

The characteristic equations are

$$\sqrt{\mathcal{H}_{kk}} \, dk \pm \sqrt{\mathcal{H}_{II}} \, dI = 0,$$

$$\frac{dx}{dt} = \mathcal{H}_{Ik} \pm \sqrt{\mathcal{H}_{kk} \mathcal{H}_{II}} \,. \tag{15.17}$$

In specific cases, other choices of variables may lead to the simplest expressions. For the Klein-Gordon example (14.26),

$$\mathcal{L} = (\omega^2 - k^2)^{1/2} F(A) - A,$$

$$F(A) = \frac{1}{2\pi} \oint \{2(A - V(\Psi))\}^{1/2} d\Psi; \tag{15.18}$$

it turns out that the phase velocity $U = \omega/k$ and A are the most convenient variables to work with. From the dispersion relation $\mathcal{L}_A = 0$,

$$k = \frac{1}{(U^2 - 1)^{1/2} F'(A)}, \qquad \omega = \frac{U}{(U^2 - 1)^{1/2} F'(A)};$$

equations 15.2 and 15.3 become

$$\frac{\partial}{\partial t} \left\{ \frac{UF}{(U^2 - 1)^{1/2}} \right\} + \frac{\partial}{\partial x} \left\{ \frac{F}{(U^2 - 1)^{1/2}} \right\} = 0,$$

$$\frac{\partial}{\partial t} \left\{ \frac{1}{(U^2 - 1)^{1/2} F'} \right\} + \frac{\partial}{\partial x} \left\{ \frac{U}{(U^2 - 1)^{1/2} F'} \right\} = 0.$$

The characteristic equations are found to be

$$\frac{dU}{U^2-1} \mp \left(-\frac{F''}{F}\right)^{1/2} dA = 0,$$

(15.19)

$$\frac{dx}{dt} = \frac{1 \pm U(-FF''/F'^2)^{1/2}}{U \pm (-FF''/F'^2)^{1/2}}.$$

(15.20)

More Dependent Variables.

When there are more dependent variables and more equations, as in (14.70)–(14.73), the number of characteristics is increased corresponding to the order of the system. The additional characteristics refer to nonlinear coupling of the wavetrain with changes in mean background quantities and are quite different from the linear group velocity. The formulas for those two velocities (associated primarily with propagation of k and A) that *do* correspond to the linear group velocity are considerably modified. In particular, in these cases, the type may be incorrectly given if the extra dependence is overlooked and the simpler formulas above are used. General formulas for the characteristics will not be developed, since the most useful choice of variables depends strongly on the particular problem at hand. Typical examples are provided by the later discussions of the Korteweg-deVries equation and Stokes waves in water of finite depth.

15.3 Type of the Equations and Stability

The type of equations can be read off according to whether the characteristics are real or imaginary. The condition for the system to be hyperbolic may be taken in any of the equivalent forms

$$r^2 - pq > 0, \qquad \omega_I J_k > 0, \qquad \mathcal{H}_{kk}\,\mathcal{H}_{II} > 0,$$

(15.21)

the second being closest in form to the near-linear result $\omega_2\omega_0'' > 0$. With opposite signs the system is elliptic.

As remarked in Section 14.2, the periodic wavetrains are unstable in a certain sense when the modulation equations are elliptic. To see this, we note that the modulation equations take the general form

$$\frac{\partial u_i}{\partial t} + a_{ij}(\mathbf{u})\frac{\partial u_j}{\partial x} = 0.$$

(15.22)

In a uniform periodic wavetrain $\mathbf{u}$ takes constant values $\mathbf{u}^{(0)}$, say. For small perturbations we take $\mathbf{u} = \mathbf{u}^{(0)} + \mathbf{u}^{(1)}$. The linearized equations for $\mathbf{u}^{(1)}$ are

$$\frac{\partial u_i^{(1)}}{\partial t} + a_{ij}^{(0)} \frac{\partial u_j^{(1)}}{\partial x} = 0, \qquad a_{ij}^{(0)} = a_{ij}(\mathbf{u}^{(0)}).$$

This system has solutions

$$\mathbf{u}^{(1)} \propto e^{i\mu(x - Ct)},$$

where

$$|a_{ij}^{(0)} - C\delta_{ij}| = 0. \tag{15.23}$$

The possible values of C are the characteristic velocities evaluated for $\mathbf{u} = \mathbf{u}^{(0)}$ [see (5.12)]. If any of the C's are complex, some of the solutions of $\mathbf{u}^{(1)}$ grow exponentially in time. Of course as usual in simple linear stability analysis this only indicates large deviations from the uniform state will occur, not necessarily that the wavetrain becomes chaotic. In the present context, the stability and the possible eventual states are considerably affected by higher order terms in the modulation approximation, as will be discussed in Section 15.5.

For the nonlinear Klein-Gordon equation (15.20), for wavetrains satisfying (14.6) with $F(A) > 0$, we have

$$\begin{aligned}
\text{Hyperbolic:} \quad & F'' < 0, \\
\text{Elliptic:} \quad & F'' > 0.
\end{aligned} \tag{15.24}$$

In particular, when $V(\Psi) = \frac{1}{2}\Psi^2 + \sigma\Psi^4$, the system is hyperbolic for $\sigma > 0$ and elliptic for $\sigma < 0$. For any even function $V(\Psi)$, the first terms of the near-linear expansion can be put in this form and the type depends in the same way on the sign of σ.

For the Sine-Gordon equation $V(\Psi) = 1 - \cos\Psi$, it may be shown that $F''(A) > 0$; the periodic wavetrains are unstable. This result applies to oscillations about $\Psi = 0$, which satisfy (14.6). We shall later note the existence of helical wavetrains in which Ψ increases or decreases monotonically. These give periodic solutions, since the same physical state is recovered after each change by 2π. They turn out to be stable in the sense considered here.

15.4 Nonlinear Group Velocity, Group Splitting, Shocks

In the hyperbolic case, the characteristic velocities are taken to define the *nonlinear group velocities*. This is the natural extension of the linear case. The splitting of the double characteristic velocity of linear theory into two different velocities is probably the most important and far-reaching result of the theory. As noted in Section 14.2, it predicts the eventual splitting of a modulation of finite extent into two separate disturbances, a result quite different from linear theory. In problems where the linear group velocity is positive, the two nonlinear group velocities will usually both be positive also. Then modulations introduced by a source at the origin will propagate on both families of characteristics as shown in Fig. 15.1. Ideally, we take a source at $x=0$ producing a highly nonlinear wavetrain up to $t=0$; it then modulates amplitude and frequency for a finite time t_0, after which it returns to producing the original wavetrain. Notice that two boundary conditions should be applied at $x=0$, so that independent distributions of a and ω may be introduced. By the usual arguments of Part I, there will be a certain interaction period, but the disturbance will ultimately separate into two simple waves on the C_+ and C_- characteristics as shown. This is analogous to Riemann's initial value problem (Section 6.12).

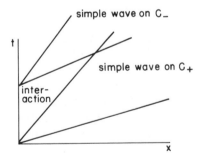

Fig. 15.1. Group splitting.

We can make an estimate of the distance to separation in terms of the difference between the characteristic velocities. We have

$$x \approx \frac{C_+ C_-}{C_+ - C_-} t_0,$$

where $C_\pm$ are typical values of the characteristic velocities and t_0 is the time the modulation lasts at the source. The velocities in the near-linear theory (15.5) provide the upper estimate

$$x \approx \frac{\omega_0'^2 t_0}{2a(\omega_2 \omega_0'')^{1/2}} .$$
(15.25)

It would be extremely valuable to have experimental evidence of this separation, since it is of fundamental importance in assessing the modulation theory. Other related nonlinear effects have been observed in nonlinear optics, but this one does not appear to have been pursued yet.

Simple wave solutions, produced as described above or otherwise, can be obtained analytically by the standard theory of Part I. One Riemann invariant is constant throughout, and the modulation variables ω, k, a remain constant along the individual characteristics of the appropriate family. In the linear theory, k remains constant but $a \propto t^{-1/2}$ along the characteristics. This difference between nonlinear and linear behavior is probably less easily detectable than the group splitting, and may be partially masked by higher order effects.

Finally, among the group of hyperbolic topics, we have the question of breaking and shocks. The dependence of the characteristic velocities on the modulation variables introduces the usual hyperbolic distortion, and "compressive" modulations in a simple wave solution will develop multivalued regions. What happens then is not clear at present. Unlike the problems treated in Part I, there is no objection to multivalued solutions as such. They would be interpreted as the superposition of two or more wavetrains with different ranges of k and a. The actual solutions would not be described correctly by (15.1)–(15.3), since these equations were derived presupposing a single phase function. But they would presumably be covered by the original equation. Certainly superposition is what happens in the linear case. Although the question was not raised in the earlier discussions, it is conceivable to set up a wavetrain with the linear group velocity $C_0(k)$ decreasing toward the front. Then, since values of k propagate with velocity $C_0(k)$, there would eventually be overlapping parts of the wavetrain. There is no problem about superposition in the linear theory and the whole process could be studied by the exact solution in Fourier integrals. Although the corresponding process would be hard to follow through analytically in the nonlinear case, this qualitative behavior seems perfectly feasible. Something like the multiphase solutions referred to in Section 14.9 would be needed within the overlap region, but the transition process poses a difficult problem.

A second possibility is that higher order terms in the modulation approximation become important near breaking and prevent the development of a multivalued solution. It is easy to see in general (and will be shown in some detail in the next section) that higher order effects typically introduce additional terms involving third derivatives into (15.2) and (15.3). They become similar in form to the Boussinesq and Korteweg-deVries equations. By analogy, it is expected that breaking is suppressed by the extra terms. Of course, just as in the case of water waves, the additional terms are introduced as small corrections for long length scales, and are the first terms in an infinite series of higher derivatives. To accept their dominating effect on breaking in all cases may be inconsistent. It is more likely that this applies to small symmetric modulations and these develop into a series of solitary waves, whereas strong unsymmetrical ones break in some sense.

Finally, we have the question of shocks. Formally, discontinuities in ω, k, A can be allowed in the solutions of (15.2) and (15.3). These would be interpreted as weak solutions and the shock conditions would be taken from appropriate conservation equations as described in Section 5.8. This is the most fascinating possibility theoretically, but it is probably least likely as a description of reality in this particular context. The unsureness as to interpretation also makes the choice of appropriate shock conditions less clear. Equations 15.2 and 15.3 are already in conservation form, but the equations for conservation of energy and momentum are also obvious candidates. The latter are

$$\frac{\partial}{\partial t}(\omega \mathcal{L}_\omega - \mathcal{L}) - \frac{\partial}{\partial x}(\omega \mathcal{L}_k) = 0, \qquad (15.26)$$

$$\frac{\partial}{\partial t}(k \mathcal{L}_\omega) - \frac{\partial}{\partial x}(k \mathcal{L}_k - \mathcal{L}) = 0. \qquad (15.27)$$

There are, in fact, an infinite number of conservation equations. However, (15.2), (15.3), (15.26), and (15.27) are the only ones with clear significance. Our system is essentially second order, [(15.1) is not a differential equation], so two conservation equations must be chosen to provide the two shock conditions. The choice is tied to what the shocks are supposed to represent. If they are taken to be approximations to solutions still covered by the original detailed equation for φ, then we should choose shock conditions from (15.26) and (15.27), the argument being that energy and momentum are conserved in the more detailed description for φ and should therefore be retained in the slowly varying approximation. The shock conditions would be

$$U_s[\omega \mathcal{L}_\omega - \mathcal{L}] + [\omega \mathcal{L}_k] = 0, \qquad (15.28)$$

$$U_s[k \mathcal{L}_\omega] + [k \mathcal{L}_k - \mathcal{L}] = 0, \qquad (15.29)$$

where U_s is the shock velocity. Conservation of phase, (15.3), and conservation of wave action, (15.2), could not then be maintained across such a shock. These were derived for a special form of solution, assuming slow variations, so there is no objection to abandoning them across abrupt shocks. These shocks would therefore represent a source of oscillations and involve jumps of the adiabatic invariants; the latter brings to mind the quantum jumps of the adiabatic invariants in quantum theory! It should be emphasized again that this is all just formalism with no positive view as to the structure of these shocks or even need for their occurrence. Moreover, these discontinuities would have irreversible properties, yet the original equation for φ is reversible. This would be another example of the long-standing problem of how systems reversible in some fine scale of description can exhibit irreversibility in "macroscopic" levels of approximation.

If, on the other hand, discontinuities are supposed to represent phenomena not covered by the original equation, but covered by some even more detailed description involving dissipation of some kind, then the choice would be different. Although momentum would probably be conserved, energy presumably would not. It is unlikely that (15.2) is the correct alternative, but one could make a case for (15.3). With dissipation, smooth oscillatory changes between different *constant* states may be constructed in dispersive models, as shown in Section 13.15. In that case the end states are constant, since the dissipation also damps out the oscillations on the two sides of the transition region. Thus it does not represent a change of state within an oscillatory wavetrain as contemplated here. But it does indicate the existence of a single-valued phase function and adds weight to the choice of (15.3) for a possible shock condition when dissipation is involved. Again this is entirely speculative and it would be pointless to go further in this direction without some more definite information and results.

15.5 Higher Order Dispersive Effects

We now consider modulation equations in the next order of approximation beyond (15.1)–(15.3). For simplicity, we work with the example

$$\varphi_{tt} - \varphi_{xx} + \varphi + 4\sigma\varphi^3 = 0 \qquad (15.30)$$

and limit the analysis to the near-linear case. But the results are typical and more specific physical applications will be shown for nonlinear optics in Section 16.4.

The variational principle for (15.30) has the Lagrangian

$$L = \frac{1}{2}\varphi_t^2 - \frac{1}{2}\varphi_x^2 - \frac{1}{2}\varphi^2 - \sigma\varphi^4, \tag{15.31}$$

and the *exact* modulation equations are given by

$$\delta \int\int \bar{L}\, dX\, dT = 0, \tag{15.32}$$

$$\bar{L} = \frac{1}{2\pi}\int_0^{2\pi}\left\{ \frac{1}{2}(-\omega\Phi_\theta + \epsilon\Phi_T)^2 - \frac{1}{2}(k\Phi_\theta + \epsilon\Phi_X)^2 - \frac{1}{2}\Phi^2 - \sigma\Phi^4 \right\} d\theta,$$

$$\tag{15.33}$$

as shown in (14.44). (We revert to the usual frequency $\omega = -\nu$.) For the near-linear theory we may use the Fourier expansion

$$\Phi = a\cos\theta + a_3\cos 3\theta + a_5\cos 5\theta + \cdots,$$

as in deriving (14.52). But we now retain also the next order terms in ϵ. The coefficients a_n are proportional to a^n, and this case is particularly simple, since to the order of approximation needed here only the term $a\cos\theta$ contributes. We have

$$\bar{L} = \frac{1}{4}(\omega^2 - k^2 - 1)a^2 - \frac{3}{8}\sigma a^4 + \frac{1}{4}\epsilon^2(a_T^2 - a_X^2) + 0(a^6, \epsilon^2 a^4). \tag{15.34}$$

This is a double expansion in which ϵ and a are assumed to be of the same order.* The linear term is $\frac{1}{4}(\omega^2 - k^2 - 1)a^2$, the first correction of nonlinearity is $-\frac{3}{8}\sigma a^4$, and the first correction of higher order dispersion is $\frac{1}{4}\epsilon^2(a_T^2 - a_X^2)$. The variational equations are

$$\delta a: \qquad (\omega^2 - k^2 - 1)a - 3\sigma a^3 - \epsilon^2(a_{TT} - a_{XX}) = 0, \tag{15.35}$$

$$\delta\Theta: \qquad\qquad \frac{\partial}{\partial T}(\omega a^2) + \frac{\partial}{\partial X}(k a^2) = 0, \tag{15.36}$$

$$\text{Consistency:} \qquad \frac{\partial k}{\partial T} + \frac{\partial\omega}{\partial X} = 0. \tag{15.37}$$

It should be noted from (15.35) that $\Theta(X, T)$ now depends also on ϵ, and

---

*Equivalently one could leave a as an $0(1)$ quantity, take σ as a measure of the nonlinearity, and use a double expansion in the small parameters σ and ϵ^2.

that a rigid separation into a hierarchy of equations for the different orders of ϵ has not been imposed. The variational principle (15.32) is exact and we have merely implemented it to the order of approximation noted in (15.34).

In this particular case, the higher order modulation equations are more complicated in form than the original equation (15.30)! Nevertheless, the behavior of modulations can be seen more easily from them than from the original equation. Of course it is usually the case that there is considerable simplification in going to the modulation equations. In any event, (15.35)–(15.37) are typical for systems in general.

When the terms in ϵ are omitted, the modulation equations are hyperbolic for $\sigma > 0$ and elliptic for $\sigma < 0$. The higher order dispersion effects introduce third derivatives of a into (15.36)–(15.37), and the modulation equations themselves become dispersive. In the case $\sigma > 0$ they have a similar structure to Boussinesq's equations. The consequences for breaking are expected to be similar, as discussed in Section 15.4. The existence of periodic solutions and solitary waves will be noted shortly. First, however, we consider how the instability found in the elliptic case $\sigma < 0$ is affected by the additional terms. A uniform wavetrain is a solution with constant values $\omega^{(0)}, k^{(0)}, a^{(0)}$, satisfying the dispersion relation (15.35). For small perturbations $\omega^{(1)}, k^{(1)}, a^{(1)}$, about these values, the linearized equations (15.35)–(15.37) are homogeneous with constant coefficients involving $\omega^{(0)}, k^{(0)}, a^{(0)}$. There are elementary solutions with $\omega^{(1)}, k^{(1)}, a^{(1)}$ all proportional to $e^{i\mu(X - CT)}$, provided C satisfies

$$\{\omega^{(0)}C - k^{(0)}\}^2 - (1-C)^2\left\{\frac{3\sigma a^{(0)2}}{2} + \frac{\epsilon^2\mu^2}{4}(1-C^2)\right\} = 0. \qquad (15.38)$$

The parameter μ determines the wave number for the modulations. For small $a^{(0)}$ and ϵ, the values for C are

$$C = \frac{k^{(0)}}{\omega^{(0)}} \pm \frac{1}{\omega^{(0)2}}\left(\frac{3\sigma a^{(0)2}}{2} + \frac{\epsilon^2\mu^2}{4\omega^{(0)2}}\right)^{1/2}. \qquad (15.39)$$

With the term in μ neglected these are just the characteristic velocities, imaginary for $\sigma < 0$. The influence of the dispersion introduced by the correction in μ is stabilizing, and the instability is now confined to the range

$$0 < \epsilon^2\mu^2 < 6|\sigma|\omega^{(0)2}a^{(0)2} \qquad (15.40)$$

in the modulation wave number $\epsilon\mu$.

For both the cases $\sigma > 0$ and $\sigma < 0$ it is important to observe that the system (15.35)–(15.37) has steady profile solutions propagating without change of shape. They are found in the usual way as solutions in which all quantities are functions of a moving coordinate $X - VT$. We have

$$\epsilon^2(1 - V^2)a'' + (\omega^2 - k^2 - 1)a - 3\sigma a^3 = 0, \qquad (15.41)$$

$$(\omega V - k)a^2 = R, \qquad (15.42)$$

$$\omega - Vk = S, \qquad (15.43)$$

where R and S are constants of integration. The last two may be used to eliminate ω and k in the first to give

$$\epsilon^2(1 - V^2)^2 a'' = \frac{R^2 - S^2 a^4 + (1 - V^2)(a^4 + 3\sigma a^6)}{a^3}. \qquad (15.44)$$

In general, there are periodic solutions (wavetrains in the envelope of the original modulated wavetrain!) in which a oscillates between two simple zeros in the numerator of the right hand side. Solitary waves will be limiting cases.

The solitary wave with $a \to 0$ as $X \to \pm \infty$ is particularly interesting; it represents a wave packet as shown in Fig. 15.2. In this case, (15.42) has $R = 0$, since $a \to 0$ at ∞. Hence

$$V = \frac{k}{\omega}; \qquad (15.45)$$

then, in turn, from (15.43), ω and k are both constant. For this example, the linear dispersion relation is $\omega_0 = (k^2 + 1)^{1/2}$ and the linear group velocity is

$$C_0 = \frac{k}{(k^2 + 1)^{1/2}} = \frac{k}{\omega_0} < 1.$$

We see that the velocity V is a nonlinear counterpart of C_0; it will be close to it for small amplitudes. Since $C_0 < 1$, we may take $V < 1$. With ω, k

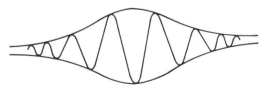

Fig. 15.2. Solitary wave modulation.

constant, (15.41) for a integrates once to

$$\epsilon^2(1-V^2)a'^2 = \frac{3}{2}\sigma a^4 - (\omega^2 - k^2 - 1)a^2. \tag{15.46}$$

For $a \to 0$ at ∞, we need

$$\omega^2 - k^2 - 1 < 0.$$

At the maximum value of a, we have

$$\omega^2 - k^2 - 1 = \frac{3}{2}\sigma a_m^2, \tag{15.47}$$

so that solitary waves of this type exist only for the elliptic case $\sigma < 0$. The velocity (15.45) of the packet is

$$V = \frac{k}{(k^2+1)^{1/2}} - \frac{3}{4}\frac{|\sigma|a_m^2}{(k^2+1)^{3/2}};$$

it moves a little slower than the linear group velocity. Ostrowskii (1967) in an analogous problem in nonlinear optics proposes that the result of instability may be a periodic solution which is essentially a sequence of such wave packets.

In the hyperbolic case $\sigma > 0$, these extreme solitary waves do not exist, but periodic wavetrains and solitary wave solutions of (15.44) with a bounded away from zero can be found. This is reasonable, since for $\sigma > 0$ the modulations distort in the hyperbolic theory ($\epsilon = 0$) but do not grow. The higher dispersion can counteract the distortion to produce steady profiles, and there is no reason for them to grow to the extreme case with $a = 0$. Their existence supports the view that in some cases breaking will not occur but rather the waveform will tend to a steady oscillatory form in the breaking region.

In near-linear theory the higher order dispersive terms arise from the quadratic part of the Lagrangian and it is easy to show that the general form corresponding to (15.34) is

$$\bar{L} = G(\omega, k)a^2 + G_2(\omega, k)a^4 + \frac{1}{2}\epsilon^2\{G_{\omega\omega}a_T^2 - 2G_{\omega k}a_T a_X + G_{kk}a_X^2\}$$

if less important terms involving derivatives of w and k are omitted.

The variational equations are similar to (15.35)–(15.37).

15.6 Fourier Analysis and Nonlinear Interactions

If the amplitudes are small and if only a few Fourier components are present, the nonlinear interactions between the components can be studied directly. This provides an alternative approach to some of the foregoing results. It was by this approach that Benjamin (1967) found an instability result of the type given in (15.40) for Stokes waves in deep water. The details for water waves by both the modulation and the interaction approaches will be discussed in Section 16.11. Here, to illustrate the method, we apply Benjamin's arguments to the Klein-Gordon equation, where the algebra is simpler.

For a finite number of Fourier components, φ can be expressed in the form

$$\varphi = \frac{1}{2} \sum \varphi_\nu(t) e^{i\kappa_\nu x}, \tag{15.48}$$

where ν runs over $\pm 1, \pm 2, \ldots, \pm N$ and we take $\kappa_{-n} = -\kappa_n$, $\phi_{-n}(t) = \varphi_n^*(t)$, $n = 1, \ldots, N$, to ensure a real φ. For the equation

$$\varphi_{tt} - \varphi_{xx} + \varphi = -4\sigma\varphi^3, \tag{15.49}$$

the linear solution (which neglects the right hand side) is

$$\varphi_\nu(t) = A_\nu e^{-i\omega_\nu t}, \tag{15.50}$$

where

$$\omega_n = (\kappa_n^2 + 1)^{1/2}, \qquad \omega_{-n} = -\omega_n, \qquad A_{-n} = A_n^*, \tag{15.51}$$

and the A_ν are constant. The near-linear theory can be developed by assuming σ to be a small parameter.

A naive perturbation expansion would express φ as a power series

$$\varphi = \varphi^{(0)} + \sigma\varphi^{(1)} + \sigma^2\varphi^{(2)} + \cdots,$$

to give the hierarchy

$$\varphi_{tt}^{(0)} - \varphi_{xx}^{(0)} + \varphi^{(0)} = 0, \tag{15.52}$$

$$\varphi_{tt}^{(1)} - \varphi_{xx}^{(1)} + \varphi^{(1)} = -4\varphi^{(0)^3}, \tag{15.53}$$

and so on. The solution of the linear equation (15.52) would be taken in the form

$$\varphi^{(0)} = \frac{1}{2} \sum A_\nu e^{i\kappa_\nu x - i\omega_\nu t}$$

and substituted in the right hand side of the equation for $\varphi^{(1)}$. However, resonance produces secular terms in $\varphi^{(1)}$ and the expansion is not uniformly valid. This is, in fact, just a more general version of the situation noted while discussing the Stokes expansion for periodic solutions in Section 13.13. A uniformly valid solution is obtained by including any third order resonant terms at the earlier level in the equation for $\varphi^{(0)}$. It amounts to adopting a more definite Fourier analysis viewpoint and grouping terms according to their contributions to the different components $e^{i\kappa_\nu x}$. That is, we substitute (15.48) in (15.49), and for each of the original components $e^{i\kappa_\nu x}$ we have

$$\frac{d^2\varphi_\nu}{dt^2} + \omega_\nu^2 \varphi_\nu = -\sigma \sum_{\kappa_\alpha + \kappa_\beta + \kappa_\gamma = \kappa_\nu} \varphi_\alpha \varphi_\beta \varphi_\gamma, \qquad (15.54)$$

where ω_ν is the same as in (15.51). The cubic term will also generate new components to be added to (15.48), but they do not resonate (at least at the cubic order) and may be neglected in the first approximation. We therefore consider the solutions of (15.54).

To take out the main oscillation, we introduce

$$\varphi_\nu(t) = A_\nu(t) e^{-i\omega_\nu t},$$

with the important difference from linear theory that $A_\nu(t)$ is still a function of t. We have

$$\frac{d^2 A_\nu}{dt^2} - 2i\omega_\nu \frac{dA_\nu}{dt} = -\sigma \sum_{\kappa_\alpha + \kappa_\beta + \kappa_\gamma = \kappa_\nu} A_\alpha A_\beta A_\gamma e^{i(\omega_\nu - \omega_\alpha - \omega_\beta - \omega_\gamma)t}.$$

The time scale of concern is now $0(\sigma^{-1})$ and each t derivative introduces an order σ. Hence it is sufficient to drop the second derivative of A_ν and take

$$\frac{dA_\nu}{dt} = -\frac{i\sigma}{2\omega_\nu} \sum_{\kappa_\alpha + \kappa_\beta + \kappa_\gamma = \kappa_\nu} A_\alpha A_\beta A_\gamma e^{i(\omega_\nu - \omega_\alpha - \omega_\beta - \omega_\gamma)t}. \qquad (15.55)$$

True resonance occurs when

$$\kappa_\alpha + \kappa_\beta + \kappa_\gamma = \kappa_\nu,$$

$$\omega_\alpha + \omega_\beta + \omega_\gamma = \omega_\nu. \tag{15.56}$$

The self-interaction

$$(\kappa_\nu, \kappa_\nu, -\kappa_\nu) \rightarrow \kappa_\nu$$

(and its permutations) are always of this type. They produce the Stokes effect on the frequency. If only one mode κ_0 were present, we should have

$$\frac{dA_0}{dt} = -\frac{3i\sigma}{2\omega_0} A_0^2 A_0^*, \qquad \omega_0 = (\kappa_0^2 + 1)^{1/2},$$

with solution

$$A_0 = a_0 e^{-(3i\sigma/2\omega_0)a_0^2 t}. \tag{15.57}$$

Then

$$\varphi = a_0 \cos\left\{\kappa_0 x - \left(\omega_0 + \frac{3}{2}\frac{\sigma a_0^2}{\omega_0}\right)t\right\} + \cdots.$$

This is the Stokes result (14.12).

In his stability discussion, Benjamin considers the effect of close "side bands" with wave numbers $\kappa_0 \pm \tilde{\mu}$ on a main wave κ_0. That is the set of κ_n, $n = 1, \ldots, N$, is $\{\kappa_0, \kappa_0 - \tilde{\mu}, \kappa_0 + \tilde{\mu}\}$, and their negative values are added to make up the full set κ_ν. We denote the corresponding A_n by A_0, A_-, A_+; their conjugates appear as in (15.51). The equations (15.55) become

$$\frac{dA_0}{dt} = -\frac{i\sigma}{2\omega_0}\left\{3A_0^2 A_0^* + 6A_0 A_+ A_+^* + 6A_0 A_- A_-^* + 6A_0^* A_+ A_- e^{-i\Omega t}\right\},$$

$$\tag{15.58}$$

$$\frac{dA_-}{dt} = -\frac{i\sigma}{2\omega_-}\left\{6A_0 A_0^* A_- + 3A_0^2 A_+^* e^{i\Omega t} + 6A_+ A_+^* A_- + 3A_-^2 A_-^*\right\}, \tag{15.59}$$

where

$$\Omega = \omega_+ + \omega_- - 2\omega_0; \tag{15.60}$$

the equation for A_+ is obtained by interchanging plus and minus subscripts. The interactions involved can be seen from the A's that contribute.

The first term in (15.58) is the self-interaction (Stokes) term; the second term comes from $(\kappa_0, \kappa_0 + \tilde{\mu}, -\kappa_0 - \tilde{\mu}) \to \kappa_0$, and so on. The terms with the factor $e^{i\Omega t}$ do not exactly satisfy the resonance condition on the frequencies. But if $\tilde{\mu}$ is small, so is Ω, and the factor is kept to preserve uniformity as $\tilde{\mu} \to 0$. The numerical coefficients in front of each term give the number of permutations in any particular interaction.

In a stability analysis, it is assumed that $A_\pm \ll A_0$, and the equations are linearized to

$$\frac{dA_0}{dt} = -\frac{3i\sigma}{2\omega_0} A_0^2 A_0^*,$$

$$\frac{dA_-}{dt} = -\frac{i\sigma}{2\omega_-} \left\{ 6A_0 A_0^* A_- + 3A_0^2 A_+^* e^{i\Omega t} \right\}.$$

The effects on A_0 are second order. We take

$$A_0 = a_0 e^{-i\rho t}, \qquad \rho = \frac{3}{2} \frac{\sigma a_0^2}{\omega_0}, \tag{15.61}$$

as in (15.57), with a_0 real. For small $\tilde{\mu}$ it is sufficient to take $\omega_\pm \simeq \omega_0$ in the coefficients of the equations for $A_\pm$ and to approximate Ω by

$$\Omega \simeq \omega_0''(\kappa_0)\tilde{\mu}^2. \tag{15.62}$$

The linear equations for $A_\pm$ then become

$$\frac{dA_-}{dt} = -i\rho \left\{ 2A_- + A_+^* e^{i(\Omega - 2\rho)t} \right\},$$

$$\frac{dA_+}{dt} = -i\rho \left\{ 2A_+ + A_-^* e^{i(\Omega - 2\rho)t} \right\}.$$

These have solutions in the form $A_\pm = a_\pm e^{i\lambda_\pm t}$, where $\lambda_\pm$ satisfy

$$\lambda^2 + (2\rho - \Omega)\lambda + \rho(\rho - 2\Omega) = 0. \tag{15.63}$$

The small side band perturbations grow if the roots of (15.63) are complex; that is, if

$$\left(\rho + \frac{\Omega}{4} \right)\Omega < 0. \tag{15.64}$$

For this particular example, $\omega_0 = (\kappa_0^2 + 1)^{1/2}$, $\omega_0'' = \omega_0^{-3}$, so (15.64) gives

$$\left(\frac{3}{2} \sigma a_0^2 + \frac{\tilde{\mu}^2}{4\omega_0^2} \right) < 0.$$

This agrees with (15.39) and (15.40), since $\tilde{\mu} = \epsilon\mu$ in the comparison. For $\sigma > 0$ the side bands always remain small. For $\sigma < 0$, there is instability for the range

$$\tilde{\mu}^2 < 6|\sigma|\omega_0^2 a_0^2.$$

Again this is a linear instability theory; the nonlinear equations (15.58)–(15.59) conserve energy, which oscillates between the modes. This agrees with the proposal that the end result is a solution with finite amplitude oscillations.

It should be fairly clear that the preceding analysis, although developed for a specific example, is general. In fact, we can recognize from (15.61) that ρ is always the Stokes correction to the frequency; in Section 14.2 it was denoted by $\omega_2 a^2$. The expression (15.62) for Ω is already in a general form. Therefore the criterion (15.64) may be written

$$\left(\omega_2 a^2 + \frac{1}{4} \omega_0'' \tilde{\mu}^2 \right) \omega_0'' \tilde{\mu}^2 < 0.$$

This should be compared with the radical in the characteristic velocity (14.21); the extra term in $\tilde{\mu}^2$ arises from the higher order dispersion effects.

The interaction and modulation approaches may be compared by noting that the modulation $k = k^{(0)} + k^{(1)}$, $a = a^{(0)} + a^{(1)}$, used to obtain (15.39), may be expressed approximately as

$$\varphi = \frac{1}{2} \{ a^{(0)} + a^{(1)} \} \exp i \{ \theta^{(0)} + \theta^{(1)} \} + \text{complex conjugate}$$

$$\simeq \frac{1}{2} a^{(0)} \exp i\theta^{(0)} + \frac{1}{2} a^{(1)} \exp i\theta^{(0)} + \frac{1}{2} i\theta^{(1)} a^{(0)} \exp i\theta^{(0)} + \text{complex conjugate}.$$

$$(15.65)$$

If we now take

$$a^{(0)} = a_0, \qquad \theta^{(0)} = \kappa_0 x - (\omega_0 + \rho)t$$

for the basic wavetrain, and express the perturbations $a^{(1)}, \theta^{(1)}$ as appropriate linear combinations of $e^{\pm i\tilde{\mu}x}$, we obtain the side band description.

The interaction theory can be carried through effectively only for the near-linear case and only for modulations consisting of a finite number of Fourier components. Even so, we note the algebraic complexity compared with the modulation approach. This can be mitigated to some extent by appealing again to the variational principle. If the expression

$$\varphi = \frac{1}{2} \sum A_\nu(t) e^{i\kappa_\nu x - i\omega_\nu t}$$

is substituted into the Lagrangian, all terms except the resonant ones will be oscillatory in x. If these are averaged out, we may use the variational principle to obtain the equations for the A_n. In the simpler cases, such as the one above, the saving is not very great. We fairly easily obtain the average Lagrangian

$$\hat{L} = \frac{1}{4} \sum \left\{ \dot{A}_n \dot{A}_n^* - i\omega_n A_n \dot{A}_n^* + i\omega_n \dot{A}_n A_n^* \right\}$$

$$- \frac{\sigma}{16} \left\{ 6 \sum A_n^2 A_n^{*2} + 24 \sum_{m \neq n} \sum A_m A_m^* A_n A_n^* \right\}$$

$$- \frac{\sigma}{16} \left\{ 12 A_0^2 A_+^* A_-^* e^{i\Omega t} + 12 A_0^{*2} A_+ A_- e^{-i\Omega t} \right\}, \tag{15.66}$$

where the summations are over A_0, A_+, A_-. But the remaining analysis of the variational equations is much the same. The main advantage is the canonical form.

The interaction theory is not restricted to neighboring sets of wave numbers. With sufficiently general dispersion relations, one can have resonances satisfying (15.56) for widely differeing κ_ν. There is then strong energy transfer between these modes. Such cases are discussed by Phillips (1967) and references given there. A problem in nonlinear optics (Section 16.5) will provide an illustrative example. This type of interaction between widely different wave numbers ties in closely with the multiple phase solutions indicated in Section 14.9.

CHAPTER 16

Applications of the Nonlinear Theory

NONLINEAR OPTICS

16.1 Basic Ideas

One of the most interesting areas for the study of nonlinear dispersive effects is in the field of nonlinear optics. The theoretical ideas fit in naturally with the experimental situation and are involved in the development of practical devices. Modulation theory is the natural approach to a number of phenomena in view of the high frequencies and wave numbers of the basic wavetrains. The self-focusing and stability of beams are studied in this way. Nonlinear interactions for the production or amplification of sum and difference frequencies are important and can be displayed dramatically by changing the color of a laser beam on passage through a nonlinear crystal. The experiments generally seem to be more easily and precisely controllable than is possible, for example, in water waves, where the many modes of fluid motion make it difficult to isolate the particular effects desired.

In the simplest formulations of the theory, the analysis is very close to that of the Klein-Gordon example and results can be taken over by analogy from that case. We start with the classical model in which the electric polarization of the material is represented as the displacement of bound electrons by the electric field. The results can be interpreted in a broader way later. We consider the basic one dimensional wavetrain and take the propagation to be in the x direction with field components E and B in the z and y directions, respectively. The electron displacement is in the z direction and we describe it by the function $r(x,t)$. Maxwell's equations reduce to

$$B_t - E_x = 0,$$

$$E_t + \frac{qN}{\epsilon_0} r_t = c_0^2 B_x,$$

(16.1)

where q is the electronic charge, N is the number of electrons per unit volume, c_0 is the speed of light in vacuum, and ϵ_0 is the permittivity of free space. To complete the system, we need the relation of r to E. We suppose that an electron driven by the field E is effectively in a potential well that provides a nonlinear restoring force. Accordingly, the equation is taken in the form

$$mr_{tt} + U'(r) = qE. \tag{16.2}$$

If the polarization $P = Nqr$ is introduced and B is eliminated from Maxwell's equations, we have

$$E_{tt} + \frac{1}{\epsilon_0} P_{tt} = c_0^2 E_{xx}, \tag{16.3}$$

$$P_{tt} + V'(P) = \epsilon_0 \nu_p^2 E, \tag{16.4}$$

where

$$V(P) = \frac{qN}{m} U(r), \qquad \nu_p^2 = \frac{Nq^2}{\epsilon_0 m} \; ; \tag{16.5}$$

ν_p is the plasma frequency.

Each oscillator responds also to the cumulative effect of all other oscillators. In the elementary treatment, this is incorporated by replacing E on the right hand sides of (16.2) and (16.4) by $E + P/3\epsilon_0$, which is the field inside a spherical cavity surrounded by dielectric with polarization P. For our purposes we can suppose the extra term is absorbed into $V'(P)$.

There are important simplifications in the analysis if the potential well is symmetric and $V'(P)$ is an odd function of P. We shall take this to be the case for the most part and indicate at appropriate places what is involved in the more general case.

Uniform Wavetrains.

In a uniform wavetrain $E = E(\theta), P = P(\theta), \theta = kx - \omega t$. Equation 16.3 then integrates to

$$(\omega^2 - c_0^2 k^2)\epsilon_0 E = -\omega^2 P + \text{constant}. \tag{16.6}$$

If $V'(P)$ is an odd function of P, we may take the constant of integration to be zero without loss of generality. In other cases, however, the constant may be required and plays an important role. If $V'(P)$ has a term in P^2, for example, we see from (16.4) that the mean values of E and P differ by a term quadratic in the amplitude; hence a constant proportional to the amplitude squared is required in (16.6). If $V'(P)$ is odd, however, we note that it is consistent to take zero mean values for E and P, and with this assumption the constant in (16.6) must be zero. Then we have

$$\epsilon_0 E = - \frac{\omega^2}{\omega^2 - c_0^2 k^2} P, \tag{16.7}$$

$$\omega^2 P_{\theta\theta} + V'(P) + \frac{\omega^2 \nu_p^2}{\omega^2 - c_0^2 k^2} P = 0. \tag{16.8}$$

In the linear case, $V'(P) = \nu_0^2 P$, the solution of (16.8) is sinusoidal in θ and we obtain the dispersion relation

$$n_0^2 = \frac{c_0^2 k^2}{\omega^2} = 1 - \frac{\nu_p^2}{\omega^2 - \nu_0^2}. \tag{16.9}$$

Damping becomes important in the absorption band around the resonant frequency ν_0 and eliminates the singular behavior there. Away from the resonant frequency, however, its effect is small and can be neglected in the first instance.

In the near-linear case, we may take

$$V'(P) = \nu_0^2 P - \alpha P^3 + \cdots,$$

$$P = b \cos\theta + b_3 \cos 3\theta + \cdots, \tag{16.10}$$

$$E = a \cos\theta + a_3 \cos 3\theta + \cdots,$$

and deduce the dispersion relation

$$n^2 = \frac{c_0^2 k^2}{\omega^2} = 1 - \frac{\nu_p^2}{\omega^2 - \nu_0^2} + \frac{3}{4} \frac{\alpha \epsilon_0^2 \nu_p^6}{(\omega^2 - \nu_0^2)^4} a^2 + \cdots. \tag{16.11}$$

In the fully nonlinear case, (16.8) has oscillatory solutions, as discussed for the Klein-Gordon equation (14.3); the close similarity between these two cases becomes apparent.

The Average Lagrangian.

A variational principle may be formulated in terms of a potential ψ, the z component of the vector potential. The field components are given by $E = -\psi_t, B = -\psi_x$, and a suitable Lagrangian is

$$L = \frac{1}{2}\epsilon_0(\psi_t^2 - c_0^2\psi_x^2) - Nq\psi_t r + N\left\{\frac{1}{2}mr_t^2 - U(r)\right\}$$

$$= \frac{1}{2}\epsilon_0(\psi_t^2 - c_0^2\psi_x^2) - \psi_t P + \frac{1}{\epsilon_0\nu_p^2}\left\{\frac{1}{2}P_t^2 - V(P)\right\}. \qquad (16.12)$$

If $V'(P)$ is an odd function of P, it is sufficient for the uniform wavetrain to take ψ and P as periodic functions of θ. Corresponding to (16.7) and (16.8), we then have

$$E = \omega\psi_\theta = -\frac{\omega^2}{\omega^2 - c_0^2k^2}\frac{P}{\epsilon_0}, \qquad (16.13)$$

$$\frac{1}{2}\omega^2 P_\theta^2 + V(P) + \frac{1}{2}\frac{\omega^2\nu_p^2}{\omega^2 - c_0^2k^2}P^2 = A, \qquad (16.14)$$

where the second equation has been integrated and the constant of integration A enters as the amplitude parameter. The average Lagrangian is then obtained by substituting (16.13)–(16.14) in (16.12) and performing the usual manipulations. The result is

$$\mathcal{L}(\omega, k, A) = \frac{1}{\epsilon_0\nu_p^2}\left(\frac{\omega}{2\pi}\oint\left\{2A - 2V(P) - \frac{\omega^2\nu_p^2}{\omega^2 - c_0^2k^2}P^2\right\}^{1/2}dP - A\right). \qquad (16.15)$$

Again the similarity with the Klein-Gordon case, (14.26), may be noted.

If $V'(P)$ is not odd, the more general form

$$\psi = \beta x - \gamma t + \Psi(\theta) \qquad (16.16)$$

is required. The parameters β and γ give nonzero mean values of B and E, and they must be treated as a pseudo-wave number and a pseudo-frequency, as explained in Section 14.7. The second constant of integration, $\tilde{A}$ say, must now be allowed in (16.6) and the triad $(\gamma, \beta, \tilde{A})$ is analogous to the main triad (ω, k, A). In the modulation theory the coupling of the changes in (ω, k, A) with changes in the mean field parameters $(\gamma, \beta, \tilde{A})$ is a crucially important effect. We shall not pursue the details here; they are similar to the case in water waves which will be discussed later.

General results can be derived from (16.15) as described in the previous chapters. However, most of the results in nonlinear optics have been found for the near-linear case. It has the advantage in this particular context that although a specific model may be used to motivate the theory, a broader interpretation of the formulas is fairly clear. The near-linear form of $\mathcal{L}$ is most easily obtained directly by substituting the expansions (16.10) into (16.12), rather than approximating (16.15). The calculation of $\mathcal{L}$ up to the fourth order in a is particularly simple since the coefficients a_3, b_3, which are third order in a, do not contribute until the sixth order terms. [Compare the derivation of (14.52).] To this order, then, we have

$$\mathcal{L} = \frac{1}{4}\left(1 - \frac{c_0^2 k^2}{\omega^2}\right)\epsilon_0 a^2 + \frac{1}{2}ab + \frac{\omega^2 - v_0^2}{4\epsilon_0 v_p^2}b^2 + \frac{3}{32}\frac{\alpha}{\epsilon_0 v_p^2}b^4. \quad (16.17)$$

The variational equation $\mathcal{L}_b = 0$ can be used to determine b in terms of a:

$$b = -\frac{\epsilon_0 v_p^2}{\omega^2 - v_0^2}a + \frac{3}{4}\frac{\alpha\epsilon_0^3 v_p^6}{\left(\omega^2 - v_0^2\right)^4}a^3. \quad (16.18)$$

This result may be resubstituted in the expression for $\mathcal{L}$ to give

$$\mathcal{L} = \frac{1}{4}\left(1 - \frac{c_0^2 k^2}{\omega^2} - \frac{v_p^2}{\omega^2 - v_0^2}\right)\epsilon_0 a^2 + \frac{3}{32}\frac{\alpha\epsilon_0^3 v_p^6}{\left(\omega^2 - v_0^2\right)^4}a^4. \quad (16.19)$$

As a check we note that the dispersion relation $\mathcal{L}_a = 0$ agrees with (16.11), as it should.

The first term in (16.12), which is also equal to $\frac{1}{2}\epsilon_0(E^2 - c_0^2 B^2)$, is the basic wave operator for electromagnetic waves in free space. It always leads to the term

$$\frac{1}{4}\left(1 - \frac{c_0^2 k^2}{\omega^2}\right)\epsilon_0 a^2$$

in $\mathcal{L}$, where a is the amplitude of the electric field. The other terms in (16.19) incorporate the response of the medium to the oscillating electric field. For other models, or to represent known material properties, it seems reasonable by analogy to suppose that

$$\mathcal{L} = \frac{1}{4}\left(1 - \frac{c_0^2 k^2}{\omega^2}\right)\epsilon_0 a^2 + g_2(\omega)\epsilon_0 a^2 + g_4(\omega)\epsilon_0 a^4 \quad (16.20)$$

for some functions $g_2(\omega), g_4(\omega)$. Then the dispersion relation $\mathcal{L}_a = 0$ allows the identification of g_2, g_4 with the behavior of the refractive index. If the dispersion relation is required to be

$$n = \frac{c_0 k}{\omega} = n_0(\omega) + \frac{1}{2} n_2(\omega) a^2, \qquad (16.21)$$

then we must choose

$$\mathcal{L} = \frac{1}{4} \left\{ n_0^2(\omega) - \frac{c_0^2 k^2}{\omega^2} \right\} \epsilon_0 a^2 + \frac{1}{8} n_0(\omega) n_2(\omega) \epsilon_0 a^4. \qquad (16.22)$$

The coefficient $n_0(\omega)$ is the linear refractive index and more realistic forms may now be taken for it in place of (16.9). For example, to include more resonant frequencies ν_j, we have

$$n_0^2(\omega) = 1 - \nu_p^2 \sum_j \frac{f_j}{\omega^2 - \nu_j^2}, \quad \sum_j f_j = 1,$$

where $f_j = N_j/N$ is the proportion of electrons with resonant frequency ν_j. This also fits the quantum theoretic description in which the ν_j are transition frequencies and the f_j are transition probabilities. Similarly, the nonlinear coefficient $n_2(\omega)$ can be chosen to represent other models or known material properties. However, the overall restriction to cases in which quadratic mean fields do not arise must be carefully noted.

16.2 One Dimensional Modulations

In the near-linear theory we can proceed simply as in Section 14.2. In optics it is more usual to consider the dispersion relation as expressing k or n as a function of ω, so we work with ω and a as the basic variables in the modulation theory. The dispersion relation is written*

$$k = k_0(\omega) + k_n(\omega) a^2, \qquad (16.23)$$

where

$$k_0(\omega) = \frac{\omega n_0(\omega)}{c_0}, \qquad k_n(\omega) = \frac{\omega n_2(\omega)}{2c_0}. \qquad (16.24)$$

*We use k_n rather than k_2 to denote the nonlinear coefficient, since k_2 will be needed later for the x_2 component of the vector $\mathbf{k}$.

To lowest order the modulation equations, corresponding to (14.18)–(14.19), are

$$\frac{\partial a^2}{\partial x} + \frac{\partial}{\partial t}\left\{ k_0'(\omega)a^2 \right\} = 0.$$

$$\frac{\partial \omega}{\partial x} + k_0'(\omega)\frac{\partial \omega}{\partial t} + k_n(\omega)\frac{\partial a^2}{\partial t} = 0,$$

(16.25)

The characteristic velocities are found to be

$$\frac{1}{C} = k_0'(\omega) \pm \left\{ k_n(\omega)k_0''(\omega) \right\}^{1/2} a.$$

(16.26)

The equations are hyperbolic if $k_n k_0'' > 0$ and elliptic if $k_n k_0'' < 0$. For (16.11), the sign of k_0'' is the same as $\nu_0^2 - \omega^2$ and the sign of k_n is the same as α. Hence we have

$$\text{Hyperbolic:} \quad \alpha(\nu_0^2 - \omega^2) > 0,$$

$$\text{Elliptic:} \quad \alpha(\nu_0^2 - \omega^2) < 0.$$

(16.27)

These results were first obtained by Ostrowskii (1967). The normal case for optics is $\omega^2 < \nu_0^2, \alpha > 0$, and the equations are hyperbolic. However, Ostrowskii (1968) reports on experiments at radiofrequencies with ferrites and semiconductor diodes, where both cases can be obtained. The hyperbolic distortion and elliptic instability are both found, and stable modulated forms seem to result, in line with the discussion of higher order effects given in Section 15.5.

The higher order effects lead to quadratic terms in the derivatives of a and b in the expression (16.17) for $\mathcal{L}$. The modulation equations are then similar in structure to those discussed in Section 15.5. Qualitatively, the phenomena are the same and the details will not be given for this case. The main results were obtained by Ostrowskii (1967), and Small (1972) shows how the variational approach may be used. We shall, however, take up in the next section the analogous problem of spatial modulations and the self-focusing of beams. Higher order effects are important for these and we outline the theory.

The fully nonlinear results corresponding to (16.26)–(16.27) can be taken from Section 15.2 using the appropriate Lagrangian. For the simple model discussed earlier this is given by (16.15).

16.3 Self-Focusing of Beams

If the nonlinear term in the dispersion relation (16.21) is positive, with $n_2(\omega) > 0$, then the phase speed $c = \omega/k = c_0/n$ increases as the amplitude falls off from the center toward the outskirts of the beam. Intuitively this suggests the beam will tend to focus. Of course the argument is only rough and we now consider such questions in more detail.

For spatial modulations, we suppose that locally the wave can be described by a periodic wavetrain propagating in the direction of the vector wave number $\mathbf{k}$. The average Lagrangian is determined from this, and in the simplest cases when pseudo-frequencies are not involved, it takes the form $\mathcal{L}(\omega, \mathbf{k}, a)$ where ω is the frequency and a is the amplitude of the electric field. In the near-linear case, $\mathcal{L}(\omega, \mathbf{k}, a)$ is given by (16.22) with $k = |\mathbf{k}|$.

For solutions in which $\omega, \mathbf{k}, a$ are independent of t, we have $\omega = $ constant and the modulation equations deduced from the average variational principle are

$$\mathcal{L}_a = 0, \tag{16.28}$$

$$\frac{\partial}{\partial x_i} \mathcal{L}_{k_i} = 0, \qquad \frac{\partial k_i}{\partial x_j} - \frac{\partial k_j}{\partial x_i} = 0. \tag{16.29}$$

If the medium is isotropic so that $\mathcal{L}$ depends only on the magnitude k of the vector $\mathbf{k}$, (16.29) reduce to

$$\frac{\partial}{\partial x_i}(k_i \rho) = 0, \qquad \frac{\partial k_i}{\partial x_j} - \frac{\partial k_j}{\partial x_i} = 0, \tag{16.30}$$

where

$$\rho = -k^{-1} \mathcal{L}_k. \tag{16.31}$$

The dispersion relation (16.28) provides a relation between a and k; hence, in principle, ρ can be taken as a function of k (although it may not always be convenient to do the actual elimination of a).

It is interesting to note that the equations in (16.30) are the same as the equations for compressible irrotational steady flow of a gas, with the wave number $\mathbf{k}$ replacing the fluid velocity vector and ρ replacing the density. The relation of ρ to k provided by (16.31) corresponds to the Bernoulli equation between density and speed in the fluid flow problem (see Section 6.17). In this analogy, the beam corresponds to a fluid jet in a gas. But care is needed in taking qualitative results directly from fluid flow, because ρ is usually an increasing function of k in optics, whereas density and velocity have the opposite variation in a gas. However, the various techniques of finding solutions may be usefully taken over.

The type of equations (16.30)–(16.31) governs their mathematical structure. This is separate from the question of whether the original time-dependent equations are elliptic. Moreover, the ellipticity of the steady equations does not indicate instability; the type affects only the properties of the solution and the form of the boundary conditions.

For two dimensional or axisymmetric beams, we shall take x in the axial direction and r in the transverse or radial direction. The wave number vector will have corresponding components (k_1, k_2), and (16.30)–(16.31) become

$$\frac{\partial k_2}{\partial x} - \frac{\partial k_1}{\partial r} = 0, \qquad (16.32)$$

$$\frac{\partial}{\partial x}(\rho k_1) + \frac{\partial}{\partial r}(\rho k_2) + \frac{m\rho k_2}{r} = 0, \qquad (16.33)$$

$$\rho = \rho(k), \qquad (16.34)$$

where $m = 0$ for the two dimensional beam and $m = 1$ for the axisymmetric beam.

The Type of the Equations.

The characteristics are most easily found by temporarily reintroducing the phase θ, with

$$k_1 = \theta_x, \qquad k_2 = \theta_r,$$

and writing (16.33) as

$$\left(1 + \frac{k_1^2}{k}\frac{\rho'}{\rho}\right)\theta_{xx} + \frac{2k_1 k_2}{k}\frac{\rho'}{\rho}\theta_{xr} + \left(1 + \frac{k_2^2}{k}\frac{\rho'}{\rho}\right)\theta_{rr} + \frac{m}{r}\theta_r = 0.$$

Then characteristic curves in the (x, r) plane must satisfy

$$\left(1 + \frac{k_1^2}{k}\frac{\rho'}{\rho}\right)dr^2 - \frac{2k_1 k_2}{k}\frac{\rho'}{\rho}dr\,dx + \left(1 + \frac{k_2^2}{k}\frac{\rho'}{\rho}\right)dx^2 = 0.$$

These are imaginary and the equation is elliptic if

$$1 + k\frac{\rho'}{\rho} > 0;$$

they are real and the equation is hyperbolic if this quantity is negative.

For the near-linear theory described by the Lagrangian (16.22), we have

$$\rho = -k^{-1}\mathcal{L}_k = \frac{\epsilon_0 c_0^2}{2\omega^2}a^2 \tag{16.35}$$

and therefore

$$\frac{c_0 k}{\omega} = n_0(\omega) + \frac{1}{2}n_2(\omega)a^2$$

$$= n_0(\omega) + \frac{\omega^2 n_2(\omega)}{\epsilon_0 c_0^2}\rho. \tag{16.36}$$

If $n_2(\omega) > 0$, then $\rho'(k) > 0$ and the steady equations are elliptic. We shall consider only this case. We note from (16.26) that the original time-dependent equations are hyperbolic if, in addition, $k_0''(\omega) > 0$.

Focusing.

The "streamlines" determined by the vector field $\mathbf{k}$ are the orthogonal trajectories of the family of phase surfaces $\theta = $ constant. In isotropic media, the group velocity has the same direction as $\mathbf{k}$ and these streamlines are rays. We can picture the phase surfaces moving out along these rays and the question of focusing is related to convergence of the rays. To analyze this, it is convenient to transform (16.32)–(16.34) and introduce coordinates (ξ, η) based on the rays and phase surfaces. If we introduce

$$k_1 = \xi_x, \qquad k_2 = \xi_r, \qquad \theta = \xi(x, r) - \omega t,$$

$$\rho k_1 r^m = \eta_r, \qquad \rho k_2 r^m = -\eta_x, \tag{16.37}$$

the equations (16.32)–(16.33) are satisfied identically; the successive positions of a phase surface are given by $\xi = $ constant and the streamlines by $\eta = $ constant. The relations (16.37) may be written in inverse form as

$$x_\xi = \frac{\cos\chi}{k}, \qquad x_\eta = -\frac{\sin\chi}{\rho k r^m},$$

$$r_\xi = \frac{\sin\chi}{k}, \qquad r_\eta = \frac{\cos\chi}{\rho k r^m},$$

where χ is the inclination of $\mathbf{k}$ to the x axis. Thus the consistency relations may be written

$$\frac{\partial \chi}{\partial \xi} = \frac{\rho r^m}{k} \frac{\partial k}{\partial \eta}, \qquad \frac{\partial \chi}{\partial \eta} = k \frac{\partial}{\partial \xi} \left(\frac{1}{\rho k r^m} \right). \qquad (16.38)$$

If ρ decreases away from the axis, then so does k, from (16.36) and the assumption that $n_2 > 0$. Therefore, from the first equation in (16.38), $\partial \chi / \partial \xi < 0$; this shows that the rays bend toward the axis and the beam focuses. If $n_2 < 0$, there is a corresponding defocusing.

Thin Beams.

Some interesting solutions of the equations have been constructed by Akhmanov et al. (1966) under the further assumption that the beam is thin. It is assumed that the nonlinear effects provide a small correction to a linear plane wave of constant wave number K and the r derivatives are of greater order than x derivatives due to the thinness of the beam. We take

$$\theta = -\omega t + Kx + Ks(x, r),$$

where s_x and s_r are both small but s_x and s_r^2 are comparable. [This can be done formally by choosing

$$\theta = -\omega t + Kx + K\epsilon^2 s \left(x, \frac{r}{\epsilon} \right),$$

where ϵ is an amplitude parameter.] With this approximation we have

$$k \simeq K \left(1 + s_x + \frac{1}{2} s_r^2 \right)$$

and the near-linear dispersion relation $n = c_0 k / \omega = n_0 + \frac{1}{2} n_2 a^2$ gives

$$s_x + \frac{1}{2} s_r^2 = \frac{1}{2} \frac{n_2}{n_0} a^2, \qquad (16.39)$$

assuming that $c_0 K / \omega = n_0(\omega)$. Since $\rho \propto a^2$ from (16.35), ρ may be replaced by a^2 in (16.33). Under the approximation $s_x \ll s_r$, (16.33) may then be taken as

$$\frac{\partial a^2}{\partial x} + s_r \frac{\partial a^2}{\partial r} + \left(s_{rr} + \frac{m}{r} s_r \right) a^2 = 0. \qquad (16.40)$$

Equations 16.39 and 16.40 are to be solved for s and a^2.

Near the axis we may expect s to be given by

$$s = \sigma(x) + \frac{1}{2} \frac{r^2}{R(x)} + O(r^4), \tag{16.41}$$

where $R(x)$ is the radius of curvature of the phase surface at the axis. Surprisingly, there is an exact solution in which $s(x,r)$ has just these two terms. From (16.39) we see that a^2 must also be quadratic in r, and (16.40) provides relations for the various coefficients. The results are

$$a^2 = \frac{a_0^2}{f^{m+1}(x)} \left\{ 1 - \frac{r^2}{r_0^2 f^2(x)} \right\}, \tag{16.42}$$

$$\sigma' = \frac{1}{2} \frac{n_2}{n_0} \frac{a_0^2}{f^{m+1}(x)}, \qquad \frac{1}{R(x)} = \frac{f'(x)}{f(x)}, \tag{16.43}$$

where

$$f'^2 = \frac{2n_2 a_0^2}{(m+1)n_0 r_0^2} \left(\frac{1}{f^{m+1}} - 1 \right) + \frac{1}{R_0^2}, \qquad f(0) = 1; \tag{16.44}$$

r_0 is the initial radius of the beam, a_0 the initial amplitude on the axis, and R_0 the initial radius of curvature of the phase surface.

If the phase surfaces at $x=0$ are plane ($R_0^{-1}=0$), the solutions of (16.44) are

$$m=0: \qquad x = \left(\frac{n_0}{2n_2} \right)^{1/2} \frac{r_0}{a_0} \left\{ f^{1/2}(1-f)^{1/2} - \sin^{-1} f^{1/2} + \frac{\pi}{2} \right\}, \tag{16.45}$$

$$m=1: \qquad x = \left(\frac{n_0}{n_2} \right)^{1/2} \frac{r_0}{a_0} (1 - f^2)^{1/2}. \tag{16.46}$$

The beam focuses and the solution becomes singular at the point where $f(x) \to 0$; at this point the radius of curvature $R(x) \to 0$ and the amplitude $a \to \infty$. The distance to the point of focus is

$$x_f = \frac{\pi}{2} \left(\frac{n_0}{2n_2} \right)^{1/2} \frac{r_0}{a_0}, \qquad \text{for } m=0, \tag{16.47}$$

$$x_f = \left(\frac{n_0}{n_2} \right)^{1/2} \frac{r_0}{a_0}, \qquad \text{for } m=1. \tag{16.48}$$

In the neighborhood of the singularity, higher derivatives of a become important and must be added to (16.39)–(16.40). As we shall see, the extra terms introduce dispersive effects which counteract the focusing and allow continuous solutions.

Before doing this, we mention an ingenious solution of the two dimensional equations ($m = 0$) that was found by Akhmanov et al. If we introduce

$$v = s_y, \qquad \tau = a^2,$$

then (16.39)–(16.40) are equivalent to

$$v_x + vv_y - \gamma \tau_y = 0, \qquad \gamma = \frac{n_2}{2n_0},$$

$$\tau_x + v\tau_y + \tau v_y = 0,$$

where we have replaced r by y for the two dimensional case. These are similar to the equations of unsteady one dimensional gas dynamics, apart from the change in sign of γ. They may be linearized *exactly* by the hodograph transformation to

$$y_\tau - vx_\tau - \gamma x_v = 0,$$

$$y_v - vx_v + \tau x_\tau = 0.$$

Since they are linear, these equations offer the possibility of general solutions by superposition. However, in the particular solution referred to here, Akhmanov et al. presumably noted the extra convenience of variables

$$p = x\tau, \qquad q = y - vx,$$

to write the hodograph equations

$$q_\tau - \frac{\gamma}{\tau} p_v = 0, \qquad q_v + p_\tau = 0,$$

but they then retransformed them to

$$v_p - \frac{\gamma}{\tau} \tau_q = 0, \qquad \tau_p + v_q = 0.$$

With $v = -\Phi_p$, $\tau = \Phi_q$, we have

$$\Phi_q \Phi_{pp} + \gamma \Phi_{qq} = 0.$$

This has a separable solution

$$\Phi = \left(ha_0^2 + \frac{\gamma}{h}p^2\right)\tanh\frac{q}{h},$$

which represents a beam. In the original variables the solution is given in terms of parameters p and q by

$$s_y = v = -\frac{2\gamma p}{h}\tanh\frac{q}{h}, \qquad a^2 = \tau = \left(a_0^2 + \frac{\gamma}{h^2}p^2\right)\text{sech}^2\frac{q}{h},$$

$$x = \frac{p}{\tau}, \qquad y = \frac{vp}{\tau} + q.$$

In the initial plane $x = 0$, we have

$$s_y = 0, \qquad a^2 = a_0^2\,\text{sech}^2\frac{y}{h}.$$

The beam starts with rays parallel to the x axis and has a realistic amplitude distribution. It may be shown that the beam focuses at the point

$$x_f = \frac{1}{2\gamma^{1/2}}\frac{h}{a_0} = \left(\frac{n_0}{2n_2}\right)^{1/2}\frac{h}{a_0}. \tag{16.49}$$

For the present solution to agree with (16.42) near the axis, we must take $h = r_0$, and we see that the focal point compares well with (16.47).

16.4 Higher Order Dispersive Effects

We return temporarily to the specific model with Lagrangian (16.12) to see the effect of higher order terms in the modulation approximation. In the near-linear theory the expansions (16.10) are substituted as before, but now derivatives of the coefficients $a, a_3, \ldots, b, b_3, \ldots,$ are retained as explained in Section 15.5. For steady beams, these modulation parameters are functions of position $\mathbf{x}$ only and it is clear from (16.12) that $\mathbf{x}$ derivatives arise only from the term $-c_0^2\psi_{x_i}^2$ (generalized to more space dimensions). On substituting

$$\psi = \frac{a}{\omega}\sin\theta + \frac{a^3}{3\omega}\sin 3\theta + \ldots,$$

chosen to agree with (16.10), we see that the expression for the average

Lagrangian in (16.17) acquires the additional term

$$-\frac{1}{4}\frac{\epsilon_0 c_0^2}{\omega^2}a_{x_i}^2.$$

The elimination of b via (16.18) does not involve this term. Therefore it is also added to (16.19). Finally, interpreting the form of (16.19) more generally as before, we have

$$\mathcal{L} = \frac{1}{4}\left\{n_0^2(\omega) - \frac{c_0^2 k^2}{\omega^2}\right\}\epsilon_0 a^2 + \frac{1}{8}n_0(\omega)n_2(\omega)\epsilon_0 a^4 - \frac{1}{4}\frac{\epsilon_0 c_0^2}{\omega^2}a_{x_i}^2. \quad (16.50)$$

The variational equations are now

$$\delta a: \qquad \left(n_0^2 - \frac{c_0^2 k^2}{\omega^2}\right)a + n_0 n_2 a^3 + \frac{c_0^2}{\omega^2}a_{x_i x_i} = 0, \qquad (16.51)$$

$$\delta\theta: \qquad \frac{\partial}{\partial x_i}(k_i a^2) = 0, \qquad (16.52)$$

$$\text{Consistency:} \qquad \frac{\partial k_i}{\partial x_j} - \frac{\partial k_j}{\partial x_i} = 0. \qquad (16.53)$$

The extra term in (16.51) is dispersive in nature and counteracts the focusing. To focus initially, it will be necessary for the nonlinear term $n_0 n_2 a^3$ to dominate the dispersion term $c_0^2 a_{x_i x_i}/\omega^2$. If the beam has initial radius r_0 this gives the estimate

$$a_0^2 \gtrsim \frac{n_0}{n_2}\frac{1}{k^2 r_0^2} \qquad (16.54)$$

for the critical strength required. As the beam focuses the reduced transverse scale increases the effect of $a_{x_i x_i}$ and the focusing to a singular point is prevented in some cases. The beam might also oscillate in thickness under the fluctuating dominance of the nonlinearity and the dispersion.

    As a particular case, we should expect that there is a solution representing a uniform beam with all quantities independent of distance along it. For a two dimensional or axisymmetric beam, the equations would reduce to

$$k_2 = 0, \qquad k_1 = k = \text{constant},$$

$$\frac{c_0^2}{\omega^2}\left(a_{rr} + \frac{m}{r}a_r\right) + \left(n_0^2 - \frac{c_0^2 k^2}{\omega^2}\right)a + n_0 n_2 a^3 = 0.$$

If a^* denotes the amplitude at which

$$\frac{c_0^2 k^2}{\omega^2} = n_0^2 + n_0 n_2 a^{*2}, \qquad (16.55)$$

we have

$$a_{rr} + \frac{m}{r} a_r = 2\gamma K^2 a (a^{*2} - a^2), \qquad (16.56)$$

where $\gamma = n_2/2n_0$, $K^2 = \omega^2 n_0^2/c_0^2$, as before. In the two dimensional case $m = 0$, the equation can be integrated further to

$$a_y^2 = \gamma K^2 a^2 (a_0^2 - a^2), \qquad (16.57)$$

where $a_0 = a^* \sqrt{2}$ is the maximum amplitude; a^* is the amplitude at the point of inflexion in the profile. The solution is

$$a = a_0 \operatorname{sech}(\gamma^{1/2} K a_0 y). \qquad (16.58)$$

This is the anolog of the solitary wave in unsteady problems. If a constant of integration is allowed in (16.57) one can obtain solutions which are oscillatory and periodic in y. These are the analogs of cnoidal waves.

For the axisymmetric beam, $m = 1$, solutions of (16.56) have been found numerically by Chiao, Garmire, and Townes (1964) and by Haus (1966). The former calculate the solution corresponding to (16.58) and it has monotonic decay from a_0 at $r = 0$ to zero as $r \to \infty$. Haus finds oscillatory solutions of successively decaying amplitude and these represent a beam surrounded by diffraction rings. Small (1972) notes that (16.56) in normalized form, with $\bar{a} = a/a^*, \bar{r} = rKa^*(2\gamma)^{1/2}$, is associated with the variational principle

$$\delta \int_0^\infty \bar{r} \left(\bar{a}_{\bar{r}}^2 + \bar{a}^2 - \frac{1}{2} \bar{a}^4 \right) d\bar{r} = 0,$$

and he uses a Rayleigh-Ritz procedure to show that

$$\bar{a} = 0.8488 e^{-0.2495 \bar{r}^2} + 1.3156 e^{-1.1810 \bar{r}^2}$$

is a good approximation to the Chiao, Garmire, Townes solution. In these dimensionless variables the strength required is

$$P = \int_0^\infty \bar{a}^2 \bar{r} \, d\bar{r} \simeq 1.86.$$

Thin Beams.

The thin beam approximation to (16.51)–(16.52) is obtained under the same assumptions that led to (16.39)–(16.40). There is now an extra second derivative term added to (16.39) and we have

$$K^2(2s_x + s_r^2)a = \frac{n_2}{n_0}K^2a^3 + \left(a_{rr} + \frac{m}{r}a_r\right), \tag{16.59}$$

$$\frac{\partial a^2}{\partial x} + s_r\frac{\partial a^2}{\partial r} + \left(s_{rr} + \frac{m}{r}s_r\right)a^2 = 0. \tag{16.60}$$

If we take

$$\Psi = ae^{iKs},$$

the two equations combine into

$$2iK\frac{\partial \Psi}{\partial x} + \nabla_\perp^2\Psi + \frac{n_2}{n_0}K^2|\Psi|^2\Psi = 0, \tag{16.61}$$

where

$$\nabla_\perp^2 = \frac{\partial^2}{\partial r^2} + \frac{m}{r}\frac{\partial}{\partial r}.$$

This nonlinear Schrodinger equation has a certain canonical structure, in the same sense as the Korteweg-deVries equation, and arises in a variety of different problems. Surprisingly, a wide class of exact solutions can be derived for the two dimensional beam ($m = 0$) by the same method developed by Gardner, Greene, Kruskal, and Miura (1967) for the Korteweg-deVries equation. This was pointed out by Zakharov and Shabat (1972), who go on to give a thorough analysis of the equation. An account will be given in Chapter 17.

By the time the various approximations have been made it is simpler to derive (16.61) directly from Maxwell's equations with an assumed nonlinear relation between **P** and **E**. For the two dimensional beam, we have

$$E_{tt} + \frac{1}{\epsilon_0}P_{tt} = c_0^2(E_{xx} + E_{yy})$$

and add

$$P = (n_0^2 - 1)\epsilon_0 E + n_0 n_2\epsilon_0 a^2 E$$

to give (to a sufficient approximation)

$$n_0^2E_{tt} + n_0 n_2 a^2 E_{tt} = c_0^2(E_{xx} + E_{yy}).$$

Then if

$$E = \frac{1}{2}\Psi(x,y)e^{iKx - i\omega t} + \frac{1}{2}\Psi^*(x,y)e^{-iKx + i\omega t},$$

we have

$$\frac{n_2}{n_0}K^2|\Psi|^2\Psi = -2iK\Psi_x - \Psi_{xx} - \Psi_{yy}.$$

If Ψ_{xx} is neglected, then (16.61) follows.

16.5 Second Harmonic Generation

One of the spectacular experiments in nonlinear optics is the production of a blue beam from a red beam on passage through a nonlinear crystal. This is a good example of the production of second harmonics due to nonlinear effects and the theory is in the spirit of the general discussion in Section 15.6. The experiment was first performed by Franken, Hill, Peters, and Weinreich (1961). A full account of the theory is given by Yariv (1967, Chapter 21). We note the main points briefly.

In this case the appropriate nonlinear effect is a *quadratic* dependence of **P** on **E**, and it is assumed that the components are given by

$$P_i = (n_0^2 - 1)\epsilon_0 E_i + d_{ijk} E_j E_k. \tag{16.62}$$

Ammonium dihydrogen phosphate, for example, exhibits this effect with d_{ijk} nonzero when i, j, and k are unequal. The anisotropy of the relation corresponds to the anisotropy of the crystal. It would be modeled by an unsymmetric potential well with a term

$$V(\mathbf{P}) \propto P_i P_j P_k$$

in the three dimensional version of (16.4). In general, n_0 depends on the frequency ω due to the dispersion, but in many cases the d_{ijk} are independent of ω.

Maxwell's equations may be reduced to

$$\frac{\partial^2 E_i}{\partial t^2} + \frac{1}{\epsilon_0}\frac{\partial^2 P_i}{\partial t^2} = c_0^2 \nabla^2 E_i, \tag{16.63}$$

but a little care is needed in using (16.62) directly when a number of interacting modes of different frequencies are involved, because of the dependence of n_0 on ω. However, if P_i is split into two parts $P_i = P_i' + P_i''$, where P_i' refers to the linear part and P_i'' to the nonlinear, we know that for any particular frequency

$$\frac{1}{c_0^2}\left(\frac{\partial^2 E_i}{\partial t^2} + \frac{1}{\epsilon_0}\frac{\partial^2 P_i'}{\partial t^2}\right) = -\frac{n_0^2\omega^2}{c_0^2}E_i = -k^2 E_i, \qquad (16.64)$$

where $k(\omega)$ is the corresponding wave number for linear waves. We now take a number of interacting plane waves whose y and z components are given by

$$E_i = \frac{1}{2}\sum_\alpha A_i^{(\alpha)}(x)\exp(ik_\alpha - i\omega_\alpha t),$$

where $\alpha = \pm 1, \pm 2, \ldots,$

$$k_{-n} = k_n, \qquad \omega_{-n} = -\omega_n, \qquad \mathbf{A}^{(-n)} = \mathbf{A}^{(n)*},$$

and k_n, ω_n satisfy the linear dispersion relation. On substitution in (16.63), we have

$$\sum_\alpha\left\{ik_\alpha\frac{dA_i^{(\alpha)}}{dx} + \frac{1}{2}\frac{d^2A_i^{(\alpha)}}{dx^2}\right\}\exp(ik_\alpha x - i\omega_\alpha t) = \mu_0\frac{\partial^2}{\partial t^2}P_i''.$$

The nonlinear term $\mathbf{P}''$ produces modulations of the amplitudes $\mathbf{A}^{(\alpha)}$. The modulations are assumed to be slow compared with the wavelength and the second derivatives with respect to x are neglected. Assuming that

$$P_i'' = d_{ijk}E_jE_k,$$

we have

$$\sum_\alpha ik_\alpha\frac{dA_i^{(\alpha)}}{dx}\exp\left(ik_\alpha x - i\omega_\alpha t\right)$$

$$= -\frac{\mu_0}{4}\sum_{\beta,\gamma}(\omega_\beta + \omega_\gamma)^2 d_{ijk}A_j^{(\beta)}A_k^{(\gamma)}\exp\left\{i(k_\beta + k_\gamma)x - i(\omega_\beta + \omega_\gamma)t\right\}.$$

$$(16.65)$$

For three interacting waves with frequencies satisfying the resonance condition

$$\omega_1 + \omega_2 = \omega_3, \qquad (16.66)$$

the terms proportional to $e^{i\omega_1 t}$ give

$$\frac{dA_i^{(1)}}{dx} = \frac{i\mu_0 c_0}{2n_0} \omega_1 d_{ijk} A_j^{(3)} A_k^{(2)*} \exp\{i(k_3 - k_1 - k_2)x\}.$$

Similarly,

$$\frac{dA_i^{(2)}}{dx} = \frac{i\mu_0 c_0}{2n_0} \omega_2 d_{ijk} A_j^{(3)} A_k^{(1)*} \exp\{i(k_3 - k_1 - k_2)x\},$$

$$\frac{dA_i^{(3)}}{dx} = \frac{i\mu_0 c_0}{2n_0} \omega_3 d_{ijk} A_j^{(1)} A_k^{(2)} \exp\{-i(k_3 - k_1 - k_2)x\}.$$

If we assume that $\mathbf{A}^{(3)} = 0$ initially at $x = 0$ and that the primary waves $\mathbf{A}^{(1)}, \mathbf{A}^{(2)}$ are very little depleted by the interaction, we may take $\mathbf{A}^{(1)}, \mathbf{A}^{(2)}$ constant in the equation for $dA^{(3)}/dx$ and obtain

$$A_i^{(3)} = \frac{\mu_0 c_0}{2n_0} \omega_3 d_{ijk} A_j^{(1)} A_k^{(1)} \frac{(1 - e^{-ix\Delta k})}{\Delta k}, \qquad \Delta k = k_3 - k_1 - k_2.$$

The amplitude is proportional to $(\sin\frac{1}{2}x\Delta k)/\frac{1}{2}\Delta k$. If the interacting waves satisfy the resonance condition

$$\Delta k = k_3 - k_1 - k_2 = 0 \tag{16.67}$$

exactly, $A^{(3)}$ increases linearly with x at first, but the other interaction equations (reducing $\mathbf{A}_1, \mathbf{A}_2$ as $\mathbf{A}_3$ builds up) must then be included. The energy oscillates between the interacting modes. Eventually, loss of energy to higher harmonics and dissipation of energy must be incorporated.

In the case of second harmonic generation, the second harmonic is produced by self-interaction in which

$$\omega_1 = \omega_2 = \omega, \qquad \omega_3 = 2\omega.$$

In normal circumstances, however, $\Delta k = k(2\omega) - 2k(\omega) \neq 0$ due to the dispersion, and the second harmonic generation would remain fairly small. To impove on this and obtain true resonance, Giordmaine (1962) and Maker et al. (1962) proposed the ingenious device of using birefractive crystals (described in Section 12.8) to match an ordinary ray at frequency ω with an extraordinary ray at frequency 2ω. The matching condition

$$k^{(e)}(2\omega) - 2k^{(0)}(\omega) = 0$$

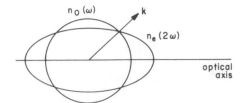

Fig. 16.1   Index matching of ordinary and extraordinary rays.

is equivalent to

$$n^{(e)}(2\omega) - n^{(0)}(\omega) = 0,$$

where subscripts e and 0 indicate extraordinary and ordinary rays, respectively. The variation of $n^{(e)}(2\omega)$ and $n^{(0)}(\omega)$ with the angle from the optical axis is as shown in Fig. 16.1. The vector $\mathbf{k}$ as shown gives the required direction for resonance. For a ruby laser beam ($\lambda = 6940$ Å) in a crystal of potassium dihydrogen phosphate the angle is $50.4°$. Full details as well as alternative possibilities are given by Yariv (1967).

With these improvements of the resonance condition, the experiments gave beautiful confirmation of the theory. A striking photograph taken by R. W. Terhune is reproduced in the frontispiece of Yariv's book.

WATER WAVES

16.6 The Average Variational Principle for Stokes Waves

We now apply the variational approach to some problems in the theory of water waves. The variational principle is given in (13.16)–(13.17) of Section 13.2, and the approximate developments of Stokes and Korteweg-deVries, backed by subsequent mathematical existence proofs, assure the existence of periodic dispersive wavetrains. Since φ is a potential and only its derivatives appear in the Lagrangian, the most general form for a periodic wavetrain is

$$\varphi = \beta x - \gamma t + \Phi(\theta, y), \qquad \theta = kx - \omega t,$$

$$\eta = N(\theta),$$

(16.68)

where $\Phi(\theta, y)$ and $N(\theta)$ are periodic functions of θ. The parameter β is the mean of the horizontal velocity φ_x, and γ is related to the mean height of the waves. In the uniform case, a frame of reference can be chosen in which $\beta = 0$ and the mean height is zero. This was done in the earlier discussion [see (13.120) and (13.121)]. Notice that $\gamma \neq 0$ even for this choice. In the modulation theory, changes in the mean velocity and the mean height are coupled with changes in the amplitude. Accordingly β, γ, and a related parameter for the mean height must be left open. The nonlinear coupling of amplitude modulations with mean velocity and height is an important physical effect and the mathematics fits it quite naturally. It is the prime example of the situation noted in (14.62) and the following discussion of Section 14.7.

In the lowest order modulation approximation, the average Lagrangian is found by substituting the periodic solution (16.68) into the expression (13.17). In the first instance we shall consider a horizontal bottom and choose the origin $y = 0$ at the bottom so that $h_0 = 0$. We have

$$\mathcal{L} = \frac{1}{2\pi} \int_0^{2\pi} L \, d\theta, \tag{16.69}$$

where

$$L = \int_0^{N(\theta)} \rho \left\{ \gamma + \omega \Phi_\theta - \frac{1}{2} (\beta + k\Phi_\theta)^2 - \frac{1}{2} \Phi_y^2 - gy \right\} dy$$

$$= \rho \left(\gamma - \frac{1}{2} \beta^2 \right) N - \frac{1}{2} \rho g N^2 + (\omega - \beta k) \rho \int_0^N \Phi_\theta \, dy$$

$$- \rho \int_0^N \left(\frac{1}{2} k^2 \Phi_\theta^2 + \frac{1}{2} \Phi_y^2 \right) dy. \tag{16.70}$$

Since exact expressions for $\Phi(\theta, y)$ and $N(\theta)$ are not known, further progress is made by adopting either the near-linear expansions of Stokes or the long wave theory of Boussinesq and Korteweg-deVries. We pursue the Stokes development. The periodic functions $\Phi(\theta, y)$ and $N(\theta)$ are expanded as Fourier series in the form

$$\Phi(\theta, y) = \sum_1^\infty \frac{A_n}{n} \cosh nky \sin n\theta \tag{16.71}$$

$$N(\theta) = h + a \cos \theta + \sum_2^\infty a_n \cos n\theta. \tag{16.72}$$

The main parameters will ultimately be the triads (ω, k, a) and (γ, β, h); a is the amplitude parameter and h is the mean height of the surface. It may be assumed in advance that the coefficients a_n, A_n will be $0(a^n)$ for small amplitudes; (16.71)–(16.72) are substituted in (16.69)–(16.70) to obtain an expression for $\mathcal{L}$ to any desired order in a. The main interest is in the first nonlinear effects, which are of order a^4 in $\mathcal{L}$, so it is convenient to calculate $\mathcal{L}$ up to this order. The coefficients A_1, A_2, a_1 appear in the expression, in addition to the two main triads, but they may be eliminated by solving the variational equations

$$\mathcal{L}_{A_1} = 0, \qquad \mathcal{L}_{A_2} = 0, \qquad \mathcal{L}_{a_1} = 0$$

for A_1, A_2, a_1 and resubstituting the results in $\mathcal{L}$. These steps are tedious but unavoidable, whatever approach is used. It is, on the whole, a little simpler to obtain the relations for A_1, A_2, a_2 from the variational principle, rather than directly from the equations as outlined in Section 13.13. This route also has the advantage that $\mathcal{L}$ is determined once and for all, and all other quantities such as mass, momentum, and energy flux are simply derived from it without repetition of similar algebraic manipulations.

The eventual expression for $\mathcal{L}$ is

$$\mathcal{L} = \rho\left(\gamma - \frac{1}{2}\beta^2\right)h - \frac{1}{2}\rho g h^2 + \frac{1}{2}E\left\{\frac{(\omega - \beta k)^2}{gk\tanh kh} - 1\right\}$$
$$-\frac{1}{2}\frac{k^2 E^2}{\rho g}\left\{\frac{9T^4 - 10T^2 + 9}{8T^4}\right\} + O(E^3), \qquad (16.73)$$

where

$$E = \frac{1}{2}\rho g a^2, \qquad T = \tanh kh.$$

It will be seen below that E is the energy density for linear waves moving into still water; it becomes a convenient amplitude parameter in place of a. In general, changes of the mean quantities (γ, β, h) are coupled to the wavemotion and it will be seen that the changes are $O(a^2)$. It is therefore consistent to replace h by the undisturbed depth h_0 in the coefficient of the term in a^4 and replace T by

$$T_0 = \tanh kh_0$$

in that term. It is important, however, to keep h in the earlier terms. In the original derivation of the expression for $\mathcal{L}$ (Whitham, 1967), it was supposed that γ and β were $O(a^2)$, since the case of interest there was propagation into initially still water. But (16.73) is in fact true without that restriction. The extension allows the study of waves on currents, for example, where the *changes* in β due to the waves will be $O(a^2)$, but β itself will include the nonzero undisturbed value of the stream velocity.

There is some slight difference in the form of $\mathcal{L}$ depending on the choice of the zero level of the potential energy. In deriving (16.73) from (13.17) the zero level was taken at the bottom, assumed to be horizontal. If, more generally, the mean surface is $y = b$ and the bottom is $y = -h_0$, the term $\frac{1}{2}\rho g h^2$ in (16.73) is replaced by

$$\frac{1}{2}\rho g b^2 - \frac{1}{2}\rho g h_0^2; \tag{16.74}$$

the other terms are the same with $h = h_0 + b$. This modification must be used when the bottom is not horizontal and therefore no longer available as the reference level for potential energy. We shall take the bottom to be horizontal and use (16.73) unless the contrary is stated specifically.

16.7 The Modulation Equations

For a modulated wavetrain the term $\beta x - \gamma t$ in (16.68) must be replaced by a pseudo-phase $\psi(x,t)$ and γ, β defined by

$$\gamma = -\psi_t, \qquad \beta = \psi_x,$$

just as $kx - \omega t$ is replaced by the phase $\theta(x,t)$ (see Section 14.7). The average variational principle

$$\delta \int \int \mathcal{L}(\omega, k, E; \gamma, \beta, h)\, dx\, dt = 0 \tag{16.75}$$

is to be used for variations in $\delta E, \delta\theta, \delta h, \delta\psi$, and we have

$$\delta E: \qquad \mathcal{L}_E = 0, \tag{16.76}$$

$$\delta\theta: \qquad \frac{\partial}{\partial t}\mathcal{L}_\omega - \frac{\partial}{\partial x}\mathcal{L}_k = 0, \qquad \frac{\partial k}{\partial t} + \frac{\partial\omega}{\partial x} = 0, \tag{16.77}$$

$$\delta h: \qquad \mathcal{L}_h = 0, \tag{16.78}$$

$$\delta\psi: \qquad \frac{\partial}{\partial t}\mathcal{L}_\gamma - \frac{\partial}{\partial x}\mathcal{L}_\beta = 0, \qquad \frac{\partial\beta}{\partial t} + \frac{\partial\gamma}{\partial x} = 0. \tag{16.79}$$

The dispersion relation $\mathcal{L}_E = 0$ gives

$$\frac{(\omega - \beta k)^2}{gk \tanh kh} = 1 + \frac{9T_0^4 - 10T_0^2 + 9}{4T_0^4}\frac{k^2 E}{\rho g} + O(E^2), \tag{16.80}$$

in agreement with (13.123). The companion relation $\mathcal{L}_h = 0$ gives

$$\gamma = \frac{1}{2}\beta^2 + gh + \frac{1}{2}\left(\frac{1-T_0^2}{T_0}\right)\frac{kE}{\rho} + O(E^2).$$

Since $\gamma = -\psi_t, \beta = \psi_x$, this is a Bernoulli type of equation for the mean flow potential ψ, modified by the wave contribution proportional to a^2. In such relations it seems to be convenient to express coefficients depending on T_0 in terms of

$$\omega_0(k) = (gk\tanh kh_0)^{1/2}, \qquad c_0(k) = (gk^{-1}\tanh kh_0)^{1/2},$$

$$C_0(k) = \frac{1}{2}c_0(k)\left\{1 + \frac{2kh_0}{\sinh 2kh_0}\right\},$$

the linear values for waves moving into still water of depth h_0. We have

$$\gamma = \frac{1}{2}\beta^2 + gh + \frac{1}{2}\left(\frac{2C_0}{c_0} - 1\right)\frac{E}{\rho h_0} + O(E^2). \tag{16.81}$$

16.8 Conservation Equations

The wave action density and flux in (16.77) are

$$\mathcal{L}_\omega = \frac{E(\omega - \beta k)}{gk\tanh kh} = \frac{E}{\omega_0} + O(E^2), \tag{16.82}$$

$$-\mathcal{L}_k = \frac{E\beta(\omega - \beta k)}{gk\tanh kh} + \frac{1}{2}\frac{E(\omega - \beta k)^2}{(gk\tanh kh)^2}\frac{d}{dk}\omega_0^2 + O(E^2)$$

$$= \frac{E}{\omega_0}(\beta + C_0) + O(E^2). \tag{16.83}$$

These take the usual form in terms of the energy density E.

Mass Conservation.

The companion quantities in (16.79) are

$$\mathcal{L}_\gamma = \rho h, \qquad -\mathcal{L}_\beta = \rho h\beta + \frac{E}{c_0} + O(E^2). \tag{16.84}$$

We see therefore that the first equation in (16.79) is the equation for conservation of mass and that the waves add a net contribution E/c_0 to the mass flow. It is particularly valuable, then, to introduce the mass transport velocity defined by

$$U = \beta + \frac{E}{\rho c_0 h}.$$ (16.85)

Energy and Momentum.

The energy density and flux, defined in (14.74), are found to be

$$\omega \mathcal{L}_\omega + \gamma \mathcal{L}_\gamma - \mathcal{L} = \frac{1}{2}\rho h U^2 + \frac{1}{2}\rho g h^2 + E + O(E^2),$$ (16.86)

$$-\omega \mathcal{L}_k - \gamma \mathcal{L}_\beta = \rho h U\left(\frac{1}{2}U^2 + gh\right) + U\left(\frac{2C_0}{c_0} - \frac{1}{2}\right)E + (U + C_0)E + O(E^2).$$

(16.87)

The momentum density and flux [see (14.75)] are found to be

$$k\mathcal{L}_\omega + \beta \mathcal{L}_\gamma = \rho h \beta + \frac{E}{c_0} = \rho h U + O(E^2),$$ (16.88)

$$-k\mathcal{L}_k - \beta \mathcal{L}_\beta + \mathcal{L} = \rho h U^2 + \frac{1}{2}\rho g h^2 + \left(\frac{2C_0}{c_0} - \frac{1}{2}\right)E + O(E^2).$$ (16.89)

The simplicity of form obtained by introducing U in place of β is to be particularly noted. For then the contributions of the mean flow, the waves, and their interactions are clearly seen. From (16.86), E is confirmed as the energy density contributed by the waves. The wave momentum E/c_0 in (16.88) takes the usual form, but the complete term is also conveniently written as $\rho h U$. The expression for the momentum flux (16.89) contains the interesting term

$$S = \left(\frac{2C_0}{c_0} - \frac{1}{2}\right)E,$$ (16.90)

which was first pointed out and exploited by Longuet-Higgins and Stewart (1960, 1961), who refer to it as radiation stress. It should then be noted that it contributes a rate of working US in the energy flux (16.87); this is a wave interaction term in addition to the usual energy flux $(U + C_0)E$. Whenever a mechanical system is described relative to a frame moving with its center of mass, the total energy is

$$\frac{1}{2}\sum m(U+v)^2 = \frac{1}{2}U^2\sum m + U\sum mv + \frac{1}{2}\sum mv^2,$$

the middle term being U times the relative momentum; the three terms in (16.87) are the counterparts of this simpler case.

From the various expressions obtained above, the original equations (16.77)–(16.79) may be written

$$\frac{\partial}{\partial t}\left(\frac{E}{\omega_0}\right) + \frac{\partial}{\partial x}\left\{(\beta + C_0)\frac{E}{\omega_0}\right\} = 0, \qquad \frac{\partial k}{\partial t} + \frac{\partial \omega}{\partial x} = 0, \qquad (16.91)$$

$$\frac{\partial}{\partial t}(\rho h) + \frac{\partial}{\partial x}\left(\rho h\beta + \frac{E}{c_0}\right) = 0, \qquad \frac{\partial \beta}{\partial t} + \frac{\partial \gamma}{\partial x} = 0, \qquad (16.92)$$

where ω and γ are given by (16.80)–(16.81) and terms $O(E^2)$ are omitted. An alternative set, introducing the energy and momentum equations in place of wave action and the consistency equation between β and γ, is

$$k_t + \omega_x = 0, \qquad (16.93)$$

$$(\rho h)_t + (\rho h U)_x = 0, \qquad (16.94)$$

$$(\rho h U)_t + \left(\rho h U^2 + \frac{1}{2}\rho g h^2 + S\right)_x = 0, \qquad (16.95)$$

$$\left(\frac{1}{2}\rho h U^2 + \frac{1}{2}\rho g h^2 + E\right)_t + \left\{\rho h U\left(\frac{1}{2}U^2 + gh\right) + US + (U + C_0)E\right\}_x = 0.$$

$$(16.96)$$

In the early work concerning waves on currents there was some question about the correct form for a "wave energy" equation for E. One way of deducing the correct form is to eliminate h and U as much as possible in (16.96) from the preceding equations. It is easily found that

$$E_t + \{(U + C_0)E\}_x + SU_x = 0 \qquad (16.97)$$

is correct. The need for the extra term SU_x was pointed out by Longuet-Higgins and Stewart (1961). Of course (16.97) is equivalent to the wave action equation in (16.91), which appears to be more fundamental in the present approach. To see the equivalence we note that the wave action equation can be expanded to

$$E_t + \{(\beta + C_0)E\}_x + \left[\frac{k}{\omega_0}\frac{\partial \omega_0}{\partial k} + \frac{h}{\omega_0}\frac{\partial \omega_0}{\partial h}\right]E\beta_x = 0.$$

The coefficient in square brackets is equal to $(2C_0/c_0 - \frac{1}{2})$ and $\beta = U + O(E)$, so the two agree.

16.9 Induced Mean Flow

Equations 16.92, or alternatively the pair (16.94)–(16.95), can be viewed as determining the changes in h and U induced by the waves. These are the long wave equations [see (13.79)] with the additional wave term S. We are concerned here with the changes in h and U induced by the wavetrain. For waves moving into still water of depth h_0, we may suppose U and $b = h - h_0$ to be small and linearize the equations to

$$b_t + h_0 U_x = 0,$$

$$U_t + gb_x = -\frac{S_x}{\rho h_0}.$$

(16.98)

For many purposes it is sufficient to take S as a known forcing term already determined from the linear dispersive theory for the distributions of k and E. Since

$$k_t + C_0(k)k_x = 0, \qquad E_t + (C_0 E)_x = 0$$

in that theory, and since S is of the form $f(k)E$, it follows that

$$\{g(k)S\}_t + \{g(k)C_0 S\}_x = 0$$

for any function $g(k)$. It is then easily verified that a solution of (16.98) is

$$b = h - h_0 = -\frac{h_0}{gh_0 - C_0^2(k)} \cdot \frac{S}{\rho h_0},$$

$$U = \beta + \frac{E}{\rho c_0 h_0} = -\frac{C_0(k)}{gh_0 - C_0^2(k)} \cdot \frac{S}{\rho h_0}.$$

(16.99)

To these may be added the solutions of the homogeneous equations, that is, functions of $x \pm \sqrt{gh_0}\, t$. It is clear from (16.99) that the group velocity and the long wave velocity $\sqrt{gh_0}$ should not be too close compared with a^2. But this is required for the validity of the Stokes expansion; in the limit $C_0^2 \to gh_0$, the Korteweg-deVries development is needed.

In starting up a wavetrain, transient long waves will be propagated with velocities $\pm \sqrt{gh_0}$, but (again assuming C_0 and $\sqrt{gh_0}$ are well enough separated) the mean flow and mean height *accompanying* the wavetrain are given by (16.99). The starting transients set up on moving an obstacle through water are discussed in detail by Benjamin (1970).

16.10 Deep Water

For deep water, $kh_0 \gg 1$, the induced changes in h and β become negligible. This should be expected in advance, but it is confirmed explicitly by (16.99). The average Lagrangian (16.73) becomes

$$\mathcal{L}_W = \frac{1}{2}\left(\frac{\omega^2}{gk} - 1\right)E - \frac{1}{2}\frac{k^2 E^2}{\rho g} + O(E^3). \tag{16.100}$$

There is no interaction between mean flow and the waves described by $\mathcal{L}_W$. As far as the waves are concerned, we may work entirely with $\mathcal{L}_W$. It fits the simple form of earlier problems, where pseudo-frequencies do not arise, and ω, k, E are the only wave parameters. The dispersion relation from $\mathcal{L}_E = 0$ is

$$\omega^2 = gk\left(1 + \frac{2k^2 E}{\rho g} + \cdots\right)$$
$$= gk(1 + k^2 a^2 + \cdots), \tag{16.101}$$

in agreement with earlier results. The modulation equations for E and k are given by (16.77).

On a given current U_0, the preceding arguments refer to $\beta - U_0$ rather than β itself and, in the limit of deep water, $\mathcal{L}_W$ is modified to

$$\mathcal{L}_W = \frac{1}{2}\left\{\frac{(\omega - U_0 k)^2}{gk} - 1\right\}E - \frac{1}{2}\frac{k^2 E^2}{\rho g} + O(E^3). \tag{16.102}$$

16.11 Stability of Stokes Waves

For deep water waves, the simple theory of Section 14.2 applies. From (16.101),

$$\omega_0(k)=(gk)^{1/2}, \qquad \omega_2(k)=\frac{1}{2}(gk)^{1/2}k.$$

The quantity $\omega_0''\omega_2<0$; therefore modulations grow in time. For finite depth, the coupling with the induced mean flow becomes important and has a stabilizing effect.

The stability is decided by the type of the full set (16.91)–(16.92), which is a fourth order system for k,E,β,h. The type in turn is decided from the characteristics. It is straightforward, but lengthy, to find the characteristic velocities directly by the standard method. The analysis can be simplified and given a more significant form by breaking the argument into two parts. First, (16.99) is a sufficient approximation for the relations of h,β to E. At the same time, this first step brings out the result that two of the characteristic velocities are $\pm\sqrt{gh_0}$. The expressions for h,β may be used in the equations (16.91) for k and E to determine the other two characteristic velocities.

With $b=h-h_0$ and β taken to be $O(E)$, the dispersion relation (16.80) may be approximated by

$$\omega=\omega_0+k\beta+k\left(C_0-\frac{1}{2}c_0\right)\frac{b}{h_0}+\frac{9T_0^4-10T_0^2+9}{8T_0^3}\frac{k^2E}{\rho c_0}+O(E^2).$$

From (16.99), this reduces to

$$\omega=\omega_0(k)+\Omega_2(k)\frac{k^2E}{\rho c_0}+O(E^2), \tag{16.103}$$

where

$$\Omega_2=\frac{9T_0^4-10T_0^2+9}{8T_0^3}-\frac{1}{kh_0}\left\{\frac{(2C_0-(1/2)c_0)^2}{gh_0-C_0^2}+1\right\}.$$

Since b,β have been eliminated, we now have the simple modulation equations for k,E of the type discussed in Section 14.2 and can read off the result for the characteristic velocities with no further calculation! The characteristic velocities are

$$C_0 \pm \left(\frac{\omega_0'' \Omega_2 k^2 E}{\rho c_0} \right)^{1/2}.$$

(16.104)

Here $\omega_0 = (gk \tanh kh_0)^{1/2}$ and ω_0'' is always negative. Thus the characteristics are imaginary for $\Omega_2 > 0$ and real for $\Omega_2 < 0$. The formula for Ω_2 shows clearly the stabilizing effect of the mean flow as kh_0 decreases from the deep water limit. The critical value for stability is determined by the value of kh_0 for which $\Omega_2 = 0$. This value is found numerically to be $kh_0 = 1.36$. For $kh_0 > 1.36$ modulations grow; for $kh_0 < 1.36$ they propagate in typical hyperbolic fashion.

The instability for deep water waves was first deduced by Benjamin (1967) by the Fourier mode analysis described in Section 15.6. This was then realized to be the significance of the elliptic modulation equations and the critical value $kh_0 = 1.36$ was deduced for the finite depth case. The value was then confirmed by Benjamin using his Fourier mode approach. This sequence shows the valuable interplay between the two approaches.

It should again be remarked that the "unstable case" refers to growth of the modulations and not necessarily to chaotic motion. To assess the eventual behavior, higher order dispersive terms must be included as described in Section 15.5. From that analysis we infer that the next stage would be the development of modulations where the envelope of the wavetrain is a sequence of solitary waves.

16.12 Stokes Waves on a Beach

For a wavetrain approaching a beach, we may take the modulation parameters to be independent of t. From the modulation equations, we then have the four relations

$$\omega, \qquad -\mathcal{L}_k, \qquad \gamma, \qquad -\mathcal{L}_\beta = \text{constants}$$

to determine $k(x)$, $E(x)$, $\beta(x)$, $h(x)$ in terms of their original constant values out at sea and the depth distribution $h_0(x)$. To the lowest approximation, the first two relations are

$$\omega = \omega_0 = (gk \tanh kh_0)^{1/2} = \text{constant},$$

$$-\mathcal{L}_k = \frac{E}{\omega_0} C_0 = \text{constant},$$

(16.105)

which are sufficient to determine the distributions of $k(x)$, $E(x)$ in terms of the depth distributions $h_0(x)$. Since ω_0 is constant, the relation for E can also be interpreted as one of constant energy flux EC_0, but it now seems that wave action is more fundamantal for such "adiabatic" processes. The relations $\gamma = $ constant, $-\mathcal{L}_\beta = $ constant determine the accompanying small changes in $h - h_0$ and β. The results are

$$b = h - h_0 = -\frac{1}{2}\left(\frac{2C_0}{c_0} - 1\right)\frac{E}{\rho g h_0},$$

$$\beta = -\frac{E}{\rho c_0 h_0}.$$

(16.106)

[The slight modification noted in (16.74) is used in calculating γ.] There is a depression of the mean surface and a countercurrent to balance the mass flow induced by the waves.

On a shelving beach the amplitude increases as the depth decreases. At sufficiently high amplitudes, which can be variously estimated as $a/\lambda = 0.142$ from Michell's deep water calculations or $a/h_0 = 0.78$ from McCowan's solitary wave estimates, the waves peak and the Stokes theory ceases to apply.

16.13 Stokes Waves on a Current

A similar discussion applies to waves propagating along a nonuniform current $U_0(x)$, which may be assumed to be due to variations in the depth $h_0(x)$ or fed by upwelling from below. In this case, we have

$$\omega = kU_0(x) + \omega_0(k) = kU_0(x) + \left\{ gk \tank k h_0(x) \right\}^{1/2} = \text{constant}$$

(16.107)

for the determination of $k(x)$, and

$$-\mathcal{L}_k = \frac{E}{\omega_0}\left\{ U_0(x) + C_0(x) \right\} = \text{constant}$$ (16.108)

for the determination of amplitude. For deep water

$$\omega_0 = (gk)^{1/2}, \qquad c_0 = \left(\frac{g}{k}\right)^{1/2}, \qquad C_0 = \frac{1}{2}c_0:$$

we may express the results in the form

$$kU_0 + (gk)^{1/2} = \text{constant},$$

$$Ec_0(2U_0 + c_0) = \text{constant}. \tag{16.109}$$

These were first found by Longuet-Higgins and Stewart (1961) after detailed direct analysis of the equations of motion. For waves moving against a current the result is singular, predicting $E \to \infty$, when the magnitude of the group velocity equals the stream velocity. At this stage the next order terms of order E^2 become crucial and ensure a finite result. This question has been studied by Crapper (1972) and Holliday (1973).

KORTEWEG-DEVRIES EQUATION

To conclude this chapter, the modulation theory is developed for the Korteweg-deVries equation. There are a number of special features in the derivation, and some nontrivial tricks were needed to get exact formulas for the characteristic relations. Because of the central position of the equation in the subject and the possible tie with further developments in the exact analysis of the equation, it seems worthwhile to document these.

16.14 The Variational Formulation

It is convenient to choose the particular normalization in which the equation becomes

$$\eta_t + 6\eta\eta_x + \eta_{xxx} = 0. \tag{16.110}$$

There is no variational principle for the equation as it stands and some potential representation is required. The simplest choice is $\eta = \varphi_x$; the equation becomes

$$\varphi_{xt} + 6\varphi_x\varphi_{xx} + \varphi_{xxxx} = 0 \tag{16.111}$$

and a suitable Lagrangian is

$$L = -\frac{1}{2}\varphi_t\varphi_x - \varphi_x^3 + \frac{1}{2}\varphi_{xx}^2. \tag{16.112}$$

One could also introduce $\chi = \varphi_{xx}$ in (16.111), work with a pair of functions φ, χ, and use a Lagrangian involving only first derivatives. That has the advantage of fitting the general discussion in Section 14.7, but it is simpler to work with (16.112) and add a few special manipulations.

The need for a potential representation is basic to the structure of the problem and affects the number of parameters in the uniform wavetrain. We must take

$$\varphi = \psi + \Phi(\theta), \qquad \psi = \beta x - \gamma t, \qquad \theta = kx - \omega t.$$

Then

$$\eta = \beta + k\Phi_\theta, \tag{16.113}$$

and the parameter β refers to the mean value of η. In terms of η the uniform wavetrain solution of (16.110) is given by

$$k^2 \eta_{\theta\theta\theta} + 6\eta\eta_\theta - \frac{\omega}{k}\eta_\theta = 0,$$

and there are two immediate integrals

$$k^2 \eta_{\theta\theta} + 3\eta^2 - \frac{\omega}{k}\eta + B = 0,$$

$$k^2 \eta_\theta^2 + 2\eta^3 - \frac{\omega}{k}\eta^2 + 2B\eta - 2A = 0, \tag{16.114}$$

where A, B are constants of integration. For this solution, the Lagrangian (16.112) may first be expressed in terms of η as

$$L = \frac{1}{2}\left(\gamma - \frac{\omega}{k}\beta\right)\eta + \frac{1}{2}\frac{\omega}{k}\eta^2 - \eta^3 + \frac{1}{2}k^2\eta_\theta^2,$$

and then, from (16.114), in the equivalent form

$$L = k^2\eta_\theta^2 + \left\{ B + \frac{1}{2}\left(\gamma - \frac{\omega}{k}\right)\beta \right\}\eta - A. \tag{16.115}$$

We now require the average Lagrangian

$$\mathcal{L} = \frac{1}{2\pi}\int_0^{2\pi} L\, d\theta.$$

From (16.113),

$$\frac{1}{2\pi}\int_0^{2\pi} \eta\, d\theta = \bar{\eta} = \beta;$$

from (16.114),

$$\frac{1}{2\pi}\int_0^{2\pi}k^2\eta_\theta^2\,d\theta = \frac{1}{2\pi}\oint k^2\eta_\theta\,d\eta = kW,$$

where*   (16.116)

$$W(A,B,U)=\frac{1}{2\pi}\oint\{2A-2B\eta+U\eta^2-2\eta^3\}^{1/2}\,d\eta,$$

$$U=\frac{\omega}{k}.$$

Finally, then,

$$\mathcal{L}=kW(A,B,U)+\beta B+\frac{1}{2}\beta\gamma-\frac{1}{2}U\beta^2-A. \qquad (16.117)$$

The variational equations for the triad (γ,β,B) are

$$\delta B:\qquad \beta=-kW_B,$$

$$\delta\psi:\qquad \frac{\partial}{\partial t}\left(\frac{1}{2}\beta\right)+\frac{\partial}{\partial x}\left(U\beta-\frac{1}{2}\gamma-B\right)=0,$$

$$\text{Consistency:}\qquad \frac{\partial\beta}{\partial t}+\frac{\partial\gamma}{\partial x}=0.$$

From the last two we may take $\gamma=U\beta-B$, without loss of generality, so we have

$$\beta=-kW_B,\qquad \gamma=-kUW_B-B,$$

and

$$\frac{\partial}{\partial t}(kW_B)+\frac{\partial}{\partial x}(kUW_B+B)=0. \qquad (16.118)$$

For the triad (ω,k,A) it is convenient to replace the variational equation for $\delta\theta$ by the momentum equation

$$\frac{\partial}{\partial t}(k\mathcal{L}_\omega+\beta\,\mathcal{L}_\gamma)+\frac{\partial}{\partial x}(\mathcal{L}-k\mathcal{L}_k-\beta\,\mathcal{L}_\beta)=0.$$

We have

$$\delta A:\qquad kW_A=1, \qquad (16.119)$$

---

*The symbol U is now used for the nonlinear phase velocity as opposed to its use for the mass flow velocity in Sections 16.9–16.13.

Momentum: $\dfrac{\partial}{\partial t}(kW_U) + \dfrac{\partial}{\partial x}(kUW_U - A) = 0,$ (16.120)

Consistency: $\dfrac{\partial k}{\partial t} + \dfrac{\partial}{\partial x}(kU) = 0,$ $\omega = kU.$ (16.121)

Equations 16.118, 16.120, and 16.121 may be viewed as three equations for A, B, U, with k given by (16.119). A more symmetric equivalent form is

$$\frac{\partial W_B}{\partial t} + U\frac{\partial W_B}{\partial x} + W_A\frac{\partial B}{\partial x} = 0,$$

$$\frac{\partial W_U}{\partial t} + U\frac{\partial W_U}{\partial x} - W_A\frac{\partial A}{\partial x} = 0, \qquad (16.122)$$

$$\frac{\partial W_A}{\partial t} + U\frac{\partial W_A}{\partial x} - W_A\frac{\partial U}{\partial x} = 0.$$

In terms of these variables, the wave number, frequency, and mean value $\bar{\eta} = \beta$ are given by

$$k = \frac{1}{W_A}, \qquad \omega = \frac{U}{W_A}, \qquad \beta = -\frac{W_B}{W_A}. \qquad (16.123)$$

The amplitude a is obtained by relating the zeros of the cubic in W to the coefficients A, B, U. The natural choice of basic parameters in the physical description would be k, β, a; the trio A, B, U is an equivalent set. The dispersion relation $\omega = \omega(k, \beta, a)$ is provided implicitly by the second expression in (16.123).

The equations in (16.122) have a reasonably symmetric appearance, whereas the original variational equations in β and γ look awkward. This seems to be associated with the hybrid nature of the Korteweg-deVries equation as an approximation to the original water waves formulation. In deriving the approximation, the fluid velocity is expressed in terms of the depth [see (13.102)], so that the triad γ, β, B becomes intermingled in an unsymmetrical way. For instance, β is introduced as the mean height in (16.113), yet its natural role is that of the mean fluid velocity. This duality is smoothed out in the more symmetric form (16.122). A related point is that because of the potential representation, we have to work at first with a fourth order system; the more balanced form is recovered only when we revert to a third order system.

16.15 The Characteristic Equations

The system (16.122) is hyperbolic in general and we now consider the characteristic equations. The function W and its derivatives W_A, W_B, W_U can all be expressed in terms of complete elliptic integrals and the characteristic form of the equations could be ground out directly, but with considerable labor. Surprisingly, however, if the zeros p, q, r of the cubic

$$\eta^3 - \frac{1}{2} U\eta^2 + B\eta - A = 0 \tag{16.124}$$

are used as new variables in place of A, B, U, and if various (nontrivial) identities among the second derivatives of W are introduced, the equations may be put in a simple form from which the characteristic relations and velocities are seen immediately. It turns out that the equations may be written

$$(q+r)_t + P(q+r)_x = 0,$$

$$P = 2\left\{ (p+q+r) - \frac{p(W_q - W_r) + q(W_r - W_p) + r(W_p - W_q)}{W_q - W_r} \right\}, \tag{16.125}$$

plus similar equations for $r+p$ and $p+q$ in cyclic permutations. Thus the Riemann invariants are

$$q+r, \qquad r+p, \qquad p+q, \tag{16.126}$$

simply, and the corresponding characteristic velocities P, Q, R are the coefficient in (16.125) and its permutations.

At this point it is useful to express the quantities concerned in terms of elliptic integrals. We introduce

$$a = \frac{p-q}{2}, \qquad s^2 = \frac{p-q}{p-r}, \qquad p > q > r; \tag{16.127}$$

a is an amplitude variable and s is the modulus of the elliptic integrals. Then it may be shown that

$$\beta = \bar{\eta} = p - 2a\frac{D(s)}{K(s)},$$

where $D(s), K(s)$ are the complete elliptic integrals in standard notation

(Jahnke and Emde, 1945). If we rather use β, a, s as basic variables, we have

$$p = \beta + 2a\frac{D}{K}, \qquad q = \beta + 2a\left(\frac{D}{K} - 1\right), \qquad r = \beta + 2a\left(\frac{D}{K} - \frac{1}{s^2}\right). \quad (16.128)$$

The wave number and phase velocity are given by

$$k = \frac{1}{W_A} = \frac{\pi a^{1/2}}{sK}, \quad (16.129)$$

$$U = \frac{\omega}{k} = 2(p + q + r) = 6\beta + 4a\left(\frac{3D}{K} - \frac{1 + s^2}{s^2}\right). \quad (16.130)$$

The Riemann invariants and characteristic velocities are as follows:

Riemann Invariant	Characteristic Velocity
$q + r$	$P = U - \dfrac{4aK}{s^2 D}$
$r + p$	$Q = U - \dfrac{4a(1 - s^2)K}{s^2(K - D)}$
$p + q$	$R = U - \dfrac{4a(1 - s^2)K}{s^2(s^2 D - K)}$

In general the velocities P, Q, R are distinct and $P < Q < R$. Thus the system is hyperbolic. The limits $s^2 \to 0$ and $s^2 \to 1$ are both singular in that two of the velocities become equal. The limiting equations are then not strictly hyperbolic, although, due to the uncoupling of one of the equations, they may still be solved by integration along characteristics. This is the situation encountered earlier in the linear theory, to which the limit $s^2 \to 0$ corresponds.

Small Amplitude Case.

If we take $a \to 0, s^2 \to 0$ but keep the wave number k given by (16.129) finite and nonzero, we have

$$s \sim \frac{2a^{1/2}}{k}.$$

In the extreme limit $s^2 \to 0$, we find

$$P, Q \to 6\beta - 3k^2, \qquad R \to 6\beta.$$

The linear theory would neglect changes in β and we would have the linear group velocity $-3k^2$ as a double characteristic. In the next order correction, the near-linear theory, we find

$$P \sim 6\beta - 3k^2 - 3a + O\!\left(\frac{a^2}{k^2}\right),$$

$$Q \sim 6\beta - 3k^2 + 3a + O\!\left(\frac{a^2}{k^2}\right), \qquad (16.131)$$

$$R \sim 6\beta + O\!\left(\frac{a^2}{k^2}\right).$$

The corresponding approximation in the original equations gives the set

$$\beta_t + 6\beta\beta_x + \left[\frac{3}{2}a^2\right]_x = 0,$$

$$k_t + \left(6\beta k - k^3 + \left[\frac{3}{2}\frac{a^2}{k}\right]\right)_x = 0, \qquad (16.132)$$

$$(a^2)_t + \{(6\beta - 3k^2)a^2\}_x + 6a^2\beta_x = 0.$$

The terms in square brackets are the near-linear corrections to linear theory. In the linear theory, the equation for β uncouples and may be solved independently; it provides the characteristic velocity $R = 6\beta$. Usually, however, the solution $\beta = 0$ is appropriate and we have the usual modulation equations for a and k. The near-linear corrections introduce the important qualitative changes that make the system genuinely hyperbolic and split the remaining group velocities. The modification of the equation for β is crucially important. If only the correction to the frequency (in the second equation) were introduced, the pair of equations would have imaginary characteristics and would appear to present a case of instability. In the terminology of Section 14.2, we have $\omega_0 = -k^3$, $\omega_2 = 3/2k$, $\omega_0''\omega_2 < 0$. But the coupling with β stabilizes the modulations and the full system is hyperbolic. We can simply calculate the characteristics of (16.132) and check (16.131), but it is again instructive to take the approach used in Section 16.11. If β is entirely induced by the wave motion, we may use the second and third equations in (16.132) to show that $(a^2)_x = (a^2/3k^2)_t$, to lowest order. Then from the first equation

$$\beta = -\frac{a^2}{2k^2}.$$

After this is substituted in the second equation the effective change in frequency is

$$\Omega_2 = 6\beta k + \frac{3}{2}\frac{a^2}{k} = -\frac{3}{2}\frac{a^2}{k}.$$

The equations for a and k are now hyperbolic with characteristic velocities $-3k^2 \pm 3a$ in agreement with (16.131).

16.16 A Train of Solitary Waves

In the other limit $s^2 \to 1$, the wavetrain becomes a sequence of near-solitary waves. In this case, K and D are asymptotically given by

$$K = \Lambda + O(1-s^2), \qquad D = \Lambda - O(1-s^2), \qquad \Lambda = \log\frac{4}{(1-s^2)^{1/2}}.$$

In the solitary wave limit it is natural to take the amplitude to be the height of the crests above the troughs and to take the wave number to apply to the number per unit length (rather than number per 2π). Accordingly we introduce

$$a_1 = 2a, \qquad k_1 = \frac{k}{2\pi}.$$

Then from (16.129) we have

$$\Lambda = \frac{(2a_1)^{1/2}}{4k_1}. \tag{16.133}$$

Errors of order $1 - s^2$ are exponentially small in $(2a_1)^{1/2}/k_1$ and we work to that order. Then

$$p \sim \beta + a_1 - 2k_1(2a_1)^{1/2}; \qquad q, r \sim \beta - 2k_1(2a_1)^{1/2}; \tag{16.134}$$

$$U \sim 6\beta + 2a_1 - 12k_1(2a_1)^{1/2}; \tag{16.135}$$

$$P \sim 6\beta - 16k_1(2a_1)^{1/2}\left\{\frac{1 - 3k_1(2a_1)^{-1/2}}{1 - 4k_1(2a_1)^{-1/2}}\right\};$$

$$Q, R \sim 6\beta + 2a_1 - 12k_1(2a_1)^{1/2} \tag{16.136}$$

We note that

$$q = r + O\left\{\exp\left(\frac{-a_1^{1/2}}{k_1}\right)\right\}, \quad Q = R + O\left\{\exp\left(\frac{-a_1^{1/2}}{k_1}\right)\right\}.$$

In the near-linear limit changes in β propagate primarily on the fastest characteristic (velocity R), whereas a and k are carried primarily by the two slower ones. By contrast, in this limit β is carried primarily by the slowest characteristic (P), and a_1 and k_1 propagate out ahead. In the forward region we may integrate along the P characteristic and deduce that $q + r$ remains equal to its initial value. But in this limit $q \approx r$ throughout so that q and r individually remain equal to the initial value. The normalization would usually be such that $q = r = 0$; hence they remain zero. Then from (16.134) we have

$$\beta \approx 2k_1(2a_1)^{1/2} \tag{16.137}$$

and

$$p \sim a_1; \quad q, r \sim 0; \quad U \sim 2a_1; \quad Q, R \sim 2a_1.$$

The corresponding approximate equations can be shown to be

$$k_{1t} + (2a_1k_1)_x = 0,$$
$$a_{1t} + 2a_1a_{1x} = 0. \tag{16.138}$$

In this approximation, the system is not strictly hyperbolic, but a_1 may be found first by integration along the characteristics $dx/dt = 2a_1$, and *then* k_1 can be found by integration along the same characteristics. The structure is similar to the one found in the linear theory. However, this time it is a_1 that remains constant on the characteristics and k_1 decreases like $1/t$.

As in the case of the linear limit, the next order approximation beyond (16.137)–(16.138) modifies the structure of the equations; the characteristics are separated and the system becomes truly hyperbolic.

The equations in (16.138) are so simple by now that one would assume there is some direct derivation without going through the general case first, and this is indeed true. The Korteweg-deVries equation may be written in conservation form as

$$\eta_t + (3\eta^2 + \eta_{xx})_x = 0, \tag{16.139}$$

and if it is averaged over a number of waves we have

$$(\bar{\eta})_t + \left(\overline{3\eta^2}\right)_x = 0. \tag{16.140}$$

Now a solitary wave with $q = r = 0$ and $p = a_1$ is given by

$$\eta = a_1 \operatorname{sech}^2 \left\{ \left(\frac{a_1}{2}\right)^{1/2} x - 4\left(\frac{a_1}{2}\right)^{3/2} t \right\}. \tag{16.141}$$

If the average values are calculated from

$$\bar{\eta} = k_1 \int_{-\infty}^{\infty} \eta \, dx, \qquad \overline{\eta^2} = k_1 \int_{-\infty}^{\infty} \eta^2 \, dx,$$

using this solution, we have

$$\bar{\eta} = 4k_1 \left(\frac{a_1}{2}\right)^{1/2}, \qquad \overline{\eta^2} = \frac{16k_1}{3}\left(\frac{a_1}{2}\right)^{3/2}.$$

Equation 16.140 becomes

$$(k_1 a_1^{1/2})_t + (2k_1 a_1^{3/2})_x = 0. \tag{16.142}$$

From (16.141), the phase velocity $U = 2a_1$; therefore the consistency equation $k_{1t} + (k_1 U)_x = 0$ becomes

$$k_{1t} + (2a_1 k_1)_x = 0. \tag{16.143}$$

This pair for k_1 and a_1 is equivalent to the pair in (16.138). Notice that in this derivation it is implicitly assumed that q and r remain equal to zero in the modulation.

An important particular solution of the equations is

$$a_1 = \frac{x}{2t}, \qquad k_1 = \frac{1}{2t} f\left(\frac{x}{2t}\right), \tag{16.144}$$

where f is an arbitrary function. This is easy to interpret. A solitary wave of amplitude a_1 moves with velocity $2a_1$. Therefore (16.144) represents a sequence of solitary waves each retaining a constant amplitude and moving on the path $x = 2a_1 t$. The decrease in k_1 is due to the divergence of solitary waves of different amplitude. The solution is represented in Fig. 17.1, where it is deduced from the discussion of exact solutions. However, it is completed there by a discontinuity in a_1 and k_1. We therefore consider what the jump conditions are for our equations.

There is the usual question of which conservation equations should be maintained across the discontinuity. If we accept (16.142)–(16.143), the jump conditions are

$$- V [k_1 a_1^{1/2}] + [2 k_1 a_1^{3/2}] = 0,$$

$$- V [k_1] + [2 a_1 k_1] = 0,$$

where V is the velocity of the discontinuity. A jump from $a_1 = 0$ to a nonzero value $a_1^{(0)}$ would therefore have $V = 2 a_1^{(0)}$. This is the phase velocity and the result indicates that the solution (16.144) may be cut off at any one of the solitary waves in the sequence. It is this choice that is confirmed by the exact solution in Section 17.5. The function f and the amplitude $a_1^{(0)}$ can be determined only from the initial conditions, which are provided in Section 17.5. Of course the exact analysis is superior to the modulation theory of solitary waves in the case of the Korteweg-deVries equation. But the confirmation of the results in this case substantiates similar uses of the modulation theory in problems where exact solutions are not known.

It might be noted finally that if the solitary wave solution is written

$$\eta = a_1 \operatorname{sech}^2 \left(\frac{a_1}{2} \right)^{1/2} \left(x - \frac{\omega_1}{k_1} t \right)$$

and it is used to calculate an average Lagrangian defined by

$$\mathcal{L} = k_1 \int_{-\infty}^{\infty} L \, dx,$$

we find

$$\mathcal{L} \propto \omega_1 a_1^{3/2} - \frac{6}{5} k_1 a_1^{5/2}.$$

The variational equations are

$$\delta a: \quad \omega_1 = 2a_1 k_1,$$

$$\delta \theta: \quad (a_1^{3/2})_t + \left(\frac{6}{5} a_1^{5/2}\right)_x = 0,$$

$$\text{Consistency:} \quad k_{1t} + \omega_{1x} = 0.$$

These are equivalent to (16.142)–(16.143).

CHAPTER 17

Exact Solutions; Interacting Solitary Waves

17.1 Canonical Equations

One of the most remarkable developments in recent work on nonlinear dispersive waves is the discovery of a variety of explicit exact solutions for some of the simple canonical equations of the subject. The main equations concerned are the following:

1. The Korteweg-deVries equation, now normalized to

$$\eta_t + \sigma\eta\eta_x + \eta_{xxx} = 0, \qquad (17.1)$$

where σ is a constant,

2. The cubic Schrödinger equation

$$iu_t + u_{xx} + \nu|u|^2 u = 0, \qquad (17.2)$$

3. The Sine-Gordon equation

$$\varphi_{tt} - \varphi_{xx} + \sin\varphi = 0. \qquad (17.3)$$

Explicit solutions representing the interaction of any number of solitary waves can be constructed, and a precise prediction can be made of the number of solitary waves that will eventually emerge from any finite initial disturbance.

These equations are canonical for the subject in that they combine some of the simplest types of dispersion with the simplest types of nonlinearity. The Korteweg-deVries equation combines linear dispersion

$$\omega = -\kappa^3 \qquad (17.4)$$

with a typical nonlinear convection operator. Equation 17.2 combines dispersion represented by

$$\omega = \kappa^2 \qquad (17.5)$$

577

with a simple cubic nonlinearity. In both cases, the dispersion relation can be viewed as a Taylor series approximation to a more general dispersion relation, (17.4) referring to those odd in κ and (17.5) to those even in κ. [A term proportional to κ added to (17.4) or (17.5) can be normalized out by choosing a moving frame of reference.] For this reason the equations are not mere models but frequently can be derived as valid approximations for long waves. The linear dispersion relation $\omega^2 = \kappa^2 + 1$ in (17.3) is still a fairly obvious general form, apart from its original relevance to relativistic particles in the Klein-Gordon equation; a number of problems in which the appropriate nonlinear term is $\sin\varphi$ were noted in Section 14.1.

Ever since the Cole-Hopf solution of Burgers' equation, countless people must have tried similar tricks to solve (17.1), but the eventual method of solution requires much more than a simple trick. Gardner, Greene, Kruskal, and Miura (1967) developed an ingenious series of steps to tie the equation to an inverse scattering problem. The end result is the transformation of (17.1) to a linear integral equation, but it seems inconceivable that anyone would discover it without intermediate steps. By hindsight, one can see that the substitution

$$\sigma\eta = 12(\log F)_{xx}, \tag{17.6}$$

which is a reasonable generalization of the substitution

$$c = -2\nu(\log\varphi)_x$$

for Burgers' equation, offers an easy route to the special solutions representing the interaction of solitary waves. The equation for F is *not* linear but it has some special structure, and solutions in series of exponentials lead to solitary waves. However, it is not clear how to extract more general information from the equation for F.

Equation 17.2 was solved by Zakharov and Shabat (1971) by a similar inverse scattering technique, relying to some extent on the general ideas of Lax (1968).

The explicit solution of (17.3) for two interacting solitary waves was first noticed by Perring and Skyrme (1962), apparently on the basis of their numerical computations. Lamb (1967, 1971) subsequently showed how a Bäcklund transformation could be used consistently to produce further solutions. Recently Lamb (1973) and Ablowitz et al. (1973) have shown how inverse scattering theory can be used.

The surprising overall result is that if a number of initially well-spaced solitary waves are allowed to interact, they will eventually emerge from the interaction and recover their original shapes and velocities. The only relic of their interaction is a constant displacement from the positions they would otherwise have had. The analogy with the collision of particles is intriguing. The main solutions and methods of derivation will be described in this chapter.

We add two related topics. Toda (1967a, 1967b) considers mass-spring chains which are discrete versions of some of our problems. If the extension of the nth spring from its equilibrium length is $r_n(t)$, the equations may be written

$$m\ddot{r}_n = 2f(r_n) - f(r_{n+1}) - f(r_{n-1}), \tag{17.7}$$

where $f(r)$ is the force law for each spring. The continuous limit of this difference equation would be a wave equation, nonlinear if $f(r)$ is a nonlinear function of r. In the nonlinear case the existence of uniform wavetrain solutions of (17.7) can be shown on the basis of a Stokes type of expansion in powers of amplitude. But Toda found ingenious exact expressions in elliptic functions for the case

$$f(r) = -\alpha(1 - e^{-\beta r}). \tag{17.8}$$

Moreover, these solutions have solitary waves as limiting cases and Toda was able to find solutions representing interactions with behavior similar to that of the continuous ones.

Finally, the equation

$$(1 - \varphi_t^2)\varphi_{xx} + 2\varphi_x\varphi_t\varphi_{xt} - (1 + \varphi_x^2)\varphi_{tt} = 0 \tag{17.9}$$

was proposed by Born-Infeld (1934) via the variational principle

$$\delta \int \int \{1 - \varphi_t^2 + \varphi_x^2\}^{1/2} \, dx \, dt = 0. \tag{17.10}$$

The idea was to extend the simple wave equation (with Lagrangian $\frac{1}{2}\varphi_t^2 - \frac{1}{2}\varphi_x^2$) and introduce nonlinear effects while preserving the Lorentz invariance properties. This equation admits waves of arbitrary shape moving with velocities of $+1$ or -1. They may be chosen as "solitary waves" but lack the specific intrinsic structure of the previous cases. However, Barbishov and Chernikov (1967) have shown that explicit solutions for interactions can be found, again with shape-preserving properties and a shift in position due to the interaction. A brief description is added even though there may be no deep relation with the others and they are not dispersive waves.

KORTEWEG-DEVRIES EQUATION

17.2 Interacting Solitary Waves

We first give the solutions derivable from the transformation (17.6). The parameter σ can be normalized out of both (17.1) and (17.6), of course, but different choices corresponding to $\sigma = 1, 6, -6$ have all been used in the literature and it is convenient to leave it open here for ease in making cross-references. The derivation of (17.1) for water waves and its more general significance through (17.4) as an approximation for long waves in other contexts have been explained in Section 13.11.

The transformation from η to F is most easily made in two stages. First $\eta = p_x$ is introduced and the equation integrated to

$$p_t + \frac{1}{2}\sigma p_x^2 + p_{xxx} = 0;$$

then the nonlinear transformation

$$\sigma p = 12(\log F)_x$$

is made. Terms up to the fourth degree in F and its derivatives arise, but the special feature of the transformation is that the terms of third and fourth degree cancel. The result is the quadratic equation

$$F(F_t + F_{xxx})_x - F_x(F_t + F_{xxx}) + 3(F_{xx}^2 - F_x F_{xxx}) = 0. \quad (17.11)$$

One notes the appearance of the basic operator

$$\frac{\partial}{\partial t} + \frac{\partial^3}{\partial x^3}$$

and a certain balance to the equation. A possible (but rather tenuous) motivation for the transformation starts from the solution for a single solitary wave, which may be written

$$\sigma\eta = 3\alpha^2 \operatorname{sech}^2 \frac{\theta - \theta_0}{2}, \qquad \theta = \alpha x - \alpha^3 t, \quad (17.12)$$

where α and θ_0 are parameters. This is the x derivative of

$$6\alpha\left\{\tanh\left(\frac{\theta - \theta_0}{2}\right) - 1\right\},$$

which is in turn the x derivative of $12 \log F$ where

$$F = 1 + \exp\{-(\theta - \theta_0)\}$$

$$= 1 + \exp\{-\alpha(x - s) + \alpha^3 t\}, \qquad s = \frac{\theta_0}{\alpha}. \qquad (17.13)$$

This "derivation" focuses on one working rule for finding exact solutions in this area: to consider transformations which make the special solitary wave solutions appear as simple exponentials. One might note for comparison that the transformation $c = -2\nu(\log\varphi)_x$ for Burgers' equation puts the steady shock solution (4.23) into the form

$$\varphi = \exp(-\alpha_1 x + \nu\alpha_1^2 t) + \exp(-\alpha_2 x + \nu\alpha_2^2 t), \qquad \alpha_i = \frac{c_i}{2\nu}. \qquad (17.14)$$

Whatever motivation is used, we see immediately that (17.13) is a solution of (17.11) for any α and s. It is a solution corresponding to the operator

$$\frac{\partial}{\partial t} + \frac{\partial^3}{\partial x^3};$$

it satisfies $F_t + F_{xxx} = 0$, and the third pair of terms in (17.11) vanishes due to the homogeneity in the derivatives.

If (17.11) were linear, we could superpose solutions with different α and s, but due to the nonlinearity there will be interaction terms. A normal interaction approach would be to take

$$F = 1 + F^{(1)} + F^{(2)} + \cdots$$

with the hierachy

$$\left\{ F_t^{(1)} + F_{xxx}^{(1)} \right\}_x = 0,$$

$$\left\{ F_t^{(2)} + F_{xxx}^{(2)} \right\}_x = -3 \left\{ F_{xx}^{(1)^2} - F_x^{(1)} F_{xxx}^{(1)} \right\},$$

and so on, to solve. Let us take two terms like (17.13) for $F^{(1)}$:

$$F^{(1)} = f_1 + f_2, \qquad f_j = \exp\left\{-\alpha_j(x - s_j) + \alpha_j^3 t\right\}, \qquad j = 1, 2. \qquad (17.15)$$

The equation for $F^{(2)}$ is

$$\left\{ F_t^{(2)} + F_{xxx}^{(2)} \right\}_x = 3\alpha_1\alpha_2(\alpha_2 - \alpha_1)^2 f_1 f_2, \qquad (17.16)$$

and its solution is

$$F^{(2)} = \frac{(\alpha_2 - \alpha_1)^2}{(\alpha_2 + \alpha_1)^2} f_1 f_2. \tag{17.17}$$

The surprising thing is that all the remaining equations in the hierarchy then have zero on the right hand side, so that

$$F = 1 + f_1 + f_2 + \frac{(\alpha_2 - \alpha_1)^2}{(\alpha_2 + \alpha_1)^2} f_1 f_2 \tag{17.18}$$

is *an exact solution* of (17.11).

The significant point in this solution is that the interaction terms produce only the product term $f_1 f_2$ on the right of (17.16) and not terms in f_1^2 and f_2^2, which could also be expected. This result generalizes to higher orders and the nonlinear terms in the equation never produce products of f's containing a repeated subscript. Thus with an input of only two terms as in (17.15) the only needed combinations are $f_1, f_2, f_1 f_2$, and we have an exact solution. If we start with

$$F^{(1)} = \sum_{j=1}^{N} f_j,$$

then $F^{(2)}$ contains all terms $f_j f_k$ with $j \neq k$ but *not* f_j^2; $F^{(3)}$ contains all terms $f_j f_k f_l$, with $j \neq k \neq l$ but *not* f_j^3 or $f_j^2 f_k$; and so on. Thus the sequence terminates at

$$F^{(N)} \propto f_1 f_2 \cdots f_N$$

(having covered all products without repeated f's), and there is an exact solution in the form

$$F = 1 + \sum_j f_j + \sum_{j \neq k} a_{jk} f_j f_k + \sum_{j \neq k \neq l} a_{jkl} f_j f_k f_l + \cdots + a_{12 \cdots N} f_1 f_2 \cdots f_N.$$

As if this were not amazing enough, it can also be shown that the solution may be written

$$F = \det |F_{mn}|, \tag{17.19}$$

where*

$$F_{mn} = \delta_{mn} + \frac{2\alpha_m}{\alpha_m + \alpha_n} f_m. \tag{17.20}$$

This result was first found by the more general approach referred to above and described in the next section, but it may be verified directly in (17.11) (see Hirota, 1971).

With N modes f_j, the solution represents the interaction of N solitary waves. We discuss the case $N = 2$. The solution for F is given in (17.18) and the corresponding expression for η given by (17.6) is

$$\frac{\sigma}{12}\eta = \frac{\alpha_1^2 f_1 + \alpha_2^2 f_2 + 2(\alpha_2 - \alpha_1)^2 f_1 f_2 + \{(\alpha_2 - \alpha_1)/(\alpha_2 + \alpha_1)\}^2 (\alpha_2^2 f_1^2 f_2 + \alpha_1^2 f_1 f_2^2)}{(1 + f_1 + f_2 + \{(\alpha_2 - \alpha_1)/(\alpha_2 + \alpha_1)\}^2 f_1 f_2)^2},$$

$$\tag{17.21}$$

where

$$f_j = \exp\{-\alpha_j(x - s_j) + \alpha_j^3 t\}.$$

A single solitary wave (17.12) may be written in terms of f as

$$\frac{\sigma}{12}\eta = \frac{\alpha^2 f}{(1 + f)^2}, \tag{17.22}$$

where the maximum of $\sigma\eta$ occurs for $f = 1$. We note that

Maximum amplitude of $\sigma\eta = 3\alpha^2$,

Position of maximum $= s + \alpha^2 t$, $\qquad\qquad$ (17.23)

Velocity of the wave $= \alpha^2$.

The solution (17.21) is approximately a solitary wave with parameter α_1 for regions of the (x, t) plane where $f_1 \simeq 1$ and f_2 is *either* large or small. To

---

*There are a number of equivalent forms for the F_{mn} which lead to the same final expression for η.

verify this, observe the following:

1. For $f_1 \simeq 1$, $f_2 \ll 1$,

$$\frac{\sigma}{12}\eta \simeq \frac{\alpha_1^2 f_1}{(1+f_1)^2}.$$

2. For $f_1 \simeq 1$, $f_2 \gg 1$,

$$\frac{\sigma}{12}\eta \simeq \frac{\{(\alpha_2-\alpha_1)/(\alpha_2+\alpha_1)\}^2 \alpha_1^2 f_1 f_2^2}{\left(f_2 + \{(\alpha_2-\alpha_1)/(\alpha_2+\alpha_1)\}^2 f_1 f_2\right)^2}$$

$$= \frac{\alpha_1^2 \tilde{f}_1}{\left(1+\tilde{f}_1\right)^2}, \qquad \tilde{f}_1 = \left(\frac{\alpha_2-\alpha_1}{\alpha_2+\alpha_1}\right)^2 f_1.$$

The latter is the solitary wave α_1 with s_1 replaced by

$$\tilde{s}_1 = s_1 - \frac{1}{\alpha_1}\log\left(\frac{\alpha_2+\alpha_1}{\alpha_2-\alpha_1}\right)^2; \qquad (17.24)$$

this represents a finite displacement of the profile in the x direction. Similarly, where $f_2 \simeq 1$ and f_1 is either large or small, we have the solitary wave α_2 with or without a shift in s_2. Where $f_1 \simeq 1$ and $f_2 \simeq 1$, we have the interaction region; where f_1 and f_2 are both small or both large we have $\sigma\eta \simeq 0$.

The behavior of the interacting solitary waves described by (17.21) can now be seen. We take $\alpha_2 > \alpha_1 > 0$ for definiteness and note from (17.23) that the solitary wave α_2 is stronger and moves faster than the wave α_1. As $t \to -\infty$, there is no interaction region in which $f_1 \simeq 1$, $f_2 \simeq 1$, and (17.21) describes

Solitary wave α_1 on $x = s_1 + \alpha_1^2 t$, $f_1 \simeq 1$, $f_2 \ll 1$,

Solitary wave α_2 on $x = s_2 - \frac{1}{\alpha_2}\log\left(\frac{\alpha_2+\alpha_1}{\alpha_2-\alpha_1}\right)^2 + \alpha_2^2 t$, $f_1 \gg 1$, $f_2 \simeq 1$;

elsewhere $\sigma\eta \simeq 0$ (f_1 and f_2 both large or small).

This represents a larger solitary wave α_2 overtaking a smaller one α_1.

As $t \to +\infty$, we have:

Solitary wave α_1 on $x = s_1 - \dfrac{1}{\alpha_1} \log\left(\dfrac{\alpha_2 + \alpha_1}{\alpha_2 - \alpha_1}\right)^2 + \alpha_1^2 t, \quad f_1 \simeq 1, \quad f_2 \gg 1,$

Solitary wave α_2 on $x = s_2 + \alpha_2^2 t, \quad f_1 \ll 1, \quad f_2 \simeq 1;$

elsewhere $\sigma\eta \simeq 0$.

The remarkable result is that the solitary waves emerge unchanged in form with the original parameters α_1 and α_2, the faster wave α_2 now being ahead. The only remnant of the collision process is a forward shift

$$\frac{1}{\alpha_2} \log\left(\frac{\alpha_2 + \alpha_1}{\alpha_2 - \alpha_1}\right)^2 \qquad \text{for the wave } \alpha_2,$$

and a backward shift

$$\frac{1}{\alpha_1} \log\left(\frac{\alpha_2 + \alpha_1}{\alpha_2 - \alpha_1}\right)^2 \qquad \text{for the wave } \alpha_1.$$

The interaction occurs in the neighborhood of

$$t = -\frac{s_2 - s_1}{\alpha_2^2 - \alpha_1^2}, \qquad x = \frac{\alpha_2^2 s_1 - \alpha_1^2 s_2}{\alpha_2^2 - \alpha_1^2}.$$

In this region $f_1 \simeq 1, f_2 \simeq 1$, and (17.21) describes how the two peaks merge into a single peak and then reemerge in the reversed order.

Similar results can be inferred from (17.19)–(17.20) for the case of N waves. The eventual behavior as $t \to \infty$ consists of N solitary waves lined up in order with strength and velocity increasing toward the front and separating further as t increases.

17.3 Inverse Scattering Theory

In setting up the theory, we temporarily let

$$u = -\frac{\sigma}{6}\eta \tag{17.25}$$

to agree with the original papers; the equation is then

$$u_t - 6uu_x + u_{xxx} = 0. \tag{17.26}$$

From the analogy with Burgers' equation, the substitution

$$u \propto \frac{\psi_{xx}}{\psi}$$

is a popular first attempt toward finding solutions, but this alone does not lead far. Gardner, Greene, Kruskal, and Miura (1967) took it further with the choice

$$u = \frac{\psi_{xx}}{\psi} + \lambda$$

and noted that rewritten this is the reduced Schrödinger equation

$$\psi_{xx} + (\lambda - u)\psi = 0. \tag{17.27}$$

At this point, presumably the emphasis changed and (17.27) was considered not as a transformation that would produce a simpler equation for ψ, but rather as an associated scattering problem from which information on ψ could be used to diagnose properties of u. In this view, the wave profile $u(x,t)$ provides the scattering potential. The time t appears as a parameter; there is a different scattering problem for each t. This time parameter is completely separate from the time τ that might have been eliminated in reducing the wave equation

$$\varphi_{xx} - \varphi_{\tau\tau} - u(x,t)\varphi = 0 \tag{17.28}$$

to (17.27), via $\varphi(x,\tau,t) = \psi(x,t)e^{i\sqrt{\lambda}\tau}$. We shall return to (17.28) later, but for the present we adopt the reduced version (17.27).

To use (17.27), we have to find the equation for ψ from (17.26). Since values of λ would belong to the spectrum of the scattering problem (17.27) and the problem changes with t, it is appropriate in the first instance to let λ be a function of t. When (17.27) is substituted in (17.26), it is found after a certain amount of ingenuity that the equation may be taken in the form

$$\psi^2 \frac{d\lambda}{dt} + \frac{\partial}{\partial x}\{\psi Q_x - \psi_x Q\} = 0, \tag{17.29}$$

$$Q = \psi_t + \psi_{xxx} - 3(u + \lambda)\psi_x. \tag{17.30}$$

We now restrict the discussion to solutions of (17.26) with $u \to 0$ as $|x| \to \infty$ and u integrable. Under this condition the spectrum of (17.27) is discrete in $\lambda < 0$ and continuous in $\lambda > 0$. For the point eigenvalues, $\lambda = -\kappa_n^2$, the corresponding eigenfunctions ψ_n satisfy

$$|\psi_n| \to 0, \qquad |x| \to 0,$$

$$\int_{-\infty}^{\infty} \psi_n^2 \, dx = \text{finite} > 0.$$

Therefore integrating (17.29) from $-\infty$ to ∞, we deduce that λ_n is independent of t. For the continuous spectrum we may choose a $\lambda > 0$ independent of t and consider the behavior of the corresponding solutions ψ with t. In either case, we deduce from (17.29), with $d\lambda/dt = 0$, that

$$Q \equiv \psi_t + \psi_{xxx} - 3(u + \lambda)\psi_x = C\psi, \tag{17.31}$$

where C is independent of x. We now know that any solution ψ of (17.27) for fixed λ will develop in time according to (17.31). In analyzing the eigenvalue problem (17.27), it is convenient to introduce $\lambda = \mu^2$ and to work with a function χ which satisfies (17.27) and the condition

$$\chi \sim e^{i\mu x}, \qquad \Im \mu \geqslant 0, \qquad \text{as} \quad x \to +\infty. \tag{17.32}$$

In order that this function, introduced for $t = 0$ say, should retain the same normalization (17.32) as it develops in time, we require (17.32) to be an asymptotic solution of (17.31) for all t. This requires the choice $C = -4i\mu^3$ and our pair of equations becomes

$$\psi_{xx} + (\mu^2 - u)\psi = 0, \tag{17.33}$$

$$\psi_t + \psi_{xxx} - 3(u + \mu^2)\psi_x + 4i\mu^3\psi = 0. \tag{17.34}$$

We now have to explain how this curious formulation allows $u(x,t)$ to be calculated.

The method depends on the result that the scattering potential u in (17.33) can be constructed from knowledge of the reflection coefficient for waves incident from $x = +\infty$, together with certain information about the point spectrum. This is the inverse scattering problem, the original problem being to determine an unknown scatterer from its reflection properties. In the present context the required information on solutions ψ is determined not from experiment but from the second equation (17.34). To be specific, we consider the problem of finding $u(x,t), t>0$, given $u(x,0)$. The procedure is as follows. For the given $u(x,0)$, we first solve the eigenvalue problem (17.33), and determine the point eigenvalues $\mu = i\kappa_n$, the corresponding eigenfunctions ψ_n, and the reflection coefficient β for incoming waves. One choice for the eigenfunctions is

$$\psi_n(x) = \chi(i\kappa_n, x),$$

where χ is specified by (17.32); the normalization constants are then given by

$$\gamma_n = \left\{ \int_{-\infty}^{\infty} \psi_n^2 \, dx \right\}^{-1}.$$

As regards the reflection coefficient β, we determine the solution Ψ of (17.33) with $u(x,0)$, which has the following behavior at $\pm\infty$:

$$\Psi(k,x) \sim \begin{cases} e^{-ikx} + \beta(k)e^{ikx}, & x \to +\infty, \\ \alpha(k)e^{-ikx}, & x \to -\infty, \end{cases}$$

for k real and positive. This determines the reflection coefficient $\beta(k)$ and the transmission coefficient $\alpha(k)$. This is the direct scattering problem: finding $\kappa_n, \gamma_n, \beta(k)$ for the given $u(x,0)$. The inverse problem would be to determine $u(x,0)$ from the knowledge of $\kappa_n, \gamma_n, \beta(k)$.

We now turn to the development of these solutions in time. We know that the κ_n are unchanged. From (17.34) and (17.33),

$$\frac{d}{dt} \int_{-\infty}^{\infty} \chi^2 \, dx = \left[-2\chi\chi_{xx} + 4\chi_x^2 + 6\mu^2\chi^2 \right]_{-\infty}^{\infty} - 8i\mu^3 \int_{-\infty}^{\infty} \chi^2 \, dx.$$

For the eigenfunctions, $\mu = i\kappa_n$, and $\psi_n(x,t) = \chi(x,t,i\kappa_n) \to 0$ as $x \to \pm\infty$; therefore the normalization constants are

$$c_n(t) = \left\{ \int_{-\infty}^{\infty} \psi_n^2 \, dx \right\}^{-1} = \gamma_n e^{8\kappa_n^3 t}.$$

The solution $\Psi(k,x,t)$ for scattered waves will have some behavior

$$\Psi(k,x,t) \sim f(k,t)e^{-ikx} + g(k,t)e^{ikx}, \qquad x \to \infty,$$

but this must be an asymptotic solution of (17.34) with $\mu = k$. On substitution we deduce that

$$f(k,t) = e^{-8ik^3t}, \qquad g(k,t) = \beta.$$

The reflection coefficient is

$$b(k,t) = \frac{g(k,t)}{f(k,t)} = \beta(k)e^{8ik^3t}.$$

The inverse scattering theory provides the construction of $u(x,t)$ from

$$\kappa_n, \qquad c_n(t), \qquad b(k,t).$$

In summary then, the direct scattering problem for $u(x,0)$ determines $\kappa_n, c_n(0), b(k,0)$; (17.34) provides their development in time; the inverse problem determines $u(x,t)$ from these quantities.

Of course a major input is now the solution of the inverse problem. This is provided by the famous Gelfand-Levitan paper (1951) and its various extensions. Their paper is phrased in terms of determining the scattering potential u from the spectral function $\rho(\lambda)$, which has jumps of magnitude c_n at the point eigenvalues $\lambda = -\kappa_n^2$ and a continuous spectrum $0 < \lambda < \infty$ related to b. Kay and Moses (1956) and Marchenko (1955) provided direct constructions from κ_n, c_n, b; a thorough review is given by Faddeyev (1959). The result is that

$$u(x,t) = -2\frac{d}{dx}K(x,x,t), \tag{17.35}$$

where the function $K(x,y,t)$ satisfies the linear integral equation

$$K(x,y,t) + B(x+y,t) + \int_x^\infty K(x,z,t)B(z+y,t)dz = 0, \qquad y > x, \tag{17.36}$$

in which

$$B(x+y,t) = \sum c_n(t)\exp\{-\kappa_n(x+y)\}$$

$$+ \frac{1}{2\pi}\int_{-\infty}^\infty b(k,t)\exp\{ik(x+y)\}dk$$

$$= \sum \gamma_n\exp\{-\kappa_n(x+y) + 8\kappa_n^3 t\}$$

$$+ \frac{1}{2\pi}\int_{-\infty}^\infty \beta(k)\exp\{ik(x+y) + 8ik^3t\}dk; \tag{17.37}$$

the initial function $u(x,0)$ provides the appropriate $\kappa_n, \gamma_n, \beta(k)$.

The derivation of (17.35)–(17.37) from the spectral approach requires more extensive discussion than would be appropriate here. But Balanis (1972) has shown that, formally at least, the results can be obtained very simply by working with the "unreduced" equation (17.28) and diagnosing the potential $u(x,t)$ from its reflection properties for an incident δ function instead of an incident periodic wave. Essentially the idea is to work in an equivalent time domain instead of the frequency domain. Balanis' treatment of the inverse scattering can be included in a reappraisal of the method of solution from this point of view.

An Alternative Version.

We consider (17.33) and (17.34) as the Fourier transforms of

$$\varphi_{xx} - \varphi_{\tau\tau} - u\varphi = 0,$$

$$\varphi_t + \varphi_{xxx} - 3u\varphi_x + 3\varphi_{x\tau\tau} - 4\varphi_{\tau\tau\tau} = 0,$$

coupling the function $u(x; t)$ with a function $\varphi(x, \tau; t)$ such that

$$\varphi(x, \tau; t) = \int_{-\infty}^{\infty} \psi(x, \mu; t) e^{i\mu\tau} d\mu.$$

This pair of equations may be written more symmetrically as

$$M\varphi \equiv \varphi_{xx} - \varphi_{\tau\tau} - u\varphi = 0, \tag{17.38}$$

$$N\varphi \equiv \varphi_t + \partial^3\varphi - 3u\partial\varphi = 0, \tag{17.39}$$

where

$$\partial \equiv \frac{\partial}{\partial x} - \frac{\partial}{\partial \tau}.$$

It is straightforward to show that

$$(NM - MN)\varphi = -(u_t - 6uu_x + u_{xxx})\varphi + 3u_x M\varphi.$$

Therefore (17.38)–(17.39) imply the Korteweg-deVries equation and we have a direct argument for their adoption. We now proceed by analogy with the previous approach. The behavior of φ at $x = +\infty$ determines the scattering potential u in (17.38) from Balanis' version of the inverse scattering problem in the (x, τ) plane. The development of φ with t is provided by (17.39).

We now give Balanis' argument for (17.38) and for the present do not display the parameter t. Consider an incident wave $\varphi = \delta(x + \tau)$ from $x = +\infty$ and let the reflected wave be $B(x - \tau)$. That is,

$$\varphi \sim \varphi_\infty = \delta(x + \tau) + B(x - \tau) \qquad \text{as} \quad x \to +\infty. \qquad (17.40)$$

We propose that the corresponding complete solution of (17.38) may be written

$$\varphi(x,\tau) = \varphi_\infty(x,\tau) + \int_x^\infty K(x,\xi)\varphi_\infty(\xi,\tau)\,d\xi. \qquad (17.41)$$

(This is equivalent to a crucial step in the Gelfand-Levitan work.) By direct substitution in (17.38) we verify that there is such a solution provided

$$K_{\xi\xi} - K_{xx} + u(x)K = 0, \qquad \xi > x,$$

$$u(x) = -2\frac{d}{dx}K(x,x), \qquad (17.42)$$

$$K, K_\xi \to 0, \qquad \xi \to \infty.$$

This is a well-posed problem, therefore K exists. From the causality property of the wave equation (17.38) we know that φ must be equal to zero for $x + \tau < 0$. Hence

$$\varphi_\infty(x,\tau) + \int_x^\infty K(x,\xi)\varphi_\infty(\xi,\tau)\,d\xi = 0, \qquad x + \tau < 0.$$

Introducing the expression for φ_∞ in (17.40), we have

$$B(x - \tau) + K(x, -\tau) + \int_x^\infty K(x,\xi)B(\xi - \tau)\,d\xi = 0, \qquad x + \tau < 0.$$

With $\tau = -y$, this is the Gelfand-Levitan equation (17.36).

To incorporate this into the solution of the Korteweg-deVries equation, we note that the development of the function B with t is given by (17.39). But at $x = +\infty$, $u \to 0$; therefore B satisfies

$$B_{xx} - B_{\tau\tau} = 0,$$

$$B_t + \partial^3 B = 0. \qquad (17.43)$$

For $t = 0$, B is determined from the direct scattering problem for (17.38) in terms of $u(x,0)$ as

$$B(x - \tau) = \sum \gamma_n \exp\{ -\kappa_n(x - \tau) \} + \frac{1}{2\pi} \int_{-\infty}^{\infty} \beta(k) \exp\{ ik(x - \tau) \} \, dk.$$

The solution of (17.43) for $t > 0$ is

$$B(x - \tau, t) = \sum \gamma_n \exp\{ -\kappa_n(x - \tau) + 8\kappa_n^3 t \}$$

$$+ \frac{1}{2\pi} \int_{-\infty}^{\infty} \beta(k) \exp\{ ik(x - \tau) + 8ik^3 t \} \, dk.$$

With $\tau = -y$, again, this is exactly (17.37).

Apart from its speed, this version gives a more symmetrical look to (17.39) as opposed to the rather awkward (17.34), and it brings out more clearly the mapping to a simpler linear problem (17.43). The basic dispersive operator

$$\frac{\partial}{\partial t} + \frac{\partial^3}{\partial x^3}$$

appears in (17.39) and (17.43) but in an extended form

$$\frac{\partial}{\partial t} + \left(\frac{\partial}{\partial x} - \frac{\partial}{\partial \tau} \right)^3.$$

The extended form shows why the factor 8 appears in the t dependence in (17.37).

We refer to (17.35)–(17.37) as the solution even though the linear integral equation (17.36) is still difficult to handle in general. However, various results can be obtained. First, the special case $\beta(k) = 0$ can be solved explicitly and gives the interaction of solitary waves discussed in the last section; each point eigenvalue corresponds to a solitary wave. Second, the number of solitary waves ultimately emerging from an arbitrary initial disturbance $u(x,0)$ can be determined from its spectrum. Third, it may be shown (Segur, 1973) that the contributions of the continuous spectrum to $u(x,t)$ die out for large t. The first two topics are considered in the next sections.

17.4 Special Case of a Discrete Spectrum Only

In this case (17.37) may be written

$$B(x+y) = \sum g_n(x) h_n(y),$$

where the explicit dependence on t is not displayed. The factor $\exp 8\kappa_n^3 t$ may be taken in g_n or h_n or divided between the two. The solution of (17.36) may then be taken in the form

$$K(x,y) = \sum w_n(x) h_n(y),$$

and we have

$$w_m(x) + g_m(x) + \sum_n w_n(x) \int_x^\infty g_m(z) h_n(z)\, dz = 0.$$

If the matrix $P(x)$ is defined by

$$P_{mn}(x) = \delta_{mn} + \int_x^\infty g_m(z) h_n(z)\, dz,$$

and f, g, h denote column vectors with components f_m, g_m, h_m, we have

$$w(x) = -P^{-1}(x) g(x)$$

and

$$K(x,x) = h^T(x) w(x) = -h^T(x) P^{-1}(x) g(x).$$

Since

$$\frac{d}{dx} P_{mn}(x) = -g_m(x) h_n(x),$$

$K(x,x)$ may be expressed as

$$K(x,x) = \operatorname{Trace} \left\{ P^{-1} \frac{dP}{dx} \right\}$$

$$= \sum_{1=n} \sum_m \frac{\mathcal{P}_{ml}}{|P|} \frac{dP_{mn}}{dx}$$

$$= \frac{1}{|P|} \frac{d}{dx} |P|$$

where $|P|$ denotes the determinant of P and $\mathscr{P}_{ml}$ is the cofactor of P_{ml}. Therefore

$$u = -2\frac{d}{dx}K(x,x) = -2\frac{d^2}{dx^2}\log|P|.$$

Since $u = -\sigma\eta/6$, this is the transformation (17.6) and it remains to verify that $|P|$ agrees with the function F quoted in (17.19)–(17.20).

There are various choices that lead to the same expression for u. If we express B in (17.37) by taking

$$g_m(x) = \gamma_m\exp(-\kappa_m x + 8\kappa_m^3 t), \qquad h_n(x) = \exp(-\kappa_n x),$$

we have

$$P_{mn} = \delta_{mn} + \frac{\gamma_m\exp\{-(\kappa_m + \kappa_n)x + 8\kappa_m^3 t\}}{\kappa_m + \kappa_n}.$$

Exponential factors may be taken in or out of the determinant without affecting the final expression for u; $\log|P|$ converts them into additive terms linear in x, which are eliminated by the double derivative in deducing u. Therefore, if each column of P is multiplied by $e^{\kappa_n x}$ and each row by $e^{-\kappa_m x}$, we have the equivalent form

$$|P| \propto \left|\delta_{mn} + \frac{\gamma_m\exp(-2\kappa_m x + 8\kappa_m^3 t)}{\kappa_m + \kappa_n}\right|.$$

This agrees with (17.20), with

$$\alpha_m = 2\kappa_m, \qquad \gamma_m = \alpha_m\exp\alpha_m s_m.$$

A symmetrical form for P is obtained from

$$g_m(x) = h_m(x) = \gamma_m^{1/2}\exp(-\kappa_m x + 4\kappa_m^3 t),$$

leading to

$$P_{mn} = \delta_{mn} + \frac{(\gamma_m\gamma_n)^{1/2}}{\kappa_m + \kappa_n}\exp\{-(\kappa_m + \kappa_n)x + 4(\kappa_m^3 + \kappa_n^3)t\}.$$

17.5 The Solitary Waves Produced by an Arbitrary Initial Disturbance

To determine the solitary waves emerging from an initial distribution $\eta = \eta_0(x)$ we have merely to find the point eigenvalues of the Schrödinger equation

$$\psi_{xx} + \{\lambda - u_0(x)\}\psi = 0, \tag{17.44}$$

where

$$\frac{\sigma \eta_0(x)}{6} = -u_0(x).$$

After emerging from the interactions the solitary wave corresponding to $\lambda = -\kappa_n^2$ will be given, according to (17.12), by

$$-u = \frac{\sigma \eta}{6} = a_n \operatorname{sech}^2\left(\kappa_n x - 4\kappa_n^3 t + \text{constant}\right), \qquad a_n = 2\kappa_n^2. \tag{17.45}$$

Some specific examples will now be noted, quoting results for the eigenvalue problem that may be found in most standard books on quantum theory.

  1. $u_0(x) = -Q\delta(x)$.

If $Q > 0$, there is one point eigenvalue $\kappa = Q/2$. Hence a single solitary wave is produced. The amplitude of u in (17.45) is $Q^2/2$. If $Q < 0$, there are no point eigenvalues and no solitary waves.

  2.   Rectangular Well.

If $u_0(x)$ is a rectangular well of width l and depth A, the eigenvalues must satisfy (Landau and Lifshitz, 1958, p. 63)

$$\sin \xi = \pm \frac{2\xi}{S}, \qquad \tan \xi < 0, \tag{17.46}$$

or

$$\cos \xi = \pm \frac{2\xi}{S}, \qquad \tan \xi > 0, \tag{17.47}$$

where

$$S = A^{1/2}l, \qquad \xi = \frac{1}{2}S\sqrt{1 - \frac{\kappa^2}{A}} \geqslant 0.$$

The number of eigenvalues is controlled by the parameter S. As S increases, corresponding to stronger initial disturbances, the number of solitary waves increases. For all $S > 0$ there is at least one solution of the eigenvalue equations; hence there is always one solitary wave. For small S, it is a solution of (17.47) with

$$\xi \simeq \frac{S}{2}, \qquad \kappa \simeq \frac{1}{2}SA^{1/2}, \qquad a \simeq \frac{1}{2}S^2A, \qquad S \ll 1.$$

As S increases a second solitary wave is produced when S reaches π, and a solution of (17.46) first appears at $\xi = \pi/2$. At this value

$$\left.\begin{array}{lll} \xi_1 = 0.934, & \kappa_1 = 0.804A^{1/2}, & a_1 = 1.30A, \\ \xi_2 = \frac{\pi}{2}, & \kappa_2 = 0, & a_2 = 0, \end{array}\right\} S = \pi.$$

More solitary waves come in as S increases; the number N is given by

$$N = \text{largest integer} \leqslant \frac{S}{\pi} + 1.$$

The dependence of S on A and l for the rectangular well suggests that, more generally,

$$Z = \int_{-\infty}^{\infty} |u|^{1/2} dx$$

will be an interesting measure of disturbances and wave shapes in this context. Indeed, if we calculate this quantity for the single solitary wave (17.45), the parameter κ drops out and we have

$$Z_s = \int_{-\infty}^{\infty} |u|^{1/2} dx = 2^{1/2}\pi \tag{17.48}$$

independent of amplitude. In this measure, therefore, solitary waves have a unit size. For a train of N solitary waves we have $Z = 2^{1/2}\pi N$; there is a Planck's constant for solitary waves!

The parameter S is the value of the integral in the initial disturbance. For large S, we have $S \sim \pi N$ from the above results, so that in the large time behavior the value of Z in the train of solitary waves is

$$Z = 2^{1/2}\pi N \sim 2^{1/2} S. \tag{17.49}$$

This shows the close relation between the ultimate Z and the initial S. But we also see that the "action" $\int |u|^{1/2} dx$ is not conserved. This is also seen for small S: there is always one solitary wave produced, even when S is smaller than the unit required in (17.48). The required value must be built up in the initial separation process, possibly with compensating drain on the continuous part of the spectrum. We shall find, however, that the result $N \sim S/\pi$, for large S, is a general one for initial disturbances $u_0(x)$ consisting of a single well and S defined by

$$S = \int_{-\infty}^{\infty} |u_0|^{1/2} dx. \tag{17.50}$$

For the δ function case in the first example we may take u_0 as the limit of $-Q(m/\pi)^{1/2} e^{-mx^2}$ as $m \to \infty$. For this, $S \to 0$ as $m \to \infty$ and the production of only one solitary wave fits in with the results for the other examples.

3. $u_0 = -A \operatorname{sech}^2 x/l.$

In this case the eigenvalues are given (Landau and Lifshitz, 1958, p. 70) by

$$\kappa_n = \frac{1}{2l} \left\{ (1 + 4Al^2)^{1/2} - (2n-1) \right\} \geqslant 0.$$

The value of S defined by (17.50) is

$$S = \pi A^{1/2} l.$$

The number of solitary waves is given by

$$N = \text{Largest integer} \leqslant \frac{1}{2} \left\{ \left(1 + \frac{4S^2}{\pi^2} \right)^{1/2} + 1 \right\}.$$

As before, there is always one solitary wave for small S, more come in as S increases; as $S \to \infty$ we again have

$$N \sim \frac{S}{\pi},$$

and (17.49) holds.

4. Continuous Distribution of Solitary Waves.

When the initial disturbance is large ($S \to \infty$), there are many closely spaced eigenvalues which satisfy the Bohr-Sommerfeld rule

$$\oint p \, dx = \oint \sqrt{\lambda - u_0(x)} \, dx = 2\pi \left(n + \frac{1}{2} \right) \qquad (17.51)$$

(see Landau and Lifshitz, 1958, p. 162). Hence the number of solitary waves (the largest value of n for $\lambda = 0$) is

$$N \sim \frac{1}{\pi} \int_{-\infty}^{\infty} |u_0|^{1/2} \, dx = \frac{S}{\pi} . \qquad (17.52)$$

This proves that the result found in the last two examples is general.

The largest value of $|\lambda|$ for the bound states in (17.51) is u_m, where $u_m = |u_0|_{max}$, so the range of κ is $0 < \kappa < u_m^{1/2}$ and the range of amplitudes in (17.45) is

$$0 < a < 2u_m. \qquad (17.53)$$

The number of eigenvalues in $(\lambda, \lambda + d\lambda)$ is approximately

$$\frac{1}{4\pi} \oint \frac{dx}{\sqrt{\lambda - u_0(x)}} \, d\lambda.$$

Therefore the number of solitary waves with amplitudes in $(a, a + da)$ is approximately $f(a) \, da$ where

$$f(a) = \frac{1}{8\pi} \oint \frac{dx}{\sqrt{|u_0| - a/2}} , \qquad (17.54)$$

a result first found by Karpman (1967) (see also Karpman and Sokolov, 1968). This distribution is over the range $0 < a < 2u_m$ and the total number is

$$N = \int_0^{2u_m} f(a) \, da = \frac{1}{\pi} \int_{-\infty}^{\infty} |u_0|^{1/2} dx,$$

in agreement with (17.52).

After the initial interaction each solitary wave of amplitude a moves with velocity $2a$ and it is found at

$$x = 2at \quad \text{as} \quad t \to \infty.$$

Therefore the distribution of amplitude is given by

$$a = \frac{x}{2t}, \qquad 0 < \frac{x}{2t} < 2u_m. \tag{17.55}$$

We have the triangular distribution noted in Fig. 17.1 and discussed in Section 16.16.

The number of waves $k(x,t)$ in $(x, x+dx)$ is given by

$$k\,dx = f(a)\,da;$$

hence

$$k(x,t) = \frac{1}{2t} f\left(\frac{x}{2t}\right), \tag{17.56}$$

where f is given by (17.54). This fixes the arbitrary function (16.144).

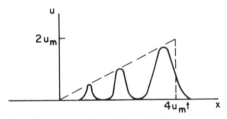

Fig. 17.1   Series of solitary waves in solution of the Korteweg-deVries equation.

17.6 Miura's Transformation and Conservation Equations

The path to finding the preceding solutions of the Korteweg-deVries equation was very much stimulated by the existence of an infinite number of conservation equations. One way of obtaining these is by Miura's transformation, which is of independent interest. If one substitutes

$$u = -\frac{\sigma\eta}{6} = v^2 + v_x \tag{17.57}$$

in the Korteweg-deVries equation, the result may be written

$$\left(2v + \frac{\partial}{\partial x}\right)\left(v_t - 6v^2 v_x + v_{xxx}\right) = 0.$$

Hence the equation

$$v_t - 6v^2 v_x + v_{xxx} = 0 \tag{17.58}$$

may also be studied by relating it to the Korteweg-deVries equation.

The infinite number of conservation laws can be generated by a modification of this. If the substitution

$$\sigma \eta = w + i\epsilon w_x + \frac{1}{6}\epsilon^2 w^2 \tag{17.59}$$

is made, then

$$\left(1 + i\epsilon \frac{\partial}{\partial x} + \frac{1}{3}\epsilon^2 w^2\right) \left\{ w_t + \left(w + \frac{1}{6}\epsilon^2 w^2\right) w_x + w_{xxx} \right\} = 0.$$

We choose w to satisfy

$$w_t + \left(w + \frac{1}{6}\epsilon^2 w^2\right) w_x + w_{xxx} = 0.$$

A simple conservation equation for w is

$$w_t + \left(\frac{1}{2}w^2 + \frac{1}{18}\epsilon^2 w^3 + w_{xx}\right)_x = 0. \tag{17.60}$$

If now w is solved recursively in terms of $\zeta = \sigma \eta$ using (17.59), we have formally

$$w = \sum_0^\infty \epsilon^n w_n \{ \zeta \},$$

where the w_n depend on ζ and its x derivatives. When this is substituted in (17.60), each coefficient of ϵ^n provides a conservation law. The first few conserved densities are

$$\zeta, \qquad \frac{1}{2}\zeta^2, \qquad \frac{1}{3}\zeta^3 - \zeta_x^2, \qquad \frac{1}{4}\zeta^4 - 3\zeta\zeta_x^2 + \frac{9}{5}\zeta_{xx}^2,$$

$$\frac{1}{5}\zeta^5 - 6\zeta^2\zeta_x^2 + \frac{36}{5}\zeta\zeta_{xx}^2 - \frac{108}{35}\zeta_{xxx}^2.$$

The existence of an infinite number of conserved quantities

$$\int_{-\infty}^\infty w_n \{ \zeta \}\, dx$$

evidently added confidence that explicit solutions would be found.

CUBIC SCHRÖDINGER EQUATION

17.7 Significance of the Equation

The relevance of the equation

$$iu_t + u_{xx} + \nu |u|^2 u = 0$$

as an approximation for modulated beams in nonlinear optics was explained in Section 16.4. Here we remark on its general significance for time-dependent dispersive waves. The general solution for a linear dispersive mode is

$$\int_{-\infty}^{\infty} F(k) e^{ikx - i\omega(k)t} \, dk, \tag{17.61}$$

where $\omega = \omega(k)$ is the dispersion relation. For a modulated wavetrain with most of the energy in wave numbers close to some value k_0, $F(k)$ is concentrated near $k = k_0$ and (17.61) may be approximated by

$$\Phi = \int_{-\infty}^{\infty} F(k) \exp\left(ikx - \left\{ \omega_0 + (k - k_0)\omega_0' + \frac{1}{2}(k - k_0)^2 \omega_0'' \right\} t \right) dk,$$

where $\omega_0 = \omega(k_0)$, $\omega_0' = \omega'(k_0)$, ... This in turn may be written

$$\Phi = \varphi \exp\{ i(k_0 x - \omega_0 t) \}, \tag{17.62}$$

where

$$\varphi = \int_{-\infty}^{\infty} F(k_0 + \kappa) \exp\left\{ i\kappa x - i\left(\kappa \omega_0' + \frac{1}{2}\kappa^2 \omega_0'' \right) t \right\} d\kappa$$

and we have substituted $k = k_0 + \kappa$. The function φ describes the modulations in (17.62); it satisfies the equation

$$i(\varphi_t + \omega_0' \varphi_x) + \frac{1}{2} \omega_0'' \varphi_{xx} = 0, \tag{17.63}$$

and it corresponds to the dispersion relation

$$W = \kappa \omega_0' + \frac{1}{2} \kappa^2 \omega_0''. \tag{17.64}$$

The equation for Φ corresponds to the original expansion

$$\omega = \omega_0 + (k - k_0)\omega_0' + \frac{1}{2}(k - k_0)^2\omega_0'';$$

it is

$$i\Phi_t - \left(\omega_0 - k_0\omega_0' + \frac{1}{2}k_0^2\omega_0''\right)\Phi + i(\omega_0' - k_0\omega_0'')\Phi_x - \frac{1}{2}\omega_0''\Phi_{xx} = 0.$$

The extra terms are eliminated by the transformation (17.62).

If this approximation to the linear dispersion is combined with a cubic nonlinearity we have

$$i(\varphi_t + \omega_0'\varphi_x) + \frac{1}{2}\omega_0''\varphi_{xx} + q|\varphi|^2\varphi = 0. \qquad (17.65)$$

Since $\varphi = ae^{i\kappa x - iWt}$ is still a solution, we see that the nonlinear correction to the dispersion relation modifies (17.64) to

$$W = \kappa\omega_0' + \frac{1}{2}\kappa^2\omega_0'' - qa^2.$$

Therefore the modulations are stable or unstable in the sense of Sections 14.2 and 15.3 according to

$$q\omega_0'' < 0: \quad \text{Stable,}$$

$$q\omega_0'' > 0: \quad \text{Unstable.}$$

Equation 17.65 can be normalized by first choosing a frame of reference moving with the linear group velocity ω_0' to eliminate the term in φ_x and then rescaling the variables to obtain

$$iu_t + u_{xx} + \nu|u|^2u = 0, \qquad (17.66)$$

where the sign of ν is the same as $q\omega_0''$.

17.8 Uniform Wavetrains and Solitary Waves

These are found as usual by looking for solutions depending on a moving coordinate $X = x - Ut$, with the slight extension here that we allow

$$u = e^{irx - ist}v(X), \qquad X = x - Ut,$$

where r and s are constants. This may be interpreted as a slight flexibility in the choice of the exponential factor in (17.62). On substitution, the ordinary differential equation for v is

$$v'' + i(2r - U)v' + (s - r^2)v + \nu|v|^2 v = 0.$$

We now choose

$$r = \frac{U}{2}, \qquad s = \frac{U^2}{4} - \alpha,$$

the first being the important one to eliminate the term in v'. Then v may be taken to be real and

$$v'' - \alpha v + \nu v^3 = 0.$$

This gives a typical cnoidal wave equation for v. It may be integrated once to

$$v'^2 = A + \alpha v^2 - \frac{\nu}{2} v^4,$$

which can be solved in elliptic functions. The limiting case of the solitary wave is possible when $\nu > 0$; we take $A = 0$, $\alpha > 0$, and the solution is

$$v = \left(\frac{2\alpha}{\nu} \right)^{1/2} \operatorname{sech} \alpha^{1/2}(x - Ut). \tag{17.67}$$

This solution represents a wave packet for u similar to that shown in Fig. 15.2; it propagates unchanged in shape with constant velocity.

It is interesting that $|u|^2$ is proportional to sech^2, the same function that describes solitary waves for the Korteweg-deVries equation. The important difference here, however, is that the amplitude and the velocity are independent parameters.

It should be particularly noted that the solutions (17.67) are possible only in the unstable case $\nu > 0$. This suggests again that the end result of an unstable wavetrain subject to small modulations is a series of solitary waves. It is confirmed by the analysis of Zakharov and Shabat described in the next section.

17.9 Inverse Scattering

Zakharov and Shabat (1972) in one more ingenious paper in this field, show how, following Lax (1968), the inverse scattering approach may be applied in analogy with the method for the Korteweg-deVries solution. The method may be used on any equation

$$u_t = Su$$

where the equation is equivalent to the relation

$$\frac{\partial L}{\partial t} = i[L,A] = i(LA - AL), \tag{17.68}$$

where L and A are linear differential operators including the function $u(x,t)$ in the coefficients and $\partial L/\partial t$ refers to differentiating u with respect to t in the expression for L. Once this factorization is obtained (a highly nontrivial step), the method proceeds as follows.

Consider the eigenvalue problem

$$L\psi = \lambda\psi. \tag{17.69}$$

On differentiation with respect to t we have

$$i\psi\frac{d\lambda}{dt} + i\lambda\frac{\partial\psi}{\partial t} = iL\psi_t + i\frac{\partial L}{\partial t}\psi$$

$$= iL\psi_t - (LA - AL)\psi$$

$$= L(i\psi_t - A\psi) + \lambda A\psi.$$

Therefore

$$i\psi\frac{d\lambda}{dt} = (L - \lambda)(i\psi_t - A\psi).$$

If ψ satisfies $L\psi = \lambda\psi$ initially and is allowed to develop with t according to

$$i\psi_t = A\psi, \tag{17.70}$$

then ψ continues to satisfy $L\psi = \lambda\psi$ with unchanged λ. Equations 17.69 and 17.70 are the pair of equations coupling the function $u(x,t)$ in the coefficients with a scattering problem. The solution proceeds as before. The behavior of ψ determines the scattering potential in (17.69); the development of ψ in time is given by (17.70).

The crucial step is still to factor L according to (17.68). Zakharov and

Shabat note, presumably by inspection, that matrix operators

$$L = i \begin{bmatrix} 1+p & 0 \\ 0 & 1-p \end{bmatrix} \frac{\partial}{\partial x} + \begin{bmatrix} 0 & u^* \\ u & 0 \end{bmatrix}, \qquad \nu = \frac{2}{1-p^2},$$

$$A = -p \begin{bmatrix} 1 & 0 \\ 0 & 1 \end{bmatrix} \frac{\partial^2}{\partial x^2} + \begin{bmatrix} \dfrac{|u|^2}{1+p} & iu_x^* \\ -iu_x & \dfrac{-|u|^2}{1-p} \end{bmatrix}$$

will do it!

From this point their analysis parallels the Korteweg-deVries discussion, although major modifications are introduced in handling the inverse scattering problem for (17.69), since a non self-adjoint operator is introduced. In nonself-adjoint problems, the direct eigenfunction expansions may be obtained from the singularities of the Green's function in the complex λ plane. Zakharov and Shabat use the development of ψ according to (17.70) to obtain information on the singularities of various related functions, and then construct $u(x,t)$ from them. However the alternative version working with the transformed equations analogous to (17.38) − 17.39) may be much simpler.

The results are similar qualitatively to those for the Korteweg-deVries equation. The solutions for interacting solitary waves are derived explicitly and are obtained when only the point spectrum contributes. The expression for $|u|^2$ is again in the form

$$|u|^2 \propto \frac{d^2}{dx^2} \log |P|,$$

where $|P|$ is a determinant of exponentials, related this time to the operator

$$i\frac{\partial}{\partial t} + \frac{\partial^2}{\partial x^2}.$$

The solution again confirms that solitary waves retain their structure and emerge in exactly their original form with possible delays due to the interaction.

The solution of the initial value problem is found as before and it seems clear that for large times the contributions of the point spectrum dominate. That is to say, the disturbance tends to become a series of solitary waves. The analysis is confined to solutions for which $|u| \to 0$ as $|x| \to \infty$, but it seems fair to deduce that a series of solitary waves is the end result of the instability of wavetrains to modulations.

SINE-GORDON EQUATION

The physical problems in which the Sine-Gordon equation occurs were described in Section 14.1. The class of solutions with φ oscillating periodically about $\varphi = 0$ is included in the discussion there. We now consider more general solutions. In particular, since φ is an angular variable, solutions in which φ increases by 2π in each cycle represent periodic solutions physically. Thus helical waves in which φ continually increases are included as periodic wavetrains. A limiting case is a single kink with φ changing by 2π from $x = -\infty$ to $x = \infty$; this is a solitary wave. Solutions representing interacting solitary waves show the same preservation properties of the previous cases, and the topological interpretation of the conservation of kinks is particularly appealing.

17.10 Periodic Wavetrains and Solitary Waves

The equation is taken in normalized form

$$\varphi_{tt} - \varphi_{xx} + \sin\varphi = 0,$$

and the steady profile solutions $\varphi = \Phi(X), X = x - Ut$, satisfy

$$\frac{1}{2}(U^2 - 1)\Phi_X^2 + 2\sin^2\frac{1}{2}\Phi = A,$$

where A is a constant of integration related to the amplitude. The following cases can be distinguished.

1. $0 < A < 2$, $U^2 - 1 > 0$. These are periodic solutions with Φ oscillating about $\Phi = 0$ over the range $-\Phi_0 < \Phi < \Phi_0$, where

$$\Phi_0 = 2\sin^{-1}\left(\frac{A}{2}\right)^{1/2}.$$

2. $0 < A < 2$, $U^2 - 1 < 0$.   These are periodic solutions with Φ oscillating about $\Phi = \pi$ in the range

$$\pi - \Phi_0 < \Phi < \pi + \Phi_0.$$

3. $A < 0$, $U^2 - 1 < 0$.   These are helical waves with

$$\Phi_X = \pm \left\{ \frac{2}{1 - U^2} \left(|A| + 2 \sin^2 \frac{1}{2} \Phi \right) \right\}^{1/2},$$

Φ monotonically increasing or decreasing.

4. $A > 2$, $U^2 - 1 > 0$.   These are also helical waves with

$$\Phi_X = \pm \left\{ \frac{2}{U^2 - 1} \left(A - 2 \sin^2 \frac{1}{2} \Phi \right) \right\}^{1/2}.$$

5.   *Limiting case $A = 0$, $U^2 - 1 < 0$.*   Solutions are

$$\tan \left(\frac{\Phi}{4} \right) = \pm \exp \left\{ \pm (1 - U^2)^{-1/2} (X - X_0) \right\}.$$

They represent single kinks of magnitude 2π. With both of the signs chosen positive it is a positive kink from $\Phi = 0$ at $x = -\infty$ to $\Phi = 2\pi$ at $x = +\infty$; with both signs chosen negative it is still a positive kink but from $\Phi = -2\pi$ at $x = -\infty$ to $\Phi = 0$ at $x = +\infty$. Opposite signs give negative kinks.

6.   *Limiting case $A = 2$, $U^2 - 1 > 0$.*   The solution is

$$\tan \left(\frac{\Phi + \pi}{4} \right) = \exp \left\{ \pm (U^2 - 1)^{-1/2} (X - X_0) \right\}$$

and this represents a kink between $\Phi = -\pi$ and $\Phi = \pi$.

The stability arguments of Section 14.2 and Section 15.3 showed that the periodic waves in case 1 above are unstable to modulations. These arguments can be extended to the other types of solution and show that cases 1 and 2 are unstable, whereas the helical waves 3 and 4 are stable. It should be remembered that these stability arguments apply only to relatively long modulations. They do not give a complete discussion of stability nor do they cover the limiting cases 5 and 6. We should expect that case 6 is completely unstable since it requires $\Phi = \pm \pi$ at infinity. In Scott's pendulum model, for example, this would correspond to the pendula in the upright vertical position. We concentrate now on the solitary waves (case 5).

17.11. The Interaction of Solitary Waves

Perring and Skyrme (1962) apparently guessed that their numerical solutions for two interacting solitary waves could be fitted by

$$\psi = \tan \frac{\varphi}{4} = U \frac{\sinh x (1 - U^2)^{-1/2}}{\cosh Ut (1 - U^2)^{-1/2}} \qquad (17.71)$$

and then verified that this is an exact solution! To see that it represents the interaction of two solitary waves, notice that its behavior as $t \to \pm \infty$ is given by

$$t \to -\infty: \quad \psi \sim U \exp\left(\frac{x + Ut}{\sqrt{1 - U^2}}\right) - U \exp\left(-\frac{x - Ut}{\sqrt{1 - U^2}}\right),$$

$$t \to +\infty: \quad \psi \sim -U \exp\left(-\frac{x + Ut}{\sqrt{1 - U^2}}\right) + U \exp\left(\frac{x - Ut}{\sqrt{1 - U^2}}\right).$$

Each of these represents solitary waves moving in opposite directions. The positive kink moving with velocity U is incident from $x = -\infty$ and emerges as a positive kink. The factor U outside the exponentials may be absorbed into the exponential as a displacement in x. The positive kink from $-\infty$ is displaced an amount

$$2\sqrt{1 - U^2} \, \log \frac{1}{U}$$

by the interaction.

The form involving $\psi = \tan(\varphi/4)$ suggests that the transformation of the equation to one for ψ may be useful in general. The equation satisfied by ψ is

$$(1 + \psi^2)(\psi_{tt} - \psi_{xx} + \psi) - 2\psi(\psi_t^2 - \psi_x^2 + \psi^2) = 0. \qquad (17.72)$$

This has a balanced structure and brings out the most nearly related linear operator

$$\frac{\partial^2}{\partial t^2} - \frac{\partial^2}{\partial x^2} + 1. \qquad (17.73)$$

It is reminiscent in these respects of the role played by the F equation (17.11) in the discussion of the Korteweg-deVries equation. The single solitary wave solutions of (17.72) are given by

$$\psi = \pm \exp\left(\pm \frac{x - Ut}{\sqrt{1 - U^2}} \right), \tag{17.74}$$

so that the transformation may again be stimulated by the rough working rule of finding the transformation for which solitary wave solutions become exponentials, and in fact satisfy (17.73).

The Perring-Skyrme solutions can be found by separation of variables, that is,

$$\psi = f(x)g(t),$$

even though the equation for ψ is nonlinear. The equation is satisfied provided

$$f'^2 = \mu f^4 + (1 + \lambda)f^2 - \nu,$$

$$g'^2 = \nu g^4 + \lambda g^2 - \mu,$$

where λ, μ, ν are separation constants. These have solutions in elliptic functions, of which the Perring-Skyrme solutions are special cases.

However, a more consistent approach to interacting solitary waves was not based on (17.72), but was developed by Lamb (1967, 1971) using Bäcklund transformations.

17.12 Bäcklund Transformations

These transformations were introduced originally as generalizations of contact transformations and were associated particularly with studies of the geometry of surfaces. As remarked earlier (Section 14.1) the Sine-Gordon equation arises in connection with surfaces of Gaussian curvature -1. An account of Bäcklund transformations and their uses is given in Forsyth (1959, Vol. VI, Chapter 21). In the application to the Sine-Gordon equation, it is convenient to choose the normalized form

$$\frac{\partial^2 \varphi}{\partial \xi \partial \eta} = \sin \varphi; \qquad \xi = \frac{x - t}{2}, \qquad \eta = \frac{x + t}{2}. \tag{17.75}$$

In general a Bäcklund transformation for a second order equation for

$\varphi(\xi, \eta)$ is of the form

$$\varphi'_\xi = P(\varphi', \varphi, \varphi_\xi, \varphi_\eta, \xi, \eta),$$

$$\varphi'_\eta = Q(\varphi', \varphi, \varphi_\xi, \varphi_\eta, \xi, \eta).$$

Consistency of the two expressions leads to a new differential equation for $\varphi'(\xi, \eta)$. The philosophy seems to be to find interesting equivalent equations in this way. Indeed, the reduction of Burgers' equation

$$\varphi_\eta + \varphi\varphi_\xi - \nu\varphi_{\xi\xi} = 0$$

to the heat equation

$$\varphi'_\eta - \nu\varphi'_{\xi\xi} = 0$$

can be written as a Bäcklund transformation

$$\varphi'_\xi = -\frac{\varphi\varphi'}{2\nu}, \qquad \varphi'_\eta = -\left(2\nu\varphi_\xi - \varphi^2\right)\frac{\varphi'}{4\nu}.$$

In general, however, the reduction of the original equation to a linear one is perhaps too much to hope for. But another use is to find equations that can be mapped into themselves, so that any known solution for φ (even a trivial one) may provide a new solution φ'. The determination of the transformation which maps (17.75) into itself is set as a problem in Forsyth with a reference to Bianchi and Darboux. It is easily shown that the appropriate Bäcklund transformation is

$$\frac{\partial \varphi'}{\partial \xi} = \frac{\partial \varphi}{\partial \xi} + 2\lambda \sin \frac{\varphi' + \varphi}{2},$$

$$\frac{\partial \varphi'}{\partial \eta} = -\frac{\partial \varphi}{\partial \eta} + \frac{2}{\lambda} \sin \frac{\varphi' - \varphi}{2}, \tag{17.76}$$

where λ is an arbitrary parameter. We note that they give, respectively,

$$\frac{\partial}{\partial \eta} \varphi'_\xi = \varphi_{\xi\eta} + \lambda(\varphi'_\eta + \varphi_\eta) \cos \frac{\varphi' + \varphi}{2}$$

$$= \varphi_{\xi\eta} + 2 \sin \frac{\varphi' - \varphi}{2} \cos \frac{\varphi' + \varphi}{2},$$

$$\frac{\partial}{\partial \xi} \varphi'_\eta = -\varphi_{\eta\xi} + \frac{1}{\lambda}(\varphi'_\xi - \varphi_\xi) \cos \frac{\varphi' - \varphi}{2}$$

$$= -\varphi_{\eta\xi} + 2 \sin \frac{\varphi' + \varphi}{2} \cos \frac{\varphi' - \varphi}{2}.$$

The two expressions for $\varphi'_{\xi\eta}$ are equal (and therefore consistent) if

$$\varphi_{\xi\eta} = \sin\varphi.$$

Moreover, adding the two, we see that

$$\varphi'_{\xi\eta} = \sin\varphi'.$$

Lamb's procedure is to generate new solutions successively from (17.76). First, since $\varphi_0 = 0$ is a solution, another solution φ_1 may be found from

$$\frac{\partial\varphi_1}{\partial\eta} = 2\lambda\sin\frac{\varphi_1}{2}, \qquad \frac{\partial\varphi_1}{\partial\xi} = \frac{2}{\lambda}\sin\frac{\varphi_1}{2}.$$

This is easily shown to be the solitary wave

$$\tan\frac{\varphi_1}{4} = C\exp\left(\lambda\eta + \frac{\xi}{\lambda}\right) = C\exp\left(\frac{x - Ut}{\sqrt{1 - U^2}}\right),$$

$$\lambda = \sqrt{\frac{1 - U}{1 + U}}.$$

Next, if φ is taken to be φ_1 in (17.76), the solution $\varphi' = \varphi_2$ is found to be the Perring-Skyrme result for two interacting solitary waves. In general the solution φ_{n-1} for $n-1$ solitary waves generates the solution φ_n for n. In terms of $\psi = \tan(\varphi/4)$, $\psi' = \tan(\varphi'/4)$ the transformation (17.76) is

$$\psi'_{\xi} = (1 + \psi^2)^{-1}\left\{(1 + \psi'^2)\psi_{\xi} + \lambda(1 - \psi^2)\psi' + \lambda\psi(1 - \psi'^2)\right\},$$

$$\psi'_{\eta} = (1 + \psi^2)^{-1}\left\{-(1 + \psi'^2)\psi_{\eta} + \frac{1}{\lambda}(1 - \psi^2)\psi' - \frac{1}{\lambda}\psi(1 - \psi'^2)\right\}.$$

Either one is a Riccati equation for ψ', or, alternatively, ψ'^2 can be eliminated between the two to give a linear first order partial differential equation for ψ'. The solution for the latter can always be found in principle but the actual expressions for the φ_n become successively more complicated.

17.13 Inverse Scattering for the Sine-Gordon Equation

Recently Lamb (1973) and Ablowitz et al. (1973) have shown how the inverse scattering approach may be used. The key step is the factorization into a scattering problem involving the desired solution φ and an evolution equation for the eigenfunctions. If the Sine-Gordon equation is taken in the normalized form

$$\varphi_{xt} = \sin \varphi$$

(reverting to x and t to make the analogy with the previous cases more apparent), the appropriate scattering equations are

$$\frac{\partial v_1}{\partial x} + i\lambda v_1 = -\frac{1}{2}\varphi_x(x,t)v_2,$$

$$\frac{\partial v_2}{\partial x} - i\lambda v_2 = \frac{1}{2}\varphi_x(x,t)v_1,$$

and the evolution equations for the vector eigenfunction (v_1, v_2) are

$$\frac{\partial v_1}{\partial t} = \frac{i}{4\lambda}(v_1\cos\varphi + v_2\sin\varphi),$$

$$\frac{\partial v_2}{\partial t} = \frac{i}{4\lambda}(v_1\sin\varphi - v_2\cos\varphi).$$

The results of Zakharov and Shabat may be applied directly to reconstruct $\varphi(x,t)$. Again the alternative approach corresponding to (17.38)–(17.39) appears to be simpler.

TODA CHAIN

Systems of mass points with nonlinear forces between nearest neighbors are of interest as models for lattice vibrations in crystals, with questions of the partition of energy among the various modes of vibrations, of thermal expansion under excitation, and the like, being of physical interest. They may be viewed as spatially discrete analogs of the continuous systems considered in this book. A single chain described by (17.7) is the simplest case for wave propagation. It is written in terms of a mass-spring system, but other interpretations are possible, such as propagation in a lumped transmission line as discussed by Hirota and Suzuki (1970, 1973). The latter authors have also performed experiments verifying the predictions on solitary waves and their interactions.

In the linearized limit $f(r) = -\gamma r$, (17.7) becomes

$$m\ddot{r}_n = \gamma(r_{n+1} + r_{n-1} - 2r_n).$$

The traveling wave solution

$$r_n = a\cos\theta, \qquad \theta = \omega t - pn$$

is well known. It works because on substitution we have

$$-\frac{m\omega^2}{\gamma}\cos\theta = \cos(\theta - p) + \cos(\theta + p) - 2\cos\theta, \qquad (17.77)$$

and the right hand side reduces to a multiple of $\cos\theta$ from the sum and difference formulas for the trigonometric functions. We have the "dispersion relation"

$$\frac{m\omega^2}{\gamma} = 2(1 - \cos p) = 4\sin^2\frac{p}{2}. \qquad (17.78)$$

The parameter p is analogous to the wave number for continuous lines.
Stokes-type expansions of the form

$$r_n = a\cos\theta + a_2\cos 2\theta + \cdots$$

can be developed for small amplitudes a; they provide nonlinear corrections to (17.78) and modulation theory can be developed as in the continuous case (Lowell, 1970). However, to obtain fully nonlinear solutions and, in particular, solitary waves is a much more difficult problem than in the continuous case. One would expect they exist but explicit examples would be welcome. The difficulty is that one requires functions and their addition formulas to handle the right hand side corresponding to (17.77). Toda (1967a, 1967b) has provided such solutions for the case

$$f(r) = -\alpha(1 - e^{-\beta r}). \qquad (17.79)$$

17.14 Toda's Solution for the Exponential Chain

It is convenient to convert the difference equation (17.7) into an equivalent form in which $\dot{s}_n = f(r_n)$ is introduced. First we have the system

$$\dot{s}_n = f(r_n),$$
$$m\dot{r}_n = 2s_n - s_{n+1} - s_{n-1}. \qquad (17.80)$$

For the exponential force (17.79),

$$\ddot{s}_n = f'(r_n)\dot{r}_n = -\alpha\beta e^{-\beta r_n}\dot{r}_n$$

$$= -\beta(\alpha + \dot{s}_n)\dot{r}_n.$$

Therefore (17.80) may be combined into

$$\frac{m}{\beta}\frac{\ddot{s}_n}{\alpha + \dot{s}_n} = s_{n+1} + s_{n-1} - 2s_n. \tag{17.81}$$

For a steady progressing wave

$$s_n = S(\theta), \qquad \theta = \omega t - pn,$$

the function $S(\theta)$ must satisfy the ordinary differential difference equation

$$\frac{m\omega^2}{\beta}\frac{S''}{\alpha + \omega S'} = S(\theta + p) + S(\theta - p) - 2S(\theta). \tag{17.82}$$

According to Toda, we now remember that

$$\mathrm{dn}^2(\theta + p) - \mathrm{dn}^2(\theta - p) = -2k^2\frac{d}{dp}\left(\frac{\mathrm{sn}\,\theta\,\mathrm{cn}\,\theta\,\mathrm{dn}\,\theta\,\mathrm{sn}^2 p}{1 - k^2\,\mathrm{sn}^2\theta\,\mathrm{sn}^2 p}\right),$$

where sn, cn, dn are the Jacobian elliptic functions and k is the modulus of these functions. If

$$E(\zeta) = \int_0^\zeta \mathrm{dn}^2 z\, dz,$$

the integral of this equation with respect to p is

$$E(\theta + p) + E(\theta - p) - 2E(\theta) = -2k^2\frac{\mathrm{sn}\,\theta\,\mathrm{cn}\,\theta\,\mathrm{dn}\,\theta\,\mathrm{sn}^2 p}{1 - k^2\,\mathrm{sn}^2\theta\,\mathrm{sn}^2 p}. \tag{17.83}$$

Moreover,

$$E'(\theta) = \mathrm{dn}^2\theta = 1 - k^2\,\mathrm{sn}^2\theta,$$

$$E''(\theta) = -2k^2\,\mathrm{sn}\,\theta\,\mathrm{cn}\,\theta\,\mathrm{dn}\,\theta.$$

Therefore the right hand side of (17.83) is

$$\frac{E''(\theta)}{q + E'(\theta)}, \qquad q = \frac{1}{\mathrm{sn}^2 p} - 1.$$

We see that the equation for $E(\theta)$ is essentially (17.82). The function $E(\theta)$ is not periodic in θ, but the related Jacobian zeta function

$$Z(\theta) = E(\theta) - \theta \frac{E(K)}{K} \qquad (17.84)$$

has period $2K$, and the change from $E(\theta)$ to $Z(\theta)$ modifies the equation trivially. The solution for s_n may be taken in the form

$$s_n = S(\theta) = bZ(2K\theta), \quad \theta = \omega t - pn, \qquad (17.85)$$

where

$$b = \left(\frac{m\alpha}{\beta}\right)^{1/2} \left(\frac{1}{\mathrm{sn}^2 2Kp} - 1 + \frac{E}{K}\right)^{-1/2}, \qquad (17.86)$$

$$\omega = \frac{1}{2K}\left(\frac{\alpha\beta}{m}\right)^{1/2} \left(\frac{1}{\mathrm{sn}^2 2Kp} - 1 + \frac{E}{K}\right)^{-1/2}. \qquad (17.87)$$

The expression for r_n is determined by

$$-\alpha(1 - e^{-\beta r_n}) = \dot{s}_n = 2K\omega b\left(\mathrm{dn}^2 2K\theta - \frac{E}{K}\right). \qquad (17.88)$$

The functions $Z(\zeta)$ and $\mathrm{dn}^2\zeta$ have period $2K$; here the phase has been normalized so that one period corresponds to unit increase in θ rather than the 2π of linear theory. The amplitude of $Z(\zeta)$ is a function of k so the amplitude of s_n is a function $A(k,p)$. Combined with (17.87), we have a dispersion relation between ω, p, A expressed in the parametric form

$$A = A(k,p), \qquad \omega = \omega(k,p).$$

In the linear limit $k \to 0$, we have

$$\mathrm{sn}^2\zeta \sim \sin^2\zeta, \qquad K, E \sim \frac{\pi}{2}$$

$$\mathrm{dn}^2\zeta \sim 1 - k^2 \sin^2\zeta, \qquad \frac{E}{K} \sim 1 - \frac{k^2}{2},$$

$$Z(\zeta) \sim \frac{k^2}{4} \sin 2\zeta.$$

Hence

$$s_n \sim \frac{bk^2}{4} \sin 2\pi\theta, \qquad \theta = \omega t - pn,$$

$$b \sim \left(\frac{m\alpha}{\beta}\right)^{1/2} \sin \pi p, \qquad \omega \sim \frac{1}{\pi}\left(\frac{\alpha\beta}{m}\right)^{1/2} \sin \pi p,$$

and

$$-\alpha(1 - e^{-\beta r_n}) \sim -\gamma r_n \sim \frac{\pi\omega b k^2}{2} \cos 2\pi\theta.$$

These may be rewritten

$$s_n \sim A \sin 2\pi(\omega t - pn), \qquad -\gamma r_n \sim 2\pi\omega A \cos 2\pi(\omega t - pn),$$

$$\frac{m(2\pi\omega)^2}{\gamma} \sim 4 \sin^2 \pi p, \qquad k^2 \sim 4\left(\frac{\beta}{m\alpha}\right)^{1/2} \frac{A}{\sin \pi p} \ll 1;$$

apart from minor renormalizations, the results reproduce the linear solution.

When $k \to 1$, $K \to \infty$ and we must take the case of finite limits for

$$2K\omega \to \Omega, \qquad 2Kp \to P.$$

The elliptic functions have the limiting forms

$$\text{sn}\,\zeta \to \tanh\zeta, \qquad \text{dn}\,\zeta \to \text{sech}\,\zeta, \qquad Z(\zeta) \to \tanh\zeta;$$

the relations (17.86)–(17.87) become

$$b \to \left(\frac{m\alpha}{\beta}\right)^{1/2} \sinh P, \qquad \Omega = \left(\frac{\alpha\beta}{m}\right)^{1/2} \sinh P.$$

Hence

$$s_n \sim \frac{m\Omega}{\beta} \tanh(\Omega t - Pn),$$

$$-\alpha(1 - \beta e^{-r_n}) \sim \frac{m\Omega^2}{\beta} \text{sech}^2(\Omega t - Pn).$$

These are the solitary waves.

On the basis of various approximate forms and special cases, Toda builds up convincing evidence that these solitary waves interact as in the continuous cases with the original forms emerging after interactions.

BORN-INFELD EQUATION

The Born-Infeld equation

$$(1-\varphi_t^2)\varphi_{xx}+2\varphi_x\varphi_t\varphi_{xt}-(1+\varphi_x^2)\varphi_{tt}=0$$

has solitary wave solutions of a kind, but they are very different from those previously discussed. It is a simple matter to check that either

$$\varphi=\Phi(x-t)\qquad\text{or}\qquad\varphi=\Phi(x+t)$$

are exact solutions for any function Φ. In particular the function Φ may be chosen as a single hump to give a solitary wave appearance. But there is no natural structure involved. The equation is hyperbolic for solutions with

$$1+\varphi_x^2-\varphi_t^2>0,$$

and perhaps some aspects properly belong in Part I. Notice that the solitary waves have constant characteristic velocities ±1 and avoid the usual breaking expected for nonlinear hyperbolic waves.

17.15 Interacting Waves

The solutions of Barbishov and Chernikov (1967) can be obtained quite naturally by a hodograph transformation, although the simplicity of the transformed equation is unexpected. First, if new variables

$$\xi=x-t,\qquad \eta=x+t,$$

$$u=\varphi_\xi,\qquad v=\varphi_\eta$$

are introduced, we may take the equivalent system

$$u_\eta-v_\xi=0,$$

$$v^2u_\xi-(1+2uv)u_\eta+u^2v_\eta=0. \tag{17.89}$$

The roles of the dependent and independent variables are then interchanged to give

$$\xi_v-\eta_u=0,$$

$$v^2\eta_v+(1+2uv)\xi_v+u^2\xi_u=0, \tag{17.90}$$

or, equivalently, the single equation

$$u^2\xi_{uu} + (1+2uv)\xi_{uv} + v^2\xi_{vv} + 2u\xi_u + 2v\xi_v = 0. \tag{17.91}$$

Assuming that the relevant solutions are to be found in the hyperbolic regime, it is now a natural step to find the characteristics for the linear system (17.90) and the linear equation (17.91). They are the integral curves of the differential form

$$u^2\,dv^2 - (1+2uv)\,du\,dv + v^2\,du^2 = 0,$$

and they are found to be curves $r = \text{constant}$, $s = \text{constant}$, where

$$r = \frac{\sqrt{1+4uv}\ -1}{2v}, \qquad s = \frac{\sqrt{1+4uv}\ -1}{2u}. \tag{17.92}$$

If r, s are introduced as new variables to replace u, v, the equations (17.90) become

$$r^2\xi_r + \eta_r = 0,$$
$$\xi_s + s^2\eta_s = 0. \tag{17.93}$$

The surprising result is that on elimination of η to deduce the new form of (17.91), we have simply

$$\xi_{rs} = 0. \tag{17.94}$$

The hodograph transformation guarantees a linear equation, but it could well have been an impossible one for practical purposes.

The general solution may be taken as

$$x - t = \xi = F(r) - \int s^2 G'(s)\,ds, \tag{17.95}$$
$$x + t = \eta = G(s) - \int r^2 F'(r)\,dr, \tag{17.96}$$

where $F(r), G(s)$ are arbitrary functions. Since

$$\varphi_r = u\xi_r + v\eta_r = \frac{r}{1-rs}\xi_r + \frac{s}{1-rs}\eta_r$$
$$= rF'(r),$$

and, similarly,

$$\varphi_s = sG'(s),$$

the corresponding expression for φ is

$$\varphi = \int rF'(r)\,dr + \int sG'(s)\,ds. \qquad (17.97)$$

Finally, it is convenient to introduce

$$F(r) = \rho, \qquad G(s) = \sigma,$$

$$r = \Phi_1'(\rho), \qquad s = \Phi_2'(\sigma),$$

and take the solution in the form

$$\varphi = \Phi_1(\rho) + \Phi_2(\sigma), \qquad (17.98)$$

$$x - t = \rho - \int_{-\infty}^{\sigma} \Phi_2'^2(\sigma)\,d\sigma, \qquad (17.99)$$

$$x + t = \sigma + \int_{\rho}^{\infty} \Phi_1'^2(\rho)\,dp. \qquad (17.100)$$

If $\Phi_1(\rho)$ and $\Phi_2(\sigma)$ are localized, say they are nonzero in

$$-1 < \rho < 0, \qquad 0 < \sigma < 1,$$

respectively, then

$$\varphi = \Phi_1(x - t) + \Phi_2(x + t) \qquad \text{for } t < 0.$$

The wave Φ_1 is incident from $x = -\infty$ and the wave Φ_2 from $x = +\infty$. As $t \to +\infty$, the solution approaches

$$\varphi = \Phi_1\left\{ x - t + \int_{-\infty}^{\infty} \Phi_2'^2(\sigma)\,d\sigma \right\} + \Phi_2\left\{ x + t - \int_{-\infty}^{\infty} \Phi_1'^2(\rho)\,d\rho \right\}. \qquad (17.101)$$

Each wave receives a displacement in the direction opposite to its direction of propagation equal to

$$\int_{-\infty}^{\infty} \Phi_i'^2(\tau)\,d\tau.$$

In this respect the interaction is similar in its residual effect to the other examples of interacting solitary waves. But in most other respects, these solutions and the Born-Infeld equation seem to belong to a different class.

One final comment on the actual solution given in (17.98)–(17.100): the mapping from the (x, t) plane to the (ρ, σ) plane may become singular in the course of the interaction, even though (17.101) shows that the solution is again single-valued in the aftermath. This would be interpreted as shock formation and tied to the fact that during the interaction the characteristic velocities depart from the values ± 1 and may form envelopes. In such cases the subsequent behavior would require modification and (17.101) should not be accepted as it stands. However, breaking would require sufficiently large amplitudes and there will be a range of solutions for which this does not occur.

In this particular example, the linear superposition performed in the transformed equation (17.94) makes the preservation of the identity of the waves obvious. Indeed in all cases the preservation of structure can presumably be traced to linear superposition in some appropriately transformed space. But the nature of the mapping and the linear solutions involved is crucial. The solution for the confluence of shocks in Section 4.7 corresponds to the superposition of solutions of the linear heat equation. But, because the latter solutions are exponential and not localized, the shocks combine to form a single new shock instead of passing through each other.

In concluding this chapter one can only comment again on the remarkable ingenuity of the various investigators involved in these recent developments. The results have given a tremendous boost to the study of nonlinear waves and nonlinear phenomena in general. Doubtless much more of value will be discovered, and the different approaches have added enormously to the arsenal of "mathematical methods." Not least is the lesson that exact solutions are still around and one should not always turn too quickly to a search for the ϵ.

References

Abbott, M. R. 1956. A theory of the propagation of bores in channels and rivers. *Proc. Camb. Phil. Soc.* **52**, 344–362.

Ablowitz, M. J. 1971. Applications of slowly varying nonlinear dispersive wave theories. *Studies Appl. Math.* **50**, 329–344.

Ablowitz, M. J. and D. J. Benney. 1970. The evolution of multiphase modes for nonlinear dispersive waves. *Studies Appl. Math.* **49**, 225–238.

Ablowitz, M. J., D. J. Kaup, A. C. Newell, and H. Segur. 1973. The initial value solution for the Sine-Gordon equation. To be published.

Akhmanov, S. A., A. P. Sukhorukov, and R. V. Khokhlov. 1966. Self-focusing and self-trapping of intense light beams in a nonlinear medium. *Soviet Physics J.E.T.P.* **23**, 1025–1033.

Balanis, G. N. 1972. The plasma inverse problem. *J. Math. Phys.* **13**, 1001–1005.

Barbishov, B. M. and N. A. Chernikov. 1966. Solution of the two plane wave scattering problem in a nonlinear scalar field theory of the Born-Infeld type. *Soviet Physics J.E.T.P.* **24**, 437–442.

Barcilon, V. 1968. Axisymmetric inertial oscillations of a rotating ring of fluid. *Mathematika* **15**, 93–102.

Barone, A., F. Esposito, C. J. Magee, and A. C. Scott. 1971. Theory and applications of the Sine-Gordon equation. *Rivista del Nuovo Cimento (2) 1*, 227–267.

Bateman, H. 1915. Some recent researches on the motion of fluids. *Monthly Weather Review* **43**, 163–170.

Benjamin, T. B. 1967. Instability of periodic wavetrains in nonlinear dispersive systems. *Proc. Roy. Soc. A* **299**, 59–75.

Benjamin, T. B. 1970. Upstream influence. *J. Fluid Mech.* **40**, 49–79.

Benjamin, T. B., J. L. Bona, and J. J. Mahony. 1972. Model equations for long waves in nonlinear dispersive systems. *Phil. Trans. Roy. Soc. A* **272**, 47–78.

Benjamin, T. B. and M. J. Lighthill. 1954. On cnoidal waves and bores. *Proc. Roy. Soc. A* **224**, 448–460.

Born, M. and L. Infeld. 1934. Foundations of a new field theory. *Proc. Roy. Soc. A* **144**, 425–451.

Boussinesq, J. 1871. Théorie de l'intumescence liquide appelée onde solitaire ou de translation se propageant dans un canal rectangulaire. *Comptes Rendus* **72**, 755–759.

Broderick, J. B. 1949. Supersonic flow round pointed bodies of revolution. *Q. J. Mech. Appl. Math.* **2**, 98–120.

Bryson, A. E. and R. W. F. Gross. 1961. Diffraction of strong shocks by cones, cylinders and spheres. *J. Fluid Mech.* **10**, 1–16.

621

Burgers, J. M. 1948. A mathematical model illustrating the theory of turbulence. *Adv. Appl. Mech.* **1**, 171–199.

Butler, D. S. 1954. Converging spherical and cylindrical shocks. Report No. 54/54 Armament Research and Development Establishment, Ministry of Supply. Fort Halstead, Kent.

Butler, D. S. 1955. Symposium on blast and shock waves. Armament Research and Development Establishment, Ministry of Supply. Fort Halstead, Kent.

Chandler, R. E., R. Herman, and E. W. Montroll. 1958. Traffic dynamics: Studies in car-following. *Oper. Res.* **6**, 165–184.

Chester, W. 1954. The quasi-cylindrical shock tube. *Phil. Mag. (7)* **45**, 1293–1301.

Chiao, R. Y., E. Garmire, and C. H. Townes. 1964. Self-trapping of optical beams. *Phys. Rev. Lett.* **13**, 479–482.

Chisnell, R. F. 1955. The normal motion of a shock wave through a nonuniform one-dimensional medium. *Proc. Roy. Soc. A* **232**, 350–370.

Chisnell, R. F. 1957. The motion of a shock wave in a channel, with applications to cylindrical and spherical shock waves. *J. Fluid Mech.* **2**, 286–298.

Cohen, D. S. and J. W. Blum. 1971. Acoustic wave propagation in an underwater sound channel. *J. Inst. Math. Applns.* **8**, 186–220.

Cole, J. D. 1951. On a quasilinear parabolic equation occurring in aerodynamics. *Q. Appl. Math.* **9**, 225–236.

Cornish, V. 1934. *Ocean waves and kindred geophysical phenomena*. Cambridge University Press.

Courant, R. and K. O. Friedrichs. 1948. *Supersonic flow and shock waves*. Interscience, New York-London.

Courant, R. and D. Hilbert. 1962. *Methods of mathematical physics*. Vol. II. Interscience, New York-London.

Crapper, G. D. 1972. Nonlinear gravity waves on steady non-uniform currents. *J. Fluid Mech.* **52**, 713–724.

Delaney, M. E. 1971. On the averaged Lagrangian technique for nonlinear dispersive waves. Ph.D. Thesis California Institute of Technology.

Donnelly, R. J., R. Herman, and I. Prigogine. 1966. *Non-equilibrium thermodynamics*. University of Chicago Press.

Dressler, R. F. 1949. Mathematical solution of the problem of roll waves in inclined open channels. *Comm. Pure Appl. Math.* **2**, 149–194.

Dressler, R. F. 1952. Hydraulic resistance effect upon the dam-break functions. *J. Res. Nat. Bur. Stand.* **49**, 217–225.

Earnshaw, S. 1858. On the mathematical theory of sound. *Phil. Trans.* **150**, 133–148.

Faddeyev, L. D. 1959. The inverse problem in the quantum theory of scattering. *Uspekhi Matem. Nauk* **14**, 57 (Transl. in *J. Math. Phys.* **4**, 72–104, 1963.)

Favre, H. 1935. *Etude théorique et expérimentale des ondes de translation dans les canaux decouverts*. Dunod et Cie, Paris.

Finsterwalder, S. 1907. Die Theorie der Gletscherschwankungen. *Z. Gletscherkunde* **2**, 81–103.

Forsyth, A. R. 1959. *Theory of differential equations*, Vol. VI. Dover Publications, New York.

Franken, P. A., A. E. Hill, C. W. Peters, and G. Weinreich. 1961. Generation of optical harmonics. *Phys. Rev. Lett.* **7**, 118–119.

Franklin, J. N. 1972. Axisymmetric inertial oscillations of a rotating fluid. *J. Math. Anal. Appl.* **39**, 742–760.

Friedman, M. P. 1960. An improved perturbation theory for shock waves propagating through non-uniform regions. *J. Fluid Mech.* **8**, 193–209.

Friedman, M. P., E. J. Kane, and A. Signalla. 1963. Effects of atmosphere and aircraft motion on the location and intensity of a sonic boom. *A.I.A.A.J.* **1**, 1327–1335.

Gardner, C. S., J. M. Greene, M. D. Kruskal, and R. M. Miura. 1967. Method for solving the Korteweg-deVries equation. *Phys. Rev. Lett.* **19**, 1095–1097.

Gelfand, I. M. and B. M. Levitan. 1951. On the determination of a differential equation from its spectral function. *Am. Math. Transl. (2)* **1**, 253–304.

Gelfand, I. M. and S. V. Fomin. 1963. *Calculus of variations.* Prentice-Hall, Englewood Cliffs, N. J.

Giordmaine, J. A. 1962. Mixing of light beams in crystals. *Phys. Rev. Lett.* **8**, 19–20.

Goldstein, S. 1953. On the mathematics of exchange processes in fixed columns. Parts I and II. *Proc. Roy. Soc. A* **219**, 151–185.

Goldstein, S. and J. D. Murray. 1959. On the mathematics of exchange processes in fixed columns. Parts III, IV, V. *Proc. Roy Soc. A* **252**, 334–375.

Greenberg, H. 1959. An analysis of traffic flow. *Oper. Res.* **7**, 79–85.

Greenspan, H. P. 1968. *The theory of rotating fluids.* Cambridge University Press.

Griffith, W. C., D. Brickl, and V. Blackman. 1956. Structure of shock waves in polyatomic gases. *Phys. Rev.* **102**, 1209–1216.

Griffith, W. C. and A. Kenny. 1957. On fully dispersed shock waves in carbon dioxide. *J. Fluid Mech.* **3**, 286–288.

Guderley, G. 1942. Starke kugelige und zylindrische Verdichtungsstösse in der Nähe des Kugelmittelpunktes bzw der Zylinderachse. *Luftfahrtforschung* **19**, 302–312.

Haus, H. A. 1966. Higher order trapped light beam solutions. *Appl. Phys. Lett.* **8**, 128–129.

Hayes, W. D. 1968. Self-similar strong shocks in an exponential medium. *J. Fluid Mech.* **32**, 305–315.

Hayes, W. D. 1973. Group velocity and nonlinear dispersive wave propagation. *Proc. Roy. Soc. A* **332**, 199–221.

Herman, R., E. W. Montroll, R. B. Potts, and R. W. Rothery. 1959. Traffic dynamics: Analysis of stability in car following. *Oper. Res.* **7**, 86–106.

Hirota, R. 1971. Exact solution of the Korteweg-deVries equation for multiple collisions of solitons. *Phys. Rev. Lett.* **27**, 1192–1194.

Hirota, R. and K. Suzuki. 1970. Studies on lattice solitons by using electrical networks. *J. Phys. Soc. Japan* **28**, 1366–1367.

Hirota, R. and K. Suzuki. 1973. Theoretical and experimental studies of lattice solitons in nonlinear lumped networks. To appear in *I.E.E.E. Proceedings.*

Hoffman, A. L. 1967. A single fluid model for shock formation in MHD shock tubes. *J. Plasma Phys.* **1**, 193–207.

Holliday, D. 1973. Nonlinear gravity-capillary surface waves in a slowly varying current. *J. Fluid Mech.* **57**, 797–802.

Hopf, E. 1950. The partial differential equation $u_t + uu_x = \mu u_{xx}$. *Comm. Pure Appl. Math.* **3**, 201–230.

Hugoniot, H. 1889. Sur la propagation du mouvement dans les corps et spécialement dans les gaz parfaits. *J. l'Ecole Polytech.* **58**, 1–125.

Huppert, H. E. and J. W. Miles. 1968. A note on shock diffraction by a supersonic wedge. *J. Fluid Mech.* **31**, 455–458.

Jahnke, E. and F. Emde. 1945. *Tables of functions.* Dover Publications, New York.

Jeffreys, H. 1925. The flow of water in an inclined channel of rectangular section. *Phil. Mag. (6)* **49**, 793–807.

Jeffreys, H. and B. S. Jeffreys. 1956. *Methods of mathematical physics,* 3rd ed. Cambridge University Press.

Jimenez, J. 1972. Wavetrains with small dissipation. Ph.D. Thesis, California Institute of Technology.

von Karman, Th., and N. B. Moore. 1932. Resistance of slender bodies moving with supersonic velocities with special reference to projectiles. *Trans. Am. Soc. Mech. Engrs.* **54**, 303–310.

Karpman, V. I. 1967. An asymptotic solution of the Korteweg-deVries equations. *Phys. Lett.* **25A**, 708–709.

Karpman, V. I. and V. P. Sokolov. 1968. On solitons and the eigenvalues of the Schrödinger equation. *Soviet Physics J.E.T.P.* **27**, 839–845.

Kay, I. and J. B. Keller. 1954. Asymptotic evaluation of the field at a caustic. *J. Appl. Phys.* **25**, 876–883.

Kay, I. and H. E. Moses. 1956. The determination of the scattering potential from the spectral measure function. *Nuovo Cimento (10)* **3**, 276–304.

Komentani, E. and T. Sasaki. 1958. On the stability of traffic flow. *Oper. Res. (Japan)* **2**, 11–26.

Korteweg, D. J. and G. deVries. 1895. On the change of form of long waves advancing in a rectangular channel, and on a new type of long stationary waves. *Phil. Mag. (5)* **39**, 422–443.

Kynch, G. F. 1952. A theory of sedimentation. *Trans. Faraday Soc.* **48**, 166–176.

Lamb, G. L., Jr. 1967. Propagation of ultrashort optical pulses. *Phys. Lett.* **25A**, 181–182.

Lamb, G. L., Jr. 1971. Analytical descriptions of ultrashort optical pulse propagation in a resonant medium. *Rev. Mod. Phys.* **43**, 99–124.

Lamb, G. L., Jr. 1973. Coherent optical pulse propagation as an inverse problem. To be published.

Lamb, H. 1932. *Hydrodynamics,* 6th ed. Cambridge University Press.

Landau, L. D. 1945. On shock waves at large distances from the place of their origin. *Soviet Journal of Physics* **9**, 496–500.

Landau, L. D. and E. M. Lifshitz. 1958. *Quantum mechanics—Nonrelativistic theory.* Pergamon Press Addison-Wesley Publishing Co., Reading, Mass.

Landau, L. D. and E. M. Lifshitz. 1959. *Theory of elasticity.* Pergamon Press Addison-Wesley Publishing Co., Reading, Mass.

Landau, L. D. and E. M. Lifshitz. 1960. *Electrodynamics of continuous media.* Pergamon Press Addison-Wesley Publishing Co., Reading, Mass.

Landau, L. D. and E. M. Lifshitz. 1960. *Mechanics.* Pergamon Press Addison-Wesley Publishing Co., Reading, Mass.

Laporte, O. 1954. On the interaction of a shock with a constriction. Rep. No. LA-1740 Los Alamos Scientific Laboratory.

Lax, P. D. 1968. Integrals of nonlinear equations of evolution and solitary waves. *Comm. Pure Appl. Math.* **21**, 467–490.

Lighthill, M. J. 1945. Supersonic flow past bodies of revolution. R&M 2003 Aeronautical Research Council Ministry of Supply H. M. Stationery Office, London.

Lighthill, M. J. 1948. The position of the shock wave in certain aerodynamic problems. *Q. J. Mech. Appl. Math.* **1**, 309–318.

Lighthill, M. J. 1949. A technique for rendering approximate solutions to physical problems uniformly valid. *Phil. Mag. (7)* **44**, 1179–1201.

Lighthill, M. J. 1949. The diffraction of blast I. *Proc. Roy. Soc. A* **198**, 454–470.

Lighthill, M. J. 1956. Viscosity effects in sound waves of finite amplitude. In *Surveys in mechanics*, Edited by G. K. Batchelor and R. M. Davies. Cambridge University Press.

Lighthill, M. J. 1957. River waves. Naval hydrodynamics publication 515 National Academy of Sciences-National Research Council.

Lighthill, M. J. and G. B. Whitham. 1955. On kinematic waves: I. Flood movement in long rivers; II. Theory of traffic flow on long crowded roads. *Proc. Roy. Soc. A* **229**, 281–345.

Longuet-Higgins, M. S. and R. W. Stewart. 1960. Changes in the form of short gravity waves on long waves and tidal currents. *J. Fluid Mech.* **8**, 565–583.

Longuet-Higgins, M. S. and R. W. Stewart. 1961. The changes in amplitude of short gravity waves on steady nonuniform currents. *J. Fluid Mech.* **10**, 529–549.

Lowell, S. C. 1970. Wave propagation in monatomic lattices with anharmonic potential. *Proc. Roy. Soc. A* **318**, 93–106.

Luke, J. C. 1966. A perturbation method for nonlinear dispersive wave problems. *Proc. Roy. Soc. A* **292**, 403–412.

Luke, J. C. 1967. A variational principle for a fluid with a free surface. *J. Fluid Mech.* **27**, 395–397.

Maker, P. D., R. W. Terhune, M. Nisenoff, and C. M. Savage. 1962. Effects of dispersion and focusing on the production of optical harmonics. *Phys. Rev. Lett.* **8**, 21–22.

Marchenko, V. A. 1955. *Dokl. Acad. Nauk SSSR* **104**, 433.

Marshall, W. 1955. The structure of magnetohydrodynamic shock waves. *Proc. Roy. Soc. A* **233**, 367–376.

McCowan, J. 1894. On the highest wave of permanent type. *Phil. Mag. (5)* **38**, 351–358.

Michell, A. G. M. 1893. The highest waves in water. *Phil. Mag. (5)* **36**, 430–437.

Moeckel, W. E. 1952. Interaction of oblique shock waves with regions of variable pressure, entropy and energy. Tech. Note 2725 Nat. Adv. Comm Aero, Washington.

Mowbray, D. E. and B. S. H. Rarity. 1967a. A theoretical and experimental investigation of the phase configuration of internal waves of small amplitude in a density stratified liquid. *J. Fluid Mech.* **28**, 1–16.

Mowbray, D. E. and B. S. H. Rarity. 1967b. The internal wave pattern produced by a sphere moving vertically in a density stratified liquid. *J. Fluid Mech.* **30**, 489–496.

Newell, G. F. 1961. Nonlinear effects in the dynamics of car-following. *Oper. Res.* **9**, 209–229.

Nigam, S. D. and P. D. Nigam. 1962. Wave propagation in rotating liquids. *Proc. Roy. Soc. A* **266**, 247–256.

Nye, J. F. 1960. The response of glaciers and ice-sheets to seasonal and climatic changes. *Proc. Roy. Soc. A* **256**, 559–584.

Nye, J. F. 1963. The response of a glacier to changes in the rate of nourishment and wastage. *Proc. Roy. Soc. A* **275**, 87–112.

Ostrowskii, L. A. 1967. Propagation of wave packets and space-time self-focusing in a nonlinear medium. *Soviet Physics J.E.T.P.* **24**, 797–800.

Ostrowskii, L. S. 1968. The theory of waves or envelopes in nonlinear media. U.R.S.I. Symposium on electromagnetic waves VI, Stresa, Italy.

Penney, W. G. and A. T. Price. 1952. Finite periodic stationary gravity waves in a perfect liquid. *Phil. Trans. Roy. Soc. A* **244**, 254–284.

Perring, J. K. and T. H. R. Skyrme. 1962. A model unified field equation. *Nucl. Phys.* **31**, 550–555.

Perry, R. W. and A. Kantrowitz. 1951. The production and stability of converging shock waves. *J. Appl. Phys.* **22**, 878–886.

Petrovsky, I. G. 1954. *Lectures on partial differential equations.* Interscience, New York-London.

Phillips, O. M. 1967. Theoretical and experimental studies of gravity wave interactions. *Proc. Roy. Soc. A* **299**, 104–119.

Poisson, S. D. 1807. Memoire sur la theorie du son. *J. l'Ecole Polytech.* **7**, 319–392.

Rankine, W. J. M. 1870. On the thermodynamic theory of waves of finite longitudinal disturbance. *Phil. Trans.* **160**, 277–288.

Rarity, B. S. H. 1967. The two-dimensional wave pattern produced by a disturbance moving in an arbitrary density stratified liquid. *J. Fluid Mech.* **30**, 329–336.

Rayleigh, Lord, 1910. Aerial plane waves of finite amplitude. *Proc. Roy. Soc. A* **84**, 247–284. (*Papers* **5**, 573–610.)

Rayleigh, Lord, 1876. On waves. *Phil. Mag. (5)* **1**, 257–279. (*Papers* **1**, 251–271.)

Richards, P. I. 1956. Shock waves on the highway. *Oper. Res.* **4**, 42–51.

Riemann, B. 1858. *Uber die Fortpflanzung ebener Luftwellen von endlicher Schwingungsweite.* Göttingen Abhandlungen, Vol. viii, p. 43. (*Werke*, 2te Aufl., Leipzig, 1892, p. 157.)

Sakurai, A. 1960. On the problem of a shock wave arriving at the edge of a gas. *Comm. Pure Appl. Math.* **13**, 353–370.

Schiff, L. I. 1951. Nonlinear meson theory of nuclear forces. *Phys. Rev.* **84**, 1–11.

Scott, A. C. 1970. *Active and nonlinear wave propagation in electronics.* Wiley-Interscience, New York.

Scott Russell, J. 1844. Report on waves. Brit. Assoc. Rep.

Seddon, J. A. 1900. River hydraulics. *Trans. Am. Soc. Civ. Engrs.* **43**, 179–243.

Sedov, L. I. 1959. *Similarity and dimensional methods in mechanics.* Academic Press, New York, London.

Segur, H. 1973. The Korteweg-deVries equation and water waves. I. Solutions of the equation. *J. Fluid Mech.* **59**, 721–736.

Seliger, R. L. 1968. A note on the breaking of waves. *Proc. Roy. Soc. A* **303**, 493–496.

Seliger, R. L. and G. B. Whitham. 1968. Variational principles in continuum mechanics. *Proc. Roy. Soc. A* **305**, 1–25.

Shercliff, J. A. 1969. Anisotropic surface waves under a vertical magnetic force. *J. Fluid Mech.* **38**, 353–364.

Skews, B. W. 1967. The shape of a diffracting shock wave. *J. Fluid Mech.* **29**, 297–304.

Small, R. D. 1972. Nonlinear dispersive waves in nonlinear optics. Ph.D. Thesis, California Institute of Technology.

Snodgrass, F. E., et al. 1966. Propagation of ocean swell across the pacific. *Phil. Trans. Roy. Soc. A* **259**, 431–497.

Sommerfeld, A. 1954. *Optics.* Academic Press, New York.

Stokes, G. G. 1847. On the theory of oscillatory waves. *Camb. Trans.* **8**, 441–473. (*Papers* **1**, 197–229.)

Stokes, G. G. 1848. On a difficulty in the theory of sound. *Phil. Mag. (3)* **23**, 349–356. (*Papers* **2**, 51–58.)

Taylor, G. I. 1910. The conditions necessary for discontinuous motion in gases. *Proc. Roy. Soc. A* **84**, 371–377.

Taylor, G. I. 1922. The motion of a sphere in a rotating liquid. *Proc. Roy. Soc. A* **102**, 180–189.

Taylor, G. I. 1946. The air wave surrounding an expanding sphere. *Proc. Roy. Soc. A* **186**, 273–292.

Taylor, G. I. 1950. The formation of a blast wave by a very intense explosion. I. Theoretical discussion. *Proc. Roy. Soc. A* **201**, 159–174.

Taylor, G. I. 1953. An experimental study of standing waves. *Proc. Roy. Soc. A* **218**, 44–59.

Taylor, G. I. 1959. The dynamics of thin sheets of fluid. II. Waves on fluid sheets. *Proc. Roy. Soc. A* **253**, 296–312.

Taylor, G. I. and J. W. Maccoll. 1933. The air pressure on a cone moving at high speeds. *Proc. Roy. Soc. A* **139**, 278–311.

Thomas, H. C. 1944. Heterogeneous ion exchange in a flowing system. *J. Am. Chem. Soc.* **66**, 1664–1666.

Toda, M. 1967a. Vibration of a chain with nonlinear interaction. *J. Phys. Soc. Japan* **22**, 431–436.

Toda, M. 1967b. Wave propagation in anharmonic lattices. *J. Phys. Soc. Japan* **23**, 501–506.

Ursell, F. 1960a. On Kelvin's ship wave pattern. *J. Fluid Mech.* **8**, 418–431.

Ursell, F. 1960b. Steady wave patterns on a non-uniform steady fluid flow. *J. Fluid Mech.* **9**, 333–346.

Weertman, J. 1958. Union géodesique and geophysique internationale, association internationale d'hydrologie scientifique. *Symposium de Chamonix*, Sept. 1958, pp. 162–168.

Whitham, G. B. 1950a. The behaviour of supersonic flow past a body of revolution, far from the axis. *Proc. Roy. Soc. A* **201**, 89–109.

Whitham, G. B. 1950b. The propagation of spherical blast. *Proc. Roy. Soc. A* **203**, 571–581.

Whitham, G. B. 1952. The flow pattern of a supersonic projectile. *Comm. Pure Appl. Math.* **5**, 301–348.

Whitham, G. B. 1955. The effects of hydraulic resistance in the dam-break problem. *Proc. Roy. Soc. A* **227**, 399–407.

Whitham, G. B. 1956. On the propagation of weak shock waves. *J. Fluid Mech.* **1**, 290–318.

Whitham, G. B. 1957. A new approach to problems of shock dynamics. Part I. Two-dimensional problems. *J. Fluid Mech.* **2**, 146–171.

Whitham, G. B. 1958. On the propagation of shock waves through regions of non-uniform area or flow. *J. Fluid Mech.* **4**, 337–360.

Whitham, G. B. 1959a. Some comments on wave propagation and shock wave structure with application to magnetohydrodynamics. *Comm. Pure Appl. Math.* **12**, 113–158.

Whitham, G. B. 1959b. A new approach to problems of shock dynamics. Part II. Three-dimensional problems. *J. Fluid Mech.* **5**, 369–386.

Whitham, G. B. 1965. A general approach to linear and nonlinear dispersive waves using a Lagrangian. *J. Fluid Mech.* **22**, 273–283.

Whitham, G. B. 1967. Nonlinear dispersion of water waves. *J. Fluid Mech.* **27**, 399–412.

Whitham, G. B. 1968. A note on shock dynamics relative to a moving frame. *J. Fluid Mech.* **31**, 449–453.

Whitham, G. B. 1970. Two-timing, variational principles and waves. *J. Fluid Mech.* **44**, 373–395.

Wu, T. T. 1961. A note on the stability condition for certain wave propagation problems. *Comm. Pure Appl. Math.* **14**, 745–747.

Yariv, A. 1967. *Quantum electronics.* John Wiley & Sons, New York.

Zakharov, V. E. and A. B. Shabat. 1972. Exact theory of two-dimensional self focusing and one-dimensional self modulation of waves in nonlinear media. *Soviet Physics J.E.T.P.* **34**, 62–69.

Zeldovich, Ya. B. and Yu. P. Raizer. 1966. *Physics of shock waves and high temperature hydrodynamic phenomena.* Academic Press, New York.

INDEX

When appropriate, bold face type singles out the main reference in a sequence.